The Insects and Arachnids of Canada

Part 1. Collecting, Preparing, and Preserving Insects, Mites, and Spiders, compiled by J.E.H. Martin, Biosystematics Research Institute, Ottawa, 1978. 182 pp. Cat. No. A42-42/1977-1.

Partie 1. Récolte, préparation et conservation des Insectes, des Acariens et des Araignées, compilé par J.E.H. Martin, Institut de recherches biosystématiques, Ottawa, 1983. 205 pp. N^{o} de cat. A42-42/1977-1F.

Part 2. The Bark Beetles of Canada and Alaska (Coleoptera: Scolytidae), by D.E. Bright, Jr., Biosystematics Research Institute, Ottawa, 1976. 241 pp. Cat. No. A42-42/1976-2.

Part 3. The Aradidae of Canada (Hemiptera: Aradidae), by R. Matsuda, Biosystematics Research Institute, Ottawa, 1977. 116 pp. Cat. No. A42-42/1977-3.

Part 4. The Anthocoridae of Canada and Alaska (Heteroptera: Anthocoridae), by L.A. Kelton, Biosystematics Research Institute, Ottawa, 1977. 101 pp. Cat. No. A42-42/1977-4.

Part 5. The Crab Spiders of Canada and Alaska (Araneae: Philodromidae and Thomisidae), by C.D. Dondale and J.H. Redner, Biosystematics Research Institute, Ottawa, 1978. 255 pp. Cat. No. A42-42/1978-5.

Part 6. The Mosquitoes of Canada (Diptera: Culicidae), by D.M. Wood, P.T. Dang, and R.A. Ellis, Biosystematics Research Institute, Ottawa, 1979. 390 pp. Cat. No. A42-42/1979-6.

Partie 7. Genera des Trichoptères du Canada et des États adjacents, par F. Schmid, Institut de recherches biosystématiques, Ottawa, 1980. 296 pp. N^{o} de cat. A42-42/1980-7F. (Available in French only.)

Part 8. The Plant Bugs of the Prairie Provinces of Canada (Heteroptera: Miridae), by L.A. Kelton, Biosystematics Research Institute, Ottawa, 1980. 408 pp. Cat. No. A42-42/1980-8.

Part 9. The Sac Spiders of Canada and Alaska (Araneae: Clubionidae and Anyphaenidae), by C.D. Dondale and J.H. Redner, Biosystematics Research Institute, Ottawa, 1982. 194 pp. Cat. No. A42-42/1982-9E.

Part 10. The Spittlebugs of Canada (Homoptera: Cercopidae), by K.G.A. Hamilton, Biosystematics Research Institute, Ottawa, 1982. 102 pp. Cat. No. A42-42/1982-10E.

Part 11. The Genera of Larval Midges of Canada (Diptera: Chironomidae), by D.R. Oliver and M.E. Roussel, Biosystematics Research Institute, Ottawa, 1983. 263 pp. Cat. No. A42-42/1983-11E.

Part 12. The Families and Subfamilies of Canadian Chalcidoid Wasps (Hymenoptera: Chalcidoidea), by C.M. Yoshimoto, Biosystematics Research Institute, Ottawa, 1984. 149 pp. Cat. No. A42-42/1983-12E.

Part 13. The Carrion Beetles of Canada and Alaska (Coleoptera: Silphidae and Agyrtidae), by R.S. Anderson and S.B. Peck, Biosystematics Research Institute, Ottawa, 1985. 121 pp. Cat. No. A42-42/1985-13E.

Part 14. The Grasshoppers, Crickets, and Related Insects of Canada and Adjacent Regions (Ulonata: Dermaptera, Cheleutoptera, Notoptera, Dictuoptera, Grylloptera, and Orthoptera), by V.R. Vickery and D.K.McE. Kevan, Biosystematics Research Institute, Ottawa, 1986. 918 pp. Cat. No. A42-42/1985-14E.

Part 15. The Metallic Wood-boring Beetles of Canada and Alaska (Coleoptera: Buprestidae), by D.E. Bright, Biosystematics Research Centre, Ottawa, 1987. 335 pp. Cat. No. A42-42/1986-15E.

Part 16. The Horse Flies and Deer Flies of Canada and Alaska (Diptera: Tabanidae), by H.J. Teskey, Biosystematics Research Centre, Ottawa, 1990. 381 pp. Cat. No. A42-42/1990-16E.

Part 17. The Wolf Spiders, Nurseryweb Spiders, and Lynx Spiders of Canada and Alaska (Araneae: Lycosidae, Pisauridae, and Oxyopidae), by C.D. Dondale and J.H. Redner, Biosystematics Research Centre, Ottawa, 1990. 383 pp. Cat. No. A42-42/1990-17E.

Part 18. The Flower Flies of the Subfamily Syrphinae of Canada, Alaska, and Greenland (Diptera: Syrphidae), by J.R. Vockeroth, Centre for Land and Biological Resources Research, Ottawa, 1992. 456 pp. Cat. No. A42-42/1991-18E.

Part 19. The Ground Spiders of Alaska and Canada (Araneae: Gnaphosidae), by N.I. Platnick and C.D. Dondale, Centre for Land and Biological Resources Research, Ottawa, 1992. 297 pp. Cat. No. A42-42/1991-19E.

THE INSECTS AND ARACHNIDS OF CANADA

PART 18

The Flower Flies of the Subfamily Syrphinae of Canada, Alaska, and Greenland

Diptera: Syrphidae

J.R. Vockeroth (retired)

Centre for Land and Biological Resources Research
Ottawa, Ontario

Research Branch
Agriculture Canada

Publication 1867 1992

Available in Canada through

Associated Bookstores
and other booksellers

or by mail from

Canada Communication Group—Publishing
Ottawa, Canada K1A 0S9

Cat. No. A42-42/1991-18E
ISBN 0-660-13830-1

Canadian Cataloguing in Publication Data

Vockeroth, J.R.

The flower flies of the subfamily Syrphinae of Canada, Alaska, and Greenland : Diptera, Syrphidae

(The Insects and arachnids of Canada, 0706-7313 ;
pt. 18)
(Publication ; 1867)

Includes bibliographical references and index.
Cat. No. A42-42/1991-18E
ISBN 0-660-13830-1

1. Syrphidae–Canada. 2. Syrphidae–Alaska. 3. Syrphidae–Greenland. I. Canada. Agriculture Canada. II. Title. III. Title: Diptera, Syrphidae. IV. Series. V. Series: Publication (Canada. Agriculture Canada). English ; 1867.

QL537.S9.V6 1991 595.7'5 C91–099104–9

Staff editor: Jane T. Buckley

Contents

Acknowledgments

I obtained distribution records over many years from specimens submitted to the Biosystematics Research Centre[1] for identification; these sources are too numerous to mention individually. I am grateful to the following institutions and individuals from whom I borrowed specimens or in whose collections I examined types:

Academy of Natural Sciences, Philadelphia, Pa. (via H.V. Weems)
American Museum of Natural History, New York, N.Y. (P. Wygodzinsky)
Bishop's University, Lennoxville, Que. (D.J. Hilton)
British Museum (Natural History), London, England (K.G.V. Smith)
California Academy of Sciences, San Francisco (P.H. Arnaud, Jr.)
California Department of Food and Agriculture, Sacramento (via H.V. Weems)
Carnegie Museum, Pittsburgh, Pa. (G. Ekis)
Cornell University, Ithaca, N.Y. (L.L. Pechuman)
Dartmouth College, Hanover, N.H. (via H.V. Weems)
Dayton Museum of Natural History, Dayton, Ohio (G.A. Coovert)
Florida State Collection of Arthropods, Gainesville (H.V. Weems)
Hope Museum, Oxford, England (G.C. Varley, C. O'Toole)
Illinois State Natural History Survey, Urbana (D.W. Webb)
Institute of Biology, Novisibirsk, Russia (A.V. Barkalov)
Kansas State University, Manhattan (via H.V. Weems)
Lyman Entomological Museum, McGill University, Ste. Anne de Bellevue, Que. (D.K.McE. Kevan)
Memorial University, St. John's, Nfld. (D.J. Larson)
Museum of Comparative Zoology, Harvard University, Cambridge, Mass. (Karen Jepson)
Natal Museum, Pietermaritzburg (B.R. Stuckenberg)
National Museum of Natural History, Smithsonian Institution, Washington, D.C. (F.C. Thompson)
Newfoundland Forestry Centre, Forestry Canada, St. John's, Nfld. (K.E. Pardy)
New York State Museum, Albany (J.K. Barnes)
North Dakota State University, Fargo (via H.V. Weems)
Nova Scotia Museum, Halifax (B. Wright)
Ohio State University, Columbus (G.A. Dahlem)
Ohio State University Experiment Station, Wooster (via H.V. Weems)
Oregon State University, Corvallis (Kathryn A. Phillips)
Peabody Museum, Yale University, New Haven, Conn. (C.L. Remington)
Research Station, Agriculture Canada, Charlottetown, P.E.I. (Mary Smith)
Research Station, Agriculture Canada, St. John's, Nfld. (R.F. Morris)

---

[1] Now the Biological Resources Division of the Centre for Land and Biological Resources Research.

Royal British Columbia Museum, Victoria (R.A. Cannings)
Royal Ontario Museum, Toronto (G.B. Wiggins)
San Jose State University, San Jose, Calif. (via H.V. Weems)
Snow Museum of Entomology, University of Kansas, Lawrence (G.M. Byers)
Texas A&M University, College Station (Sammie J. Merritt)
Université de Montréal, Montreal, Que. (Monique Coulloudon)
University of Alberta, Edmonton (D. Shpeley)
University of Arkansas, Fayetteville (via H.V. Weems)
University of British Columbia, Vancouver (S.G. Cannings)
University of California, Riverside (S. Frommer)
University of Guelph, Guelph, Ont. (D. Pengelly)
University of Idaho, Moscow (J.B. Johnson)
University of Michigan, Ann Arbor (via H.V. Weems)
University of Minnesota, St. Paul (P.J. Clausen)
University of Nebraska, Lincoln (via H.V. Weems)
University of New Hampshire, Durham (D.C. Chandler)
University of Prince Edward Island, Charlottetown (L. Drake)
University of Saskatchewan, Saskatoon (D.M. Lehmkuhl)
Utah State University, Logan (W.J. Hanson)
Washington State University, Pullman (R. Zack)
Zoological Institute, Academy of Sciences, St. Petersburg, Russia (A.A. Stackelberg)
Zoological Institute, Lund University, Lund, Sweden (H. Andersson)
Zoologisk Museum, Universitets Copenhagen, Copenhagen, Denmark (L. Lyneborg)
K. Barber, Sault Ste. Marie, Ont.
W. Barkemayer, Oldenburg, Germany
D.J. Bickel, Sydney, Australia
K.G. Capelle, Brigham City, Utah
C. Claussen, Flensburg, Germany
F.D. Fee, State College, Pa.
P. Hallett, Toronto, Ont.
R. Hurley, Arcata, Calif.
W. Krivda, The Pas, Man.
D.W. Langor, Edmonton, Alta.
P. Láska, Olomouc, Czechoslavakia
J. Lucas, Rotterdam, Netherlands
V.A. Mutin, Vladivostok, Russia
T. Nielsen, Sandnes, Norway
K. Ôhara, Fukuoka, Japan
K. Rognes, Madla, Norway
M.C.D. Speight, Bray, Ireland
H.V. Weems, Jr., Gainesville, Fla.

Particular thanks are due to L.V. Knutson, U.S. Department of Agriculture, Washington, for permission to use the original drawings of Figs. 22–31 and 173–199. I wish to thank also R. Idema who prepared

Figs. 1, 2, 4–6, 17–21, 36–46, 104, and 124; B. Flahey who prepared the many other figures; H.C. Walther who arranged, mounted, and labeled the illustrations, prepared the distribution maps, and helped in many other ways; and Marie D. Landry, who translated the keys into French.

Introduction

The Syrphidae, flower flies or hover flies, are among the most common, most brightly colored and conspicuous, and most frequently collected of Diptera. They occur on almost all land areas except Antarctica; they occur from the northern end of Greenland and Ellesmere Island southward. Because of their diverse larval habits, they occur in almost all habitats, from rain forests to arctic or torrid deserts, and from marine salt marshes to the upper limits of vegetation on mountains.

Adults of the two larger subfamilies, Syrphinae and Eristalinae (=Milesiinae), which together make up more than 95% of the Nearctic species (the third subfamily, Microdontinae, has less than 5% of the Nearctic species) feed on nectar and pollen. They are conspicuous visitors to flowers. Almost all are diurnal and are most active during bright sunlight. Perhaps because of their exposed feeding habits they are particularly subject to predation. Apparently for protection, they show a general or sometimes closely specific resemblance to bees and wasps, which exceeds that of any other large family of Diptera (see "Mimicry").

Adults of Syrphinae and Eristalinae are of importance as pollinators. Larvae of Syrphinae and of the tribe Pipizini (of doubtful position but here referred to the Eristalinae) are probably all predators of Homoptera (particularly aphids) and of a few other groups of insects. Only larvae of a few species of Eristalinae are harmful (see sections on "Economic importance" and "Biology").

This work is concerned primarily with identification of adult Syrphinae of Canada, Alaska, and Greenland. Although the subfamily has received much attention from professional and amateur entomologists, both generic and specific taxonomy of Syrphinae, until recently, has been unsatisfactory. The many Nearctic revisions by Curran, Fluke, and Shannon between 1921 and 1954 were commonly based on scanty material; their overemphasis on the importance of color pattern led to recognition, in many genera, of far too many apparent species. The major treatments by Sack (1928–1932) and Séguy (1961) of the Palaearctic and western European species, respectively, suffered from the same defects.

A landmark publication was that of Coe (1953) on the British species. Although he included part or all of 10 genera in *Syrphus* Fabricius, for the first time almost all the species of a considerable fauna could be reliably identified. Later, the generic classification of most Syrphinae of the Holarctic region was revised independently by Dušek and Láska (1967), Hippa (1968), and Vockeroth (1969). These classifications were undoubtedly all influenced by studies of the male terminalia by Fluke (1950, 1957) and differed only in minor details. Subsequent publication of keys to the species of European USSR by Stackelberg (1970) and studies of individual genera by several European workers led to the publication of keys to most northwestern European

Syrphidae (Goot 1981). A recent identification guide to the British Syrphidae (Stubbs and Falk 1983) includes excellent color illustrations of many species. Violovitsh (1983) published a treatment of the species of Siberia; an English translation of the keys is available. Torp (1984) has published an excellent account of the Danish Syrphidae. It includes keys to species, color photographs of specimens of 112 species, and extensive sections on immature forms, cytotaxonomy, migration, mimicry, and habitats. Of particular interest is a detailed account of changes in the Danish syrphid fauna during this century. In North America far less attention has been paid to the Syrphidae, and particularly to the Syrphinae. Knutson (1973) published an excellent revision of *Sphaerophoria* Lepeletier & Serville, and Vockeroth (1980, 1983, 1986*a*, 1990) has published reviews or revisions of several genera and subgenera.

The 22 genera treated here have a total of 166 species in Canada, Alaska, and Greenland. For all but five of these genera, all species known to occur north of Mexico are included in the keys. For two of these genera, *Chrysotoxum* Meigen and *Eupeodes* Osten Sacken, the number of species north of Mexico is still in doubt. The other three genera (*Allograpta* Osten Sacken, *Ocyptamus* Macquart, and *Toxomerus* Macquart) are primarily neotropical; they have two, nine, and nine species, respectively, occurring in the United States but not in Canada (F.C. Thompson, written communication). Eleven species of the genera *Dasysyrphus* Enderlein, *Leucozona* Schiner, *Paragus* Latreille, *Platycheirus* Lepeletier & Serville, and *Syrphus*, known from the conterminous United States but not from Canada, Alaska, or Greenland, are included; four of these species occur in states bordering on Canada and may occur here. Five genera with a total of seven species in the southern United States but almost certainly not occurring in Canada are not treated in the text; they are *Dideomima* Vockeroth, *Leucopodella* Hull, *Pseudodoros* Becker, *Salpingogaster* Schiner, and *Xanthandrus* Verrall. A key to all Nearctic genera and subgenera of Syrphinae is given so the work can be used more reliably in the United States. The five genera not known from Canada, Alaska, or Greenland are enclosed in square brackets in the key to genera.

Only the first reference to each specific name is given, but a complete catalog of Nearctic Syrphidae is being prepared by F.C. Thompson. Distribution outside Canada, Alaska, and Greenland is outlined briefly; records from the literature are followed by an asterisk. Because I have examined most available material from these three areas, and because I have seen many misidentified specimens in collections, the distribution maps in this book are based only on specimens examined by me. The number of specimens examined, plus the distribution maps, give a reasonable estimate of the relative abundance of the various species within the area covered. The figures for other areas (United States, Mexico, Central America, Europe, and Asia) are usually much less significant as they are based in large part on specimens in the Canadian National Collection (CNC). The exceptions are for the species of *Paragus* and of *Platycheirus*; most available Nearctic material of these groups was studied. Seasonal distribution is given following geographic distribution; it is based on months of collection of adult specimens, usually for British Columbia, and for Ontario and Quebec together, but sometimes for other areas. As an example, B.C., V, VII; Ont., Que., VI–IX means specimens were collected in British Columbia in May and July and in Ontario and Quebec in all months from June to September. Keys, unless otherwise indicated, are to the species of Canada, Alaska, and Greenland. Keys to Nearctic species include only those north of Mexico. For some genera, keys to all New World species are given.

In some cases males and females are treated separately in the keys, or characters of one or other sex are mentioned. In males of all Nearctic Syrphinae the eyes meet on the frons or are separated by at most the width of two ommatidia, and the apex of the abdomen is asymmetrical with the genital capsule usually conspicuous in ventral view. In females the eyes are widely separated on the frons and the apex of the abdomen is symmetrical.

Dušek and Láska (1974*b*), after observing that early spring adult specimens of some species were much darker than summer specimens, showed that temperature during pupation in two species of *Eupeodes* (as *Metasyrphus*), in one of *Episyrphus* Matsumura, and in one of *Sphaerophoria* had a very marked effect on the color pattern of the adults. In specimens developing at low temperatures, the abdomen was mostly black; in those developing at high temperatures, it was extensively yellow. Intermediate temperatures produced intermediate coloring. Geography as well as season probably has a strong influence; e.g., northern Canadian specimens of *Sphaerophoria philanthus* (Meigen) are usually darker than southern ones. Some of the abdominal color variations shown in single species of *Sphaerophoria* by Knutson (1973) are probably produced by differences in temperature. I found that specimens of *Eupeodes fumipennis* (Thompson) reared from a single egg batch at 16° and 24°C produced adults which, because of differences in hair and body color, would key to three different species in the key given by Fluke (1952). Specimens of *Syrphus opinator* Osten Sacken reared the same way showed almost no variation in color. Keep in mind this temperature effect when identifying specimens.

Eggs, larvae, and pupae are not treated here, because these stages of the Nearctic species are poorly known. References to the more important European papers are given in the section on "General biology"; at the generic level they are in large part applicable to the Canadian fauna. Of the genera known from Canada only three (*Allograpta*, *Ocyptamus*, and *Toxomerus*) do not occur in Europe.

Distribution

The distribution of each species known to occur in Canada, Alaska, and Greenland is shown in Maps 1–118. The species are divided into two groups on the basis of distribution: group 1, those with a transcontinental distribution; and group 2, those with restricted east–west distribution. Species are included in group 1 if they occur in Alaska, Yukon Territory, British Columbia, or in the Rocky Mountains of Alberta and extend at least as far east as Ontario (Maps 3, 9). They are included in group 2 either if they do not occur as far west as the Rocky Mountains (Maps 6, 50, 118), or if they do not occur east of the Manitoba–Ontario border (or east of Hudson Bay, if they are northern in distribution) (Maps 25, 31). The two groups are divided into five and four subgroups, respectively. Most species are easy to place in one of the nine subgroups. Only about 16 (10% of the total) have a distribution that is intermediate or doubtful; they are assigned arbitrarily to the subgroup in which they fit best.

The subgroups are

1a. Transcontinental, widespread; 47 species, e.g., *Chrysotoxum flavifrons* Macquart (Map 5), *Dasysyrphus venustus* (Meigen) (Map 11).

1b. Transcontinental, arctic (or arctic–alpine); 5 species, e.g., *Eupeodes nigroventris* (Fluke) (Map 26), *Parasyrphus groenlandicus* (Nielsen) (Map 53).

1c. Transcontinental, boreal (or boreoalpine); 11 species, e.g., *Didea alneti* (Fallén) (Map 8), *Melangyna arctica* (Zetterstedt) (Map 35).

1d. Transcontinental, boreal and cordilleran; 18 species, e.g., *Melangyna labiatarum* (Verrall) (Map 38), *Parasyrphus relictus* (Zetterstedt) (Map 57).

1e. Transcontinental, southern; 17 species, e.g., *Didea fuscipes* Loew (Map 12), *Epistrophe xanthostoma* (Williston) (Map 17).

2a. Cordilleran; 39 species, e.g., *Melangyna coei* Nielsen (Map 36), *Parasyrphus macularis* (Zetterstedt) (Map 55).

2b. Western; 14 species, e.g., *Chrysotoxum fasciatum* Müller) (Map 4), *Eupeodes flukei* (Jones) (Map 21).

2c. Eastern; 16 species, e.g., *Chrysotoxum pubescens* Loew (Map 6), *Syrphus rectus* Osten Sacken (Map 112).

2d. Greenland only; 2 species, *Dasysyrphus pinastri* (Degeer) (Map 10), *Sphaerophoria scripta* (Linnaeus) (Map 109).

A complete list of the occurrence of each species in each political area would occupy much space and is unnecessary. However, the number of species in each of these areas is given in Table 1; the species present can be determined from the maps. For each area, a second figure indicates the percentage of Holarctic species. Of the 166 species in the total area, 56 are presently known from Eurasia; this number will undoubtedly increase as the fauna of northern Siberia (especially that of the very large and markedly boreal genus *Platycheirus*) becomes better known. With four exceptions, the species now considered to be Holarctic are known to occur in Alaska or Yukon Territory. Two of these (*Platycheirus discimanus* Loew and *P. jaerensis* Nielsen) have been found only in southern or eastern Canada but probably occur also in northwestern North America. The other two, *Dasysyrphus pinastri* (Degeer) and *Sphaerophoria scripta* (Linnaeus), occur in southern Greenland and are unlikely to be found in Canada; they should perhaps be considered as Palaearctic rather than Nearctic species.

Table 1. Species of Syrphinae in the political divisions of the area treated here

Political division	No. of species	Holarctic %
Alaska	63	73.0
Yukon Territory	83	61.4
Northwest Territories		
District of Mackenzie	66	65.1
District of Keewatin	14	85.7
District of Franklin	9	66.7
British Columbia	116	38.8
Alberta	96	39.6
Saskatchewan	48	45.8
Manitoba	79	50.6
Ontario	86	41.9
Quebec	91	48.4
New Brunswick	54	42.6
Prince Edward Island	18	50.0
Nova Scotia	59	49.2
Labrador	51	62.7
Newfoundland	36	63.9
Greenland	9	55.6

Economic importance

Syrphidae are of direct economic importance in five ways. First, adults of probably all Syrphinae and Eristalinae (but apparently not those of Microdontinae) feed extensively on pollen and nectar; thus they are important as pollinators of many plants. Second, larvae of most or all species of Syrphinae and of species of the tribe Pipizini (Eristalinae or Syrphinae) are predators of various groups of Homoptera (particularly Aphidoidea), of eggs and larvae of Chrysomelidae, and, outside Canada, of Thysanoptera nymphs and of larvae of Tortricidae. Third, species of a few genera of Eristalinae are secondary but destructive invaders in diseased plants. Some species of *Eumerus* Meigen and *Merodon* Meigen in bulbs of ornamental Liliaceae or of onions and one or more species of *Cheilosia* Meigen in burrows of Scolytidae in timber cause disfiguration of the wood known as black check. Fourth, other plant feeders, e.g., species of *Cheilosia* Meigen in *Cirsium* and other plants, and species of *Copestylum* Macquart in Cactaceae, may be of some importance in weed control but their significance is undetermined. Fifth, larvae of a few species of Eristalinae have been reported as causing accidental myiasis in the gut of humans; these cases are almost certainly the result of ingesting larvae in contaminated water or food. There is no evidence that any species of the family seeks humans or any other vertebrate as a preferred host.

The literature includes many comments on the role of Syrphidae as pollinators. Many studies of pollination of individual plant species mention Syrphidae as significant. The use of *Eristalis* spp. as pollinators of fruits, especially in greenhouses, has been suggested in Japan. Syrphidae have been used as pollinators of carrot seed plants in cages (Lyon 1965). The importance of syrphids in pollination of *Plantago* was established by Stelleman and Meeuse (1976). In a study of pollinators of apple in Ontario, Boyle and Philogene (1983) showed that *Apis* and other bees, especially Andrenidae and Halictidae, were most important; at least 15 species of Syrphidae formed the second largest taxonomic group in terms of amount of fruit pollen on their bodies and therefore available for transfer. Unfortunately it was not possible to evaluate the effect of the Syrphidae independent of the other groups.

The literature on larvae of Syrphinae and Pipizini as predators, especially of aphids, is much more extensive than that on syrphids as pollinators. Many studies have suggested that whereas Syrphidae may destroy large numbers of aphids they often appear too late to prevent serious mechanical damage or virus transmission by the aphids (B.D. Frazer, personal communication). However, some studies indicate that Syrphidae are significant control agents, e.g., of *Myzus persicae* on peach trees (Tamaki et al. 1967) and of several species of aphids on apple trees (Schneider et al. 1957). Schneider (1969) reviewed the subject; some of that review and other information are summarized in the section on "General biology."

General biology

Syrphidae, like most Diptera of the infraorder Muscomorpha (=Cyclorrhapha), have as life stages the egg, three free-living larval instars, a pupa inside a puparium (the contracted and hardened integument of the mature third-instar larva), and the adult. This handbook is concerned primarily with identification of adults, but a brief account of other stages is desirable.

Much literature on the biology of immature Syrphinae is related to their importance as predators. Schneider (1969) presented an excellent summary of this information; a brief account of it, with some additions, follows here.

Eggs. Eggs of Syrphinae are usually laid in or very near colonies of the host, although some may be laid on uninfested plants (Chandler 1966, 1968). Most species lay single eggs but some of *Melanostoma* Schiner and *Platycheirus* lay them in overlapping rows. Eggs usually hatch within 2 days.

Larvae. First-instar larvae of Syrphinae, under favorable conditions, can travel more than 1 m in search of their first prey (Chandler 1969). After contact with prey, they continue to search randomly; older larvae can travel for more than 1 m and search among a series of branches of one plant; if they find no prey, they may move to another plant (Bansch 1966). When a larva contacts an aphid it seizes it with its mouthparts, lifts it from the substrate, and ingests the body contents. Many counts of prey consumption have been given; the highest seems to be 867 medium-sized *Aphis fabae* and *Myzus persicae* during the life of an individual larva of *Eupeodes corollae* (Fabricius) (Bombosch 1962). Also, many figures for duration of various life stages under different conditions have been given; at 22–24°C, 40–80% relative humidity, and with a 17-h photoperiod, *E. corollae* had a mean larval life of 7.5 days and pupal life of 8.0 days with males developing slightly faster than females (Barlow 1961). Lower temperatures or reduction in food supply will extend these periods (Bombosch 1962; Dušek and Láska 1974*a*, 1974*b*). Obligatory diapause in univoltine species or facultative diapause in autumn generations of multivoltine species may extend the larval period to many months (Schneider 1948, Dušek and Láska 1974*a*). Larvae of at least some Bacchini seem to develop more slowly than those of many Syrphini (Davidson 1922).

Larvae of many species are highly polyphagous; 38 aphid host species of *Episyrphus balteatus* (Degeer) have been recorded in Czechoslovakia alone (Láska and Starý 1980). Very few species seem to be restricted to a single host, although Dušek and Láska (1966) found larvae of *Melangyna cinctus* (Fallén) (as *Fagisyrphus cinctus*) only on *Phyllaphis fagi* in rolled leaves of *Fagus*. Some species seem to need aphids with a waxy secretion, or those forming galls; this requirement is more common in the Pipizini than in the Syrphinae. Dušek and Láska (1966) found that habitat and vertical position of the host colonies were important: *Syrphus torvus* Osten Sacken and *Epistrophe eligans* (Harris) [as *E. bifasciata* (Fabricius)] selected aphids on shrubs and trees, whereas species of *Sphaerophoria* preferred those on herbs. Some species did not seem to discriminate. Yakhontov (1966) reported that several species feeding on a variety of aphids in central Asia showed slight or marked preferences when offered a choice in the laboratory. It has been suggested that species of several genera, especially of *Toxomerus* and Bacchini, can use, or perhaps require, fresh or decaying plant material; the evidence was summarized by Hamrum (1966).

Schneider (1948) studied differences in development of larvae with and without obligatory diapause and suggested that development rate of imaginal eye discs might differentiate the two types. Schneider (1948), as well as Dušekand Láska (1974*a*), found that in central Europe several species overwinter as mature larvae but that *Eupeodes luniger* (Meigen) overwinters as a pupa. A few species, however, hibernate as adults (Schneider 1947).

Larvae of Syrphidae have a variety of parasites. Weems (1954) listed 65 species from North America, distributed among nine families of Hymenoptera. As many as 40% of the larvae occurring among *Aphis fabae* on beets may be parasitized (Bombosch 1959). Schneider (1950*a*, 1950*b*, 1951) studied many of the physiological aspects of host–parasite interaction.

Further information, especially concerning host records, is given in the accounts of individual genera or species.

Pupae. Pupation of Syrphinae usually takes place in or near larval habitat. Many multivoltine, predacious species pupate on leaves or stems in an exposed situation among the prey. Species going into diapause usually pupate in leaf litter, just below the surface of the ground, in crevices in bark, or in similar situations.

Adults. Adult Syrphinae, unlike those of many other aphid predators, feed not on aphids but primarily on pollen and nectar (and sometimes on honeydew). Newly emerged females have poorly developed ovaries; both Schneider (1948) and Frazer (1972) reported that freeze-dried pollen of *Corylus* and sugar water were effective in promoting development of eggs. Sturken (1964) discussed various dietary factors related to egg production.

Selectivity in pollen feeding has received much attention recently. Schneider (1958) found pollen from 14 plants in the crops of *Scaeva pyrastri* (Linnaeus) and concluded that syrphids were general feeders. Goot and Grabandt (1970) found that some, but not all, species of *Melanostoma* and *Platycheirus* feed either exclusively or mostly on pollen of anemophilous plants, (i.e., plants without nectar or showy petals and commonly thought to be wind pollinated). These include Gramineae and *Plantago*. Holloway (1976) found that, in New Zealand, larger and more hairy syrphids, especially *Eristalis tenax* (Linnaeus), picked up much pollen on their grooved or plumose hairs. They were probably much more effective as pollinators than the small, nearly bare species *Melanostoma fasciatum* (Macquart), which fed almost entirely on pollen of Gramineae and *Plantago* and picked up almost no pollen on its smooth body hairs. Stelleman and Meeuse (1976) showed, however, that some species of *Platycheirus* and *Melanostoma* can be important pollinators of *Plantago lanceolata* but found few pollen grains of Gramineae on the flies; they thought that the flies were probably less important as pollinators of this group. Leereveld (1982) analyzed gut pollen of about 160 specimens of several species of these genera. He found that many species from Europe, North America, New Zealand, and Madagascar fed extensively on pollen of *Plantago*, Gramineae, and Cyperaceae. (He cited published reports of three species of *Platycheirus* feeding on *Typha* pollen and of *Toxomerus marginatus* (Say) feeding on *Plantago* pollen). The importance of the flies as pollinators of plants other than *Plantago lanceolata* is still unsettled.

Gilbert (1981) gave an interesting account of the interaction between Syrphidae and flowers. Factors studied included relationship between body size and mouthpart size and morphology of about nine species, the morphology and nectar production of flowers visited, and the preferences of different species and sexes for pollen or nectar.

Mating of most species apparently takes place in the air; males of some species hover and wait for passing females, but others take stations and fly out at passing insects. Frazer (1972) was able to mate five species in large cages but Bombosch (1957) found that *Eupeodes corollae* will mate readily in small containers. Both authors, as well as Barlow (1961), described methods for maintaining laboratory cultures. Mating behavior, involving territoriality of males, has recently been described for several species of Eristalinae (Maier 1978, 1982; Maier and Waldbauer 1979; Wellington and Fitzpatrick 1981; Fitzpatrick and Wellington 1983). Maier (1982) has shown that many species of Eristalinae copulate near where eggs are later deposited and that its observation may lead to discovery of larval breeding sites. It is not known whether the same is true of Syrphinae.

There have been many studies of the factors involved in host finding and oviposition. Visual, chemical, and mechanical stimuli are apparently all involved, but the presence of light and of prey (or their exudates) seem to be the most important. The subject has been summarized by Schneider (1969).

The excellent flight capability of adult Syrphidae undoubtedly assists them in finding food, mates, and oviposition sites, but how far they travel for these purposes is not known. Many European workers have reported mass flights of a number of species, particularly over water or through mountain passes; hundreds of thousands of individuals have been reported in some flights. Marking experiments have shown that some individuals may travel up to 30 km/h over 5 km and, under certain conditions, up to 111 km in less than 3 days (Aubert et al. 1969). The reason for these flights is unknown, but, as some migratory species are known to hibernate as adults, they may be connected with a search for winter quarters. As far as I know no such mass flights have been reported in North America.

Schneider (1947, 1958) has studied hibernation of adults, particularly of those of *Scaeva pyrastri*. Typically only mated females hibernate. In the early spring the flies travel up to 2 km daily to and from the hibernation site to find pollen and nectar. Oviposition of hibernating species begins early on small aphid colonies. Wnuk and Fuchs (1977) showed that one larva of *Episyrphus balteatus* (Degeer) could eliminate a colony of up to 50 *Brevicoryne brassicae* but could not do so with appreciably larger colonies because of the reproduction of the aphids. Therefore hibernating species that begin laying when aphid colonies are small may be particularly important in aphid control. No Canadian species have been reported to hibernate as adults; however, three of the hibernating species mentioned by Schneider (1969) occur here, and specimens of a number of species that appear in early spring do not look as though they are newly emerged.

Many faunistic studies of Syrphidae have been published, especially in Europe, but many are little more than lists of species taken in a particular habitat, or in different habitats in one geographic or political area. The most interesting is that of Speight et al. (1975) on the Irish Syrphidae; similar studies for other areas where most or all the species can be reliably identified would be very useful.

Mimicry

There is no reasonable doubt that many species of Syrphidae are excellent Batesian mimics of species of bees and wasps. Color polymorphism in some species, especially the eristalines *Merodon equestris* (Fabricius) and *Volucella bombylans* (Linnaeus), is correlated geographically with color patterns in the local species of *Bombus* (Gabritschevsky 1924, 1926). Many species of nonpolymorphic Eristalinae show a remarkably close resemblance in body color and shape to certain species of Hymenoptera, and their behavior (e.g., waving black fore legs in front of the head to resemble long antennae of vespids, or flicking dark wings as do Pompilidae) in many cases increases the resemblance. Many examples of these similarities are discussed by Waldbauer (1970). He and his co-workers have given strong circumstantial evidence that, in Illinois, birds are the most important predators of Syrphidae and that they remember warning patterns from one year to the next (Waldbauer and Sheldon 1971, Waldbauer et al. 1977). Brower et al. (1960) and Brower and Brower (1965) found that young toads could be conditioned to recognize color patterns of species of Hymenoptera and would then avoid similarly patterned Diptera. Their results were much more conclusive with native Nearctic *Bombus* and *Laphria* Meigen (as *Bombomima* Enderlein) (Asilidae) than with *Apis* and *Eristalis* Latreille, perhaps because *Apis* is of Old World origin and the *Eristalis* used is a New World species.

Many Eristalinae resemble closely individual species or species groups of Hymenoptera, but few Syrphinae do so, particularly in the Nearctic region. The only striking exceptions among Canadian species are those of *Chrysotoxum*, which resemble species of *Vespula*, and the single species of *Doros* Meigen, which resembles Eumenidae of the genus *Ancistrocerus*. Most Nearctic Syrphini and Toxomerini, and to some extent Bacchini, have a conspicuous, black and yellow or orange, wasplike color pattern, so that mimicry of wasps seems probable. However, there is so much variation in color pattern, body shape, and body size, both inter- and intra-specifically, that specific models for most species cannot be recognized. Possibly, a general or nonspecific mimicry occurs that gives species at least slight protection from predators because of a wasplike color pattern; so far no experimental or observational evidence supports this. Why so many Eristalinae and so few Syrphinae appear to have specific mimetic models is as yet unexplained.

Anatomy

Eggs. "Typically, the eggs are about 1 mm long, white in color and elongate oval in shape, with an inconspicuous micropyle and microscopic surface sculpturing. The ventral surface of the egg (that in contact with the substratum) is flattened but the dorsal surface is strongly convex. The eggs may readily be distinguished from those of other aphid predators by these features" (Chandler 1968). The pattern of the chorion of a few eggs of Nearctic species was shown by Metcalf (1916, 1917) and of some Japanese species by Ninomiya (1959). The most extensive study of eggs of predacious Syrphidae is that of Chandler (1968); he keyed and illustrated the chorionic sculpture of 35 British species of Syrphinae and Pipizini. Species of a single genus (as now defined) commonly have very similar eggs that can be hard to distinguish; however, in several cases, eggs of congeneric species show striking differences, e.g., those of *Platycheirus scutatus* (Meigen) and *P. albimanus* (Fabricius). Kuznetzov (1988) gave keys to eggs for 37 species of 24 genera of Syrphinae and Eristalinae, illustrated with scanning electron microscope photographs.

Larvae. Larvae of Syrphidae vary greatly in habitus and habits and are difficult to characterize as a group. They have the posterior spiracles elevated on structures ranging from a short prominence (or rarely a pair of prominences) to a very long and retractile respiratory tube. Aquatic larvae of Eristalinae, with this retractile tube, are known as "rat-tailed maggots." The spiracular prominences in mature larvae are united along the median margin except in at least one species of *Toxomerus*. The spiracular openings are usually three in number and straight to strongly convoluted; but, in some Eristalinae and most or all Microdontinae, they may be numerous and very irregular, with many branches or with scattered groups of pores. The body surface has dense pubescence or systematically arranged spicules or tubercles; various segments (especially the posterior one or more) may have simple or elaborate fleshy processes. Some species, especially those of aquatic larvae of Eristalinae, may have well-developed thoracic and abdominal prolegs with strong crochets (data from Heiss 1938, Hartley 1961, Teskey 1981).

Larvae of Syrphinae are probably all partly or entirely predacious (Hamrum 1966). They are soft bodied and maggotlike or, rarely, strongly depressed and taper markedly toward the head. Heiss (1938) presented the most extensive treatment of Nearctic larvae and pupae but omitted many genera. Since then only a few isolated descriptions of Nearctic species have appeared. The European species have been more extensively studied. Dušek and Láska (1967) gave a key for larvae of most European genera. Scott (1939) and Dixon (1960) gave keys to larvae of many British species, and Goeldlin (1974) for those of many species of Switzerland. Rotheray and Gilbert (1989) presented an excellent study of larvae and puparia for most of the predacious genera in Europe. They gave a cladistic analysis of relationships, a tribal classification, generic diagnoses, a review of the biology of each genus with many aphid host records, and a key to genera. Whereas in some cases larvae or pupae of the species of one genus may be distinguished, in other cases (Dušek and Láska 1964) differences may be extremely slight. Goeldlin (1974) discussed color patterns of larvae and concluded they must be used with caution for purposes of identification.

Larvae of Eristalinae are much more varied than those of Syrphinae. They occur in water (usually with high organic content), marshy soil, rot-holes or ulcerating wounds of trees, rotting logs, fungi, stems or roots of vascular plants (usually or perhaps always as secondary invaders), nests of colonial Hymenoptera, and as predators in colonies of aphids (Pipizini). Metcalf (1916) gave a table of larval habitats. Many species are illustrated by Heiss (1938) and Hartley (1961) and a few by Teskey (1981).

Larvae of Microdontinae live in nests of ants; they are oval with a flat venter and strongly convex dorsum.

Pupae. Pupation takes place within the puparium. The distinguishing characters of the puparium are essentially those of the mature larva, but the body processes are usually shriveled or distorted.

Probably in all Eristalinae (except Pipizini) and in Microdontinae a pair of anterior pupal spiracles (respiratory horns) protrude through the integument of the puparium and offer important taxonomic characters (Heiss 1938, Hartley 1961). Anterior spiracles probably do not appear in Syrphinae. Many puparia can be identified by using keys to larvae. Thompson (1981) gave a key to, and illustrations of, puparia of many species of Nearctic *Microdon* Meigen.

Adults. The comprehensive account of the morphology of adult Diptera by McAlpine (1981) makes a detailed description superfluous here. The important anatomical features of adult Syrphinae are shown and labeled in Figs. 4, 12, 15, 26, 32, 33, 36–47, 68, 205, 211, 229, 236, 254, and 261; each is defined in the glossary.

Collection and preservation of specimens

Adult Syrphidae are best collected with an aerial net at flowers, on leaves (especially where shafts of sunlight penetrate a forest), on tree trunks or fallen logs, or while they hover under tree branches. Many species of both Syrphinae and Eristalinae congregate on open hilltops to mate; males are much more abundant than females and may hover over, or rest on, shrubs or small trees. Males of species of Syrphinae usually hover in specific locales. Some species are attracted to yellow paper flowers coated with sugar (Schneider 1958). Gentle sweeping of vegetation with a soft net, particularly in marshes, may yield many specimens of the smaller species. Pinning freshly killed specimens and, if possible, placing them in a deep-freezer for at least a month prevents discoloration during drying. Freezing also reduces distortion or shriveling of reared specimens. Even the smallest specimens, e.g., those of *Paragus*, can be pinned directly on fine pins (size 00 or 0); removal of male terminalia is much easier with pinned specimens than with those glued to a pin or to a card point. Fresh specimens can be mounted easily after many months if placed in a freezer in a watertight container with a few leaves (those of *Acer* [maple] are excellent).

Large numbers of adults can be taken in traps of many kinds (e.g., suction traps, Malaise traps, Malaise troughs, pan traps, sticky traps, car or bicycle nets). For large-scale collecting or for sampling of populations many of these methods are useful, but most of them yield poorer specimens than those obtained by netting. Specimens preserved in ethyl alcohol can be dried and pinned using either ethyl acetate (Vockeroth 1966) or a critical-point dryer.

Eggs may be laid by females in the laboratory, or they may be dissected from gravid females. Eggs can be preserved in Pampel's fluid or can be freeze-dried. A method of examination is described by Chandler (1968). Scanning electron microphotographs of freeze-dried specimens are valuable. Larvae can either be found in the field or, for some species, be reared from eggs in the laboratory on prey on potted and caged plants. The latter method (Frazer 1972) is preferable because many field-collected specimens are parasitized. Larvae can be freeze-dried and mounted on card points or killed with boiling water and preserved in 80% ethanol or Kahle's fluid.

Puparia, if adults have not emerged, can be preserved in the same way as are larvae. If adults have emerged the puparial remains, with the opercula, can be mounted on a card (or placed in a gelatin capsule) below the pinned adult.

Classification

Division of the Syrphidae into the three subfamilies proposed by Thompson (1969), namely, Syrphinae, Eristalinae, and Microdontinae, is recognized here, although previous authors recognized as few as two (Wirth et al. 1965, Vockeroth 1969) or as many as 21 subfamilies (Shiraki 1949). The subfamily position of only one small group, the Pipizini with three Nearctic genera (*Pipiza* Fallén, *Heringia* Rondani, and *Trichopsomyia* Williston) and about 30 Nearctic species, is in doubt. The adults have the diagnostic characters of the Eristalinae but the larvae, like most or all Syrphinae and unlike other Eristalinae, are predacious. They have therefore been referred to Syrphinae by some and to Eristalinae by other authors. Because they run to Eristalinae in the key to subfamilies given here, they are not treated here.

Four tribes of Syrphinae are currently recognized: the Bacchini, Paragini, Toxomerini, and Syrphini. They are distinguished by external characters and by those of the male terminalia.

The Bacchini and Melanostomatini of Wirth et al. (1965) and Vockeroth (1969) are combined as a single tribe characterized by a simple, unsegmented aedeagus. Most of the species of *Baccha* Fabricius of Wirth et al. have a two-segmented aedeagus and are therefore referred to the Syrphini as the genus *Ocyptamus* Macquart; only one Nearctic species, *B. elongata* (Fabricius), is left in the genus *Baccha*. The species referred by Wirth et al. (1965) to *Carposcalis* Enderlein and *Pyrophaena* Schiner, and many of those of *Melanostoma*, are here placed in *Platycheirus*; this generic synonymy is discussed by Vockeroth (1990).

The classification of the Syrphini is little changed from that of Vockeroth (1969). The major changes are as follows:

- the genus *Ocyptamus* is transferred from the Bacchini to the Syrphini
- *Chrysotoxum* (with male terminalia very similar to those of *Epistrophe*) is included in the Syrphini

- the name *Parasyrphus* Matsumura is used instead of *Phalacrodira* Enderlein
- the number of genera recognized is slightly reduced.

The most important change in nomenclature is the inclusion of the species of *Metasyrphus* Matsumura in the genus *Eupeodes* (Vockeroth 1986*b*).

The four tribes are not treated separately; the Nearctic genera are distributed among them as follows:

Bacchini: *Baccha*, *Leucopodella*, *Melanostoma*, *Platycheirus*, and *Xanthandrus*
Paragini: *Paragus*
Toxomerini: *Toxomerus*
Syrphini: the remaining 20 Nearctic genera.

Key to Nearctic subfamilies of Syrphidae

1. Postpronotal lobe bare (Fig. 40). Head posteriorly strongly concave, closely appressed to thorax, partly or entirely hiding postpronotal lobes. Male abdomen with tergite 5 visible in dorsal view, varying in form from subquadrate or subtriangular to short transverse sclerite (e.g., Figs. 102–111, 112–122). Male holoptic .... **Syrphinae**

   Postpronotal lobe with at least a few suberect or appressed hairs (Fig. 39). Head posteriorly less strongly concave clearly exposing postpronotal lobes. Male abdomen with tergite 5 not visible in dorsal view. Male holoptic or dichoptic 2

2. Cell r_{4+5} either with long stump vein projecting backward from R_{4+5} or with short stump vein projecting forward from M. Antenna with dorsal arista (as in Fig. 3). Postcoxal bridge present and entire. Face uniformly and densely haired, in profile usually gently convex, rarely with weak tubercle just below antenna or straight on upper one-third and evenly convex on lower two-thirds. Male broadly dichoptic .... **Microdontinae**

   Cell r_{4+5} usually without stump vein from R_{4+5} or M, rarely with very short stump vein from apical crossvein; if with stump vein from R_{4+5} then antenna with terminal style. Postcoxal bridge usually absent or incomplete, complete or nearly so only in some species with petiolate abdomen. Face variable, usually broadly bare medially, usually either with distinct tubercle at or below mid length or concave in profile, or with oral margin produced, rarely densely and uniformly haired and straight or slightly convex in profile. Male holoptic to broadly dichoptic .... **Eristalinae**

Clé des sous-familles néarctiques de Syrphidae

1. Lobe du postpronotum glabre (fig. 40). Tête fortement concave dans la région postérieure, appuyant étroitement contre le thorax, cachant ainsi partiellement ou totalement les lobes du postpronotum. Chez le mâle, tergite 5 visible en vue dorsale, variant de presque carré ou triangulaire à un court sclérite transversal (p. ex. fig. 102–111, 112–122). Mâle holoptique .... **Syrphinae**

   Lobe du postpronotum pourvu d'au moins quelques poils légèrement dressés ou affaissés (fig. 39). Tête moins fortement concave dans la région postérieure, exposant clairement les lobes du postpronotum. Chez le mâle, tergite 5 non visible en vue dorsale. Mâle holoptique ou dichoptique 2

2. Cellule r_{4+5} pourvue d'une longue nervure tronquée s'étendant vers l'arrière depuis R_{4+5} ou d'une nervure courte et épaisse s'étendant vers l'avant depuis M. Antenne portant une arista dorsale (comme dans la fig. 3). Pont postcoxal complet. Face garnie de poils uniformes et denses, de profil en général légèrement convexe, rarement pourvue d'un faible tubercule juste en dessous de l'antenne, ou droite sur le tiers supérieur et uniformément convexe sur les deux tiers inférieurs. Mâle largement dichoptique .... **Microdontinae**

   Cellule r_{4+5} habituellement dépourvue de nervure tronquée depuis R_{4+5} ou M, rarement avec une très courte nervure à partir de la nervure transversale apicale; lorsqu'une nervure part de R_{4+5}, l'antenne porte alors un stylet terminal. Pont postcoxal habituellement absent ou incomplet; il n'est complet ou presque complet que chez quelques espèces à abdomen pétiolé. Face variable, en général largement glabre sur la portion médiane, généralement pourvue d'un tubercule distinct à mi-longueur ou plus bas ou de profil concave, ou à marge buccale en saillie, rarement à poils denses et uniformes et de profil droit ou légèrement convexe. Mâle variant de holoptique à très dichoptique .... **Eristalinae**

Key to Nearctic genera and subgenera of Syrphinae[2]

1. Tergite 1 well-developed especially on disc, frequently half as long as tergite 2, always extending well beyond scutellum, sublaterally about three-quarters as long as tergite 2 (Fig. 148); tergites minutely punctate .... ***Paragus*** **Latreille** ... 2

   Tergite 1 greatly reduced, frequently almost linear on disc, practically covered by scutellum, sublaterally at most half as long as tergite 2 (e.g., Figs. 106–115, 119–140); tergites not punctate 3

2. Eye hairs arranged in three more or less vertical bands of contrasting color. Scutellum black with apex narrowly yellow or reddish .... ***Paragus (Paragus)*** **Latreille**

   Eye hairs of nearly uniform color. Scutellum entirely black .... ***Paragus (Pandasyopthalmus)*** **Stuckenberg**

3. Antenna elongate, sometimes longer than head; first flagellomere at least three times as long as wide; pedicel often longer than wide (Fig. 4). Abdomen robust, strongly convex dorsally, strongly margined, usually with posterolateral angles of tergites slightly to strongly projecting (Figs. 104, 105) .... ***Chrysotoxum*** **Meigen**

   Antenna short, shorter than head; first flagellomere at most twice as long as wide; pedicel no longer than width (Figs. 3, 5–17, 22, 27). Abdomen variable but without posterolateral angles of tergites projecting (Figs. 102, 103, 106–208) 4

4. Lower lobe of calypter with many long coarse yellow hairs above, especially on posteromedial part (Fig. 37) .... ***Syrphus*** **Fabricius**

   Lower lobe of calypter with only microscopic pile above or, rarely, with some very fine pale scattered erect hairs 5

5. Posterior margin of eye with distinct triangular emargination at or above level of insertion of antennae (Fig. 17). Facial tubercle well-developed, beginning immediately below insertion of antennae, laterally compressed in some specimens (Fig. 17). Metasternum bare. Abdomen at least weakly margined. Male with sclerotized haired triangular process, variable in length, arising from sternite 10 and projecting posteriorly between bases of surstyli .... ***Toxomerus*** **Macquart**

---

[2] Names of genera not known from Canada are enclosed in square brackets.

Posterior margin of eye with emargination usually either indistinct or shallow and rounded, or if emargination distinct and subtriangular then situated below level of insertion of antennae (Figs. 3, 5, 7–14, 24). Facial tubercle variable, indistinct or absent in some species, not compressed. Metasternum bare or haired. Abdomen margined or unmargined. Male without sclerotized process, with weak bare semimembranous process projecting between bases of surstyli (Figs. 261, 266) 6

6. Hind femur with distinct anteroventral and posteroventral spines on apical half. R_{4+5} slightly to strongly dipped into cell r_{4+5} and abdomen distinctly petiolate .... [***Salpingogaster* Schiner**] ... 7

Hind femur without spines. R_{4+5} usually straight or nearly so (Figs. 32–35); if R_{4+5} distinctly dipped into cell r_{4+5} (Figs. 18–21) then abdomen not petiolate 8

7. R_{4+5} deeply dipped. Tergite 1 produced laterally into strong spur. Upper occipital setae in one row .... [***Salpingogaster (Salpingogaster)* Schiner**] 2 spp.; Texas and Florida

R_{4+5} shallowly dipped. Tergite 1 not produced into spur. Upper occipital setae in three to four rows .... [***Salpingogaster (Eosalpingogaster)* Hull**] 1 sp., *S. nepenthe* (Hull); Florida

8. Anterior anepisternum with short to moderately long erect hairs at least posterodorsally (Fig. 40, anepst) 9

Anterior anepisternum without hairs, with only microscopic pubescence 11

9. Hind coxa with tuft of two or more hairs at posteromedial apical angle (Fig. 46). Eye bare or haired. Abdomen oval (Figs. 149–153) .... ***Parasyrphus* Matsumura**

Hind coxa without hairs at posteromedial apical angle. Eye bare. Abdomen oval, parallel-sided, or petiolate 10

10. Extreme posterior margin of wing with series of minute closely spaced rounded black dots (Fig. 36). Metepisternum bare. Abdomen suboval to parallel-sided (Fig. 147) .... ***Meliscaeva* Frey**

Extreme posterior margin of wing without dots. Metepisternum commonly with several fine hairs. Abdomen variable, parallel-sided to distinctly petiolate (as in Fig. 2) .... ***Ocyptamus* Macquart**

11. Abdomen distinctly petiolate (Fig. 2) 12

Abdomen parallel-sided to oval, never distinctly petiolate (Figs. 102, 103, 106–146, 148, 154–208) 15

12. Face produced rather strongly forward on lower half 13

Face not produced forward, without or with very weak tubercle (Fig. 3) 14

13. Antennal sockets separated by about twice length of scape. Pleura mostly yellow; scutellum yellow .... ***Allograpta* Osten Sacken** (in part)

Antennal sockets separated by less than length of scape. Pleura black; scutellum yellow with broad transverse black band ... . [***Pseudodoros* Becker**]
1 sp., *P. clavatus* (Fabricius); central USA southward

14. Face straight, without tubercle. Metepisternum with row of fine subappressed hairs .... [***Leucopodella* Hull**]
1 sp., *L. marmorata* (Bigot); southern Arizona

Face with low but distinct tubercle just above lower margin (Fig. 3). Metepisternum bare .... ***Baccha* Fabricius**

15. Face and scutellum with entirely black background. Abdomen unmargined (as in Figs. 145, 146). Metasternum and eye bare 16

Face or scutellum, or both, not entirely black, with at least partly yellow or yellow-brown background. Abdomen unmargined or margined. Metasternum and eye bare or haired 18

16. Metepisternum with several fine subappressed hairs; upper and lower katepisternal hair patches broadly separated posteriorly, joined anteriorly (as in Fig. 43). Hind coxa with two or more hairs at posteromedial apical angle (as in Fig. 46) .... [***Xanthandrus* Verrall**]
1 sp., *X. mexicanus* Curran; southern Texas

Metepisternum bare; katepisternal hair patches broadly separated throughout (as in Fig. 42). Hind coxa without hairs at posteromedial apical angle 17

17. Metasternum normal, with median portion broadly joined to lateral arms (Fig. 45). Face variable, almost straight with weak tubercle or moderately or strongly produced forward below; facial pruinosity in some species forming punctate or rippled pattern (Figs. 6–11). Legs of male slender and unornamented, with fore tibia or tarsus (or both) broadened (Figs. 47, 49–62, 65–88, 90), or with femora or tibiae with bristles, hair tufts, or

outstanding or modified hairs (Figs. 91–98) ....
.... ***Platycheirus*** **Lepeletier & Serville**

Metasternum reduced, with deep posterior incision on each side; median portion narrowly joined to lateral arms (Fig. 44). Face with small tubercle, and not produced below; facial pruinosity neither punctate nor rippled (Fig. 5). Legs of male slender, without bristles, hair tufts, or outstanding or modified hairs .... ***Melanostoma*** **Schiner**

18. Metepisternum with tuft of fine hairs below spiracle; metasternum haired. R_{4+5} distinctly dipped into cell r_{4+5} (Fig. 18). Large species (length 9.7 mm or more) with broad strongly margined abdomen (Figs. 116, 117) ....
.... ***Didea*** **Macquart**

Metepisternum bare below spiracle; metasternum haired or bare. R_{4+5} straight or dipped. Size and shape variable 19

19. Scutum with sharply defined clear yellow or whitish yellow lateral or sublateral stripe extending at least from postpronotum to transverse suture (Figs. 28, 29) 20

Scutum with at most poorly defined dull yellow pruinose lateral stripe 25

20. Abdomen margined; tergites 4 and 5 with at least weak margin; tergites 3–5 commonly with strong margin (Figs. 118, 120–123, 208) 21

Abdomen unmargined (Figs. 102, 103, 173–199) 23

21. Katepisternum entirely dark or with diffuse pruinose yellow area above ....
... ***Epistrophe (Epistrophella)*** **Dušek & Láska** (in part)

Upper part of katepisternum with sharply defined bright yellow spot 22

22. Abdomen strongly convex dorsally; tergite 5 with anterolateral angles black or very narrowly yellow (Fig. 118). Anepisternum black; scutellum uniformly brown .... ***Doros*** **Meigen**

Abdomen weakly convex dorsally; tergite 5 with anterolateral angles extensively yellow (Fig. 208). Anepisternum yellow posteriorly; scutellum bright yellow posteriorly ....
.... ***Xanthogramma*** **Schiner**

23. Ventral scutellar fringe present but sparse laterally, absent or nearly so on at least median third. Male terminalia extremely large, globose; tergite 9 as wide as abdomen ....
.... ***Sphaerophoria*** **Lepeletier & Serville**

Ventral scutellar fringe complete, well-developed, moderately dense (Fig. 38). Male terminalia small, inconspicuous; tergite 9 at most one-third as wide as abdomen 24

24. Metasternum bare .... ***Melangyna (Meligramma)*** **Frey** (in part)

Metasternum haired ..... ***Allograpta*** **Osten Sacken** (in part)

25. Wing with narrow but distinct brown band extending from anterior margin across crossvein r-m. Tergite 2 and base of tergite 3 yellow or gray; tergites otherwise black (Fig. 141) .... ***Leucozona (Leucozona)*** **Schiner**

Wing unmarked except for stigmal darkening. Tergites 2–4 each with pale band or pair of spots (Figs. 107–117, 119–140, 142–147, 154) 26

26. Metasternum bare 27

Metasternum with at least some hairs 36

27. Eye distinctly haired; hairs at least as long as distance between their bases 28

Eye bare or nearly so (some species of *Melangyna* s.s. with very short and sparse eye hairs key out to either couplet 28 or 32) 32

28. Wing membrane with very sparse and scattered microtrichia; apical one-third (Fig. 21) with extensive bare areas. Male eye with well-defined enlarged upper facets . ***Scaeva*** **Fabricius**

Wing membrane densely and uniformly trichose on at least apical one-third; veins without bare areas (Fig. 19). Male eye without well-defined enlarged upper facets 29

29. Tergite 2 with large subquadrate gray or yellow spots; spots much larger than pale markings on tergites 3 and 4 (Fig. 140) .... ***Leucozona (Ischyrosyrphus)*** **Bigot**

Tergite 2 either with oval or transverse yellow spots smaller than pale markings on tergites 3 and 4 (Figs. 106–115, 142), or entirely black 30

30. Abdomen unmargined, slender and parallel-sided or narrowly oval (Figs. 142, 143) .... ***Melangyna (Melangyna)*** **Verrall** (in part)

Abdomen weakly but distinctly margined, oval (Figs. 106–115) .... ***Dasysyrphus*** **Enderlein**

31. R_{4+5} at least moderately dipped into cell r_{4+5} (Fig. 20) ... 32

R_{4+5} straight or nearly so (as in Figs. 32–35) 33

32. R_{4+5} moderately dipped (Fig. 20). Face with black median stripe . . . . ***Eupeodes (Lapposyrphus)*** **Dušek & Láska**

R_{4+5} very strongly dipped. Face entirely yellow . . . . [***Dideomima*** **Vockeroth**]
1 sp., *D. coquilletti* (Williston); southern Arizona

33. Abdomen with at least faint trace of margin on tergites 3, 4, or 5 (Figs. 119–125); broader species with oval or suboval abdomen. Face with at most an obscure dark stripe medially 34

Abdomen entirely unmargined (Figs. 142–144); more slender species with nearly parallel-sided abdomen. Face commonly with clearly defined dark median stripe 35

34. Upper and lower katepisternal hair patches narrowly joined posteriorly (Fig. 41). Tergite 4 with entire yellow band (Figs. 119–123) . . . . ***Epistrophe (Epistrophe)*** **Walker** (in part)

Upper and lower katepisternal hair patches broadly separated (Fig. 42). Tergite 4 with yellow band divided medially (Figs. 124, 125) . . . . ***Epistrophe (Epistrophella)*** **Dušek & Láska** (in part)

35. Hind coxa with tuft of hairs at posteromedial apical angle (as in Fig. 46). Pale spots on tergites 3 and 4 transverse and separated (Figs. 142–143). Face usually with black median stripe, rarely entirely yellow . . . . ***Melangyna (Melangyna)*** **Verrall** (in part)

Hind coxa without hairs at posteromedial apical angle. Pale spots of tergites 3 and 4 at least slightly oblique (Fig. 44), in some specimens confluent medially. Face usually entirely yellow, if darkened without black median stripe . . . . ***Melangyna (Meligramma)*** **Frey** (in part)

36. R_{4+5} slightly but distinctly dipped into cell r_{4+5} (Fig. 19). Eye densely haired . . . . ***Eriozona (Megasyrphus)*** **Dušek & Láska**

R_{4+5} straight or nearly so (as in Figs. 32–35). Eye usually bare, rarely with distinct but very sparse hairs 37

37. Abdomen with strong distinct margin extending clearly from middle of tergite 2 to end of tergite 5 (Figs. 127–137). Upper and lower katepisternal hair patches nearly confluent anteriorly, distinctly separated posteriorly (Fig. 43) . . . . ***Eupeodes (Eupeodes)*** **Osten Sacken**

Abdomen with at most very weak indistinct margin from tergite 3 or 4 (Figs. 119, 123). Upper and lower katepisternal hair patches broadly separated anteriorly, narrowly joined posteriorly (Fig. 41) ....
.... ***Epistrophe (Epistrophe)*** **Walker** (in part)

Clé des genres et sous-genres néarctiques de Syrphinae[3]

1. Tergite 1 très long, en particulier sur le disque, souvent d'une longueur égale à la moitié du tergite 2 et s'étendant toujours bien au-delà du scutellum; sublatéralement, d'une longueur équivalant environ aux trois quarts du tergite 2 (fig. 148); tergites ornés de minuscules ponctuations ....
.... ***Paragus*** **Latreille** ... 2

   Tergite 1 très court, souvent presque linéaire sur le disque et presque recouvert par le scutellum; sur la face sublatérale, d'une longueur équivalant au plus à la moitié du tergite 2 (p. ex., fig. 106–115, 119–140); tergites non ponctués .. 3

2. Poils des yeux disposés en trois bandes plus ou moins verticales, de couleur contrastante. Scutellum noir, légèrement jaune ou rougeâtre à l'apex .... ***Paragus (Paragus)*** **Latreille**

   Poils des yeux de couleur presque uniforme. Scutellum tout noir ...... ***Paragus (Pandasyopthalmus)*** **Stuckenberg**

3. Antenne allongée, parfois plus longue que la tête; premier article du flagelle au moins trois fois plus long que large; pédicelle souvent plus long que large (fig. 4). Abdomen robuste, fortement convexe sur la face dorsale, fortement marginé, habituellement avec des tergites dont les angles postéro-latéraux font légèrement à fortement saillie (fig. 104, 105) .... ***Chrysotoxum*** **Meigen**

   Antenne courte, plus courte que la tête; premier article du flagelle au plus deux fois plus long que large; pédicelle pas plus long que large (fig. 3, 5–17, 22, 27). Abdomen variable, mais sans projection au niveau des angles postéro-latéraux des tergites (fig. 102, 103, 106–208) 4

4. Lobe inférieur du cuilleron garni sur le dessus de nombreux poils jaunes, longs et épais, en particulier sur la partie postéro-médiane (fig. 37) .... ***Syrphus*** **Fabricius**

[3] Les noms des genres inconnus au Canada sont indiqués entre crochets.

Lobe inférieur du cuilleron recouvert uniquement de poils microscopiques sur le dessus ou, rarement, de quelques poils très fins, pâles, dressés et clairsemés 5

5. Marge postérieure de l'œil portant une échancrure triangulaire distincte, au niveau ou au-dessus du point d'insertion des antennes (fig. 17). Tubercule facial proéminent, prenant naissance immédiatement sous le point d'insertion des antennes, comprimé sur les côtés chez certains spécimens (fig. 17). Métasternum glabre. Abdomen tout au moins faiblement marginé. Mâle pourvu d'un appendice (de longueur variable) triangulaire recouvert de poils sclérifiés, prenant naissance au sternite 10 et s'étendant vers l'arrière entre les bases des surstyli .... ***Toxomerus* Macquart**

Marge postérieure de l'œil portant une échancrure habituellement non distincte ou peu profonde et arrondie ou, s'il y a une échancrure distincte presque triangulaire, celle-ci est alors située sous le point d'insertion des antennes (fig. 3, 5, 7–14, 24). Tubercule facial de grandeur variable, non distinct ou absent chez certaines espèces, non comprimé. Métasternum glabre ou poilu. Abdomen avec ou sans marge. Mâle dépourvu d'appendice sclérifié, mais pourvu d'un faible appendice semi-membraneux glabre qui s'étend entre les bases des surstyli (fig. 261, 266) 6

6. Fémur postérieur portant des épines antéro-ventrale et postéro- ventrale sur la moitié apicale. R_{4+5} de légèrement à fortement incliné dans la cellule r_{4+5} et abdomen nettement pétiolé .... [***Salpingogaster* Schiner**] ... 7

Fémur postérieur sans épines. R_{4+5} habituellement droit ou presque (fig. 32–35); si R_{4+5} est fortement incliné dans la cellule r_{4+5} (fig. 18–21), l'abdomen n'est alors pas pétiolé 8

7. R_{4+5} fortement incliné. Projection latérale du tergite 1 formant un éperon robuste. Soies occipitales supérieures réunies en une rangée ..... [***Salpingogaster (Salpingogaster)* Schiner**]
2 espèces; Texas et Floride

R_{4+5} légèrement incliné. Tergite 1 ne faisant pas saillie. Soies occipitales supérieures réparties sur trois à quatre rangées .... [***Salpingogaster (Eosalpingogaster)* Hull**]
1 espèce, *S. nepenthe* (Hull); Floride

8. Anépisterne antérieur garni de poils dressés, de courts à modérément longs, au moins sur la partie postéro-dorsale (fig. 40 Anepst) 9

Anépisterne antérieur glabre, garni seulement d'une pubescence microscopique 11

9. Hanche postérieure portant deux poils ou plus au niveau de l'angle apical postéro-médian (fig. 46). Oeil glabre ou poilu. Abdomen ovale (fig. 149–153) .... ***Parasyrphus* Matsumura**

Hanche postérieure glabre au niveau de l'angle apical postéro-médian. Oeil glabre. Abdomen ovale, à côtés parallèles ou pétiolé 10

10. Extrémité de la marge postérieure de l'aile marquée d'une série de minuscules points noirs, arrondis et très rapprochés (fig. 36). Métépisterne glabre. Abdomen presque ovale ou à côtés parallèles (fig. 147) .... ***Meliscaeva* Frey**

Extrémité de la marge postérieure de l'aile non ponctuée. Métépisterne souvent garni de plusieurs poils fins. Abdomen variable, à côtés parallèles ou nettement pétiolé (comme dans la fig. 2) .... ***Ocyptamus* Macquart**

11. Abdomen nettement pétiolé (fig. 2) 12

Abdomen à côtés parallèles ou ovale, jamais nettement pétiolé (fig. 102, 103, 106–146, 148, 154–208) 15

12. Face s'avançant de façon assez marquée vers l'avant, sur la moitié inférieure 13

Face ne faisant pas saillie vers l'avant et pourvue ou non d'un très faible tubercule (fig. 3) 14

13. Cavités des antennes séparées par une longueur équivalant à environ deux fois celle du scape. Pleurites surtout jaunes; scutellum jaune ...... ***Allograpta* Osten Sacken** (*partim*)

Cavités des antennes séparées par une longueur inférieure à celle du scape. Pleurites noirs; scutellum jaune, orné d'une large bande transversale noire .... [***Pseudodoros* Becker**]
1 espèce, *P. clavatus* (Fabricius); du centre des États-Unis vers le sud

14. Face droite, sans tubercule. Métépisterne portant une rangée de poils fins et affaissés .... [***Leucopodella* Hull**]
1 espèce, *L. marmorata* (Bigot); sud de l'Arizona

Face pourvue d'un tubercule petit mais distinct, juste au-dessus du bord inférieur (fig. 3). Métépisterne glabre .... ***Baccha* Fabricius**

15. Face et scutellum sur fond tout noir. Abdomen sans marge (comme dans les fig. 145, 146). Métasternum et œil nus 16

Face ou scutellum, ou les deux, pas entièrement noirs, sur un fond tout au moins partiellement jaune ou brun jaune.

Abdomen avec ou sans marge. Métasternum et œil glabres ou poilus 18

16. Métépisterne orné de plusieurs poils fins et légèrement affaissés; touffes de poils inférieure et supérieure du katépisterne largement séparées sur la partie postérieure, mais réunies antérieurement (comme dans la fig. 43). Hanche postérieure portant au moins deux poils au niveau de l'angle apical postéro-médian (comme dans la fig. 46) .... [***Xanthandrus*** **Verrall**]
1 espèce, *X. mexicanus* Curran; sud du Texas

Métépisterne glabre; touffes de poils du katépisterne largement séparées partout (comme dans la fig. 42). Hanche postérieure sans poils au niveau de l'angle apical postéro-médian 17

17. Métasternum normal, sa portion médiane étant largement reliée aux bras latéraux (fig. 45). Face variable, presque droite et pourvue d'un faible tubercule, ou faisant modérément à fortement saillie vers l'avant, en dessous; pruinosité faciale à motif ponctué ou ridé (fig. 6–11). Chez le mâle, les pattes sont minces et dégarnies; tibia ou tarse antérieur (ou les deux) élargis (fig. 47, 49–62, 65–88, 90), ou fémurs ou tibias garnis de soies, de touffes de poils ou de poils différenciés ou modifiés (fig. 91–98) .... ***Platycheirus*** **Lepeletier & Serville**

Métasternum plus petit, bordé d'une profonde incision postérieure sur chaque côté; portion médiane reliée étroitement aux bras latéraux (fig. 44). Face pourvue d'un petit tubercule, ne faisant pas saillie; pruinosité faciale ni ponctuée, ni ridée (fig. 5). Chez le mâle, les pattes sont minces et dépourvues de soies, de touffes de poils ou de poils modifiés ou différenciés .... ***Melanostoma*** **Schiner**

18. Métépisterne orné d'une touffe de poils fins sous le spiracle; métasternum velu. R_{4+5} nettement incliné dans la cellule r_{4+5} (fig. 18). Chez les espèces mesurant 9,7 mm ou plus, l'abdomen est large et fortement marginé (fig. 116, 117) .... ***Didea*** **Macquart**

Métépisterne nu sous le spiracle; métasternum velu ou glabre. R_{4+5} droit ou incliné. Taille et forme variables 19

19. Scutum orné d'une rayure latérale ou sublatérale bien définie, de couleur jaune clair ou blanchâtre, qui s'étend au moins du postpronotum à la suture transversale (fig. 28, 29) 20

Scutum orné tout au plus d'une rayure latérale pruineuse et mal définie de couleur jaune terne 25

20. Abdomen avec marge; tergites 4 et 5 présentant tout au moins une faible marge; tergites 3–5 souvent fortement marginés (fig. 118, 120–123, 208) 21

Abdomen sans marge (fig. 102, 103, 173–199) 23

21. Katépisterne tout noir ou portant sur le dessus une zone pruineuse jaune et floue ....
... ***Epistrophe (Epistrophella)* Dušek & Láska** (*partim*)

Katépisterne dont la partie supérieure est ornée d'une tache jaune clair bien définie 22

22. Face dorsale de l'abdomen fortement convexe; angles antéro-latéraux du tergite 5 noirs ou avec un très étroit liséré jaune (fig. 118). Anépisterne noir; scutellum tout brun .....
.... ***Doros* Meigen**

Face dorsale de l'abdomen faiblement convexe; angles antéro-latéraux du tergite 5 largement jaunes (fig. 208). Anépisterne jaune sur sa face postérieure; scutellum jaune clair sur sa face postérieure .... ***Xanthogramma* Schiner**

23. Frange scutellaire ventrale présente mais clairsemée sur les côtés, absente ou presque sur au moins le tiers médian. Chez le mâle, appendices terminaux extrêmement larges et globuleux; tergite 9 aussi large que l'abdomen ....
.... ***Sphaerophoria* Lepeletier & Serville**

Frange scutellaire ventrale complète, bien formée et modérément dense (fig. 38). Chez le mâle, appendices terminaux petits et peu visibles; tergite 9 d'une largeur équivalant tout au plus au tiers de l'abdomen 24

24. Métasternum nu ....
.... ***Melangyna (Meligramma)* Frey** (*partim*)

Métasternum poilu ..... ***Allograpta* Osten Sacken** (*partim*)

25. Aile soulignée d'une bande brune, étroite mais distincte, qui s'étend de la marge antérieure à travers la nervure transversale r-m. Tergite 2 et base du tergite 3, jaune ou gris; ailleurs, les tergites sont noirs (fig. 141) ....
.... ***Leucozona (Leucozona)* Schiner**

Aile sans marque, à l'exception du ptérostigma. Tergites 2–4 tous ornés d'une bande pâle ou d'une paire de taches (fig. 107–117, 119–140, 142–147, 154) 26

26. Métasternum nu 27

Métasternum pourvu tout au moins de quelques poils ... 36

27. Oeil nettement poilu, garni de poils d'une longueur au moins égale à la distance entre leurs bases 28

Oeil nu ou presque (certaines espèces de *Melangyna* s.s. sont pourvues de poils très courts et clairsemés et peuvent être identifiées au couplet 28 ou 32) 32

28. Membrane alaire pourvue de microchètes très clairsemés; larges zones glabres sur le tiers apical (fig. 21). Chez le mâle, l'œil présente des facettes supérieures élargies bien définies .... ***Scaeva* Fabricius**

Membrane alaire fortement et uniformément poilue, sur au moins le tiers apical; nervures sans zones nues (fig. 19). Chez le mâle, absence de facettes supérieures élargies bien définies 29

29. Tergite 2 orné de grosses taches grises ou jaunes et presque carrées, qui sont beaucoup plus grosses que les marques pâles présentes sur les tergites 3 et 4 (fig. 140) .... ***Leucozona (Ischyrosyrphus)* Bigot**

Tergite 2 orné de taches jaunes, ovales ou transversales, plus petites que les marques pâles sur les tergites 3 et 4 (fig. 106–115, 142), ou tout noir 30

30. Abdomen sans marge, mince et à côtés parallèles ou étroitement ovale (fig. 142, 143) .... ***Melangyna (Melangyna)* Verrall** (*partim*)

Abdomen portant une marge petite mais distincte, de forme ovale (fig. 106–115) .... ***Dasysyrphus* Enderlein**

31. R_{4+5} au moins modérément incliné dans la cellule r_{4+5} (fig. 20) 32

R_{4+5} droit ou presque (comme dans les figures 32–35) ... 33

32. R_{4+5} modérément incliné (fig. 20). Face ornée d'une rayure médiane noire .... ***Eupeodes (Lapposyrphus)* Dušek & Láska**

R_{4+5} très fortement incliné. Face entièrement jaune .... [***Dideomima* Vockeroth**]
1 espèce, *D. coquilletti* (Williston); sud de l'Arizona

33. Abdomen portant tout au moins une marge peu apparente sur les tergites 3, 4 ou 5 (fig. 119–125); chez les espèces plus larges, l'abdomen est ovale ou presque. Face ornée tout au plus d'une rayure médiane noire obscure 34

Abdomen sans marge (fig. 142–144); chez les espèces plus minces, l'abdomen a presque des côtés parallèles. Face souvent ornée d'une rayure médiane noire, clairement définie ... 35

34. Plaques de poils supérieure et inférieure du katépisterne étroitement réunies sur la face postérieure (fig. 41). Tergite 4 orné d'une bande jaune complète (fig. 119–123) .... ***Epistrophe (Epistrophe)* Walker** (*partim*)

Plaques de poils inférieure et supérieure du katépisterne largement séparées (fig. 42). Tergite 4 orné d'une bande jaune séparée au milieu (fig. 124, 125) .... ***Epistrophe (Epistrophella)* Dušek & Láska** (*partim*)

35. Hanche postérieure garnie d'une touffe de poils à l'angle apical postéro-médian (comme dans la fig. 46). Taches pâles transversales et séparées sur les tergites 3 et 4 (fig. 142–143). Face ornée habituellement d'une rayure médiane noire, rarement entièrement jaune .... ***Melangyna (Melangyna)* Verrall** (*partim*)

Hanche postérieure dépourvue de poils à l'angle apical postéro-médian. Tergites 3 et 4 ornés de taches pâles au moins légèrement obliques (fig. 44), réunies au milieu chez certains spécimens. Face habituellement tout jaune; lorsqu'elle est plus foncée, il n'y a pas de bande médiane noire .... ***Melangyna (Meligramma)* Frey** (*partim*)

36. R_{4+5} légèrement mais nettement incliné dans la cellule r_{4+5} (fig. 19). Oeil fortement poilu .... ***Eriozona (Megasyrphus)* Dušek & Láska**

R_{4+5} droit ou presque (comme dans les fig. 32–35). Oeil habituellement nu, rarement orné de poils distincts mais très clairsemés 37

37. Abdomen ayant une marge bien distincte qui s'étend clairement du milieu du tergite 2 à l'extrémité du tergite 5 (fig. 127–137). Touffes de poils supérieure et inférieure du katépisterne presque réunies sur la face antérieure, nettement séparées postérieurement (fig. 43) .... ***Eupeodes (Eupeodes)* Osten Sacken**

Abdomen orné tout au plus d'une marge très faible peu apparente à partir du tergite 3 ou 4 (fig. 119, 123). Touffes de poils supérieure et inférieure du katépisterne largement séparées sur la face antérieure, étroitement reliées postérieurement (fig. 41) .... ***Epistrophe (Epistrophe)* Walker** (*partim*)

Genus *Allograpta* Osten Sacken

Diagnosis. Species slender with bright yellow markings, commonly with longitudinal submedian and oblique sublateral yellow stripes on tergite 4 (Fig. 103). Length 6.5–9.0 mm.

Description. Eye bare. Face yellow or with brown-to-black median stripe. Scutum black, shining or dull; postpronotum and broad lateral presutural stripe bright yellow. Scutellum mostly or entirely yellow; ventral scutellar fringe complete, moderately long. Pleura black, with bright or very obscure yellow markings. Anterior anepisternum, meron and metepimeron bare. Upper and lower sternopleural hair patches broadly separated posteriorly, joined or nearly so anteriorly. Metasternum with several long hairs. Wing membrane moderately to extensively bare on basal one-quarter. Hind coxa without posteromedial apical hair tuft.

Abdomen slender, parallel-sided, unmargined. Sternites yellow.

Distribution. Two species in southern Canada; 2 more in the United States; about 75 in the Neotropical region. The genus well-represented in warmer parts of Old World (12 species in Afrotropical region, 8 in Oriental and southeastern part of Palaearctic region, and several in Australian region). Many species in Neotropical region and several in New Zealand with much diversity in body form, color pattern, and structure.

Biology. Larvae of two Nearctic species, *A. obliqua* (Say) and the non-Canadian species *A. exotica* (Wiedemann), have been recorded as aphid predators; the former has also been reported as a predator of Coccidae. However, most species of the genus, because they occur in the Neotropical region where aphids are few, probably have prey other than aphids.

Key to Canadian species of *Allograpta*

1. Tergite 4 with arcuate yellow band (Fig. 102). Frons of male with large black spot above antennal bases. Southwestern British Columbia .... ***micrura* (Osten Sacken)**

   Tergite 4 with two narrow submedian yellow stripes and two broader oblique sublateral yellow stripes (Fig. 103). Frons of male yellow. Eastern Canada[4] .... ***obliqua* (Say)**

---

[4] Distribution data in the keys for species occurring in Canada refer to Canadian distribution only (most of these species occur also in the United States).

Clé des espèces canadiennes de *Allograpta*

1. Tergite 4 orné d'une bande jaune arquée (fig. 102). Chez le mâle, front marqué d'une large tache noire au-dessus de la base des antennes. Sud-ouest de la Colombie-Britannique .... ***micrura* (Osten Sacken)**

   Tergite 4 orné de deux étroites rayures jaunes submédianes et de deux rayures jaunes sublatérales, obliques et plus larges (fig. 103). Front du mâle jaune. Est du Canada .... ***obliqua* (Say)**

Allograpta micrura (Osten Sacken)

Fig. 102; Map 1

Sphaerophoria micrura Osten Sacken, 1877:330.
Sphaerophoria picticauda Bigot, 1884:102.
Sphaerophoria transversa Hull, 1943*a*:32.

Length. 6.5–8.8 mm.

Male. Frons yellow, with large black semicircular area above antennal bases. Face yellow with narrow to broad and brown to black median stripe; face slightly protruding below; subcranial cavity about 2.8 times as long as its greatest width.

Scutum subshining black with postpronotum and lateral presutural stripe dull yellow and postalar callus at most slightly yellowish; scutal hairs whitish yellow. Scutellum yellow with posterior margin narrowly blackish. Pleura black, with at most obscure yellow areas near upper margin. Wing only moderately bare basally; cells c and bm trichose on about apical half. Legs mostly yellow brown to dark brown with coxae, bases of fore and mid femora and most of hind femur usually blackish.

Abdomen (Fig. 102). Tergite 1 yellow anterolaterally; tergite 2 with narrow commonly divided transverse yellow band; tergites 3 and 4 each with slightly broader arcuate yellow band; tergite 5 with two small submedian and two larger sublateral yellow spots.

Female. Frons mostly black, yellow laterally on about anterior two-thirds. Facial stripe pale brown in some specimens. Postalar callus distinctly yellow. Scutellum blackish only laterally. Anepisternum posterodorsally, katepisternum above fore coxa and posterodorsally, upper margin of anepimeron, and most of laterotergite bright to dull yellow. Wing less trichose than in male; cells c and bm trichose only at extreme apex. Legs mostly yellow; fore and mid tarsi and most of hind leg usually brownish.

Distribution. Southwestern British Columbia (Map 1), south to California, Texas, Costa Rica, Ecuador, and Argentina (Catamara). B.C., VII–X.

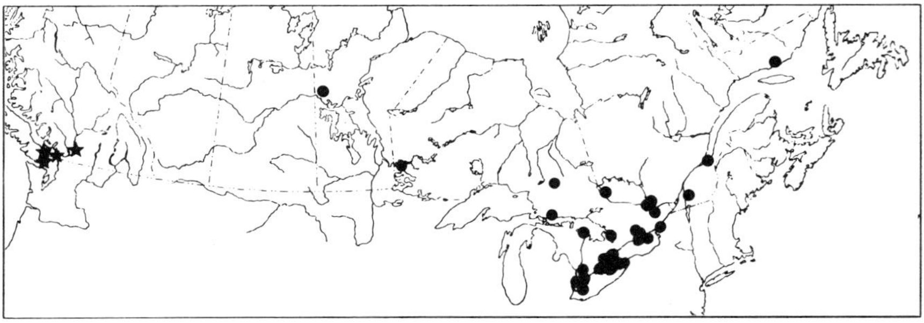

Map 1. Collection localities for *Allograpta micrura* (Osten Sacken) (★) and for *A. obliqua* (Say) (●).

Specimens identified. Canada, 1 ♂, 6 ♀♀; United States,[5] 19 ♂♂, 17 ♀♀; Neotropical region, 56 ♂♂, 108 ♀♀.

Allograpta obliqua (Say)

Fig. 103; Map 1

Scaeva obliqua Say, 1823:89.
Syrphus securiferus Macquart, 1842:160.
Syrphus bacchides Walker, 1849:594.
Syrphus dimensus Walker, 1852:235.
Syrphus signatus Wulp, 1867:144

Length. 5.4–9.0 mm.

Male. Frons yellow. Face yellow, usually with opalescent median stripe; face not produced forward below; subcranial cavity about 2.1 times as long as its greatest width.

Scutum shining black with postpronotum, broad lateral presutural stripe and postalar callus bright yellow; scutal hairs golden yellow. Scutellum yellow, obscurely reddish brown on disc in some specimens.

---

[5] Includes all mainland states except Alaska.

Pleura black with most of upper half, including anterior anepisternum and katepimeron, bright yellow. Wing extensively bare; cells sc and bmwith at most some microtrichia near apex. Legs yellow with preapical ring on hind femur; subbasal ring on hind tibia and apical one-third of hind tibia and most of tarsi brownish.

Abdomen (Fig. 103). Tergite 1 mostly yellow; tergite 2 black with yellow anterior corners and with moderately broad yellow band; tergite 3 black with broad arcuate yellow band; tergite 4 black with two narrow submedian yellow stripes and two broader oblique sublateral yellow stripes; tergite 5 with four yellow stripes.

Female. Frons mostly black, yellow laterally below level of ocelli and narrowly above antennae. Face usually with narrow pale to dark brown median stripe.

Distribution. Eastern Canada (Map 1), west to Oregon*[6] and California, south to Florida and Brazil (Santa Catarina); Hawaii. Ont., Que., V–IX.

Specimens identified. Canada, 36 ♂♂, 75 ♀♀; United States, 75 ♂♂, 74 ♀♀; Neotropical region, 17 ♂♂, 22 ♀♀; Hawaii, 1 ♂, 5 ♀♀.

Biology. Heiss (1938) listed 13 species of aphids, mostly on herbs but also on trees, among which larvae have been found and which presumably serve as hosts. Hardy (1964) reported another species, *Aphis maidis*, as a common host in Hawaii. Swezey (1926) and Hardy (1964) recorded two species of mealybugs (Coccidae) as hosts in Hawaii. Larvae occur throughout the season in the northern United States so the species is undoubtedly multivoltine. Davidson (1922) reported that larvae can survive several days after hatching on plant material without aphids.

Discussion. *Allograpta exotica* (Wiedemann), 1830, a common Neotropical species, extends as far north as Oregon* and Nebraska* but probably does not occur in Canada. The abdominal markings are similar to those of *A. obliqua* but the anterior anepisternum and the katepisternum are black rather than yellow.

---

[6] An asterisk (*) indicates a record based on published literature.

Genus *Baccha* Fabricius

Diagnosis. Species extremely slender with petiolate abdomen at least 25 times as long as its least subbasal width; wings clear or with very faint brownish markings (Fig. 2). Length 7.2–10.2 mm.

Description. Eye with extremely short sparse hairs. Frons and face black, mostly pruinose, some pruinosity yellow below or entirely yellow. Face (Fig. 3) narrow, slightly receding below, with low tubercle; margin of subcranial cavity curved strongly upward. Antenna yellow to black, very short, first flagellomere at least as deep as long.

Thorax with short hairs. Scutum mostly shining, lightly pruinose anteriorly and laterally, entirely black or with margins yellow. Scutellum black to yellow, with sparse ventral scutellar fringe only laterally. Pleura mostly weakly pruinose, black or partly or entirely yellowish. Anterior anepisternum, meron, metapleuron, and metasternum bare. Upper and lower katepisternal hair patches widely separated. Postmetacoxal bridge entire. Wing with alula much narrower than base of cell c, anal angle reduced; membrane with small bare areas near base; crossveins and extreme apex of wing faintly clouded in some specimens. Hind coxa without posteromedial apical hair tuft.

Abdomen (Fig. 2) in males about 40 times as long, in females at least 25 times as long, as least width of segment 2; dark brown to black, reddish towards apex in some females. Tergite 2 with narrow yellow anterolateral spots; tergites 3 and 4 with short basal yellow band; tergite 5 in females with very short yellow basal band or with pair of small anterolateral yellow spots.

Distribution. One widespread Holarctic species (see in the "Discussion"); one Oriental species; perhaps additional species in the eastern part of the Palaearctic region.

Biology. Láska and Starý (1980) recorded three species of aphids as larval hosts in Czechoslovakia. Rotheray and Gilbert (1989) stated that larvae of the British species are associated with ground layer aphids. Dŭsek and Láska (1960) reported that larvae collected in September in Czechoslovakia spent the winter in diapause and pupated in the spring, but Goeldlin (1974) reared larvae from eggs, with *Aphis fabae* as prey, and fund that they developed without diapause in Switzerland.

Discussion. Most European authors have recognized two species, *B. elongata* (Fabricius) and *B. obscuripennis* Meigen, but I cannot distinguish two European species on the basis of the characters given by Coe (1953) and others. The problem is discussed and left unresolved by Stubbs and Falk (1983). I believe that only one species occurs in Europe and in North America.

Baccha elongata (Fabricius)

Figs. 2, 3; Map 2

Syrphus elongatus Fabricius, 1775:768.
Baccha obscuricornis Loew, 1863*a*:15.
Baccha cognata Loew, 1863*a*:15
Baccha angusta Osten Sacken, 1877:332.
Baccha tricincta Bigot, 1883:333.

Length. 7.2–10.2 mm.

Male and female. Characters as given for the genus. Head as in Fig. 3. Abdomen as in Fig. 2.

Distribution. Alaska, Canada (Map 2), south to California and Georgia; Europe. B.C., V–X; Ont., Que., V–VII.

Specimens identified. Alaska, 3 ♀♀; Canada, 27 ♂♂, 73 ♀♀; United States, 12 ♂♂, 16 ♀♀; Europe, 7 ♂♂, 29 ♀♀.

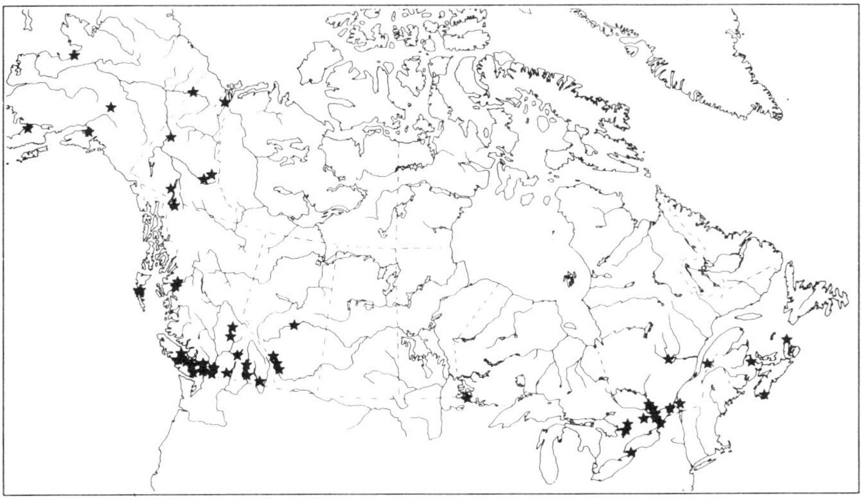

Map 2. Collection localities for *Baccha elongata* (Fabricius).

Genus *Chrysotoxum* Meigen

Diagnosis. Species medium sized to very large with bright yellow markings on thorax and abdomen. Antennae long and usually porrect. Abdomen strongly convex and oval. Length 8.9–17.4 mm.

Description. Eye with long dense hairs in males and with shorter and slightly sparser hairs in females. Face yellow with broad black to brown median stripe, with low tubercle near lower margin. Antenna (Fig. 4) long, black, usually porrect, with scape and pedicel together shorter than first flagellomere or arista.

Scutum black with pair of narrow gray pruinose submedian stripes on anterior one-third and with yellow presutural and postsutural lateral markings. Scutellum yellow with darker translucent discal spot or transverse band or rarely mostly translucent blackish. Pleura black with posterior part of anepisternum bright yellow, usually also with other yellow markings. Ventral scutellar fringe present, sparse to moderately dense. Anterior anepisternum bare or with up to 15 long fine erect pale hairs on lower half. Upper and lower katepisternal hair patches broadly joined posteriorly. Meron, metapleuron, and metasternum bare. Anterior margin of wing brownish; vein R_{4+5} moderately dipped into cell dm. Wing membrane entirely trichose or with cells bm and cu and anal lobe slightly to extensively bare. Lower squama with some erect hairs on upper surface. Hind coxa without posteromedial apical hair tuft.

Abdomen (Figs. 104, 105) strongly convex above, strongly margined laterally, usually with posterolateral angles of tergites 2–4 slightly to strongly projecting beyond margin of succeeding tergite. Tergite 1 black; tergites 2–5 black, each with extensive and variable bright yellow, narrow to broad, entire or divided, arcuate band, and usually also narrow to broad commonly strongly V-shaped posterior band. Sternite 1 entirely yellow or black medially; sternite 2 black, usually yellow anteromedially and posteriorly; sternites 3–5 black, with posterior margin and usually pair of spots on anterior half bright yellow. Surstylus narrow (Fig. 209) or broad (Fig. 210); male terminalia of species north of Mexico otherwise uniform in structure.

Distribution. In New World (North America only), about 10 species of which *C. aztec* Shannon, and two apparently undescribed species only in Mexico; in Old World, about 50 species (mostly Palaearctic, a few Oriental, one Afrotropical).

Biology. Although adults may be very common, especially in Europe and western North America, records of larvae or pupae are few. Larvae or pupae have been found in an ant nest, in a compost heap, under a stone on moist ground, and in turf. The structure of the larval mouth parts suggests they are predacious (Coe 1953, Wirth et al. 1965). A larva of the Palaearctic *C. bicinctum* (Linnaeus) has been reared on pea aphids in the laboratory (Rotheray and Gilbert 1989).

Discussion. The genus *Chrysotoxum*, with long antennae (Fig. 4) and strongly convex abdomen (Figs. 104, 105), has been treated by many authors (e.g., Hull 1949, Cole 1953, Wirth et al. 1965, Vockeroth 1969) as a subfamily of Syrphidae or as a tribe of Syrphinae. However, the male terminalia are so similar to those of *Syrphus* and *Epistrophe* that inclusion of the genus in the tribe Syrphini is warranted. The wasplike habitus of the flies is probably the result of mimicry.

The taxonomy of Nearctic *Chrysotoxum* is difficult and is not fully resolved here. Many of the color characters of both integument and hairs, used by Curran (1924*a*), Shannon (1926), and other authors to distinguish apparent species, appear to be variable although a few can be used in conjunction with other characters. The relative lengths of antennal segments are sometimes of value but are slightly variable and difficult to measure accurately; the proportions given in the key are reasonably close approximations only. The presence of fine, pale, erect hairs on the anterior portion of the anepisternum is a useful character but is variable in several species. Most species show variation in the distribution of the wing microtrichia; only *C. ypsilon* Williston, 1887, from the southwestern United States, has cell bm consistently extensively bare. The extent of the projection of the posterolateral angles of the tergites is useful but more variable within a species than indicated by Shannon. Only *C. flavifrons* Macquart has a surstylus (Fig. 210) differing slightly from that of all other species (Fig. 209) occurring north of Mexico; other structures of the male terminalia appear to be constant.

Variation in length of thoracic hairs is particularly confusing. Many females from southern Manitoba west to southern British Columbia, and from the western United States, have the hairs of the scutum and of the anterior half or more of the scutellum much shorter than usual. These specimens appear to be of a distinct species. However, because males from the same area (and in some cases with the same data) cannot be distinguished from those of *C. fasciatum* (Müller) from northern British Columbia and Yukon Territory, and because a few females from the western United States have scutal hairs of intermediate length, these females are here referred to *C. fasciatum*. Four apparently distinct species can be recognized among Canadian specimens; of these only *C. fasciatum* seems likely to be a species complex. I have examined the extant types of the New World species (except the fragmentary type of *C. villosulum* Bigot (Shannon 1926)), plus much material from the United States and Mexico. *C. ypsilon* Williston (southwestern United States) is apparently distinct; the holotypes of *C. radiosum* Shannon (Indiana) and *C. chinook* Shannon (Washington State) appear to represent additional species. The specimens from the eastern United

States in the CNC appear to be of the three species occurring in eastern Canada; those from the western United States vary considerably and may include one or more species not mentioned here.

Key to Canadian species of *Chrysotoxum*

1. Proepimeron mostly or entirely bright yellow. Hairs of scutum and pleura entirely yellow. Dorsal length of first flagellomere at most 1.2 times dorsal length of scape and pedicel combined. Tergite 2 with entire posterior margin, or at least its middle two-thirds, bright yellow. Sternite 2 with at least posterolateral angles extensively yellow. Alberta eastward .... ***pubescens* Loew**

   Proepimeron black or at most obscurely yellowish. Hairs of scutum and pleura entirely yellow or partly black. Dorsal length of first flagellomere at least 1.4 times dorsal length of scape and pedicel combined (Fig. 4). Tergite 2 and sternite 2 each with posterior margin entirely black or partly or entirely yellow 2

2. Scape at least 1.4 times as long as pedicel (measured on either dorsal or ventral surface) (Fig. 4). Tergites 3 and 4 with posterolateral angles distinctly produced beyond level of margin of following tergite, often subacute; tergite 5 with black inverted Y narrow, its greatest posterior width not more than 0.8 times its length (Fig. 104). Sternite 2 with posterior margin yellow; sternite 3 usually black with posterior margin yellow, some with pair of yellow spots on anterior half. Katepisternum usually with well-defined bright yellow spot near upper margin, some with obscure dull yellow spot. Anterior anepisternum bare or with one to three long fine pale erect hairs on lower half. Widespread .... ***derivatum* Walker**

   Scape not longer than pedicel. Tergites 3 and 4 with posterolateral angles variable, distinctly produced and subacute to scarcely produced and rounded (Fig. 105); tergite 5 with black inverted Y usually wider than long but in some specimens about 0.9 times as wide as long. Sternite 2 with posterior margin yellow or black; sternite 3 with pair of yellow spots on anterior half. Katepisternum entirely black or with obscure or distinct yellow spot near upper margin. Anterior anepisternum bare or with up to 15 long fine pale erect hairs on lower half 3

3. Posterolateral angles of tergites 3 and 4 strongly produced (as in Fig. 104). Sternite 2 with posterior margin yellow. Anterior anepisternum usually with five to 15 hairs on lower half, rarely

with fewer hairs or bare. Katepisternum black or with obscure dull yellow spot near upper margin. Lateral notopleural hairs and anepimeral hairs black. Male: surstylus broader, evenly tapered from base to apex (Fig. 210). Female: scutal hairs long, most of those on anterior half at least four times as long as longest dorsoapical hairs of pedicel; scutellar hairs long and dense, those on anterior half not distinctly shorter than those on posterior half. Widespread .... ***flavifrons*** **Macquart**

Posterolateral angles of tergites 3 and 4 at most slightly produced, usually rounded (Fig. 105). Sternite 2 with posterior margin black or narrowly yellow. Anterior anepisternum bare. Katepisterum black or with dull yellow or well-defined bright yellow spot near upper margin. Lateral notopleural hairs and anepimeral hairs black or yellow. Male: surstylus narrower, tapered on basal two-thirds, nearly parallel-sided on apical one-third (as in Fig. 209). Female: scutal hairs usually long as in *C. flavifrons*, in some very short with those on anterior half at most twice as long as longest dorsoapical hairs of pedicel; scutellum usually with hairs long and dense as in *C. flavifrons*, in some with those on anterior half or more much shorter and sparser than those near posterior margin. Manitoba westward .... ***fasciatum*** **(Müller)**

Clé des espèces canadiennes de *Chrysotoxum*

1. Proépimère surtout ou entièrement jaune clair. Poils du scutum et des pleurites tout jaunes. Sur la face dorsale, premier article du flagelle d'une longueur d'au plus 1,2 fois celle du scape et du pédicelle réunis. Tergite 2 orné d'une marge postérieure complète ou couvrant au moins les deux tiers médians, de couleur jaune clair. Sternite 2 pourvu tout au moins d'angles postéro-latéraux en grande partie jaunes. Alberta et plus à l'est .... ***pubescens*** **Loew**

Proépimère noir ou au plus d'un jaunâtre flou. Poils du scutum et des pleurites entièrement jaunes ou partiellement noirs. Sur la face dorsale, premier article du flagelle d'une longueur d'au moins 1,4 fois celle du scape et du pédicelle réunis (fig. 4). Tergite 2 et sternite 2 ornés tous deux d'une marge postérieure toute noire ou partiellement ou entièrement jaune 2

2. Scape au moins 1,4 fois plus long que le pédicelle (mesuré sur la face dorsale ou ventrale) (fig. 4). Tergites 3 et 4 pourvus d'angles postéro-latéraux souvent subaigus, qui s'avancent nettement au-delà du bord du tergite suivant; tergite 5 portant une étroite marque noire en forme de Y inversé, d'une largeur postérieure maximale ne dépassant pas 0,8 fois sa longueur

(fig. 104). Sternite 2 orné d'une marge postérieure jaune; sternite 3 habituellement noir avec une marge postérieure jaune, parfois garni d'une paire de taches jaunes sur la moitié antérieure. Katépisterne portant habituellement une tache jaune clair bien définie près du bord supérieur, parfois avec une tache jaune terne et floue. Portion antérieure de l'anépisterne glabre ou garnie, dans sa moitié inférieure, d'un à trois poils longs, fins, pâles et dressés. Espèce répandue ....
.... ***derivatum*** **Walker**

Scape pas plus long que le pédicelle. Angles postéro-latéraux des tergites 3 et 4 variant de nettement prolongés et subaigus à arrondis et à peine prolongés (fig. 105); tergite 5 orné d'un Y inversé noir habituellement plus large que long, quoique, chez certains spécimens, sa largeur soit d'environ 0,9 fois sa longueur. Marge postérieure du sternite 2 jaune ou noir; sternite 3 orné d'une paire de taches jaunes dans sa moitié antérieure. Katépisterne tout noir ou souligné d'une tache jaune nette ou floue près du bord supérieur. Anépisterne antérieur glabre ou garni d'au plus 15 poils longs, fins, pâles et dressés dans la moitié inférieure 3

3. Angles postéro-latéraux des tergites 3 et 4 faisant fortement saillie (comme dans la fig. 104). Marge postérieure du sternite 2 jaune. Anépisterne antérieur garni habituellement de 5 à 15 poils dans sa portion inférieure, rarement moins poilu ou glabre. Katépisterne noir ou portant une tache jaune, terne et floue, près du bord supérieur. Poils du notopleurite latéral et de l'anépimère noirs. Mâle : surstylus plus large, fuselé uniformément de la base à l'apex (fig. 210). Femelle : poils du scutum longs, ceux de la moitié antérieure étant pour la plupart au moins quatre fois plus longs que les poils dorso-apicaux les plus longs sur le pédicelle; poils du scutellum longs et denses, ceux de la moitié antérieure n'étant pas nettement plus courts que ceux de la portion postérieure. Espèce répandue .... ***flavifrons*** **Macquart**

Angles postéro-latéraux des tergites 3 et 4 tout au plus légèrement en saillie, habituellement arrondis (fig. 105). Marge postérieure du sternite 2 noire ou légèrement jaune. Anépisterne antérieur glabre. Katépisterne noir ou marqué d'une tache jaune terne ou jaune clair bien définie, près du bord supérieur. Poils du notopleurite latéral et de l'anépimère noirs ou jaunes. Mâle : surstylus plus étroit, fuselé sur les deux tiers basaux, à côtés presque parallèles sur le tiers apical (comme dans la fig. 209). Femelle : poils du scutum habituellement comme chez le *C. flavifrons*; chez certains, les poils sont très courts, ceux de la moitié antérieure faisant au plus deux fois la longueur des poils dorso-apicaux les plus longs du pédicelle;

scutellum habituellement garni de poils longs et denses comme chez le *C. flavifrons*; chez certains, les poils d'au moins la moitié antérieure sont beaucoup plus courts et clairsemés que ceux près du bord postérieur. Manitoba et plus à l'ouest .... ***fasciatum*** **(Müller)**

Chrysotoxum derivatum Walker

Figs. 4, 104, 209; Map 3

Chrysotoxum derivatum Walker, 1849:542.
Chrysotoxum integre Williston, 1887:16.
Chrysotoxum plumeum Johnson, 1924:99.
Chrysotoxum columbianum Curran, 1927*b*:206.
Chrysotoxum minor Curran, 1927*b*:206.

Length. 8.9–16.2 mm.

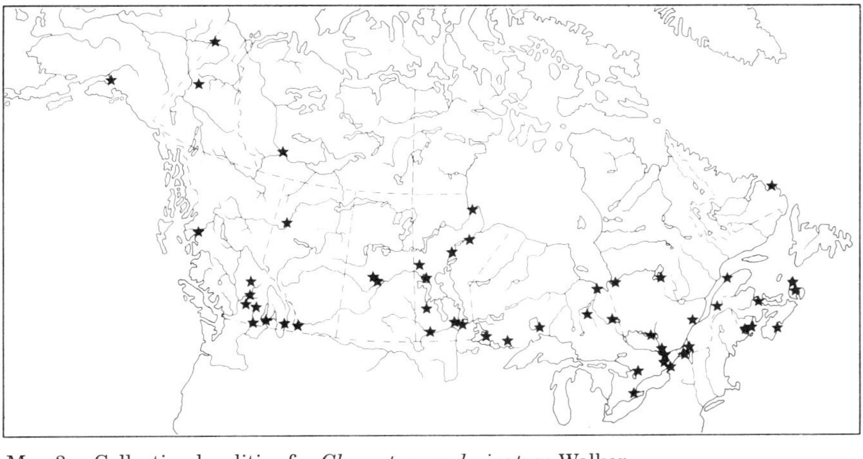

Map 3. Collection localities for *Chrysotoxum derivatum* Walker.

Male. Proportions of antennal segments about 44:27:100 (dorsal), 47:29:100 (ventral).

Hairs of scutum and of upper half of pleura mostly black to entirely yellow; hairs of lower half of pleura yellow. Anterior anepisternum bare or with one to three long fine pale erect hairs on lower half. Proepimeron black; katepisternum with bright yellow or dull yellow spot near upper margin; katatergite obscurely yellowish. Cell bm usually entirely trichose, with narrow bare median stripe in some specimens.

Posterolateral angles of tergites 3 and 4 strongly produced, subacute (Fig. 104). Tergite 2 with posterior margin usually black except laterally, rarely narrowly yellow; discal yellow bands of tergites 3 and 4 usually divided medially, rarely entire. Tergite 5 with black inverted Y narrow, its greatest posterior width at most 0.8 times its length (Fig. 104). Sternite 2 with posterior margin narrowly to broadly yellow; sternite 3 usually black with posterior margin narrowly to broadly yellow, also with pair of yellow spots on anterior half in some specimens. Surstylus (Fig. 209) rather narrow, tapered on about basal two-thirds, then almost parallel-sided to apex.

Female. Proportions of antennal segments about 32:20:100 (dorsal), 35:22:100 (ventral). Scutal and scutellar hairs long.

Distribution. Southern Alaska, Canada (Map 3), south to Arizona, South Carolina, and Mexico (Mexico). B.C., V–IX; Ont., Que., VI–VIII.

Specimens identified. Alaska, 1 ♂, 1 ♀; Canada, 60 ♂♂, 132 ♀♀; United States, 31 ♂♂, 39 ♀♀; Mexico, 24 ♂♂, 5 ♀♀.

Chrysotoxum fasciatum (Müller)

Fig. 105; Map 4

Musca arcuata Linnaeus of recent European authors, not of Linnaeus, 1758:592 (Thompson et al. 1982).
Musca fasciata Müller, 1764:85.
Chrysotoxum ventricosum Loew, 1864:72.
Chrysotoxum coloradense Greene, 1918:70.

Length. 9.6–15.0 mm.

Male. Proportions of antennal segments about 34:38:100 (dorsal), 42:48:100 (ventral).

Hairs of scutum and of upper half of pleura mostly black to entirely yellow; hairs of lower half of pleura yellow. Anterior anepisternum bare. Proepimeron black; katepisterum black or with obscure or bright yellow spot near upper margin; katatergite black or slightly to mostly yellow. Cell bm entirely trichose or with narrow bare median stripe.

Posterolateral angles of tergites 3 and 4 usually only slightly produced (Fig. 105), much less than in *C. derivatum*. Tergite 2 with posterior margin mostly black except laterally, rarely very narrowly yellowish; discal yellow bands of tergites 3 and 4 usually divided medially, rarely entire; tergite 5 with black inverted Y usually slightly wider than long but about 0.9 times as wide as long in some specimens, rarely with the stem or arms narrowly broken. Sternite 2 with posterior margin narrowly black or entirely yellow; sternite 3 with pair of yellow spots on anterior half. Surstylus as in *C. derivatum* (Fig. 209).

Female. Proportions of antennal segments about 33:36:100 (dorsal), 44:47:100 (ventral). Scutum and scutellum usually with hairs long as in other species; hairs on anterior half of scutum about four times as long as dorsoapical hairs of pedicel; hairs of scutellum dense and nearly uniform in length throughout; in some specimens from southern British Columbia to southern Manitoba and from the western United States hairs much shorter, most of those on anterior half of scutum at most twice as long as dorsoapical hairs of pedicel; hairs on anterior half or more of scutellum much shorter and sparser than those on posterior portion; these specimens with hairs of head, pleura, and abdomen also shorter than usual; some specimens from western United States with hairs intermediate in length.

Distribution. Western Canada (Map 4), south to California and New Mexico; Europe. B.C., V–VIII.

Specimens identified. Canada, 261 ♂♂, 293 ♀♀; United States, 11 ♂♂, 68 ♀♀; Europe, 10 ♂♂, 13 ♀♀.

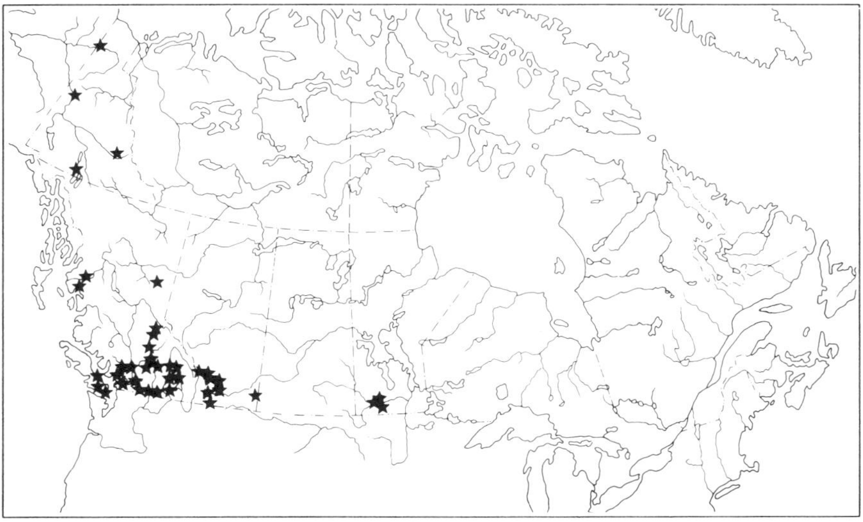

Map 4. Collection localities for *Chrysotoxum fasciatum* (Müller).

Chrysotoxum flavifrons Macquart

Fig. 210; Map 5

Chrysotoxum flavifrons Macquart, 1842:77
Chrysotoxum occidentale Curran, 1924*a*:37.

Length. 12.3–17.4 mm.

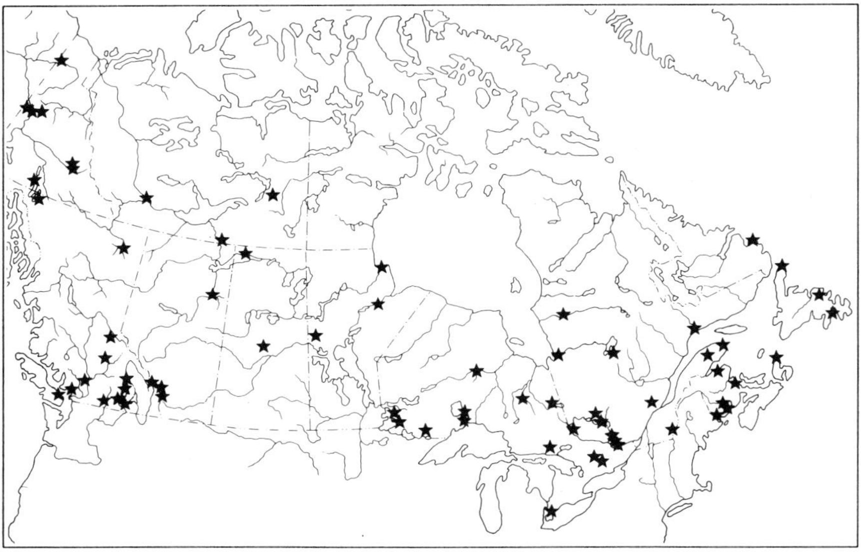

Map 5. Collection localities for *Chrysotoxum flavifrons* Macquart.

Male. Proportions of antennal segments about 30:30:100 (dorsal), 42:42:100 (ventral).

Hairs of scutum black and yellow; hairs of notopleuron always partly black; hairs of anepimeron mostly or entirely black; other pleural hairs mostly yellow. Anterior anepisternum usually with 5 to 15 long fine erect pale hairs on lower half, rarely with fewer hairs or bare. Proepimeron black; katepisterum black or obscurely yellowish near upper margin; katatergite black to extensively yellowish. Cell bm usually with distinct bare median stripe, rarely entirely trichose.

Posterolateral angles of tergites 3 and 4 strongly produced, with that of tergite 4 subacute (as in Fig. 104). Tergite 2 with posterior margin black except laterally in some specimens; discal yellow bands of tergites 3 and 4 divided; tergite 5 with black inverted Y wider than long. Sternite 2 with posterior margin bright yellow; sternite 3 with pair of yellow spots on anterior half. Surstylus rather broad, almost evenly tapered from base to apex (Fig. 210).

Female. Proportions of antennal segments 29:29:100 (dorsal), 43:43:100 (ventral). Hairs of scutum and scutellum long.

Distribution. Canada (Map 5), south to California, Michigan, and New Hampshire. B.C., VI–VIII; Ont., Que., VI–VIII.

Specimens identified. Canada, 139 ♂♂, 133 ♀♀; United States, 16 ♂♂, 46 ♀♀.

Chrysotoxum pubescens Loew

Map 6

Chrysotoxum pubescens Loew, 1860:84.
Chrysotoxum luteopilosum Curran, 1924*a*:36.

Length. 11.6–15.0 mm.

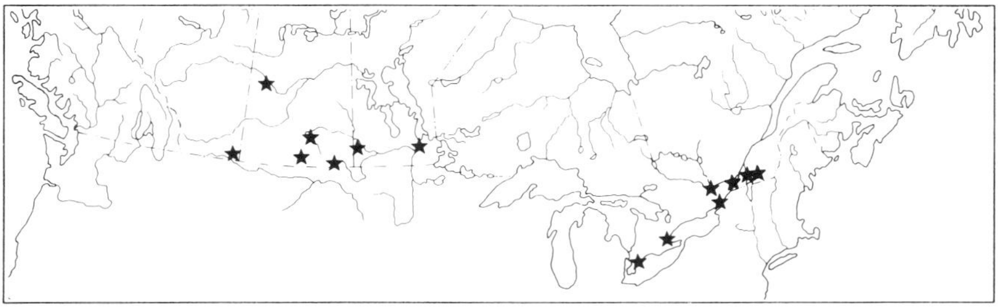

Map 6. Collection localities for *Chrysotoxum pubescens* Loew.

Male. Proportions of antennal segments about 46:50:100 (dorsal), 49:64:100 (ventral).

Hairs of scutum and pleura entirely yellow. Anterior anepisternum bare. Proepisternum mostly or entirely bright yellow; katepisternum with large bright yellow spot near upper margin; katatergite almost entirely bright yellow. Cell bm entirely trichose.

Posterolateral angles of tergites 3 and 4 strongly produced, subacute (as in Fig. 104). Tergite 2 with posterior margin or at least its middle two-thirds bright yellow; discal yellow bands of tergites 3 and 4 divided medially; tergite 5 with inverted black Y moderately broad, with its posterior width about 1.1 times its length. Sternite 2 with posterolateral angles extensively yellow; median part of posterior margin sometimes narrowly yellow; sternite 3 with pair of large yellow spots on anterior half. Surstylus as in *C. derivatum* (Fig. 209).

Female. Proportions of antennal segments about 41:44:100 (dorsal), 45:52:100 (ventral). Scutal and scutellar hairs long.

Distribution. Canada (Alberta eastward; Map 6), south to Nebraska, Kansas,* Tennessee, and Virginia. Ont., Que., VII–VIII.

Specimens identified. Canada, 13 ♂♂, 30 ♀♀; United States, 9 ♂♂, 5 ♀♀.

Genus *Dasysyrphus* Enderlein

Diagnosis. Species robust with yellow bands of tergites undivided or medially divided; half-bands oblique, usually slightly to strongly constricted or divided into pair of spots. Length 5.8–12.9 mm.

Description. Eye with dense long hairs. Frons shining black with pruinose areas. Face slightly broadened below, shining or rarely dull, yellow or rarely mostly gray, with distinct broad brown to black median stripe in some specimens.

Scutum usually shining black, rarely with obscure grayish submedian stripes and slightly pruinose notopleuron. Scutellum dull yellow to brown or gray, commonly blackish on posterior margin, translucent. Pleura black, usually shining, rarely slightly pruinose on upper half. Ventral scutellar fringe well-developed. Upper and lower katepisternal hair patches narrowly to broadly joined posteriorly, otherwise broadly separated. Anterior anepisternum, meron, metapleuron, and metasternum bare. Stigma pale brown to dark brown. Wing membrane entirely trichose or with slight to moderate bare areas on basal half. Hind coxa in some specimens with some posteromedial apical hairs. Legs with femora entirely yellow to entirely black; tibiae and tarsi mostly yellow to mostly black with those of hind leg usually darker, especially in specimens with black femora.

Abdomen narrowly to broadly oval, rather convex above in males and in some females, with distinct margin from near middle of tergite 2 to end of tergite 5 (Figs. 106–115). Tergite 1 black; tergite 2 with pair of yellow spots; tergites 3 and 4 each with entire or more often divided yellow band; each half-band oblique, slightly to very strongly constricted or even divided into pair of spots, reaching or not reaching lateral margin. Sternites yellow, most or all with distinct black bands.

Distribution. Seven Nearctic species (one only in Greenland); one undescribed species in Costa Rica and one in southern Brazil; several additional Palaearctic and Oriental species.

Biology. Goeldlin (1974) and Rotheray and Gilbert (1989) have summarized most of the scanty biological data. Larvae of a European species—probably either *D. tricinctus* (Fallén), 1817, or *D. albostriatus* (Fallén), 1817—have been reported feeding on larvae of Lepidoptera and Symphyta. Goeldlin reared to maturity larvae of *D. albostriatus* and of *D. postclaviger* Štys & Moucha, 1962 [=*D. venustus* (Meigen)] on *Aphis fabae*, but those of the latter species failed to pupate. Rotheray and Gilbert (1989) report larvae of *D. tricinctus* as a regular predator of *Drepanosiphon platanoidis* on *Acer pseudoplatanus* in Scotland. Degeer (1776) reared larvae of *D. pinastri* (Degeer) on conifer aphids and said they fed on aphids on three species of angiosperms as well. A specimen with puparium of *D. lotus* (Williston) from California, in the CNC, is labeled "feeding on membracid nymphs on *Quercus kellogi*." In Czechoslovakia a specimen of *D. venustus* overwintered as a mature larva (Dušek and Láska 1962). Goeldlin (1974) reported that newly emerged larvae of the two species he reared wrap themselves transversely partly around a bean stalk and remain in one position until mature, feeding on aphids that come within reach. Goeldlin suggested that *D. albostriatus* may be bivoltine; the collection dates of *D. pauxillus* (Williston) suggest that this species is univoltine, so there is apparently variation within the genus.

Key to Nearctic species of *Dasysyrphus*

1. Yellow bands of tergites 3 and 4 extending broadly to lateral margins of tergites, not narrowed near margins (Figs. 109, 114, 115) 2

   Either yellow bands of tergites 3 and 4 not extending to lateral margins of tergites or one or both bands extending very narrowly to margins (Figs. 106–108, 110–113) 3

2. Yellow spots of tergite 2 large, extending over one-quarter to two-fifths length of tergite and reaching its lateral margins in some specimens; yellow half-bands of tergites 3 and 4 usually strongly constricted with posterior margin strongly convex, often divided into pair of spots, rarely with posterior margin nearly straight (Figs. 114, 115). Wing with microtrichia variable—cell bm entirely trichose, bare along mid length, or trichose only along posterior margin and at base and apex. Widespread . . . . ***venustus* (Meigen)** (in part)

Yellow spots of tergite 2 small, extending over at most one-quarter length of tergite and not reaching its lateral margins; yellow half-bands of tergites 3 and 4 not distinctly constricted but narrowed toward lateral margin and with posterior margin nearly straight (Fig. 109). Wing with cell bm mostly bare, trichose only along narrow posterior margin, at base and at apex. Widespread .... ***limatus* (Hine)**

3. Yellow bands of tergite 3 or tergite 4, or both, reaching or nearly reaching lateral margins but strongly narrowed just before margins (Fig. 114) 4

Yellow bands of tergites 3 and 4 not reaching lateral margins, their outer ends subtruncate and not distinctly narrowed (Figs. 106–108, 110–113) 5

4. Yellow half-bands of tergites 3 and 4 strongly constricted, curved distinctly forward laterally (as in Figs. 114, 115). Femora broadly blackened basally. Widespread .... ***venustus* (Meigen)** (in part)

Yellow half-bands of tergites 3 and 4 weakly constricted, curved only very slightly forward laterally (as in Fig. 111). Femora entirely yellow or narrowly blackened basally. Widespread .... ***pauxillus* (Williston)** (in part)

5. Yellow bands of tergites 3 and 4 (or, if bands medially divided, inner end of each half-band) closely approximated to anterior margin of tergite for one-third to one-half of distance from mid line to lateral margin, abruptly narrowed laterally (Fig. 108). Wing membrane extensively bare, with cell c bare on at least basal one-fifth and commonly bare on basal half or more and cell bm bare at extreme base and apex and on narrow posterior margin in some specimens. Cheek narrowly to broadly black anteriorly, part yellow posteriorly. Western Canada .... ***creper* (Snow)**

Yellow bands of tergites 3 and 4 (or, if bands medially divided, inner end of each half-band) closely approximated to anterior margin of tergite for at most one-quarter of distance from mid line to lateral margin, usually not abruptly narrowed laterally (Figs. 106, 107, 110–113). Wing membrane usually extensively trichose; cell c entirely trichose or bare only at extreme base; cell bm entirely trichose or bare on at most anterior two-thirds; rarely cell c more extensively bare and cell bm almost entirely bare. Cheek usually yellow or black; rarely black anteriorly and yellow posteriorly 6

6. Yellow bands of tergites 3 and 4 entire although usually emarginate posteromedially (Figs. 110, 112) 7

Yellow band of tergite 3 and usually of tergite 4 divided medially (Figs. 106, 107, 111, 113) 8

7. Cheek entirely yellow. Scutellum entirely pale yellow. Wing with cell bm bare on anterior half to two-thirds. Female frons with pruinose band covering about three-quarters of frons and extending posteromedially to ocelli. Western United States .... ***lotus*** **(Williston)**

Cheek usually entirely black, rarely yellow posteriorly. Scutellum broadly black laterally and usually narrowly black posteriorly. Wing with cell bm usually entirely trichose but in some specimens extensively bare. Female frons with pruinose band covering about one-third of frons and well-separated from ocelli. Widespread .... ***pauxillus*** **(Williston)** (in part)

8. Yellow half-bands of tergites 3 and 4 each strongly constricted or divided into pair of spots (Figs. 106, 107). Boreal and western Canada .... ***amalopis*** **(Osten Sacken)**

Yellow half-bands of tergites 3 and 4 only slightly constricted (Figs. 111, 113) 9

9. Wing with cell bm entirely trichose or partly bare. Widespread, not in Greenland .... ***pauxillus*** **(Williston)** (in part)

Wing with cell bm entirely trichose. Greenland .... ***pinastri*** **(Degeer)**

Clé des espèces néarctiques de *Dasysyrphus*

1. Tergites 3 et 4 ornés de bandes jaunes qui s'étendent largement et uniformément jusqu'aux bords latéraux (fig. 109, 114, 115) 2

Tergites 3 et 4 ornés de bandes jaunes qui n'atteignent pas les bords latéraux ou dont l'une ou les deux se rétrécissent jusqu'aux bords (fig. 106–108, 110–113) 3

2. Tergite 2 maculé de grosses taches jaunes couvrant de un quart aux deux cinquièmes de la longueur du tergite et atteignant les bords latéraux chez certains spécimens; demi-bandes jaunes sur les tergites 3 et 4 en général très rétrécies avec bord postérieur très convexe, souvent divisées en deux taches; bord postérieur rarement presque droit (fig. 114, 115). Aile pourvue de microchètes de longueur variable – cellule bm entièrement velue, glabre à mi-longueur ou velue seulement le long de la marge postérieure ainsi qu'à la base et à l'apex. Espèce répandue .... ***venustus*** **(Meigen)** (*partim*)

Tergite 2 maculé de petites taches jaunes qui s'étendent tout au plus sur un quart de sa longueur et qui n'atteignent pas les bords latéraux; demi-bandes jaunes sur les tergites 3 et 4 sans rétrécissement distinct, mais plus étroites vers le bord latéral, ayant le bord postérieur presque droit (fig. 109). Aile pourvue d'une cellule bm presque glabre, un peu velue seulement le long de la marge postérieure ainsi qu'à la base et à l'apex. Espèce répandue .... ***limatus* (Hine)**

3. Tergites 3 ou 4, ou les deux, ornés de bandes jaunes atteignant ou presque les bords latéraux, mais très rétrécies juste avant les bords (fig. 114) 4

Tergites 3 et 4 ornés de bandes jaunes n'atteignant pas les bords latéraux, à extrémités extérieures presque tronquées et sans rétrécissement distinct (fig. 106–108, 110–113) 5

4. Demi-bandes jaunes sur les tergites 3 et 4 très rétrécies et nettement recourbées vers l'avant sur les côtés (comme dans les fig. 114, 115). Partie basilaire du fémur largement noircie. Espèce répandue .... ***venustus* (Meigen)** (*partim*)

Demi-bandes jaunes sur les tergites 3 et 4 peu rétrécies, très légèrement recourbées vers l'avant sur les côtés (comme dans la fig. 111). Fémurs tout jaunes ou légèrement noircis dans la partie basilaire. Espèce répandue .... ***pauxillu*s (Williston)** (*partim*)

5. Bandes jaunes sur les tergites 3 et 4 (ou portion inférieure de chaque demi-bande, si les bandes sont séparées au milieu) s'approchant près du bord antérieur du tergite sur une distance variant du tiers à la moitié de celle qui sépare la ligne médiane du bord latéral, brusquement rétrécies sur les côtés (fig. 108). Membrane alaire largement glabre; cellule c glabre sur au moins un cinquième de la partie basilaire et souvent sur au moins la moitié basilaire; cellule bm glabre sauf, chez certains spécimens, aux extrémités basilaire et apicale et sur le bord postérieur étroit. Joue légèrement à largement noire sur la face antérieure, en partie jaune vers l'arrière. Ouest du Canada .... ***creper* (Snow)**

Bandes jaunes sur les tergites 3 et 4 (ou portion inférieure de chaque demi-bande si les bandes sont séparées au milieu) s'approchant près du bord antérieur du tergite sur au plus un quart de la distance séparant la ligne médiane du bord latéral, habituellement sans rétrécissement brusque sur les côtés (fig. 106, 107, 110–113). Membrane alaire en général très velue; cellule c entièrement velue ou glabre seulement à l'extrémité basilaire; cellule bm entièrement velue ou glabre sur au plus les deux tiers antérieurs; cellule c rarement plus glabre et cellule bm presque entièrement glabre. Joue habituellement jaune ou noire; rarement noire sur la face antérieure et jaune sur la face postérieure 6

6. Bandes jaunes sur les tergites 3 et 4 complètes, mais habituellement échancrées sur la partie postéro-médiane (fig. 110, 112) 7

Bande jaune sur le tergite 3 et habituellement sur le tergite 4 séparées sur la partie médiane (fig. 106, 107, 111, 113) 8

7. Joue entièrement jaune. Scutellum entièrement jaune pâle. Aile pourvue d'une cellule bm glabre sur la moitié ou les deux tiers antérieurs. Chez la femelle, bande pruineuse couvrant environ les trois quarts du front et s'étendant sur la partie postéro-médiane jusqu'aux ocelles. Ouest des États-Unis .... ***lotus*** **(Williston)**

Joue en général entièrement noire, rarement jaune postérieurement. Scutellum largement noir sur les côtés et en général pourvu d'un étroit liséré noir sur la face postérieure. Aile pourvue d'une cellule bm en général entièrement velue, quoique largement glabres chez certains spécimens. Chez la femelle, bande pruineuse couvrant environ le tiers du front, bien séparée des ocelles. Espèce répandue .... ***pauxillus*** **(Williston)** (*partim*)

8. Demi-bandes jaunes sur les tergites 3 et 4, toutes deux très rétrécies ou divisées en deux taches (fig. 106, 107). Régions boréales et ouest du Canada ..... ***amalopis*** **(Osten Sacken)**

Demi-bandes jaunes des tergites 3 et 4, seulement un peu rétrécies (fig. 111, 113) 9

9. Aile pourvue d'une cellule bm entièrement velue ou partiellement glabre. Espèce répandue, absente au Groënland .... ***pauxillus*** **(Williston)** (*partim*)

Aile pourvue d'une cellule bm entièrement velue. Groënland .... ***pinastri*** **(Degeer)**

Key to males of Nearctic *Dasysyrphus*

1. Surstylus strongly compressed, without projecting ridge along posterolateral surface (Figs. 213*c*, 217*c*), in dorsal view nearly flat 2

Surstylus with weak to strong projecting ridge along most of posterolateral surface (Figs. 211*a*, 212*c*, 214*c*, 215*c*, 216*c*), in dorsal view subtriangular to subrectangular (Fig. 211*d*) 3

2. Distal segment of aedeagus long, with short stiff setulae on apical two-fifths of anterior surface; aedeagal base not strongly swollen, longer than broad (Fig. 217*e*) .... ***venustus* (Meigen)**

Distal segment of aedeagus short, without setulae; aedeagal base strongly swollen, not longer than broad (Fig. 213*e*) .... ***limatus* (Hine)**

3. Paramere in form of strong tapering curved hook (Fig. 214*f*); posterolateral ridge of surstylus weak (Fig. 214*c*) .... ***lotus* (Williston)**

Paramere in form of subtriangular or subrectangular plate with posterolateral angle projecting as short acute spine (Figs. 211*a*, 212*f*, 215*f*, 216*f*); posterolateral ridge of surstylus strong (Figs. 211*a*, 212*c*, 215*c*, 216*c*) 4

4. Aedeagal base with about 20 minute spicules on dorsoapical surface (Fig. 212*e*) .... ***creper* (Snow)**

Aedeagal base with four to six large distinct spicules on dorsoapical surface (Figs. 211*a*, 215*e*, 216*e*) 5

5. Distal segment of aedeagus with strong abrupt hump on posterior surface (Fig. 215*e*) .... ***pauxillus* (Williston)**

Distal segment of aedeagus with at most obscure hump on posterior surface (Figs. 211*a*, 216*e*) 6

6. Distal segment of aedeagus with slight but distinct hump on posterior surface (Fig. 211*a*); surstylus with posterior surface broad, in dorsal view almost subquadrate (Fig. 211*d*) .... ***amalopis* (Osten Sacken)**

Distal segment of aedeagus slightly curved but without distinct hump (Fig. 216*e*); surstylus with posterior surface narrow, in dorsal view very slender (Fig. 216*c*) .... ***pinastri* (Degeer)**

Clé des espèces néarctiques de *Dasysyrphus*

Mâles

1. Surstylus très comprimé, dépourvu d'une crête en saillie le long de la face postéro-latérale (fig. 213*c*, 217*c*); presque plat en vue dorsale 2

Surstylus pourvu d'une crête légèrement ou fortement en saillie, presque tout le long de la face postéro-latérale (fig. 211*a*, 212*c*, 214*c*, 215*c*, 216*c*); quasi triangulaire à rectangulaire en vue dorsale (fig. 211*d*) 3

2. Segment distal de l'édéage long, garni de soies courtes et raides sur les deux cinquièmes apicaux de la portion antérieure; base de l'édéage peu renflée, plus longue que large (fig. 217*e*) .... ***venustus* (Meigen)**

Segment distal de l'édéage court, dépourvu de soies; base de l'édéage très renflée, pas plus longue que large (fig. 213*e*) .... ***limatus* (Hine)**

3. Paramère en forme de crochet recourbé très fuselé (fig. 214*f*); crête postéro-latérale du surstylus peu prononcée (fig. 214*c*) .... ***lotus* (Williston)**

Paramère en forme de plaque quasi triangulaire ou rectangulaire avec un angle postéro-latéral en saillie formant une épine courte et acérée (fig. 211*a*, 212*f*, 215*f*, 216*f*); surstylus pourvu d'une forte crête postéro-latérale (fig. 211*a*, 212*c*, 215*c*, 216*c*) 4

4. Base de l'édéage garnie d'environ 20 minuscules spicules sur la face dorso-apicale (fig. 212*e*) .... ***creper* (Snow)**

Base de l'édéage garnie de quatre à six grosses spicules distinctes sur la face dorso-apicale (fig. 211*a*, 215*e*, 216*e*) 5

5. Segment distal de l'édéage orné d'une bosse raide et prononcée sur la face postérieure (fig. 215*e*) ..... ***pauxillus* (Williston)**

Segment distal de l'édéage portant tout au plus une bosse peu visible sur la face postérieure (fig. 211*a*, 216*e*) 6

6. Segment distal de l'édéage portant une bosse petite mais bien visible sur la face postérieure (fig. 211*a*); face postérieure du surstylus large, presque carrée en vue dorsale (fig. 211*d*) .... ***amalopis* (Osten Sacken)**

Segment distal de l'édéage légèrement recourbé, mais dépourvu de bosse distincte (fig. 216*e*); face postérieure du surstylus étroite, très mince en vue dorsale (fig. 216*c*) .... ***pinastri* (Degeer)**

Dasysyrphus amalopis (Osten Sacken)

Figs. 106, 107, 211; Map 7

Syrphus amalopis Osten Sacken, 1875*b*:148.

Length. 6.6–10.6 mm.

Male. Frons obscurely dark gray pruinose on about posterior half. Face dull yellow laterally; median stripe black, one-third to two-fifths as wide as face; lower facial margin and cheek black; cheek in some specimens brown-orange posteriorly.

Scutal hairs mostly black, some lateral and posterior hairs yellow; pleural hairs mostly pale but some black. Scutellum dull yellow, blackish laterally; scutellar hairs almost all black. Wing membrane entirely trichose. Fore and mid femora usually black on about basal five-sixths but rarely only on about basal half; hind femur black, obscurely yellowish only at extreme apex.

Abdomen (Figs. 106, 107). Tergite 2 with pair of well-separated subrectangular or slightly constricted yellow spots not reaching margins; tergites 3 and 4 each with broadly divided yellow band; each half-band slightly oblique, strongly constricted or completely divided, and not extending to lateral margin; tergite 5 in some specimens with pair of small yellow submarginal anterolateral spots. Terminalia (as in Fig. 211) as described in key.

Female. Frons with distinct to obscure entire yellow-gray or gray pruinose band on about middle one-third. Tergite 5 with pair of large rounded yellow submarginal anterolateral spots.

Distribution. Alaska, Canada (Map 7), south to Colorado and New Hampshire. B.C., V–VIII; Que., VII, VIII.

Specimens identified. Alaska, 6 ♀♀; Canada, 15 ♂♂, 98 ♀♀; United States, 1 ♂, 13 ♀♀.

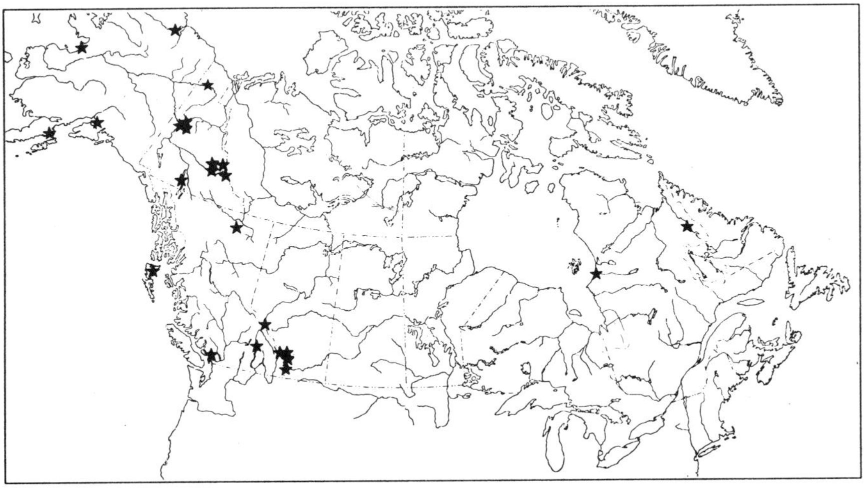

Map 7. Collection localities for *Dasysyrphus amalopis* (Osten Sacken).

Dasysyrphus creper (Snow)

Figs. 108, 212; Map 8

Syrphus creper Snow, 1895:234.

Length. 7.0–10.3 mm.

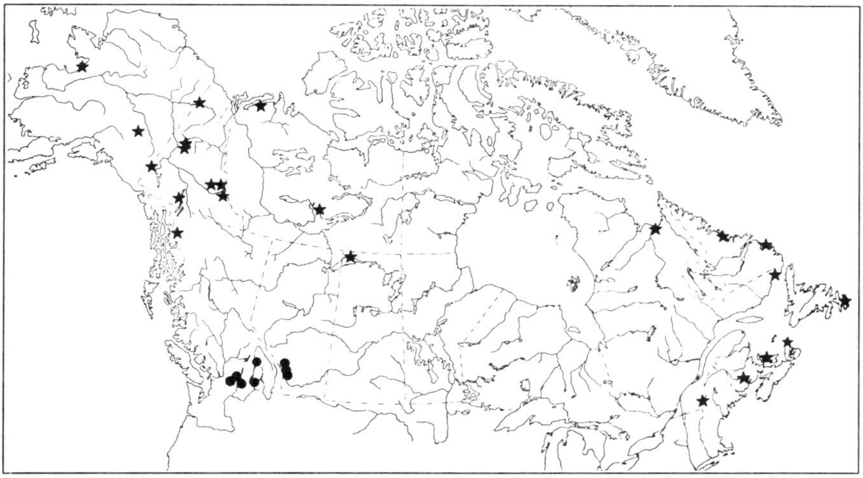

Map 8. Collection localities for *Dasysyrphus creper* (Snow) (●) and for *Didea alneti* (Fallén) (★).

Male. Frons densely yellow pruinose on about posterior half. Face bright yellow laterally; median stripe black, about one-third as wide as face; lower facial margin and anterior part of cheek black; posterior part of cheek yellow to yellow-brown.

Scutal and pleural hairs white to pale yellow. Scutellum dull yellow; scutellar hairs mostly black. Wing extensively bare on basal half; cell c bare or with sparse scattered microtrichia on basal one-eighth to three-quarters; cell bm usually bare except at extreme base and apex but some specimens with some microtrichia along posterior margin and on most of apical one-third of cell. Fore and mid femur brown to black on about basal one-third to two-fifths; hind femur brown to black on about basal seven-eighths.

Abdomen (Fig. 108). Tergite 2 with pair of large subrectangular yellow spots not reaching margins; tergites 3 and 4 each with yellow band strongly constricted posteromedially or narrowly divided; each half-band with anterior margin of inner end straight and closely approximated to anterior margin of tergite for one-third to half of distance from mid line to lateral margin, very oblique and strongly narrowed laterally but not reaching margin; tergite 5 black with broad yellow posterior margin. Terminalia (as in Fig. 212) as described in key.

Female. Frons with distinct entire yellowish pruinose band on about middle one-third. Tergite 5 with pair of rather large subtriangular submarginal anterolateral yellow spots.

Distribution. Southwestern Canada (Map 8), south to California, New Mexico, and Nebraska.* B.C., VI–IX.

Specimens identified. Canada, 8 ♂♂, 9 ♀♀; United States, 12 ♂♂, 22 ♀♀.

Dasysyrphus limatus (Hine)

Figs. 109, 213; Map 9

Syrphus limatus Hine, 1922:146.

Length. 8.2–12.0 mm.

Male. Frons narrowly and obscurely yellowish pruinose along eye margins. Face bright to dark yellow laterally; median stripe black, from four- to five-ninths as wide as face; lower facial margin broadly black; cheek black.

Scutal hairs varying from mostly pale to mostly black; pleural hairs entirely pale or partly black. Scutellum yellow brown to brown, usually with blackish posterior margin; scutellar hairs mostly black. Wing with cell c entirely trichose; cell bm bare on most of anterior two-thirds. Fore and mid femora brown to black on about basal one-third to half; hind femur dark brown to black except at extreme apex.

Abdomen in Fig. 109. Tergite 2 with small yellow spots extending over at most one-quarter length of tergite, not reaching margins; tergites 3 and 4 each with broadly divided yellow band; each half-band slightly oblique, slender, at its longest point not covering more than one-third length of tergite, narrowed about half way to margin, then parallel-sided to margin; tergite 5 with small yellow anterolateral marginal spots. Terminalia (as in Fig. 213) as described in key.

Female. Frons with distinct or obscure, broadly divided, yellowish or grayish pruinose band on about middle one-third.

Distribution. Alaska, Canada (Map 9), south to Colorado and New York. B.C., V–VIII; Que., VI, VII.

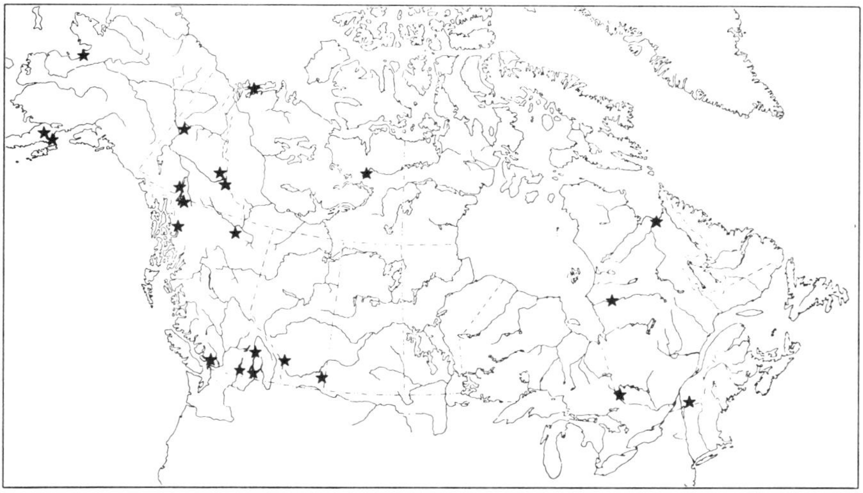

Map 9. Collection localities for *Dasysyrphus limatus* (Hine).

Specimens identified. Alaska, 1 ♂, 3 ♀♀; Canada, 28 ♂♂, 24 ♀♀; United States, 5 ♂♂, 3 ♀♀.

Dasysyrphus lotus (Williston)

Figs. 110, 214

Syrphus lotus Williston, 1887:75.

Length. 9.8–11.0 mm.

Male. Frons densely yellow pruinose on posterior half to three-fifths. Face and cheek bright yellow; median stripe pale brown to brown, in some specimens indistinct laterally, from one-quarter to two-fifths as wide as face.

Scutal and pleural hairs entirely pale yellow. Scutellum dull yellow; scutellar hairs mostly black. Wing with extreme base of cell c and most of anterior two-thirds of cell bm bare. Fore and mid femora black on about basal three-fifths.

Abdomen in Fig. 110. Tergite 2 with pair of large yellow spots not reaching margins; tergites 3 and 4 each with entire yellow band, strongly emarginate posteriorly on mid line in some specimens; each half-band slightly oblique and gradually broadened laterally, not reaching lateral margin; tergite 5 broadly yellow laterally and posteriorly. Terminalia (as in Fig. 214) as described in key.

Female. Frons with entire yellowish gray pruinose band on about middle three-quarters; band extending posteriomedially to anterior ocellus; shining area above antenna slightly reddish in some specimens. Tergite 5 with pair of large yellow triangular submarginal anterolateral spots.

Distribution. Idaho* to California, south to El Salvador. Published records from British Columbia are almost certainly based on misidentifications. Ariz., IV–IX.

Specimens identified. United States, 3 ♂♂, 7 ♀♀; Mexico and Central America, 3 ♂♂.

Biology. A specimen has been reared from larvae feeding on membracid nymphs (see "*Dasysyrphus* Enderlein, Biology").

Dasysyrphus pauxillus (Williston)

Figs. 111, 112, 215; Map 10

Syrphus pauxillus Williston, 1887:74.
Syrphus pacificus Lovett, 1919:245.
Syrphus laticaudus Curran, 1925:175.

Length. 5.8–10.2 mm.

Male. Frons obscurely gray pruinose on about posterior two-thirds. Face bright or dull yellow to grayish yellow or mostly gray laterally; median stripe black, usually about one-third as wide as face, apparently broader and very obscure when face darkened laterally; lower facial margin and cheek black.

Scutal and pleural hairs mixed black and white or mostly black. Scutellum dull yellow to grayish-yellow or mostly gray; posterior margin black, with hairs entirely white to mostly black. Wing usually entirely trichose, in some western specimens cell bm narrowly bare along middle. Fore and mid femora varying from yellow with extreme base slightly darkened to entirely black; hind femur with at least basal two-thirds black, entirely black in some specimens; tibiae entirely yellow to mostly black (legs palest in eastern specimens and darkest in northwestern specimens).

Abdomen in Figs. 111, 112. Tergite 2 with pair of well-separated usually subrectangular spots not extending to lateral margins; spots very reduced and rounded in some specimens; tergites 3 and 4 usually each with broadly divided yellow band; each half-band with anterior margin slightly convex and posterior margin straight, oblique and ending slightly before lateral margin, very rarely with one or both bands undivided (Fig. 112); tergite 5 black. Terminalia (as in Fig. 215) as described in key; parameres in some specimens directed obliquely or horizontally outward away from base of aedeagus.

Female. Frons with gray pruinose band distinct to very obscure, entire to broadly divided, on about middle one-fifth to one-quarter. Face (and scutellum and legs) only rarely as dark as in darker males; cheek in some specimens with posterior half yellow. Wing in some western specimens with cell c bare at extreme base and cell bm partly or almost entirely bare. Tergite 2 with yellow spots extending narrowly or broadly to lateral margins in some specimens; tergites 3 and 4 in many eastern specimens with yellow half-bands extending narrowly to or nearly to lateral margins, in some western specimens with bands entire but usually emarginate posteromedially; tergite 5 with pair of small rounded subtriangular or elongate submarginal anterolateral yellow spots.

Distribution. Alaska, Canada (Map 10), south to California, New Mexico, Michigan,* and New York. B.C., III–VII; Ont., Que., V–VII.

Specimens identified. Alaska, 1 ♀; Canada, 78 ♂♂, 147 ♀♀; United States, 15 ♂♂, 26 ♀♀.

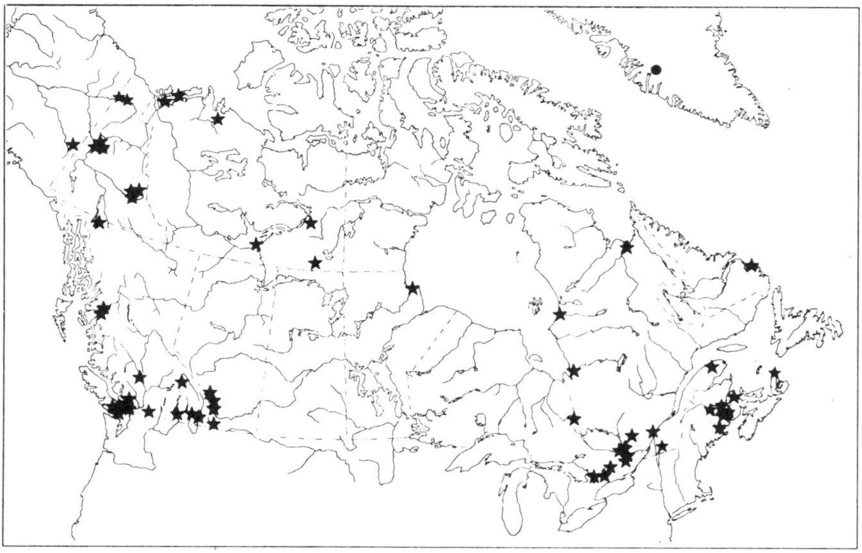

Map 10. Collection localities for *Dasysyrphus pauxillus* (Williston) (★) and for *D. pinastri* (Degeer) (●).

Dasysyrphus pinastri (Degeer)

Figs. 113, 216; Map 10

Musca pinastri Degeer, 1776, p. 113.
Syrphus lunulatus of authors (not *S. lunulatus* Meigen, 1822).

Length. 8.4–10.2 mm.

Male. Frons densely pale gray to dark gray pruinose on about posterior half. Face bright yellow laterally; median stripe black, about one-third as wide as face; lower facial margin and cheek black; cheek in some specimens yellow-brown to orange-brown posteriorly.

Scutal and pleural hairs pale yellow to yellow-brown. Scutellum dull yellow-brown; scutellar hairs mostly black. Wing membrane entirely trichose. Fore and mid femora black on about basal five-sixths.

Abdomen in Fig. 113. Tergite 2 with pair of well-separated subrectangular yellow spots, not extending to lateral margins; tergites 3 and 4 each with broadly divided yellow band; each half-band distinctly narrowed laterally, slightly to moderately constricted, ending well before lateral margin; tergite 5 black. Terminalia (as in Fig. 216) as described in key.

Female. Frons with dense entire yellowish gray pruinose band on about middle one-third. Tergite 5 with small anterolateral submarginal yellow spots.

Distribution. Greenland (Map 10); Iceland;* Europe. Greenland, VIII.

Specimens identified. Greenland, 1 ♀; Europe, 10 ♂♂, 9 ♀♀.

Dasysyrphus venustus (Meigen)

Figs. 114, 115, 217; Map 11

Syrphus arcuatus Fallén, 1817:42 (name declared unavailable by ICZN, Opinion 978, 1972).
Syrphus venustus Meigen, 1822:299.
Syrphus lunulatus Meigen, 1822:299.
Syrphus amalopis of authors (in part) (not *S. amalopis* Osten Sacken, 1875*a*).
Syrphus intrudens Osten Sacken, 1877:326.
Syrphus disgregus Snow, 1895:233.
Syrphus reflectipennis Curran, 1921*a*:157.

Syrphus laticaudatus Curran, 1925:176.
Syrphus osburni Curran, 1925:177.

Length. 8.0–12.9 mm.

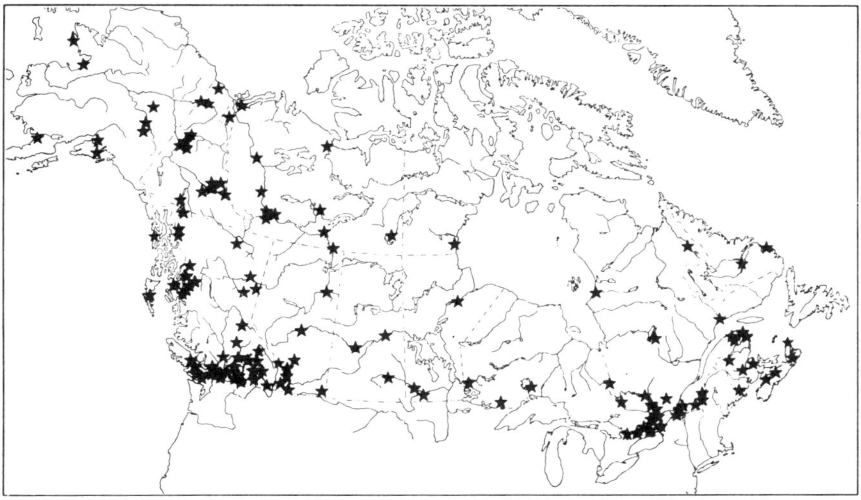

Map 11. Collection localities for *Dasysyrphus venustus* (Meigen).

Male. Frons obscurely or distinctly yellow-gray to gray pruinose on posterior one-quarter to two-thirds. Face dull to bright yellow laterally; median stripe black, from one-third to five-ninths as wide as face; lower facial margin and cheek black; cheek in some specimens brown-orange posteriorly.

Scutal hairs varying from entirely pale yellow to mostly black; pleural hairs usually pale yellow to brownish yellow, rarely partly black. Scutellum dull yellow to brownish yellow, blackish laterally in some specimens; scutellar hairs usually mostly or entirely black, rarely partly or entirely white. Wing membrane usually entirely trichose, with bare areas on basal one-third in some specimens, at most cell bm bare on much of anterior two-thirds. Fore and mid femora dark brown to black on basal one-third to half, hind femur dark brown to black on about basal five-sixths.

Abdomen in Figs. 114, 115. Tergite 2 with pair of large rounded to subrectangular yellow spots extending over one-quarter to two-fifths length of tergite, each spot with shallow rounded anterior sublateral emargination, in some specimens not extending to lateral margin; tergites 3 and 4 each with divided yellow band; each half-band slightly to strongly convex posteriorly, slightly to strongly narrowed laterally, slightly to strongly constricted or even completely divided, extending broadly to lateral margin; tergite 5 with pair of marginal anterolateral yellow spots confluent in some specimens with yellow posterior margin of tergite. Terminalia (as in Fig. 217) as described in key.

Female. Frons with band on about middle one-third entire to broadly divided, obscure to distinct, yellowish-gray to gray pruinose. Tergite 2 with yellow spots commonly extending narrowly or broadly to lateral margins; tergites 3 and 4 with yellow bands very rarely entire although strongly emarginate medially, in some western specimens each half-band strongly narrowed just before lateral margin, extending narrowly to or almost to margin; tergite 5 with anterolateral spots large, extending in some specimens almost to mid line, separated from yellow posterior margin.

Distribution. Alaska, Canada (Map 11), south to California, New Mexico, and North Carolina; Europe; Asia. B.C., IV–X; Ont., Que., IV–IX.

Specimens identified. Alaska, 7 ♂♂, 39 ♀♀; Canada, 229 ♂♂, 537 ♀♀; United States, 62 ♂♂, 104 ♀♀; Europe, 25 ♂♂, 37 ♀♀; Siberia, 2 ♂♂.

Biology. Larvae will feed on aphids but these may not be their natural prey (see "*Dasysyrphus* Enderlein, Biology").

Discussion. The abdominal pattern varies greatly, with each half-band of tergites 3 and 4 shallowly emarginate (Fig. 114), deeply emarginate (Fig. 115), or even completely divided. The male terminalia of these forms do not differ, nor am I able to find other characters to indicate that *D. venustus* is a complex of species. Several students of Palaearctic Syrphidae (e.g., Goot 1981, Stubbs and Falk 1983, Violovitsh 1983) recognized as distinct *D. friulensis* Goot, which seems to differ from *D. venustus* primarily in having more deeply emarginate half-bands on tergites 3 and 4; but I do not think a comparable species occurs in North America.

Genus *Didea* Macquart

Diagnosis. Species large with broad flat abdomen bearing strongly emarginate or narrowly divided yellow, yellow-green, or pale green bands. Length 9.7–15.0 mm.

Description. Eye with moderately long but extremely sparse hairs. Face yellow or with narrow dark median line, slightly receding below. First flagellomere rather large, twice as long as broad.

Scutum black, shining or subshining, notopleural callus and postsutural scutal margin in some specimens obscurely yellowish. Ventral scutellar fringe short and sparse. Pleura black with densely pale-pruinose areas on upper half. Anterior anepisternum bare. Upper and lower katepisternal hair patches broadly separated. Meron bare; metepisternum with some hairs; metasternum with many black hairs. Vein R_{4+5} broadly and rather deeply dipped into cell r_{4+5}. Lower calypter usually with very few fine pale erect scattered hairs on upper surface. Hind coxa with several hairs at posteromedial apical angle (as in Fig. 46).

Abdomen strongly margined. Most sternites with posterior median black triangle.

Distribution. Two Nearctic species; several Palaearctic species.

Biology and larval hosts. Larvae of *D. fuscipes* were reported by Metcalf (1916) and Heiss (1938) to feed on *Longistigma caryae* on *Tilea* and *Platanus*, and by Metcalf (1916) on *Plocomaphis flocculosa* on *Salix*.

Key to New World species of *Didea*

1. Scutellum yellow haired on basal one-third or more, otherwise black haired. Basal half of cell bm trichose only on posterior one-third. Tergite 4 with yellow band deeply emarginate but entire; tergite 5 at least slightly reddish anterolaterally in male (Fig. 117), yellow with black posteromedian triangle in female. Widespread .... ***fuscipes* Loew**

   Scutellum entirely black-haired or with yellow hairs on extreme anterior margin. Basal half of cell bm trichose on posterior two-thirds or more. Tergite 4 with yellow band narrowly divided; tergite 5 black in male (Fig. 116), black or with pair of small yellow spots in female. Widespread, boreal .... ***alneti* (Fallén)**

Clé des espèces du Nouveau Monde de *Didea*

1. Scutellum pourvu de poils jaunes sur au moins le tiers basilaire; ailleurs, les poils sont noirs. La moitié basilaire de la cellule bm n'est velue que sur le tiers postérieur. Tergite 4 orné d'une bande jaune très échancrée, mais complète; tergite 5 tout au moins légèrement rougeâtre sur la face antéro-latérale chez le mâle (fig. 117); jaune et orné d'un triangle noir postéro-médian chez la femelle. Espèce répandue .... ***fuscipes* Loew**

Scutellum pourvu de poils tout noirs ou de poils jaunes à l'extrémité du bord antérieur. Moitié basilaire de la cellule bm velue sur au moins les deux tiers postérieurs. Tergite 4 orné d'une bande jaune étroitement divisée; tergite 5 noir chez le mâle (fig. 116), noir ou marqué d'une paire de petites taches jaunes chez la femelle. Espèce répandue, régions boréales .... ***alneti* (Fallén)**

Didea alneti (Fallén)

Fig. 116; Map 8

Scaeva alneti Fallén, 1817:38.

Length. 10.5–13.8 mm.

Male. Frons yellow with large black spot above base of each antenna. Face yellow with lower margin entirely blackish, and with distinct dark brown triangular median stripe on lower half.

Scutellum dull yellow with entire posterior margin dark or rarely entirely dark, with yellow hairs on at most extreme anterior margin. Wing membrane mostly trichose, with bare areas in anterior one-third or less of cell bm, and in some specimens in very narrow strip behind CuP.

Markings of tergites yellow-green to pale green; band of tergite 4 and rarely of tergite 3 divided medially; tergite 5 black (Fig. 116).

Female. Arms of dark Y-shaped mark above antennae wider than anteromedian yellow area. Tergite 5 usually black, rarely with pair of small anterolateral yellow spots.

Distribution. Alaska, Canada (Map 8), south to Colorado; Europe; Asia.* Canada, VI–VIII.

Specimens identified. Alaska, 2 ♂♂, 4 ♀♀; Canada, 11 ♂♂, 17 ♀♀; United States, 2 ♀♀; Europe, 3 ♂♂, 3 ♀♀.

Didea fuscipes Loew

Fig. 117; Map 12

Didea fuscipes Loew, 1863*b*:318.

Didea fasciata of American authors (not *D. fasciata* Macquart, 1834).

Length. 9.7–15.0 mm.

Male. Frons yellow, usually with small dark spot above base of each antenna. Face yellow, including most of lower margin, with narrow brown median stripe on lower half indistinct in some specimens.

Scutellum dull yellow with only lateral corners darkened, usually with black hairs only on posterior half or less, rarely with some black hairs on anterior half. Wing membrane with bare areas at base of cell r, in basal two-thirds of cell br, in most of anterior two-thirds of cell bm, and in narrow strip behind CuP.

Markings of tergites yellow to yellow-green; bands of tergites 3 and 4 strongly narrowed medially but undivided; tergite 5 at least slightly yellowish to reddish anterolaterally (Fig. 117).

Female. Arms of dark Y-shaped mark above antennae narrower than anteromedian yellow area. Tergite 5 yellow with black posteromedian triangle.

Distribution. Southern Canada (Map 12), south to Arizona and Mississippi. B.C., V–IX; Ont., Que., V–VIII.

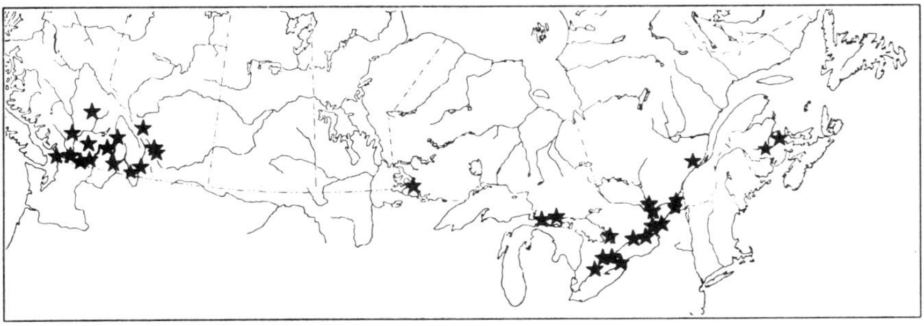

Map 12. Collection localities for *Didea fuscipes* Loew.

Specimens identified. Canada, 24 ♂♂, 36 ♀♀; United States, 20 ♂♂, 24 ♀♀.

Genus *Doros* Meigen

Diagnosis. Species elongate, robust, wasplike, with wing darkened anteriorly, with abdomen parallel-sided, and with narrow yellow spots or bands on tergites. Length 11.5–15.5 mm.

Description. Eye bare.

Scutum black, weakly pruinose, with moderately broad, well-defined shining bright yellow sublateral stripe. Scutellum dull yellow-brown, with base and anterior corners narrowly black. Ventral scutellar fringe short and sparse. Pleura black, with distinct yellow spot above fore coxa and on upper part of katepisternum. Anterior anepisternum bare. Upper and lower katepisternal hair patches broadly separated. Meron and metapleuron bare. Hind coxa without posteromedial apical hair tuft.

Abdomen strongly convex above, strongly margined. Tergite 1 black; tergite 2 with pair of narrow yellow spots; tergite 3 with narrow basal yellow band; tergite 4 with narrow basal and apical yellow bands; tergite 5 black, with long slender posterior yellow triangle. Sternites black, with yellow incisures.

Distribution. One Nearctic species; two Palaearctic species.

Biology. Unknown.

Doros aequalis Loew

Fig. 118; Map 13

Doros aequalis Loew, 1863*b*:318.

Length. 11.5–15.5 mm.

Male and female. Characters as given for genus. Abdomen as in Fig. 118. Species strongly resembling wasps of genus *Ancistrocerus* Wesmael (Eumenidae).

Distribution. Canada (Map 13), south to Oregon and North Carolina.* B.C., V–VI; Ont., Que., V–VII.

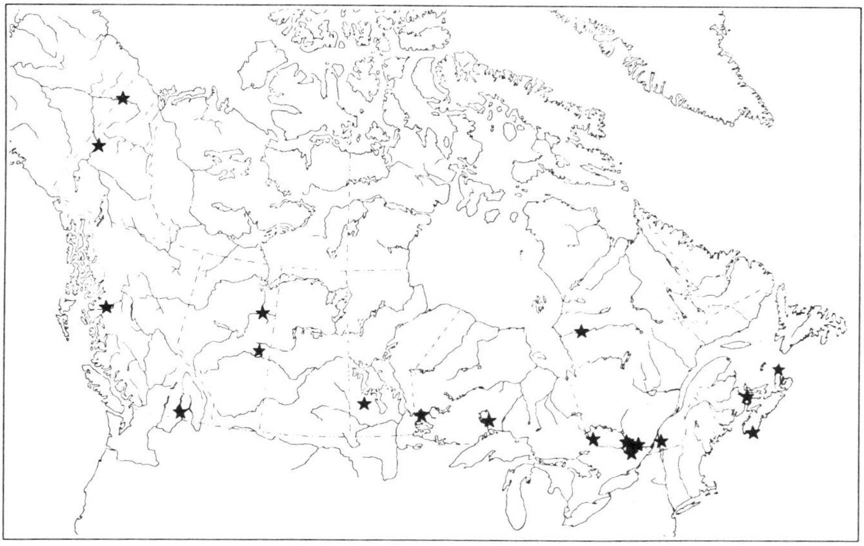

Map 13. Collection localities for *Doros aequalis* Loew.

Specimens identified. Canada, 31 ♂♂, 31 ♀♀; United States, 6 ♂♂, 3 ♀♀.

Genus *Epistrophe* Walker

Diagnosis. Species robust to rather slender, with pair of yellow spots on tergite 2 and entire or divided yellow band on each of tergites 3 and 4, with anterior corners of tergites yellow in some specimens, rarely with dark submetallic instead of yellow markings. Length 7.8–15.0 mm.

Description. Eye bare. Face usually entirely yellow, rarely with obscure dark median stripe.

Scutum usually black, shining to subshining, and yellow pruinose laterally; rarely scutum entirely black or with poorly or well-defined lateral yellow stripe. Ventral scutellar fringe complete. Pleura black, usually slightly yellow pruinose, with obscure pruinose yellow markings in some specimens. Upper and lower katepisternal hair patches narrowly joined posteriorly or widely separated. Anterior anepimeron, meron, and metapleuron bare. Metasternum haired or bare. Wing membrane entirely trichose or partly bare on basal one-third. Hind coxa with posteromedial apical hair tuft only in *E. grossulariae* (Meigen).

Abdomen broadly oval to nearly parallel sided, weakly margined on tergites 4 and 5 or strongly margined on tergites 2–5. Tergite 2 with pair of yellow spots; tergites 3 and 4 each with entire or nearly divided yellow band or with pair of rounded yellow spots, anterior corners of tergites yellow in some specimens. Sternites usually yellow or yellow with obscure dark markings; with discrete black bands, or entirely black, only in *E. metcalfi* (Fluke).

Distribution. *Epistrophe* s.s. is Holarctic and Oriental with perhaps two to three times as many species in the Old World as in the New World. *E. (Epistrophella)* occurs only in the Holarctic region. (The Oriental distribution shown by Vockeroth [1969] is an error.)

Key to subgenera of *Epistrophe*

1. Tergite 4 with undivided yellow band (Figs. 119–123). Upper and lower katepisternal hair patches joined posteriorly (Fig. 41) .... ***E.* (*Epistrophe*) Walker**

   Tergite 4 with pair of yellow spots, narrowly joined anteromedially in some specimens (Figs. 124, 125). Upper and lower katepisternal hair patches broadly separated posteriorly (Fig. 42) .... ***E.* (*Epistrophella*) Dušek & Láska**

Clé des sous-genres de *Epistrophe*

1. Tergite 4 orné d'une bande jaune continue (fig. 119–123). Touffes de poils supérieure et inférieure du katépisterne réunies sur la face postérieure (fig. 41) .... ***E.* (*Epistrophe*) Walker**

   Tergite 4 maculé d'une paire de taches jaunes, étroitement réunies chez certains spécimens sur la face antéro-médiane (fig. 124, 125). Touffes de poils supérieure et inférieure du katépisterne largement séparées dans la région postérieure (fig. 42) .... ***E.* (*Epistrophella*) Dušek & Láska**

Subgenus *Epistrophe* (*Epistrophe*) Walker

Diagnosis. Species robust; anterior corners of tergites black; tergite 4 with entire although in some specimens distinctly emarginate yellow band. Length 9.3–15.0 mm.

Description. Scutum slightly yellow pruinose laterally. Upper and lower katepisternal hair patches narrowly joined posteriorly (Fig. 41). Metasternum haired or bare. Hind coxa with or without posteromedial apical hair tuft.

Abdomen broadly to narrowly oval, weakly margined.

Biology and larval hosts. Nothing has been reported from North America. Dušek and Láska (1961, 1966) and Goeldlin (1974) reported three European species from a variety of unnamed aphids hosts; these species are said to have a larval diapause and to be univoltine. Láska and Starý (1980) gave a number of aphid host records from Czechoslovakia for the above three species. Schneider (1948) reported *E. grossulariae* as a predator of two species of aphids in Switzerland; he said it had a larval diapause and was univoltine. Collection dates of adults of Nearctic species suggest that they are also univoltine, with the possible exception of *E. grossulariae*.

Key to New World species of *Epistrophe* (*Epistrophe*)

1. Antenna black, at most scape slightly paler below. Metasternum with many hairs. Hairs of hind coxa mostly or entirely black. Male with hairs of sternites 3 and 4 mostly black. Widespread .... ***grossulariae* (Meigen)**

   Antenna yellow, first flagellomere darkened above in some species. Metasternum usually bare, rarely with up to four hairs anteromedially. Hairs of hind coxa mostly or entirely yellow. Male with hairs of sternites 3 and 4 yellow 4

2. Wing membrane entirely trichose. Fore and mid femora usually with at least some black hairs posteriorly. Scutellar hairs entirely yellow. Widespread .... ***terminalis* (Curran)**

   Wing with cell bm bare on at least anterior half. Fore and mid femora with only yellow hairs posteriorly. Scutellar hairs mostly black in some species 3

3. Anal lobe with extensive bare areas in front of and behind A_2; alula bare on at least anterior half. Tergites 3 and 4 with yellow bands not reaching margins; tergite 4 black posteriorly (Fig. 120). Sternites 2–4 each with distinct black band or mostly or entirely black. Eastern .... ***metcalfi* (Fluke)**

   Anal lobe and alula entirely trichose. Tergites 3 and 4 with yellow bands usually reaching margins; tergite 4 narrowly yellow posteriorly (Figs. 121, 123). Sternites 2–4 yellow or with irregular dark markings 4

4. Scutellar hairs mostly or entirely yellow. Fore coxa yellow; hind coxa yellow to yellow-brown. Male with yellow bands of tergites 3 and 4 deeply emarginate posteriorly (as in Fig. 123). Female with broad entirely pruinose band on frons, and with yellow spots of tergite 2 reaching or almost reaching anterior margin of tergite 2 over nearly full width (Fig. 123). Widespread .... ***xanthostoma* (Williston)**

Scutellar hairs mostly black. Fore and hind coxae dark brown to black. Male with yellow bands of tergites 3 and 4 at most slightly emarginate posteriorly (Fig. 121). Female with frons shining medially and pruinose only laterally, and with yellow spots of tergite 2 well-separated from anterior margin of tergite except at anterolateral angles in some specimens (as in Fig. 121). Widespread .... ***nitidicollis* (Meigen)**

Clé des espèces du Nouveau Monde de *Epistrophe* *(Epistrophe)*

1. Antenne noire; scape tout au plus légèrement plus pâle en dessous. Métasternum très poilu. Poils de la hanche postérieure noirs sur toute la hanche ou sa plus grande partie. Chez le mâle, poils des sternites 3 et 4 surtout noirs. Espèce répandue .... ***grossulariae* (Meigen)**

Antenne jaune; chez certaines espèces, le premier article du flagelle est plus foncé sur le dessus. Métasternum généralement glabre, ayant rarement plus de quatre poils sur la face antéro-médiane. Sur la hanche postérieure, poils surtout ou entièrement jaunes. Chez le mâle, les poils des sternites 3 et 4 sont jaunes 4

2. Aile pourvue d'une membrane entièrement velue. Fémurs antérieur et médian garnis généralement d'au moins quelques poils noirs dans la région postérieure. Poils du scutellum tous jaunes. Espèce répandue .... ***terminalis* (Curran)**

Aile pourvue d'une cellule bm glabre sur au moins la moitié antérieure. Fémurs antérieur et médian garnis uniquement de quelques poils jaunes dans la partie postérieure. Poils du scutellum surtout noirs chez certaines espèces 3

3. Lobe anal présentant de vastes zones glabres en avant et en arrière de A_2; alule glabre au moins sur la moitié antérieure. Tergites 3 et 4 ornés de bandes jaunes n'atteignant pas les bords; tergite 4 noir dans la région postérieure (fig. 120). Sternites 2–4 portant une bande noire distincte, ou dont toute la surface ou presque est noire. Est ...... ***metcalfi* (Fluke)**

Lobe anal et alule entièrement velus. Tergites 3 et 4 ornés de bandes jaunes s'étendant généralement jusqu'aux bords; tergite 4 souligné par une bande jaune étroite dans la région postérieure (fig. 121, 123). Sternites 2–4 jaunes ou ornés de marques foncées irrégulières 4

4. Scutellum pourvu de poils en majorité jaunes ou tous jaunes. Hanche antérieure jaune; hanche postérieure jaune à jaune brun. Chez le mâle, bandes jaunes des tergites 3 et 4 très échancrées dans la région postérieure (comme dans la fig. 123). Chez la femelle, large bande entièrement pruineuse sur le front et taches jaunes sur le tergite 2 atteignant ou presque le bord antérieur de ce tergite sur presque toute sa largeur (fig. 123). Espèce répandue .... ***xanthostoma* (Williston)**

Scutellum pourvu de poils surtout noirs. Hanches antérieure et postérieure brun foncé à noir. Chez le mâle, bandes jaunes des tergites 3 et 4 tout au plus légèrement échancrées dans la région postérieure (fig. 121). Le front de la femelle est brillant sur la face médiane et pruineux seulement sur les côtés; taches jaunes du tergite 2 bien séparées du bord antérieur, sauf, chez certains spécimens, aux angles antéro-latéraux (comme dans la fig. 121). Espèce répandue .... ***nitidicollis* (Meigen)**

Epistrophe (*Epistrophe*) *grossulariae* (Meigen)

Figs. 41, 119; Map 14

Syrphus grossulariae Meigen, 1822:306.
Syrphus lesueurii Macquart, 1842:152.
Epistrophe conjungens Walker, 1852:242.
Syrphus grossulariae var. *melanis* Curran, 1922*c*:96.

Length.. 10.4–15.0 mm.

Male. Frons yellow-gray pruinose on upper half to three-fifths, shining brownish black on lower half to two-fifths, usually yellowish anteromedially. Face pale yellow, usually with narrow obscure brownish median stripe on lower three-fifths. Antenna black, scape yellowish below in some specimens.

Scutellar hairs almost all black. Metasternum with many long pale hairs. Wing membrane entirely trichose. Hind coxa with posteromedial apical hair tuft and with many black hairs. Fore coxa yellow to yellow-brown, mid and hind coxae brown to black. Fore and mid femora with up to basal one-fifth, and hind femur with up to basal half, brown to black; femora otherwise yellow.

Tergite 2 with pair of large pale yellow spots regularly broadened laterally and extending over anterior half to two-thirds of margins; tergites 3 and 4 each with rather long anterior yellow band, in some specimens emarginate posteriorly and reaching the margins in its full width; tergite 4 broadly yellow posteriorly; margins of tergites 2–4 alternately yellow and black (Fig. 119). Posterior half of sternite 3 and all sternite 4 with both long suberect and short appressed hairs black.

Female. Frons mostly yellow-gray pruinose, area around ocelli and narrow to broad median stripe dark, subshining brownish black only above antennae. Hind femur commonly entirely yellow posteriorly. Markings of tergites submetallic gray rather than yellow in some specimens.

Distribution. Alaska, Canada (Map 14), south to California, North Carolina, and South Carolina;* Europe; Asia. B.C., VII–X; Ont., Que., VI–X.

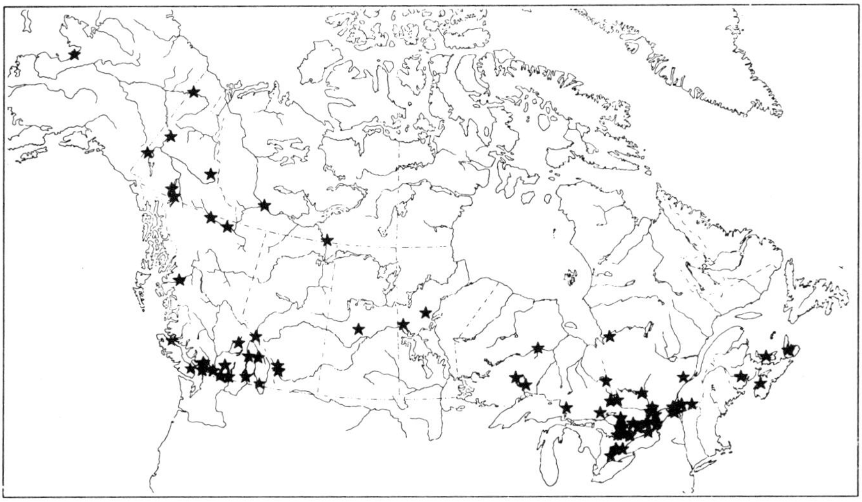

Map 14. Collection localities for *Epistrophe grossulariae* (Meigen).

Specimens identified. Alaska, 36 ♀♀; Canada, 62 ♂♂, 146 ♀♀; United States, 22 ♂♂, 88 ♀♀; Europe, 6 ♂♂, 10 ♀♀; Japan, 1 ♂, 3 ♀♀.

Biology. Schneider (1948) listed two species of aphids as larval hosts in Switzerland.

Epistrophe (*Epistrophe*) *metcalfi* (Fluke)

Fig. 120; Map 15

Metasyrphus metcalfi Fluke, 1933:119.

Length. 11.0–12.3 mm.

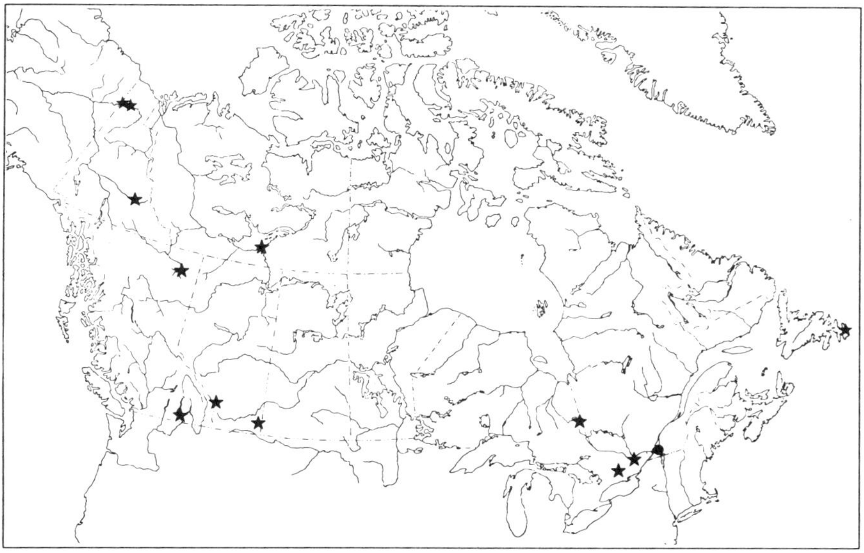

Map 15. Collection localities for *Epistrophe metcalfi* (Fluke) (●) and for *E. terminalis* (Curran) (★).

Male. Frons mostly black, gray pruinose on upper half to two-thirds, subshining to shining below, slightly yellowish immediately above antennae. Antenna orange-brown, slightly paler on lower half.

Scutellar hairs yellow. Metasternum bare. Wing membrane extensively bare at base of cell c, in most of cell br, in cell bm except narrowly posteriorly and apically, in cell *cup* basally and narrowly anteriorly and posteriorly, in anal lobe in front of and behind A_2, and in alula except posterior and apical margins. Coxae brown to black; hind coxa with yellow hairs. Fore and mid femora with basal one-third to half and hind femur with basal half to two-thirds brown to black, femora otherwise yellow; fore and mid femora without black hairs posteriorly.

Tergite 2 with pair of small triangular yellow spots not reaching lateral margins; tergites 3 and 4 each with yellow band broadly and shallowly emarginate posteriorly and not reaching lateral margins; tergite 4 without yellow posterior margin; tergite 5 usually with small indistinct orange anterolateral spots (Fig. 120). Sternites 2–4 each with distinct black band or mostly or entirely black, with yellow hairs.

Female. Frons mostly shining black, gray pruinose laterally on about lower two-thirds, slightly yellowish immediately above antennae. Femora mostly yellow; hind femur brownish anteriorly on middle half. Sternites 2–5 each with narrow to broad black band, with only yellow hairs.

Distribution. Southeastern Canada (Map 15), Wisconsin, Mississippi, and North Carolina. Wis., V; Miss., N.C., IV.

Specimens identified. Canada, 1 ♂; United States, 4 ♂♂, 1 ♀.

Epistrophe (*Epistrophe*) *nitidicollis* (Meigen)

Fig. 121; Map 16

Syrphus nitidicollis Meigen, 1822;308.
Syrphus protritus Osten Sacken, 1877:328.
Stenosyrphus hunteri Curran, 1925:171.

Length. 9.3–13.0 mm.

Male. Frons with upper one-third to half pruinose, pale yellow to almost black; lower two-thirds to half of frons shining, yellow or extensively brown to black with only narrow anterior margin yellowish. Face yellow. Antenna yellow-orange; first flagellomere brown above in some specimens.

Scutellar hairs almost all black. Metasternum bare. Wing membrane bare in most of basal half of cell br, about middle half of cell bm except posteriorly, and in narrow strip behind CuP. Fore coxa brown to black; mid and hind coxae black; latter with only yellow hairs. Femora entirely yellow or fore and mid femora obscurely or distinctly brown on about basal one-fifth and hind femur brown on up to basal one-third.

Tergite 2 with yellow spots only slightly broadened laterally over most of their width, abruptly broadened near margin, and extending over anterior two-thirds of margin; tergites 3 and 4 with yellow bands scarcely emarginate posteriorly and only slightly narrowed laterally; tergite 4 narrowly yellow posteriorly; margins of tergites 2–4 alternately black and yellow (Fig. 121). Sternites with yellow hairs.

Female. Frons mostly shining to subshining black, narrowly yellow-gray pruinose laterally on lower two-thirds, narrowly yellow anteriorly. Tergite 2 with yellow spots well-separated from anterior margin over most of their width, strongly broadened near margin.

Sternites 3 and 4 with suberect yellow hairs, with some short appressed black hairs in some specimens.

Distribution. Alaska,* Canada (Map 16), south to California, Utah, Maryland, and South Carolina;* Europe; Asia. B.C., IV–VI; Ont., Que., V–VII.

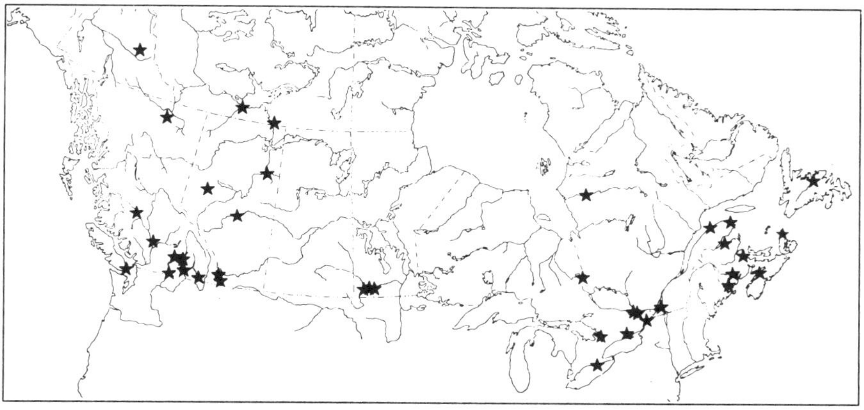

Map 16. Collection localities for *Epistrophe nitidicollis* (Meigen).

Specimens identified. Canada, 39 ♂♂, 29 ♀♀; United States, 4 ♂♂, 2 ♀♀; Europe, 7 ♂♂, 16 ♀♀; Siberia, 1♂.

Biology. Láska and Starý (1980) listed seven species of aphids as larval hosts in Czechoslovakia.

Epistrophe (*Epistrophe*) *terminalis* (Curran)

Fig. 122; Map 15

Stenosyrphus terminalis Curran, 1925:98.
Stenosyrphus submarginalis Curran, 1925:101.

Length. 9.0–12.0 mm.

Male. Frons with upper half dark gray pruinose and lower half shining dark yellow, or mostly or entirely blackish. Face yellow. Antenna yellow-orange.

Scutellar hairs yellow. Metasternum bare. Wing membrane entirely trichose. Coxae brown to blackish; hind coxa with yellow hairs. Femora entirely yellow or brownish on up to basal one-third; fore and mid femora with some black hairs posteriorly.

Tergite 2 with yellow spots strongly broadened only near lateral margins; tergites 3 and 4 each with yellow band broadly emarginate posteriorly and slightly or strongly narrowed laterally; tergite 4 with broad yellow posterior margin; margins of tergites 2–4 alternately yellow and black (Fig. 121). Sternites 3 and 4 with yellow hairs.

Female. Frons usually with upper two-thirds subshining to shining black with obscure lateral pruinose triangle and with lower one-third shining bright yellow, rarely mostly black. Fore and hind coxae, and all femora, yellow; mid coxa brown. Spots of tergite 2 well-separated from anterior margin over most of their width. Sternites 3 and 4 with suberect hairs mostly yellow, with appressed hairs partly black.

Distribution. Canada (Map 15), Maine.* B.C., V–VI; Ont., V.

Specimens identified. Canada, 24 ♂♂, 11 ♀♀.

Discussion. The characters distinguishing *E. terminalis* from the European species *E. melanostoma* (Zetterstedt), 1843, and *E. ochrostoma* (Zetterstedt), 1849, are unclear. It is possible that *E. terminalis* is a synonym of one or the other of these, or even that all three are synonymous. *E. ochrostoma* (as *Metasyrphus ochrostomus*) has been recorded from Alberta, Ontario, and New York (Wirth et al. 1965). Referral of the species to *Metasyrphus* and description of the metasternum as haired (Fluke 1933, 1952) as well as the figure of the male terminalia (Fluke 1950) indicate that Fluke, and probably Wirth et al., incorrectly used the name *ochrostoma* for a species of *Eupeodes*.

Epistrophe (*Epistrophe*) *xanthostoma* (Williston)

Fig. 123; Map 17

Syrphus xanthostoma Williston, 1887:86.

Length. 9.9–12.7 mm.

Male. Frons yellow pruinose on upper half, shining yellow on lower half, posteromedian part of shining area usually brown. Face yellow. Antenna yellow-orange, upper margin of first flagellomere brownish in some specimens.

Scutellar hairs almost all yellow, some black hairs present near posterior margin in some specimens. Metasternum bare or with some long pale hairs near mid line. Wing membrane with bare areas as in

E. nitidicollis or slightly more extensive. Fore coxa yellow; mid coxa brown; hind coxa narrowly brown basally, otherwise yellow, usually with only yellow hairs but with some black hairs in some specimens. Femora yellow.

Tergite 2 with pair of large yellow spots well-separated from anterior margin over most of their width but reaching lateral margin over their entire length; tergites 3 and 4 each with very broad yellow band strongly emarginate posteriorly and reaching margins over its entire length; tergite 4 broadly yellow posteriorly; margins of tergites 2–4 continuously and rather broadly yellow. Sternites with yellow hairs.

Female. Frons mostly yellow-gray pruinose, darker around ocelli and medially; area above antenna shining yellow, usually blackish posteromedially. Hind coxa with only yellow hairs. Tergite 2 with yellow spots very large, nearly reaching anterior margin over most of their width (Fig. 123). Sternites 3 and 4 with suberect yellow hairs, with some appressed black hairs in some specimens.

Distribution. Alaska, Canada (Map 17), south to California and North Carolina. B.C., V–VI; Ont., Que., V–VII.

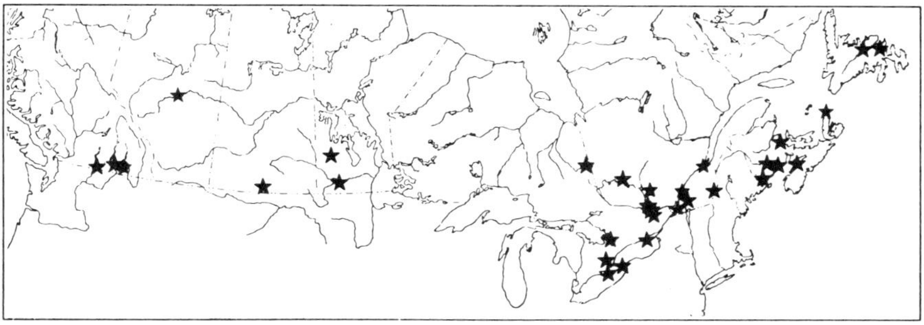

Map 17. Collection localities for *Epistrophe xanthostoma* (Williston).

Specimens identified. Alaska, 1 ♂; Canada, 30 ♂♂, 21 ♀♀; United States, 11 ♂♂, 7 ♀♀.

Subgenus *Epistrophe* (*Epistrophella*) Dušek & Láska

Diagnosis. Species rather slender; anterior corners of tergites 3 and 4 yellow in some specimens; tergites 2–4 each with pair of broadly or narrowly separated or (especially on tergite 3) medially confluent yellow spots. Length 7.8–11.7 mm.

Description. Scutum with obscure to very distinct lateral yellow stripe. Upper and lower katepisternal hair patches distinctly separated (Fig. 42). Metasternum bare. Hind coxa without posteromedial apical hair tuft.

Abdomen narrowly oval to almost parallel sided, distinctly margined from middle of tergite 2 to apex of tergite 5.

Distribution. One Nearctic species; one European species.

Biology. Larvae of the Nearctic species, *E. emarginata* (Say) have been reported to feed on a number of species of aphids (see "*E. emarginata*, Biology"). Láska and Starý (1980) reported that the European species *E. euchromus* (Kowarz) has been reared from three species of aphids in Czechoslovakia. Fluke (1931) indicated that larvae of *E. emarginata* require a winter diapause, so the species is almost certainly univoltine.

Discussion. Although I have been able to recognize only one Nearctic species, it shows so much variation that a fairly extensive description is given here.

Epistrophe (*Epistrophella*) *emarginata* (Say)

Figs. 42, 124, 125; Map 18

Scaeva emarginata Say, 1823:91.
Xanthogramma felix Osten Sacken, 1875*b*:67.
Xanthogramma divisa Williston, 1882:311.
Syrphus disjunctus Williston, 1882:314 (preoccupied Macquart, 1842).
Syrphus maculifrons Bigot, 1884:89.
Syrphus disjectus Williston, 1887:73 (new name for *disjunctus* Williston).
Xanthogramma aenea Jones, 1907:93.
Syrphus invigorus Curran, 1921*b*:171.
Xanthogramma fragila Fluke, 1922:237.
Syrphus infuscatus Fluke, 1931:297.
Syrphus weborgi Fluke, 1931:299.

Length. 7.8–11.7 mm.

Male. Frons entirely yellow or narrowly to broadly black medially. Face and cheek entirely yellow or lower margin of face and, more commonly, part of or entire cheek blackish. Antenna varying from yellow-orange with upper margin darkened to mostly black.

Scutum with obscure dull yellow to well-defined bright yellow lateral stripe. Scutellar hairs usually partly black, especially posteriorly, entirely yellow in some specimens; pleuron entirely dark or with obscure yellow pruinose areas on much of upper half. Wing with microtrichia very variable: at one extreme membrane nearly entirely trichose with only small bare area at base of cell c and near base of cell bm, at other extreme with cell c trichose only on apical one-sixth, cell br almost bare, cell bm with only very few microtrichia near apex, cell cu*p* bare on basal half or more, and bases of cells r_{2+3}, dm, and cua_1 bare (with many intermediate conditions between these extremes). Legs almost entirely yellow with only coxae and hind tarsus dark in some specimens, commonly with pale brown to dark brown ring on apical half or more of hind femur and with hind tibia brownish towards apex, rarely with bases of all femora darkened.

Abdomen in Figs. 124, 125. Tergite 2 with pair of yellow spots small and widely separated from lateral margins in some specimens or larger and commonly extending to margins, rarely narrowly confluent medially; tergite 3 with anterolateral corners or entire anterior margin narrowly yellow in some specimens, with pair of large yellow spots commonly narrowly to broadly confluent medially, reaching lateral margins narrowly to broadly or extending forward anterolaterally to join anterolateral yellow spots; some with narrow posterior margin yellow; tergite 4 similar to tergite 3 but large yellow spots rarely confluent medially and broader posterior margin yellow; tergite 5 black with anterolateral corners yellow, or yellow with broad black inverted U or V. Sternites 2–4 entirely reddish yellow or some or all with obscure to distinct entire black band on posterior half, rarely almost entirely black.

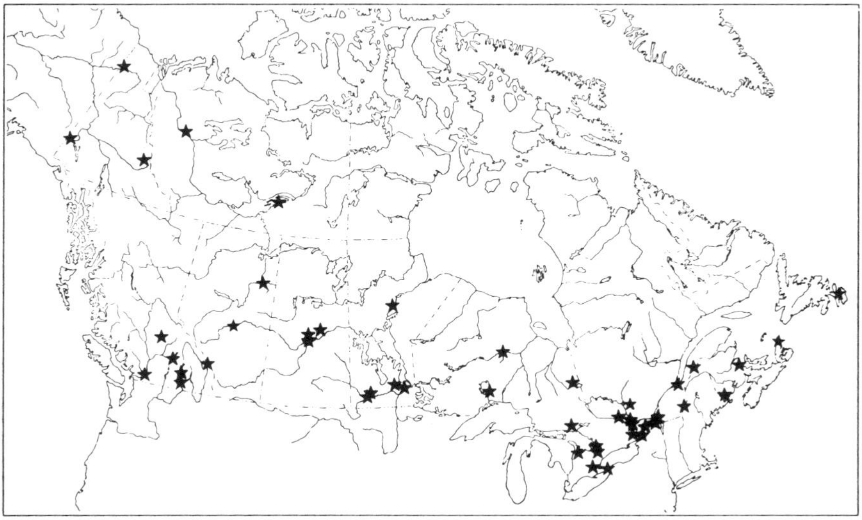

Map 18. Collection localities for *Epistrophe emarginata* (Say).

Female. Frons yellow with incomplete median black stripe extending halfway from ocelli to antennae and tapering anteriorly, or more extensively black with narrow or broad median stripe extending to antennae or broadened to cover all frons except narrow anterolateral margins.

Distribution. Canada (Map 18) south to Washington, Colorado, Texas,* Mississippi, and Florida.* B.C., VI–VIII; Ont., Que., V–VIII.

Specimens identified. Canada, 112 ♂♂, 73 ♀♀; United States, 29 ♂♂, 34 ♀♀.

Biology. Larval host records have been given by Metcalf (1917) [as *Syrphus divisus* Williston], Fluke (1931) [as *S. emarginatus* (Say) and as *S. divisus]*, and Heiss (1938) [as *"Metasyrphus" emarginatus* and as *"M." invigorus* (Curran)]. These authors recorded nine different aphids on various angiosperms as larval hosts. Fluke suggested that the species he treated as *S. divisus* is restricted to the aphid *Chaitophorus populicola* on various species of *Populus*, and that the body color and posterior spiracular process of the larva differed from those of the species he treated as *S. emarginatus*. Heiss also indicated that the posterior spiracular process of *"M." invigorus* differed from those of the two species treated by Fluke. Because of the apparent conspecificity of these three species, further studies of biology and of larval morphology are needed.

Discussion. The specimens of the species *E. emarginata*, as already defined, show more variation than those of any other Nearctic species of Syrphinae. Many characters, which in other genera are constant or nearly so within a species, show wide variation in *E. emarginata*. These characters include color of frons and cheeks, distinctness of lateral scutal stripes, color of scutellar hairs, color of legs, and, particularly, extent of wing microtrichia and markings of abdominal tergites. Specimens with an extensively trichose wing and with the tergites bearing well-separated yellow spots are more abundant in the west; those with the wing extensively bare and the abdominal spots confluent medially are almost restricted to the east. However, so many intermediates and so many different combinations of these characters can be found throughout the range of the species that to recognize more than one species appears impossible.

A series of 33 males taken 17.VII.1985 on the summit of Mt. Rigaud, Rigaud, Que., in an area no more than 10 × 10 m, run in the key of Fluke (1931) to *E. felix* (O.S.), *E. emarginata* (Say), and either *E. divisa* (Will.) or *E. fragila* (Fluke). Specimens that run to *E. felix* show moderate variation in abdominal pattern, in color of frons and of legs, and in the distinctness of the yellow lateral scutal stripe. It seems probable that the 33 specimens are of one species.

Most of the specimens of *E. emarginata* have cell c densely trichose on at least the apical half and usually on the apical five-sixths or more. Thirteen specimens, from Manitoba to New Brunswick and Massachusetts, have cell c trichose on at most the apical one-sixth. In other characters these specimens show little variation. The black frontal stripe of the female is entire, although it varies from narrow to very broad; the scutellar hairs are entirely yellow; tergites 2 and 4 each have a pair of spots and tergite 3 (with one exception) an entire yellow band, with none of these markings reaching the margins; sternite 4 of the male, and sternites 4 and 5 of the female, each have a black band. These specimens may represent a distinct species, but, as the only character not found in some other specimens of *E. emarginata* is the extensively bare cell c, their specific distinctness is doubtful.

Genus *Eriozona* Schiner

The genus includes two subgenera: *Eriozona* (*Eriozona*) Schiner has two species in the Palaearctic region; *Eriozona* (*Megasyrphus*) Dušek & Láska has one Nearctic and one Palaearctic species.

Subgenus *Eriozona* (*Megasyrphus*) Dušek & Láska

Diagnosis. Species robust with broadly oval moderately convex and strongly margined abdomen with broad deeply emarginate or divided yellow or greenish yellow bands on tergites 3 and 4. Length 10.2–15.6 mm.

Description. Eye with long dense hairs above, less dense hairs below. Frons of male black, densely yellow-gray pruinose on upper two-thirds; frons of female black with two large oblique joined or narrowly separated yellow-gray pruinose areas. Face yellow with dark brown to black median stripe.

Scutum black, shining, at most very slightly yellowish laterally. Scutellum yellow-brown. Ventral scutellar fringe complete. Pleura black, shining; posterior part of anepisternum and upper part of katepisternum distinctly pale pruinose, with slightly yellowish ground color in some specimens. Anterior anepisternum, meron, and metapleuron bare. Upper and lower katepisternal hair patches narrowly separated posteriorly because of posterior broadening of both patches, otherwise broadly separated. Metasternum with many long black hairs. Wing membrane entirely trichose; vein R_{4+5} with shallow broad symmetrical but distinct dip into cell r_{4+5} (Fig. 19). Hind coxa with tuft of strong hairs at posteromedial apical angle.

Abdomen (Fig. 126) broadly oval, usually distinctly convex above, with strong margin from two-thirds length of tergite 2 to end of tergite 5. Tergite 2 with pair of large yellow spots; tergites 3 and 4 as described in "Diagnosis." Sternites yellow, 2–4 each with broad medially expanded subapical black band.

Distribution. One Nearctic and one Palaearctic species are known. The nominotypical subgenus, *Eriozona* Schiner, has two Palaearctic species.

Biology. Goeldlin (1974) suggested that the European species of the subgenus (as *Megasyrphus annulipes* (Zetterstedt)) is univoltine and that the larva has an obligatory diapause. He reared two larvae, from eggs, on *Aphis fabae* but the larvae did not pupate. Kula (1983) found larvae of *Eriozona* (*Eriozona*) *syrphoides* (Fallén) feeding on *Cinara pineae* on spruce. They went into diapause in November and produced adults in April.

Discussion. Although the larvae of *Megasyrphus* and of *Didea* are very similar (Rotheray and Gilbert 1989), the adults are so different that I cannot accept the synonymy of the two names proposed by these authors.

Eriozona (*Megasyrphus*) *laxa* (Osten Sacken)

Figs. 19, 126; Map 19

Didea laxa Osten Sacken, 1875*b*:66.
Didea daphne Hull, 1925:280.
Syrphus catalina Curran, 1930*a*:14.
Metasyrphus nigrocomus Hull, 1943*b*:48.

Length. 10.2–15.6 mm.

Male and female. Characters as given for subgenus. Wing as in Fig. 19. Abdomen as in Fig. 126.

Distribution. Alaska, Canada (Map 19), south to California, Colorado, Mexico,* and North Carolina. B.C., V–IX; Ont., Que., V–IX.

Specimens identified. Alaska, 18 ♀♀; Canada, 105 ♂♂, 134 ♀♀; United States, 20 ♂♂, 23 ♀♀.

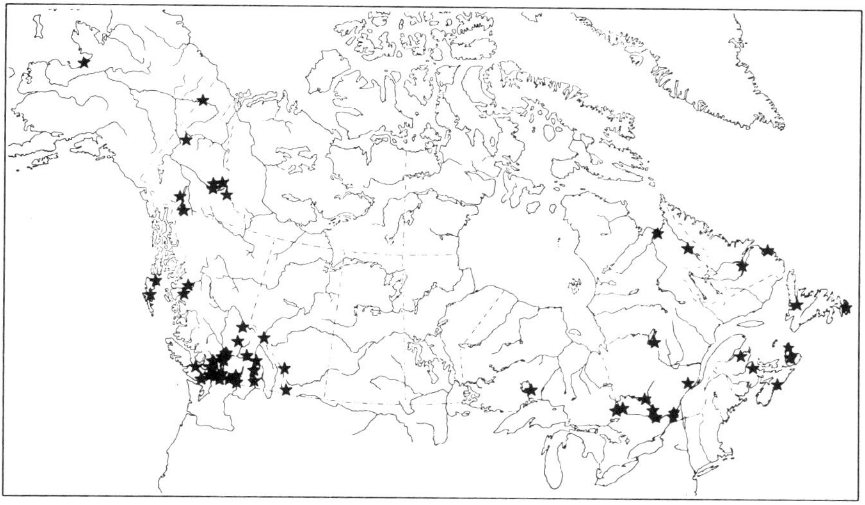

Map 19. Collection localities for *Eriozona laxa* (Osten Sacken).

Discussion. The Nearctic species *E. laxa* and the Palaearctic species *Eriozona erratica* (Linnaeus), until recently known as *E. annulipes* (Zetterstedt), are extremely similar and perhaps should be considered conspecific. However, European specimens can be distinguished from those of *E. laxa* by slight differences in the shape of the yellow bands of tergites 3 and 4; they are narrower, less strongly emarginate medially, and at most very slightly narrowed laterally. The pair of Japanese specimens I have seen agree with European rather than with Nearctic specimens; F.C. Thompson (written communication) has verified this finding from other Japanese specimens. This supports recognition of two species, although study of Siberian material is desirable.

Genus *Eupeodes* Osten Sacken

Diagnosis. Species robust with shining or subshining thorax. Abdomen oval, usually flattened, strongly margined. Tergites usually with pale yellow to reddish yellow spots commonly lunulate or with bands of similar color almost straight, wavy, or distinctly emarginate; rarely with about apical one-third of abdomen reddish above or with tergites mostly or entirely black. Length 6.3–14.1 mm.

Description. Eye bare or with very short sparse scarcely apparent hairs. Face yellow, usually with narrow but distinct brown or black median stripe; cheek yellow or blackened. Upper postocular orbit (Fig. 13) narrow to broad.

Scutum shining black, rarely slightly yellow pruinose laterally. Scutellum dull yellow, translucent, narrowly opaque posteriorly in some specimens. Ventral scutellar fringe complete, moderately dense. Pleura black or grayish black, shining or with very sparse subshining pruinosity on about upper half. Anterior anepisternum, meron, and metepisternum bare. Upper and lower katepisternal hair patches moderately separated or narrowly joined posteriorly, almost confluent anteriorly because of anterior broadening of lower patch (Fig. 43). Metasternum haired or bare. Vein R_{4+5} nearly straight or slightly but distinctly dipped into cell r_{4+5} (Fig. 20). Wing membrane commonly nearly entirely trichose with only small bare areas near base; in *E. volucris* (Osten Sacken) bare areas over most of basal half and extending along main veins almost to wing apex, only apex and posterior margin of wing densely trichose. Hind coxa without hairs at posteromedial apical angle.

Abdomen oval, usually nearly flat above, with strong margin from near middle of tergite 2 to apex of tergite 5. Tergite 2 with pair of yellow spots; tergites 3 and 4 each with entire emarginate or narrowly divided yellow band or with pair of lunulate and usually slightly oblique spots; rarely black markings of tergite 4 absent or nearly so and apical one-third of abdomen reddish above, or yellow markings reduced so tergites are nearly or entirely black. Sternites yellow; sternites 2 and 3, and usually all sternites, with median dark brown to black spot or transverse band; black markings so extensive in some specimens that sternites are almost entirely black. Male terminalia usually small and scarcely apparent from above, in *E. volucris* Osten Sacken protruding as bluntly rounded cylinder at apex of abdomen (Fig. 137).

Distribution. Predominantly Holarctic but with one or some species in all other major regions—see under *E.* (*Eupeodes*) and *E.* (*Lapposyrphus*).

Biology. Larvae of many species have been recorded as predators of a wide variety of aphids. Several species are known to be multivoltine; it is probable that all species are. Several European species, particularly *E. corollae* (Fabricius), have been observed to migrate, but this behavior has not been reported for Nearctic species.

Discussion. The species referred by recent authors (e.g., Dušek and Láska 1976, 1980, and Vockeroth 1969) to *Metasyrphus* Matsumura, 1917, are here referred to *Eupeodes* (cf. Vockeroth 1986*b*). The species show an unusual amount of intraspecific variation in color of head, legs, abdomen, and body hairs; in some and perhaps in all species, these variations seem to result from differences in temperature during the pupal stage (Dušek and Láska 1974*b*). Structural differences between the species are few; in some, as in *E. volucris* and *E. americanus*, the male terminalia are very distinctive, but in most cases the differences are slight and there is apparently moderate variation within some species. As a result, about 3% of the available specimens are unidentified. They may represent unusual variations of known species or, in a few cases, undescribed species. It seems unwise, in a genus as difficult as this one, to treat as distinct species forms known from only one or a few specimens.

Key to subgenera of *Eupeodes*

1. R_{4+5} nearly straight (as in Fig. 32). Metasternum haired. Male terminalia small, scarcely apparent from above (Figs. 127–136), or greatly enlarged, protruding as bluntly rounded cylinder at apex of abdomen (Fig. 137) .... ***Eupeodes (Eupeodes)* Osten Sacken**

   R_{4+5} distinctly dipped into cell r_{4+5} (Fig. 20). Metasternum bare. Male terminalia small, scarcely apparent from above (Figs. 138, 139) .... ***Eupeodes (Lapposyrphus)* Dušek & Láska**

Clé des sous-genres de *Eupeodes*

1. R_{4+5} presque droite (comme dans la fig. 32). Métasternum poilu. Appendices terminaux petits chez le mâle, à peine visibles du dessus (fig. 127–136), ou fortement élargis et en saillie, formant un cylindre nettement arrondi à l'apex de l'abdomen (fig. 137) .... ***Eupeodes (Eupeodes)* Osten Sacken**

   R_{4+5} nettement inclinée dans la cellule r_{4+5} (fig. 20). Métasternum glabre. Chez le mâle, petits appendices terminaux, peu visibles du dessus (fig. 138, 139) .... ***Eupeodes (Lapposyrphus)* Dušek & Láska**

Subgenus *Eupeodes* (*Eupeodes*) Osten Sacken

Diagnosis. As for genus *Eupeodes*. Length 6.3–13.0 mm.

Description. Face entirely yellow or with brown to black median stripe. Metasternum haired. R_{4+5} nearly straight (as in Fig. 32). Male terminalia small, scarcely apparent from above (Figs. 127–136), or projecting as bluntly rounded cylinder at apex of abdomen (Fig. 137).

Distribution. Fourteen species in Nearctic region north of United States; two or three Holarctic species. More species probably in United States. One or two undescribed species in Ecuador and Chile. About 20–25 species in Palaearctic region, 1 species in Afrotropical region, and 1 or more species of distinctive species group in Oriental region southeastward to Australia.

Biology. Larvae of many species have been reared from many aphids on herbs, shrubs, and both broad-leaved and coniferous trees (as described under individual species). Host records for unidentified females of *E. americanus*, *E. fumipennis*, or *E. pomus* in the CNC are *Chaitophorus vinimalis*, *Cinara banksiana*, *Cinara lanatra*, *Eriosoma lanigerum*, and *Schizolachnus* sp.

Discussion. The genus–group name *Macrosyrphus* Matsumura is available for the Oriental group should it be given subgeneric or generic rank.

Key to Canadian species of *Eupeodes* (*Eupeodes*)

1. Tergite 5 without black markings on disc, either entirely yellow to red-orange or red-orange with very narrow black lateral margins (Figs. 133, 136). Frons of male entirely yellow; frons of female yellow on anterior one- to two-thirds with at most pair of obscure brownish spots above antennae. Face usually entirely yellow or with only oral margin black, if with dark median stripe then not extending above apex of tubercle. Alula extensively bare on anterior half or more 2

   Tergite 5 usually yellow with at least median basal black spot, often with black arcuate band or mostly black, rarely entirely black; if entirely yellow or yellow-orange then alula entirely trichose. Frons of both sexes often extensively darkened above antennae. Face with dark median stripe extending well above apex of tubercle in some specimens. Alula partly bare or entirely trichose 4

2. Tergite 4 almost entirely reddish yellow to red, paler on anterior half, narrowly black laterally; if obscure black band on posterior half then much less extensive than posterior reddish band; tergite 5 usually with very narrow black lateral margins (Fig. 136). Frons of female entirely shining, without lateral pruinose spots. Western Canada .... ***snowi* (Wehr)**

   Tergite 4 with distinct broad black band on posterior half more extensive than posterior yellow or reddish band; tergite 5 entirely yellow-orange or reddish, without black lateral margins. Frons of female with pair of distinct lateral pruinose spots at about mid length 3

3. Anterior margin of bands or spots of tergites 3 and 4 nearly straight and very close to anterior margin of tergite, usually red-orange (Fig. 133). Lower facial margin narrowly to broadly black; cheek entirely black. Female usually with hind femur black on at least basal one-third but entirely yellow-orange in some specimens, usually without black hairs on disc of scutellum but some specimens with some on posterior margin. Western Canada . . . . ***montivagus* (Snow)**

Bands or spots of tergites 3 and 4 with anterior margins wavy, distinctly separated from anterior margin of tergite at least sublaterally, never reddish (Fig. 129). Lower facial margin yellow to brownish; cheek black only posteriorly. Female with hind femur yellow, with many black hairs on disc of scutellum in some specimens. Widespread . . . . ***flukei* (Jones)**

4. Wing extensively bare; cell c densely trichose on apical half or less, often nearly bare; cell bm bare except for some microtrichia at extreme apex; cells r_{2+3}, r_{4+5}, dm, and cua_1 broadly bare basally. Tergites 3 and 4 each with pair of slightly oblique lunulate spots, well-separated from lateral tergite margins, rarely narrowly joined medially 5

Wing less extensively bare; cell c densely trichose on at least apical two-thirds; cell bm with at least narrow apex entirely trichose, extensively trichose in some species; cells r_{2+3}, r_{4+5}, dm, and cua_1 entirely trichose. Tergites 3 and 4 as above or each with entire yellow band; yellow markings commonly extending to lateral margins of tergites 6

5. Scutellum with at least some and usually many black hairs. Frons of both sexes with pair of large brown to black spots above antennae; in female these spots separated from upper black part of frons by diffuse poorly defined yellowish band. Male terminalia small, normal, almost hidden below apex of abdomen (Fig. 135). Female with black band on sternites 4 and 5. Widespread . . . . ***perplexus* (Osburn)**

Scutellum with only pale hairs. Frons of male entirely yellow; frons of female with single semicircular brown to black spot above antennae; this spot separated from black upper part of frons by well-defined pale yellow band. Male terminalia very large, projecting at apex of abdomen as long bluntly rounded cylinder (Fig. 137). Female with sternites 4 and 5 yellow-orange. Widespread . . . . ***volucris* Osten Sacken**

6. Alula entirely trichose 7

Alula narrowly to broadly bare anteriorly 8

7. Yellow bands of tergites 3 and 4 entire, constricted medially, or divided into pair of spots, each band or spot on tergite 3 with anterior margin nearly straight (Fig. 131). Male frons entirely yellow. Hypandrium with short posterior projection at each side of base of aedeagus (Fig. 220*b*); aedeagal base with one short bluntly rounded tooth and one acute tooth (Fig. 220*d*). Female frons entirely shining, black on upper two-thirds, bright yellow on lower one-third. Widespread . . . . ***latifasciatus* (Macquart)**

Yellow bands of tergites 3 and 4 broadly divided medially into pair of spots, each spot of tergite 3 usually more slender than in *E. latifasciatus* with anterior margin distinctly concave (Fig. 128). Male frons entirely yellow or with pair of dark spots above antenna. Hypandrium truncate apically, without posterior projections (Fig. 219*b*); aedeagal base with one short acute tooth and one longer acute tooth (Fig. 219*d*). Female frons with pair of large lateral pruinose spots at about mid length, either black on upper two-thirds and entirely yellow on lower one-third, or with pair of dark spots above antennae. Boreal and Arctic Canada, Alaska, and Greenland . . . . ***curtus* (Hine)**

8. Male (eyes meeting on frons) 9

Female (eyes widely separated on frons) 14

9. Tergites 3 and 4 each with undivided although medially constricted yellow band in some species (Figs. 127, 130) 10

Tergites 3 and 4 each with pair of lunulate yellow spots (Figs. 132, 134) 12

10. Surstylus short, about twice as long as broad; in dry specimens distinctly shorter than exposed part of epandrium (Fig. 223*c*). Widespread . . . . ***pomus* (Curran)**

Surstylus elongate, about 2.5 times as long as broad; in dry specimens subequal in length to exposed part of epandrium (Figs. 218*a*,*c*; 221*c*,*e*) 11

11. Surstylus irregular in outline, appearing slightly twisted in dorsal view (Fig. 218*c*), slightly enlarged at apex in lateral view (Fig. 218*a*); epandrium elongate, rather slender, with membrane-filled incision ventrolaterally (Fig. 218*a*); hypandrium moderately rugose laterally, with posterolateral angles slightly rounded (Fig. 218*b*). Widespread . . . . ***americanus* (Wiedemann)**

Surstylus tapering regularly to apex (Fig. 221*c*), not enlarged apically in lateral view (Fig. 221*e*); epandrium not elongate, without ventrolateral incision (Fig. 221*e*); hypandrium strongly rugose laterally, with posterolateral angles prominent (Fig. 221*b*). Western Canada ..... ***fumipennis* (Thomson)**

12. Scutellar hairs mostly or entirely white. Frons entirely pale yellow. Upper postocular orbit moderately broad (Fig. 13). Arctic Canada, Greenland .... ***nigroventris* (Fluke)**

Scutellar hairs almost entirely black. Frons usually with two brown to blackish spots above antenna, rarely entirely yellow. Upper postocular orbit narrow (Fig. 12) 13

13. Tergite 5 yellow or yellow-orange with an anteromedian black spot. Aedeagal base with one short bluntly rounded tooth and one acute tooth (as in Fig. 220*d*). Central Canada .... ***neoperplexus* (Curran)**

Tergite 5 with black median spot reaching lateral margin, commonly mostly or entirely black. Aedeagal base with one short acute tooth and one longer acute tooth (Figs. 226*d*, 227*d*). Widespread, including Greenland .... ***luniger* (Meigen)**

14. Frons entirely shining, black on upper two-thirds, yellow on lower one-third. Tergites entirely black or with at most faint indication of pale spots. Arctic Canada, Greenland .... ***nigroventris* (Fluke)**

Frons with pair of distinct lateral pruinose spots. Tergites usually with large distinct yellow spots or bands 15

15. Tergites 3 and 4 each with entire yellow band strongly constricted medially in some specimens .... ***americanus* (Wiedemann)**
.... ***fumipennis* (Thompson)**
.... ***pomus* (Curran)**

Tergite 3 with pair of lunulate yellow spots; tergite 4 with similar pair of spots, narrowly joined medially in some specimens .... ***luniger* (Meigen)**
.... ***neoperplexus* (Curran)**

Clé des espèces canadiennes de *Eupeodes* (*Eupeodes*)

1. Tergite 5 dépourvu de marques noires sur le disque, soit tout jaune à rouge orangé, soit rouge orangé avec des bandes latérales noires très étroites (fig. 133, 136). Chez le mâle, le front est tout jaune; chez la femelle, il est jaune sur un à deux

tiers de la région antérieure avec, tout au plus, une paire de taches brunâtres floues au-dessus des antennes. La face est en général tout jaune ou ornée seulement d'un liséré buccal noir; lorsqu'il y a une rayure médiane noire, celle-ci ne s'étend pas au-delà de l'apex du tubercule. Alule largement glabre sur au moins la moitié antérieure 2

Tergite 5 habituellement jaune, pourvue au moins d'une tache médiane noire dans la partie basilaire, souvent souligné d'une bande arquée tout noire ou largement noire, rarement entièrement noir; lorsque le tergite est tout jaune ou jaune orangé, l'alule est alors entièrement velue. Chez les deux sexes, le front est souvent très foncé au-dessus des antennes. Face marquée d'une rayure médiane foncée qui s'étend bien au-delà de l'apex du tubercule chez certains spécimens. Alule partiellement glabre ou entièrement velue 4

2. Tergite 4 presque entièrement jaune rougeâtre à rouge, plus pâle sur la moitié antérieure, souligné sur les côtés d'un étroit liséré noir; lorsqu'il y a une bande noire floue dans la moitié postérieure, celle-ci est alors beaucoup moins étendue que la bande postérieure rougeâtre. Tergite 5 orné en général de liserés latéraux noirs très étroits (fig. 136). Chez la femelle, le front est entièrement luisant, sans taches pruineuses latérales. Ouest du Canada .... ***snowi* (Wehr)**

Tergite 4 orné d'une large bande noire distincte sur la moitié postérieure, plus large que la bande jaune ou rougeâtre postérieure; tergite 5 tout jaune orangé ou rougeâtre, sans bords latéraux noirs. Chez la femelle, le front est marqué d'une paire de taches pruineuses latérales distinctes, à environ mi-longueur 3

3. Sur les tergites 3 et 4, bord antérieur des bandes ou des taches presque droit et très près de la marge antérieure du tergite, habituellement rouge orangé (fig. 133). Marge faciale noire, étroite à large, dans la partie inférieure; joue tout noire. Chez la femelle, le fémur postérieur est habituellement noir, au moins sur le tiers basal, quoiqu'il soit entièrement jaune orangé chez certains spécimens; disque du scutellum sans poils noirs, bien que certains spécimens en aient quelques-uns sur le bord postérieur. Ouest du Canada .... ***montivagus* (Snow)**

Sur les tergites 3 et 4, bandes ou taches avec bords antérieurs ondulés, nettement séparées de la marge antérieure du tergite du moins sublatéralement, jamais rougeâtres (fig. 129). Liséré facial inférieur jaune à brunâtre; joue noire uniquement dans la partie postérieure. Chez la femelle, le fémur postérieur est jaune chez certains spécimens, disque du scutellum pourvu de nombreux poils noirs. Espèce répandue .... ***flukei* (Jones)**

4. Aile très glabre; cellule c fortement velue sur au plus la moitié apicale, souvent presque nue; cellule bm glabre, à l'exception de quelques microchètes à l'extrémité apicale; partie basilaire des cellules r_{2+3}, r_{4+5}, dm et cua_1 largement glabre. Tergites 3 et 4 maculés tous deux d'une paire de lunules légèrement obliques, bien séparées des bords latéraux, rarement réunies sur le plan médian 5

Aile moins glabre; cellule c fortement velue sur au moins les deux tiers apicaux; cellule bm pourvue tout au moins d'une étroite zone entièrement velue à l'apex, très velue chez certaines espèces; cellules r_{2+3}, r_{4+5}, dm et cua_1 entièrement velues. Tergites 3 et 4 similaires à la description précédente ou portant tous deux une bande jaune complète; il y a des marques jaunes s'étendant souvent jusqu'aux bords latéraux des tergites 6

5. Scutellum pourvu tout au moins de plusieurs poils noirs, habituellement nombreux. Chez les deux sexes, le front est maculé d'une paire de larges taches brunes ou noires au-dessus des antennes; chez la femelle, ces taches sont séparées de la région supérieure noire du front par une bande jaunâtre floue et mal définie. Chez le mâle, appendices terminaux petits et normaux, presque cachés sous l'apex de l'abdomen (fig. 135). Femelle marquée d'une bande noire sur les sternites 4 et 5. Espèce répandue .... ***perplexus* (Osburn)**

Scutellum pourvu seulement de poils pâles. Chez le mâle, le front est tout jaune; chez la femelle, il est orné d'une unique tache semi-circulaire brune à noire, au-dessus des antennes; cette tache est séparée de la partie supérieure noire du front par une bande jaune pâle bien définie. Appendices terminaux très gros chez le mâle, faisant saillie à l'apex de l'abdomen et formant un long cylindre nettement arrondi (fig. 137). Sternites 4 et 5 jaune orangé chez la femelle. Espèce répandue .... ***volucris* Osten Sacken**

6. Alule entièrement velue 7

Alule étroitement ou largement glabre sur la partie antérieure 8

7. Bandes jaunes complètes sur les tergites 3 et 4, étranglées en leur milieu ou séparées en une paire de taches, chaque bande ou tache sur le tergite 3 ayant un bord antérieur presque droit (fig. 131). Front du mâle tout jaune. Hypandrium pourvu d'une courte projection postérieure de chaque côté de la base de l'édéage (fig. 220*b*); base de l'édéage portant une dent courte bien arrondie et une autre acérée (fig. 220*d*). Chez la femelle, le front est entièrement luisant, noir sur les deux tiers supérieurs

et jaune clair sur le tiers inférieur. Espèce répandue .... ***latifasciatus* (Macquart)**

Bandes jaunes sur les tergites 3 et 4, largement séparées sur le plan médian en une paire de taches, chaque tache sur le tergite 3 étant habituellement plus mince que chez l'*E. latifasciatus*, avec un bord antérieur nettement concave (fig. 128). Chez le mâle, le front est tout jaune ou il est orné d'une paire de taches foncées au-dessus des antennes. Hypandrium tronqué dans la portion apicale et dépourvu de projections postérieures (fig. 219*b*); base de l'édéage garnie d'une dent courte et acérée et d'une autre plus longue (fig. 219*d*). Chez la femelle, le front porte une paire de grosses taches latérales pruineuses, à peu près à mi-longueur; il est, soit noir sur les deux tiers supérieurs et tout jaune sur le tiers inférieur, soit orné d'une paire de taches foncées au-dessus des antennes. Régions boréales et arctiques du Canada, Alaska et Groënland .... ***curtus* (Hine)**

8. Mâle (yeux contigus sur le front) 9

Femelle (yeux largement séparés sur le front) 14

9. Tergites 3 et 4 ornés tous deux d'une bande jaune continue, mais étranglée en son milieu chez certaines espèces (fig. 127, 130) 10

Tergites 3 et 4 ornés tous deux d'une paire de taches jaunes en forme de lunule (fig. 132, 134) 12

10. Surstylus court, environ deux fois plus long que large; chez les spécimens desséchés, nettement plus court que la partie exposée de l'épandrium (fig. 223*c*). Espèce répandue .... ***pomus* (Curran)**

Surstylus allongé, environ 2,5 fois plus long que large; chez les spécimens desséchés, de longueur quasi égale à la partie exposée de l'épandrium (fig. 218*a*,*c*; 221*c*,*e*) 11

11. Surstylus de profil irrégulier, paraissant légèrement tordu en vue dorsale (fig. 218*c*) et légèrement élargi à l'apex, vu de côté (fig. 218*a*); épandrium allongé et plutôt mince, avec une incision membraneuse ventro-latérale (fig. 218*a*); hypandrium modérément rugueux sur les côtés, avec angles postéro-latéraux légèrement arrondis (fig. 218*b*). Espèce répandue .... ***americanus* (Wiedemann)**

Surstylus régulièrement fuselé vers l'apex (fig. 221*c*), sans élargissement apical vu de côté (fig. 221*e*); épandrium non allongé et sans incision ventro-latérale (fig. 221*e*); hypandrium fortement rugueux sur les côtés, pourvu d'angles postéro-

latéraux proéminents (fig. 221*b*). Ouest du Canada ....
.... ***fumipennis*** **(Thomson)**

12. Poils du scutellum surtout ou totalement blancs. Front entièrement jaune pâle. Orbite postoculaire supérieur modérément large (fig. 13). Région arctique du Canada. Groënland .... ***nigroventris*** **(Fluke)**

Poils du scutellum presque tout noirs. Front marqué habituellement de deux taches brunes à noirâtres au-dessus de l'antenne, rarement tout jaune. Orbite postoculaire supérieur étroit (fig. 12) 13

13. Tergite 5 jaune ou jaune orangé, maculé d'une tache noire antéro-médiane. Base de l'édéage pourvue d'une dent courte et bien arrondie et d'une autre acérée (comme dans la fig. 220*d*). Centre du Canada .... ***neoperplexus*** **(Curran)**

Tergite 5 orné d'une tache médiane noire qui s'étend jusqu'au bord latéral, souvent largement ou tout noir. Base de l'édéage pourvue d'une dent courte et acérée et d'une dent acérée plus longue (fig. 226*d*, 227*d*). Espèce répandue, y compris au Groënland .... ***luniger*** **(Meigen)**

14. Front entièrement brillant, noir sur les deux tiers supérieur et jaune sur le tiers inférieur. Tergites tout noirs ou soulignés tout au plus de taches pâles peu apparentes. Région arctique du Canada, Groënland .... ***nigroventris*** **(Fluke)**

Front orné d'une paire de taches pruineuses distinctes sur les côtés. Tergites portant habituellement de larges taches ou bandes jaunes distinctes 15

15. Tergites 3 et 4 marqués tous deux d'une bande jaune complète fortement étranglée au milieu chez certains spécimens .....
.... ***americanus*** **(Wiedemann)**
.... ***fumipennis*** **(Thompson)**
.... ***pomus*** **(Curran)**

Tergite 3 orné d'une paire de taches jaunes en forme de lunules; tergite 4 présentant une paire similaire de taches, réunies étroitement au milieu chez certains spécimens ....
.... ***luniger*** **(Meigen)**
.... ***neoperplexus*** **(Curran)**

Eupeodes (*Eupeodes*) *americanus* (Wiedemann)

Figs. 43, 127, 218; Map 20

Syrphus americanus Wiedemann, 1830:129.
Syrphus wiedemanni Johnson, 1919:32 (unjustified new name for *americanus* Wiedemann).

Length. 7.0–11.0 mm.

Male. Frons dull yellow, usually with pair of small semicircular pale to dark brown spots above antennae. Face yellow, with narrow brown median stripe extending well above tubercle; lower facial margin yellow; cheek yellow anteriorly, blackish posteriorly. Upper postocular orbit moderately narrow (as in Fig. 12).

Scutum and pleura with yellow hairs. Scutellar hairs usually pale yellow, rarely some black hairs on disc. Wing moderately bare; cell c bare on about basal one-tenth; cell bm bare on about anterobasal two- to three-fifths; alula bare on most of anterior one-third. Fore and mid femora black on basal four-fifths; hind tibia with brownish ring at mid length in some specimens; hind tarsi brownish; legs otherwise yellow.

Abdomen in Fig. 127. Tergite 2 with pair of large yellow spots strongly narrowed medially, almost reaching margins laterally in some specimens; tergites 3 and 4 each with broad yellow band with anterior margin nearly straight and posterior margin either nearly straight or broadly and shallowly emarginate medially; bands not extending to lateral margins; tergite 4 with posterior margin broadly yellow; tergite 5 yellow with black band usually extending to narrow black lateral margins. Sternites yellow, 1–4 each with black band covering at most half sternite. Surstylus (Figs. 218*a*,*c*) elongate, irregular in outline, in dorsal view appearing slightly twisted, in lateral view with apex slightly enlarged; epandrium (Fig. 218*a*) long, slender, with membrane-filled emargination on ventrolateral margin; hypandrium (Fig. 218*b*) moderately rugose laterally, with apicolateral angles rounded or nearly so, without apical projections.

Female. Not distinguishable from those of *E. fumipennis* and *E. pomus*.

Distribution. Canada (Map 20) south to California,* Texas, Mexico,* and Florida (California and Mexico records possibly based on *E. fumipennis*). Ont., Que., V–X.

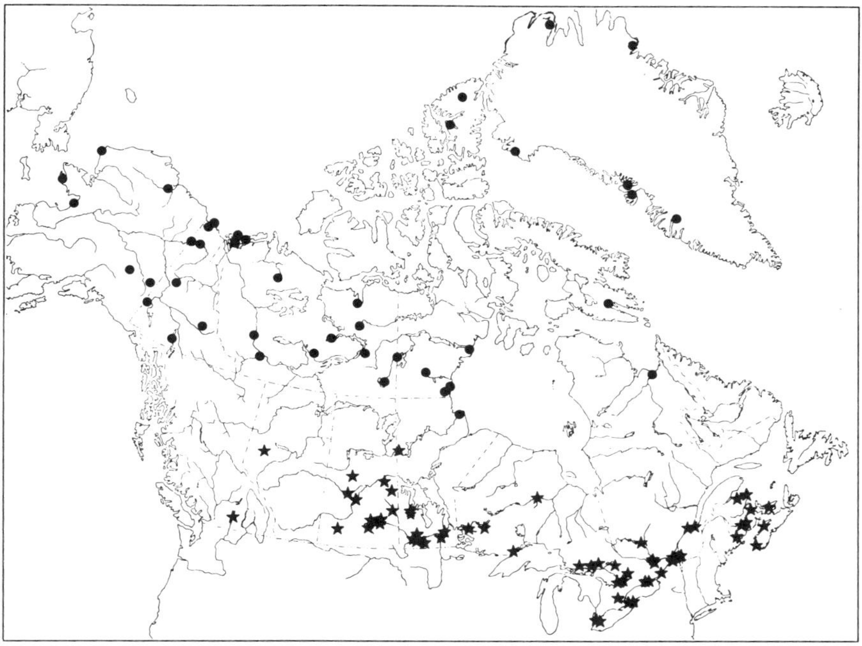

Map 20. Collection localities for *Eupeodes americanus* (Wiedemann) (★) and for *Eupeodes curtus* (Hine) (●).

Specimens identified. Canada, 158 ♂♂; United States, 65 ♂♂.

Biology. Heiss (1938) lists 22 aphid species as larval hosts from her own rearings and from literature records; most if not all these observations are probably based on correct adult syrphid identifications. Additional aphid host records from the CNC are *Cinara carolina*, *Eriosoma americanum*, and *Myzus persicae*.

Eupeodes (*Eupeodes*) *curtus* (Hine)

Figs. 128, 219; Map 20

Syrphus curtus Hine, 1922:145.

Length. 7.9–11.5 mm.

Male. Frons bright yellow to grayish yellow, with pair of obscure or distinct brown semicircular spots above antennae in some specimens. Face bright yellow, with rather broad brown to black median stripe of variable length; lower facial margin and cheek usually broadly black, cheek rarely yellow anteriorly. Upper postocular orbit moderately narrow (as in Fig. 12).

Scutal hairs pale yellow, few to many brown to black hairs on disc in some specimens; pleural hairs very pale yellow; scutellar hairs mostly black. Wing extensively trichose; cell c entirely trichose or bare on about basal one-tenth; cell bm bare on anterobasal one- to two-fifths; alula entirely trichose. Fore and mid femora black on basal two-fifths to three-quarters; hind femur black on basal three-quarters to five-sixths; hind tibia with brownish ring beyond middle in some specimens; tarsi brownish; legs otherwise dark yellow.

Abdomen in Fig. 128. Tergite 2 with pair of small to rather large yellow spots nearly reaching lateral margins in some specimens; tergites 3 and 4 each with pair of yellow spots only rarely extending narrowly to lateral margins; each spot with anterior margin rather strongly concave; tergite 4 with very narrow to broad yellow posterior margin; tergite 5 varying from yellow with small anteromedian black spot or band to black with narrow posterior margin yellowish. Sternites usually yellow with large black bands, rarely with small black bands or entirely yellow. Surstylus short; hypandrium (Fig. 219*b*) not narrowed apically, without apical projections, weakly rugose; aedeagal base with two short heavy acute teeth and two longer acute teeth.

Female. Frons black on about upper two-thirds, yellow on about lower one-third, usually with pair of brown to black spots above antennae; yellow area slightly or strongly brownish in some specimens, lower part of black area with pair of small to large pruinose lateral spots. Anterior part of cheek commonly yellow. Femora in some specimens only narrowly black at base. Yellow spots of tergites commonly extending rather broadly to lateral margins. Sternites varying from mostly yellow-orange to almost entirely black; sternite 5 mostly yellow-orange except in specimens from Greenland.

Distribution. Alaska, Canada, Greenland (Map 20); Iceland;* Finland.* Y.T., VI–VIII; Greenland, VI–VIII.

Specimens identified. Alaska, 62 ♀♀; Canada, 23 ♂♂, 95 ♀♀; Greenland, 16 ♂♂, 7 ♀♀.

Eupeodes (*Eupeodes*) *flukei* (Jones)

Fig. 129; Map 21

Syrphus flukei Jones, 1917:222.
Syrphus palliventris Curran, 1925:173.

Length. 7.3–11.6 mm.

Male. Frons yellow. Face yellow; lower facial margin and tubercle brownish in some specimens; cheek black only posteriorly. Upper postocular orbit very narrow.

Scutum with bright yellow to pale yellow hairs; pleura with white or yellow hairs. Scutellum usually with pale yellow hairs, with few or many black hairs in some specimens. Wing moderately bare; cell c bare on basal one-fifth to one-sixth; cell bm about five-sixths to seven-eighths bare; alula bare on most of anterior half. Fore and mid femora blackish on basal one-quarter or less; hind femur obscurely blackish in some specimens; rest of legs yellow-orange.

Abdomen in Fig. 129. Tergite 2 with pair of large yellow spots extending to lateral margins anterolaterally in some specimens; tergites 3 and 4 each with pair of very large yellow spots with concave anterior margin; spots in some specimens extending to lateral and anterior margins of tergites 3 and 4 and in some specimens rather broadly confluent medially; tergite 4 with rather broad yellow-orange posterior margin; tergite 5 entirely red-orange. Sternites yellow-orange, 1–4 each with small to large black band. Terminalia as in *E. montivagus* (Fig. 228).

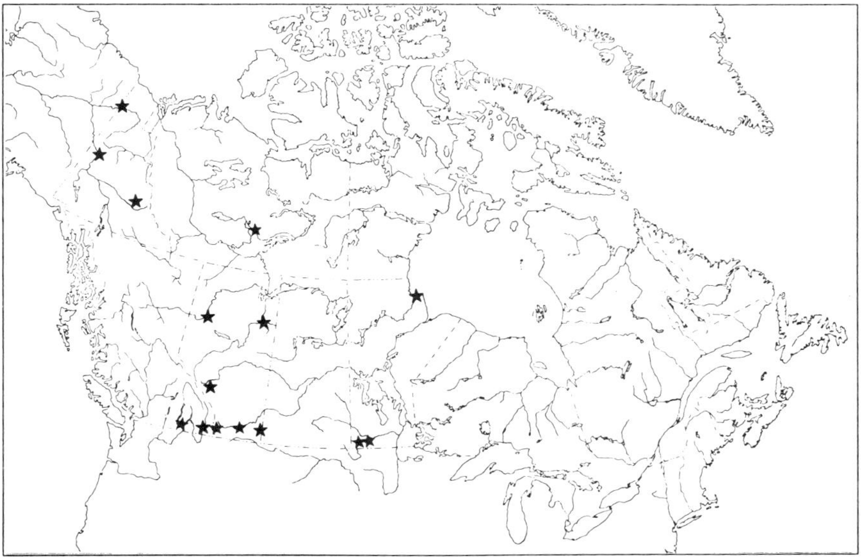

Map 21. Collection localities for *Eupeodes flukei* (Jones).

Female. Frons black on upper one-third to half, yellow below with pair of obscure brown spots above antenna and with pair of large yellow pruinose lateral spots. Scutellum with few to many black hairs on disc. Fore and mid femora blackish at extreme base in some specimens;

femora otherwise yellow-orange. Yellow spots on tergite 3 narrowly separated; spots on tergite 4 separated or narrowly joined. Sternite 4 with or without small black band; sternite 5 yellow-orange; rarely all sternites apparently entirely black.

Distribution. Western Canada (Map 21), Ontario,* south to Washington* and Colorado. B.C., IX; Alta., VI, VIII.

Specimens identified. Canada, 16 ♂♂, 14 ♀♀; United States, 1 ♂, 2 ♀♀.

Eupeodes (*Eupeodes*) *fumipennis* (Thomson)

Figs. 130, 221; Map 22

Syrphus fumipennis Thomson, 1869:499.

Length. 8.2–11.8 mm.

Male. Frons yellow, with two faint to distinct semicircular brown spots above antennae. Face yellow with narrow brown to blackish stripe extending well over tubercle; cheek brown to black posteriorly. Upper postocular orbit moderately narrow.

Scutum and pleura with white or yellowish hairs. Scutellar hairs mostly black. Wing as in *E. americanus*. Fore and mid femora black on about basal one-third; hind femur varying from slightly brownish near base to blackish on basal two-thirds; rest of legs yellow-orange with tarsi brownish above.

Abdomen in Fig. 130. Tergite 2 with pair of large yellow spots extending narrowly to lateral margins in some specimens; tergites 3 and 4 each with broad yellow band with slightly wavy anterior margin, medially emarginate posterior margin; bands extending to lateral margins in some specimens; tergite 4 with broad yellow posterior margin; tergite 5 yellow with anteromedian black spot usually joining narrow black lateral margins. Sternites yellow, 1–4 each with large black band. Surstylus (Figs. 221*c*,*e*) elongate, tapering evenly to apex, not enlarged apically in lateral view; epandrium (Fig. 221*e*) short, without notch on ventrolateral margin; hypandrium (Fig. 221*b*) strongly rugose laterally, with posterolateral angles projecting.

Female. Similar to male; not definitely distinguishable from females of *E. americanus* and *E. pomus*.

Distribution. Western Canada (Map 22), south to California, Mexico (Chiapas), and New Mexico.

Map 22. Collection localities for *Eupeodes fumipennis* (Thomson).

Specimens identified. Canada, 55 ♂♂; United States, 52 ♂♂; Mexico, 1 ♂.

Biology. Specimens in the CNC were reared from larvae found among woolly aphids and from larvae associated with sugar beet root aphids (*Pemphigus* sp.). Campbell and Davidson (1924) reported larvae of *E. meadii* Jones from 12 different aphids in California; it is probable that most or all of their specimens were of *E. fumipennis*.

Eupeodes (*Eupeodes*) *latifasciatus* (Macquart)

Figs. 131, 220; Map 23

Syrphus latifasciatus Macquart, 1829:242.
Syrphus pallifrons Curran, 1925:172.
Syrphus depressus Fluke, 1933:97.
Metasyrphus chillcotti Fluke, 1952:20.

Length. 7.0–10.1 mm.

Male. Frons bright yellow. Face bright yellow, with brown median stripe of variable width and length; lower facial margin and cheek dark brown to black. Upper postocular margin rather broad (as in Fig. 13).

Scutum and pleura with bright yellow hairs. Scutellum with bright yellow hairs, usually with few to many black hairs on disc. Wing extensively trichose; cell c entirely trichose or bare on at most basal one-tenth; cell bm usually entirely trichose, with small bare area anteriorly near base in some specimens; alula entirely trichose. Fore and mid femora with about basal one-third brown to black; hind femur with about basal half to three-fifths brown to black; tarsi brownish above; legs otherwise bright yellow.

Abdomen in Fig. 131. Tergite 2 with pair of yellow spots of varied size only rarely extending to lateral margins; tergites 3 and 4 each with broad yellow band extending rather broadly to lateral margins in some specimens; each band emarginate posteromedially or, especially in northern specimens, narrowly or broadly divided medially in two large spots with anterior margins straight or only slightly concave; tergite 4 with moderately broad yellow posterior margin; tergite 5 yellow with anteromedian black spot extending laterally as arcuate band almost to lateral margins in some specimens. Sternites yellow with small black bands; sternite 4 without black band in some specimens. Surstylus short; hypandrium (Fig. 220*b*) slightly narrowed apically, weakly rugose, with short but rather deep posterior projection on either side of base of aedeagus; aedeagal base (Fig. 220*d*) with two short bluntly rounded teeth and two longer acute teeth.

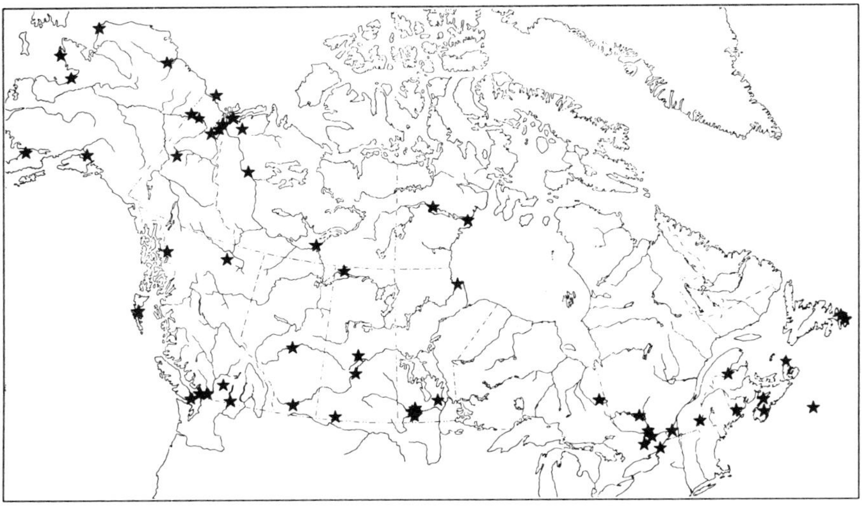

Map 23. Collection localities for *Eupeodes latifasciatus* (Macquart).

Female. Frons black on upper two-thirds, bright yellow on lower one-third, entirely shining. Face almost without median stripe in some specimens; oral margin and anterior half of cheek yellow in some specimens. Wing usually less extensively trichose than in male; cell bm commonly bare on about anterobasal one-quarter. Legs paler than in male; fore and mid femora only narrowly darkened at base, hind femur entirely bright yellow in some specimens. Tergites 3 and 4 with yellow bands or spots slightly narrower than in male. Sternite 5 usually bright yellow, rarely with small median black spot.

Distribution. Alaska, Canada (Map 23), south to California,* Utah, Texas,* Delaware, and Virginia;* Europe. B.C., IV–VII, X; Ont., Que., V–VIII.

Specimens identified. Alaska, 41 ♂♂, 10 ♀♀; Canada, 73 ♂♂, 72 ♀♀; United States, 2 ♂♂; Europe, 7 ♂♂, 6 ♀♀.

Biology. Láska and Starý (1980) listed four species of aphids as larval hosts in Czechoslovakia.

Eupeodes (*Eupeodes*) *luniger* (Meigen)

Figs. 12, 132, 226, 227; Map 24

Syrphus luniger Meigen, 1822:300.
Metasyrphus luniger astutus Fluke, 1952:15.
Metasyrphus luniger vockerothi Fluke, 1952:17.

Length. 7.8–11.5 mm.

Male. Frons dull yellow to grayish yellow with two brown to blackish spots above antenna. Face dull yellow, with distinct to diffuse brown to black median stripe extending well above tubercle; oral margin and cheeks brown to black. Upper postocular orbit rather narrow (Fig. 12).

Scutum with hairs entirely yellowish or with few to many black hairs on disc; pleura with hairs white to yellow. Scutellar hairs mostly black. Wing moderately haired; cell c bare on about basal one-eighth to one-twelfth; cell bm bare on about basal one- to two-thirds; alula very narrowly to broadly bare on basal half. Fore and mid femora brown to black on basal one- to two-thirds; hind femur blackish on about basal five-sixths; tarsi and tibia commonly brownish; rest of legs dark yellow.

Abdomen in Fig. 132. Tergite 2 with pair of slender yellow spots well removed from lateral margins; tergites 3 and 4 each with pair of slightly lunulate oblique yellow spots of varying width well-removed from lateral margins; tergite 4 with moderately broad posterior yellow margin; tergite 5 usually yellow with broad arcuate black band reaching lateral black margins; some specimens black with only narrow posterior margin yellow (in Greenland specimens yellow tergite markings greatly reduced). Sternites yellow with large black bands. Surstylus short; hypandrium variable, moderately rugose, with very short posterior processes beside aedeagus in some specimens (Figs. 226*b*, 227*b*); base of aedeagus with two acute or rarely rather blunt teeth and two longer acute teeth (Figs. 226*d*, 227*d*).

Female. Frons black on about upper two-thirds, yellow on lower one-third, with pair of faint to distinct semicircular brown to black spots above antennae or in some specimens extensively brownish below, with pair of large lateral pruinose spots at mid length; spots confluent medially in some specimens. Scutal hairs mostly white, rarely some hairs above wing bases black.

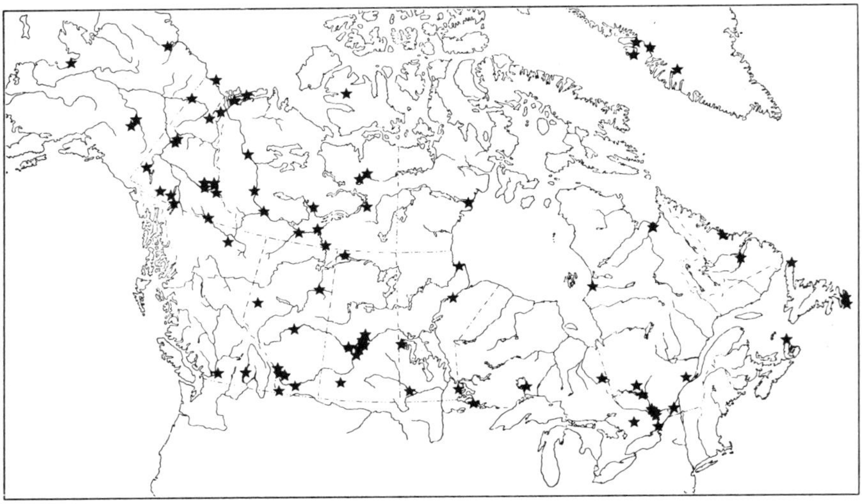

Map 24. Collection localities for *Eupeodes luniger* (Meigen).

Distribution. Alaska, Canada, Greenland (Map 24), south to California,* Utah, and Maryland; Europe; Asia.* B.C., VI–VIII; Ont., Que., V–X.

Specimens identified. Alaska, 3 ♂♂; Canada, 84 ♂♂; Greenland, 7 ♂♂; United States, 6 ♂♂; Europe, 13 ♂♂.

Biology. Láska and Starý (1980) listed 33 species of aphids as larval hosts in Czechoslovakia. Dušek and Láska (1942) found that in Czechoslovakia specimens of *E. luniger* overwinter as pupae with at least a partial diapause.

Discussion. The species shows greater variation in structure of male terminalia than do the other species of the genus (two extremes are shown in Figs. 226, 227). A number of species may be included here but correlations between color characters, structure of terminalia, and geographic distribution could not be determined.

Eupeodes (Eupeodes) montivagus (Snow)

Figs. 133, 228; Map 25

Syrphus montivagus Snow, 1895:236.

Length. 7.8–12.3 mm.

Male. Frons pale yellow, obscurely brownish above antennae in some specimens. Face yellow with lower facial margin broadly black, and with lower part of tubercle brown to black in some specimens; cheek black. Upper postocular orbit broad.

Scutum and pleura with yellow to yellow-orange hairs. Scutellum usually with only yellow hairs; some specimens with some black hairs on disc. Wing moderately bare; cell c bare on about basal one-sixth; cell bm about five-sixths bare; alula bare on most of anterior half. Fore and mid femora black on basal one-third to half; hind femur black on about basal two-thirds; rest of legs yellow-orange with tarsi brownish above.

Abdomen in Fig. 133. Tergite 1 and anterior and lateral margins of tergite 2 with slight bluish tint; tergite 2 with pair of rather large orange-red spots having inner ends broad or distinctly narrowed and broadly separated from lateral margins; tergite 3 with pair of large yellow-orange to orange-red spots with nearly straight anterior margins; spots in some specimens narrowly to broadly joined anteromedially and rarely extending narrowly to lateral margins; tergite 4 similar to tergite 3 but with broad orange-red posterior margin; tergite 5 entirely orange-red. Sternites orange-red; sternites 1–3 each with large black band; sternite 4 with small black band. Surstylus short; hypandrium slightly broadened apically, moderately rugose laterally, without posterior processes beside aedeagus (Fig. 228); aedeagal base with two very short and two long acute teeth.

Female. Frons black on upper one-third to half, yellow below with pair of large yellow pruinose lateral spots. Femora usually less extensively black than in male; hind femur entirely yellow-orange in some specimens. Reddish spots usually narrowly to broadly joined anteromedially on tergite 3; spots always joined on tergite 4. Sternite 4 with black band very small or absent in some specimens; sternite 5 orange-red.

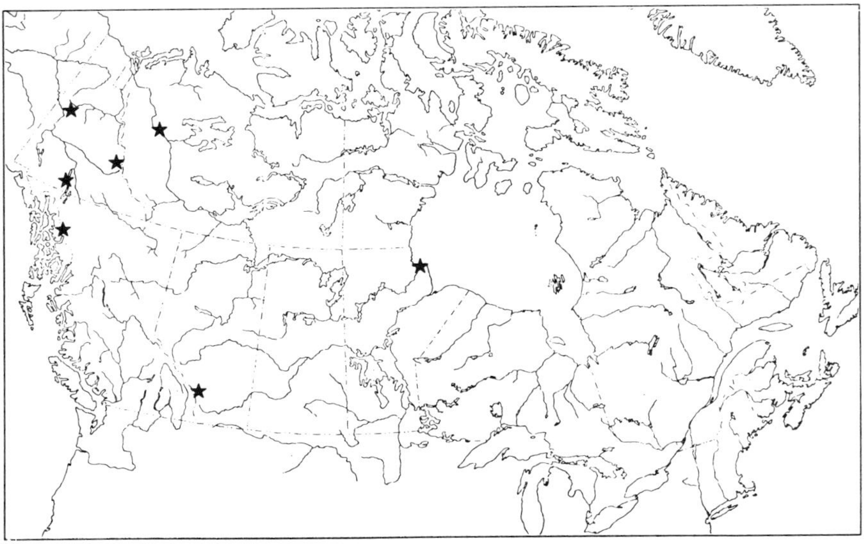

Map 25. Collection localities for *Eupeodes montivagus* (Snow).

Distribution. Western Canada (Map 25), south to California, Colorado, and New Mexico.* Y.T., VII–VIII.

Specimens identified. Canada, 3 ♂♂, 5 ♀♀; United States, 15 ♂♂, 10 ♀♀.

Eupeodes (*Eupeodes*) *neoperplexus* (Curran)

Map 26

Syrphus neoperplexus Curran, 1925:93.

Length. 10.0–11.6 mm.

Male. Very similar to that of *E. luniger* but slightly paler. Facial stripe broad, obscure, pale brown. Scutal hairs all pale. Hind femur brown to black on basal half or slightly less. Tergite 5 yellow-orange with black anteromedian spot far removed from lateral margins. Terminalia similar to those of *E. luniger* but shorter teeth of aedeagal base bluntly rounded.

Female. Apparently indistinguishable from that of *E. luniger.*

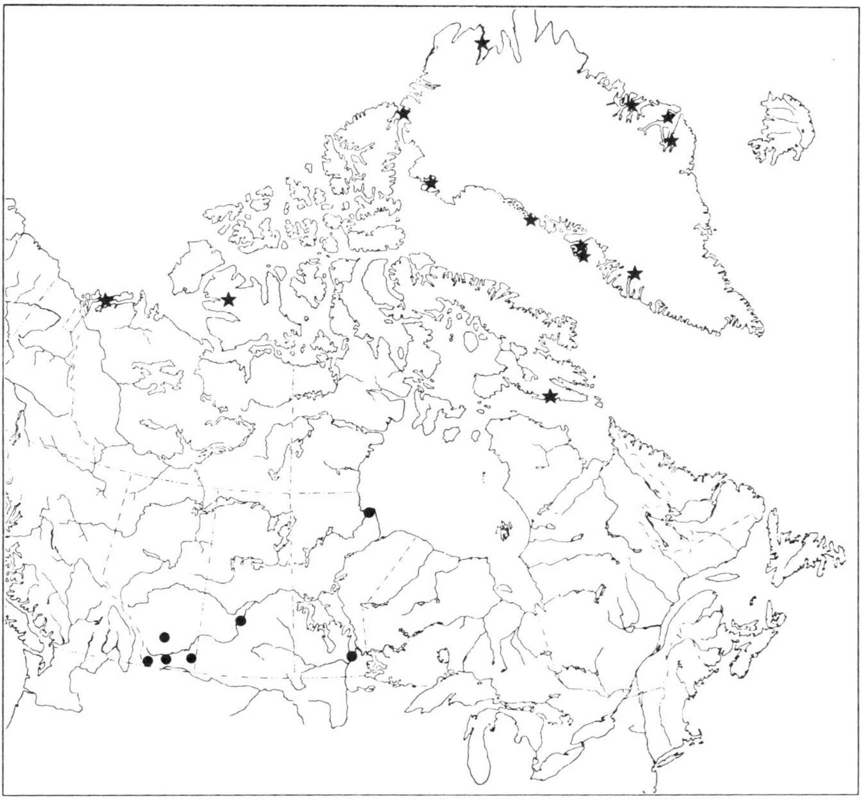

Map 26. Collection localities for *Eupeodes neoperplexus* (Curran) (●) and for *E. nigroventris* (Fluke) (★).

Distribution. Central Canada (Map 26), Ontario,* south to North Dakota.* Alta., V–VII.

Specimens identified. Canada, 13 ♂♂.

Discussion. This species, as here defined, differs from paler specimens of *E. luniger* only in the markings of tergite 5 and in the structure of the aedeagal base. It is possibly a synonym of *luniger.*

Eupeodes (*Eupeodes*) *nigroventris* (Fluke)

Fig. 134; Map 26

Syrphus nigroventris Fluke, 1933:97.

Length. 9.0–11.4 mm.

Male. Frons pale yellow. Face pale yellow; broad lower facial margin, median stripe extending to upper margin of tubercle, and cheek black. Upper postocular orbit moderately broad (Fig. 13).

Scutum and pleura with white or yellowish hairs. Scutellum with white hairs, also with some black hairs on disc in some specimens. Wing mostly trichose; cell c bare on about basal one-tenth; cell bm bare on about one-third to three-fifths; alula bare on most of anterior half. Femora black, yellowish only at extreme apex; fore and mid tibiae yellowish, brownish apically; hind tibia mostly blackish; tarsi brown to black.

Abdomen (Fig. 134) entirely black with posterior margins of tergites 4 and 5 narrowly yellowish, and with small and obscure paired yellowish spots on tergites 2–4 in some specimens. Sternites black with narrow yellowish incisures. Surstylus moderately long, evenly tapered to apex; hypandrium weakly rugose, without apical projections; aedeagal base with two short and two long acute teeth.

Female. Frons shining, black on about upper two-thirds, bright yellow on about lower one-third.

Distribution. Arctic Canada, Greenland (Map 26). Greenland, VI–VIII.

Specimens identified. Canada, 12 ♂♂, 2 ♀♀; Greenland, 1 ♂, 9 ♀♀.

Biology. A specimen from Peary Land, Greenland, was reared from a larva collected on *Salix*. Both aphids and coccids were present and the larval prey was not determined.

Eupeodes (*Eupeodes*) *perplexus* (Osburn)

Figs. 135, 222; Map 27

Syrphus perplexus Osburn, 1910:55.
Syrphus meadii Jones, 1917:223.

Length. 9.7–13.0 mm.

Male. Frons dull yellow with two large brown to black spots above antennae. Face pale yellow with distinct brown to black median stripe on lower two-thirds; lower facial margin usually brown anteriorly; cheek yellowish brown to blackish. Antenna brownish black; first flagellomere orange-brown below. Upper postocular orbit very narrow.

Scutum with pale yellow to white hairs, with some black hairs laterally behind suture; pleura with white hairs. Scutellum with few to many black hairs on disc, otherwise with pale hairs. Wing extensively bare; cell c densely trichose on apical half or less; some specimens with some scattered microtrichia on basal half; cell bm with very few microtrichia at extreme apex; cells r_{2+3}, r_{4+5}, bm, and cua_1 with bare areas at base; alula bare on most of basal two-thirds. Fore and mid femora brown to black on basal one-third to half; hind femur brown to black on about basal nine-tenths; hind tibia with brown ring beyond mid length; tarsi brownish above; legs otherwise dull yellow to yellow-brown; hairs of apical half or entire posterior surface of fore and mid femora black.

Abdomen in Fig. 135. Tergites 2–4 each with pair of yellow spots well separated from lateral margins; spots on 3 and 4 slightly oblique and slightly concave anteriorly; tergite 4 with very narrow yellow posterior margin; tergite 5 black with narrow yellow posterior margin and with pair of sublateral anterior yellowish spots in some specimens. Sternites 1–4 yellow, each with large black band. Surstylus moderately long and tapering (Fig. 222); hypandrium subquadrate, without posterior processes beside aedeagus, moderately rugose laterally; aedeagal base with two short and two long acute teeth.

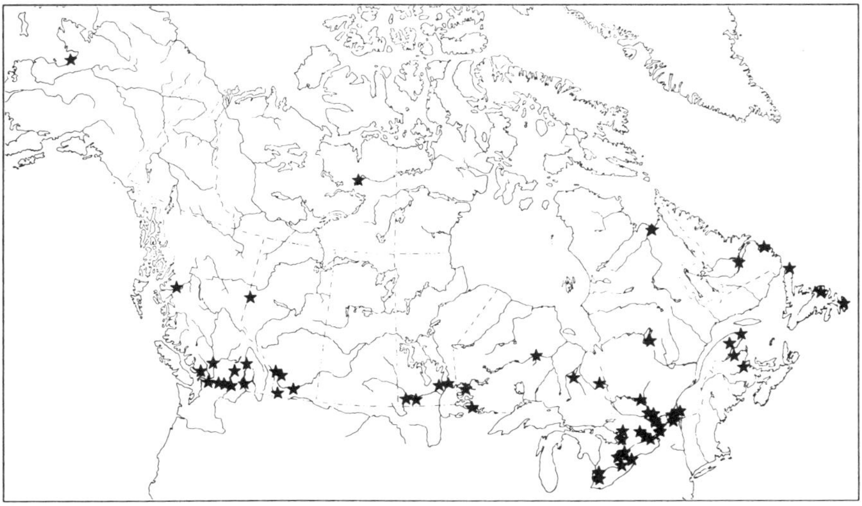

Map 27. Collection localities for *Eupeodes perplexus* (Osburn).

Female. Frons black on upper half to two-thirds, with two large semicircular brown to black spots above antennae; between these dark areas usually with irregular and poorly defined dull yellow median band but some specimens with dark areas confluent; frons with pair of large whitish pruinose lateral spots. Scutum commonly with only pale hairs. Abdomen of some specimens with yellow spots of tergites 3 and 4 narrowly joined medially. Sternites 1–5 yellow, each with large black band.

Distribution. Alaska, Canada (Map 27), south to Oregon, New Mexico,* and Mississippi. B.C., V–IX; Ont., Que., IV–IX.

Specimens identified. Alaska, 1 ♀; Canada, 40 ♂♂, 94 ♀♀; United States, 1 ♂, 12 ♀♀.

Biology. A specimen in the CNC was reared from a larva feeding on *Aphis* sp. on *Bidens* sp. Heiss (1938) lists four other aphid species as hosts; it is probable that the syrphids were correctly identified.

Eupeodes (*Eupeodes*) *pomus* (Curran)

Fig. 223; Map 28

Syrphus americanus var. *pomus* Curran, 1921*b*:172.
Syrphus americanus var. *vinelandi* Curran, 1921*b*:172.

Length. 6.8–12.0 mm.

Male. Frons yellow with two pale brown to black semicircular spots above antennae. Face yellow with narrow and obscure to broad and distinct brown median stripe; cheek yellow or rarely brownish anteriorly, brownish posteriorly.

Scutum and pleura with yellowish hairs. Scutellar hairs mostly yellow to mostly black. Wing as in *E. americanus*. Fore and mid femora brown to black on about basal one-third; hind femur brown to black on basal half to four-fifths; tibia and tarsi often brownish, rest of legs yellow-orange.

Abdomen as in *E. fumipennis*. Bands of tergites 3 and 4 in some specimens with both anterior and posterior margins nearly straight, varying to nearly divided medially. Sternites yellow with large black bands. Surstylus short (Fig. 223); epandrium without ventrolateral incision (as in Fig. 221*e*); hypandrium neither broadened posteriorly nor with posterior projections, weakly rugose (Fig. 223*b*); aedeagal base with two short and two long acute teeth (as in Fig. 219).

Female. Similar to male; not definitely distinguishable from females of *E. americanus* and *E. fumipennis*.

Distribution. Canada (Map 28), south to Oregon, Arizona, Oklahoma,* and Virginia.* B.C., V–VIII; Ont., Que., V–VIII.

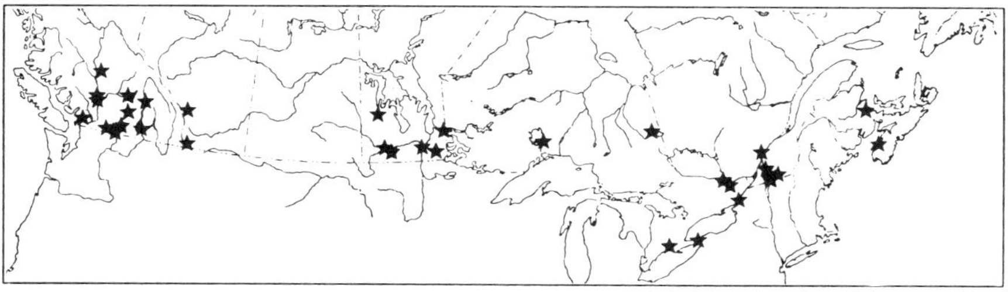

Map 28. Collection localities for *Eupeodes pomus* (Curran).

Specimens identified. Canada, 42 ♂♂; United States, 32 ♂♂.

Biology. The type specimens of *E. pomus* were reared from larvae feeding on *Aphis pomi*. Other specimens in the CNC were reared from an aphid on *Acer* sp.

Eupeodes (*Eupeodes*) *snowi* (Wehr)

Figs. 136, 224; Map 29

Syrphus ruficauda Snow, 1892:36 (preoccupied Bigot, 1883).
Syrphus snowi Wehr, 1922:137 (new name for *ruficauda* Snow).
Syrphus snowi Curran, 1925:173 (new name for *ruficauda* Snow).

Length. 7.6–11.4 mm.

Male. Frons pale yellow. Face yellow with lower margin broadly black and rarely with tubercle slightly brownish; cheek black. Upper postocular orbit narrow (as in Fig. 12).

Scutum and pleura with pale yellow hairs. Scutellum usually with only yellow hairs; some specimens with some black hairs on disc. Wing moderately bare; cell c trichose on apical half or more; cell bm about four-fifths bare, trichose near apex and along most of posterior margin; alula bare on most of anterior half. About basal half of fore and mid femora and basal three- to four-fifths of hind femur black; rest of legs bright yellow-orange.

Abdomen in Fig. 136. Tergite 2 with pair of large yellow spots well-separated from lateral margins; tergite 3 with similar spots broadly confluent medially in some specimens; tergite 4 mostly red-orange, narrowly black laterally, with narrow obscure darker arcuate band just beyond mid length in some specimens; tergite 5 red-orange, very narrowly black laterally. Sternites 1–3 yellow, each with large black band; sternite 4 red-orange. Surstylus short; hypandrium (Fig. 224) parallel-sided or slightly narrowed apically, strongly rugose laterally, with strong posterior projection on each side of base of aedeagus; aedeagal base with pair of short and pair of long acute teeth (as in Fig. 219*d*).

Female. Frons black on upper half or slightly more, yellow below, entirely shining. Scutellum with only yellow hairs. Fore and mid femora usually narrowly black at base, rarely entirely yellow-orange; hind femur usually entirely yellow-orange, rarely black on about basal half. Abdominal markings similar to those of male, but with markings of tergites 2 and 3 usually narrower and tergite 3 with almost straight transverse band in some specimens; markings of tergites 2–4 rarely extending to lateral margins; tergite 5 red-orange, rarely with narrow lateral margins darker. Sternites usually without distinct dark markings; sternites 2 and 3 mostly blackish in some specimens.

Distribution. Western Canada (Map 29), south to Oregon,* Colorado, Arizona,* and Nebraska.* B.C., V–VIII.

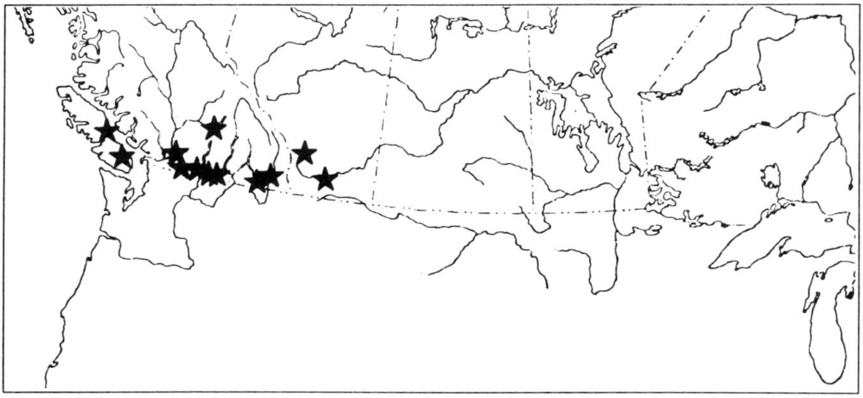

Map 29. Collection localities for *Eupeodes snowi* (Wehr).

Specimens identified. Canada, 3 ♂♂, 8 ♀♀; United States, 7 ♂♂, 4 ♀♀.

Eupeodes (*Eupeodes*) *volucris* Osten Sacken

Figs. 137, 225; Map 30

Eupeodes volucris Osten Sacken, 1877:328.
Syrphus perpallidus Bigot, 1884:90.
Eupeodes braggii Jones, 1917:221.
Eupeodes weldoni Jones, 1917:221.

Length. 6.3–9.8 mm.

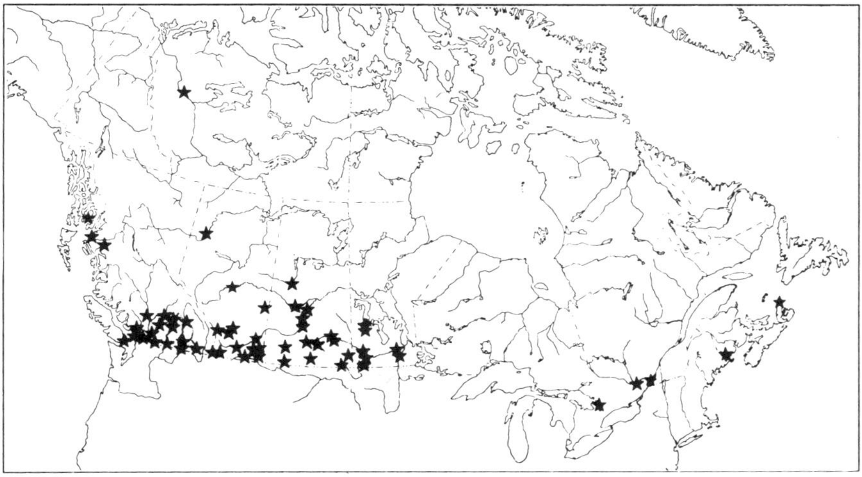

Map 30. Collection localities for *Eupeodes volucris* Osten Sacken.

Male. Frons pale yellow. Face pale yellow with distinct dark brown to black median stripe on lower three-fifths; stripe continuing on each side on anterior half of lower facial margin; cheek pale brown to black posteriorly. Antenna entirely black. Upper postocular orbit moderately broad.

Scutum, pleura, and scutellum with white to pale yellow hairs. Wing extensively bare; cells c and bm with scattered microtrichia only at extreme apex; all cells bare at base; only about apical one-quarter and posterior one-third of wing with moderately dense microtrichia. Basal one-third to three-fifths of fore and mid femora and about basal four-fifths of hind femur black; legs otherwise dark yellow; hairs of posterior surface of fore and mid femora yellow.

Abdomen in Fig. 137. Tergites 2–4 each with pair of large pale yellow spots well-separated from lateral margins; spots of tergites 3 and 4 slightly oblique and with anterior margin slightly concave; tergite 4 with posterior margin mostly narrowly pale yellow; tergite 5 black with posterior margin narrowly yellow and usually with part or all lateral margin narrowly yellow to reddish. Sternites 1–3 pale yellow, each with black band; sternite 4 yellow-orange. Terminalia very large, protruding at apex of abdomen as large shining black cylinder; upper surface composed of nearly fused sternites 7 and 8 (Fig. 137); surstyli on under side of cylinder, each very long, slender, distinctly curved and with enlarged and notched apex (Fig. 225).

Female. Frons with upper two-thirds black and with large undivided semicircular black spot above antennae, with well-defined narrow pale yellow band between two black areas; lower part of upper black area with pair of large lateral whitish pruinose spots. Femora usually less extensively black basally than in male; hind femur black on basal two-fifths to three-quarters. Sternites 4 and 5 yellow-orange.

Distribution. Canada (Map 30) south to California, Texas, Mexico (Chiapas), and Louisiana.* B.C., V–VIII; Ont., VII.

Specimens identified. Canada, 70 ♂♂, 129 ♀♀; United States, 112 ♂♂, 214 ♀♀; Mexico, 15 ♂♂, 11 ♀♀.

Biology. *E. volucris* is multivoltine. Jones (1922) gave 21 days as the average period from egg to adult in the laboratory. Heiss (1938) listed 20 species of aphids as larval hosts; an additional host record from the CNC is *Cinara ponderosae*.

Subgenus *Eupeodes* (*Lapposyrphus*) Dušek & Láska

Diagnosis. As for genus *Eupeodes*. Length 7.2–14.1 mm.

Description. Face with distinct but some species with very narrow pale brown to black median stripe. Metasternum bare. R_{4+5} distinctly dipped into cell r_{4+5} (Fig. 20); wing membrane extensively and rather variably bare on much of basal one-third. Male terminalia small, scarcely apparent from above (Fig. 139).

Distribution. One Holarctic species, one western Nearctic species.

Biology. Larvae have been reported to feed on a number of species of aphids. Most of the reared specimens in the CNC are from larvae collected on conifers but this predominance may be a collection bias. Goeldlin (1974) stated that *E. lapponicus* Zetterstedt is multivoltine in

Switzerland and overwinters as a pupa, or apparently, in the case of some females, as an imago.

Key to species of *Eupeodes* (*Lapposyrphus*)

1. Tergites 3 and 4 each with pair of well-separated yellow spots (Fig. 139) or, in some females, tergites entirely black. Widespread .... ***lapponicus* (Zetterstedt)**

   Tergites 3 and 4 each with undivided yellow band (Fig. 138). Western Canada .... ***aberrantis* (Curran)**

Clé des espèces de *Eupeodes* (*Lapposyrphus*)

1. Tergites 3 et 4 pourvus de deux taches jaunes bien distinctes (fig. 139) ou, chez certaines femelles, tergites entièrement noirs. Répandu .... ***lapponicus* (Zetterstedt)**

   Tergites 3 et 4 pourvus d'une bande jaune continue (fig. 138). Ouest du Canada .... ***aberrantis* (Curran)**

Eupeodes (*Lapposyrphus*) *aberrantis* (Curran)

Fig. 138; Map 31

Syrphus aberrantis Curran, 1925:90.

Length. 7.2–11.4 mm.

Male. Eye angle about 100°. Frons yellow, usually with semicircular brown line above antennae, weakly pale pruinose above this line. Face with narrow pale brown to blackish median stripe extending narrowly on either side of oral opening to join pale brown to blackish cheeks.

Scutum shining black; in some specimens obscurely yellowish laterally and on posterior callus. Scutellum bright to dull yellow. Pleura shining black. Legs mostly yellow; about basal one-third of fore and mid femora, and hind femur except apex, brown to blackish; hind tibia often obscurely brownish near mid length.

Abdomen in Fig. 138. Tergite 2 with pair of large yellow spots not reaching margins; tergite 3 with slightly wavy, posteriorly emarginate yellow band not reaching margins; tergite 4 with similar band and with posterior margin rather broadly yellow; tergite 5 black with pair of sublateral anterior spots and posterior margin yellow. Sternites yellow with broad black bands.

Female. Frons black on little less than upper half, otherwise yellow except for brown line above antennae, shining except for pair of rather large anteriorly oblique lateral pruinose spots at about mid length. Dark markings of legs usually less extensive and paler than in male.

Distribution. Western Canada (Map 31), south to Oregon and Idaho.* B.C., V–VIII.

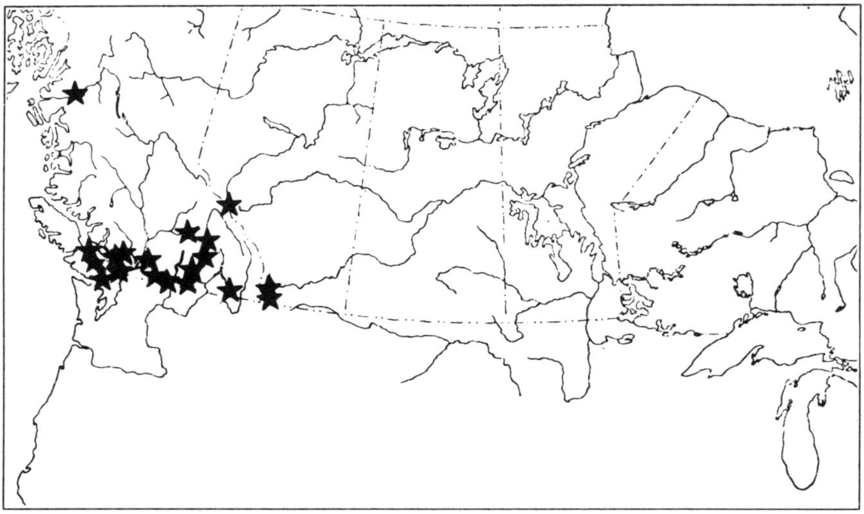

Map 31. Collection localities for *Eupeodes aberrantis* (Curran).

Specimens identified. Canada, 26 ♂♂, 16 ♀♀; United States, 8 ♂♂, 9 ♀♀.

Biology. Specimens in the CNC have been reared from larvae predacious on *Adelges piceae* on *Abies amabalis* and from an adelgid on *Pseudotsuga menziesii*. Others have been reared from larvae or pupae found on *Larix*, *Tsuga heterophylla*, and *Pinus flexilis*.

Eupeodes (Lapposyrphus) lapponicus (Zetterstedt)

Figs. 20, 139; Map 32

Scaeva lapponica Zetterstedt, 1838:598.
Syrphus agnon Walker, 1849:579.
Syrphus alcidice Walker, 1849:579.
Syrphus arcucinctus Walker, 1849:580.

Length. 8.3–14.1 mm.

Male. Similar to male of *E. aberrantis*, but with eye angle about 90°. Frons with broader semicircular dark line; upper part more densely yellowish gray pruinose. Median facial stripe usually slightly broader.

Tergites 3 and 4 each with pair of large yellow spots; each spot with distinctly concave anterior margin and with nearly straight posterior margin (Fig. 139).

Female. Similar to female of *E. aberrantis*, but with frons slightly to distinctly darker, commonly blackish anterolaterally and just above antennae, with larger pruinose spots. Legs not paler than in male. Tergites with markings similar to those of male *E. lapponicus* but yellow spots distinctly narrower or, rarely, tergites entirely black.

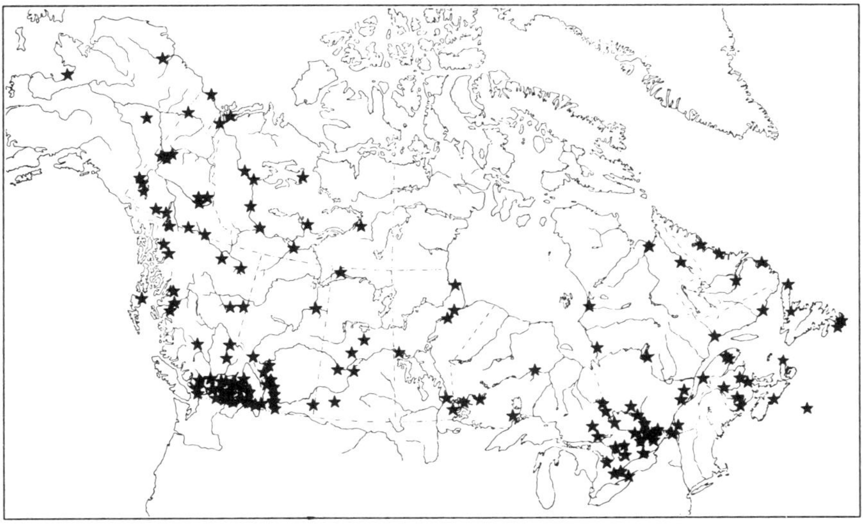

Map 32. Collection localities for *Eupeodes lapponicus* (Zetterstedt).

Distribution. Alaska, Canada (Map 32), Greenland,* south to California, Mexico (Durango), Kentucky, and Virginia;* Europe; Asia. B.C., IV–X; Ont., Que., IV–IX. (I do not know the source of the Greenland record given by Wirth et al. [1965] but think it is probably erroneous.)

Specimens identified. Alaska, 22 ♂♂, 34 ♀♀; Canada, 555 ♂♂, 571 ♀♀; United States, 280 ♂♂, 79 ♀♀; Mexico, 1 ♂, 1 ♀; Europe, 21 ♂♂, 22 ♀♀; Siberia, 1 ♂, 1 ♀.

Biology. Specimens from Canada in the CNC have been reared from larvae predacious on *Adelges piceae*, *Cinara braggii*, and *Cinara lasiocarpae* and on aphids on *Picea glauca* and *Picea mariana*. Others have been reared from larvae or pupae found on *Abies balsamea*, *Larix* sp., *Gleditsia triacanthos*, and *Rhododendron* sp. Specimens from Germany were also reared from *Adelges piceae*. Láska and Starý (1980) recorded as hosts in Czechoslovakia three species of aphids on angiosperms.

Genus *Leucozona* Schiner

Diagnosis. Species slender to robust; some specimens densely pilose. Tergite 2 mostly or entirely pale; following tergites mostly or entirely dark. Wing unmarked or with oblique brown anterior spot at mid length. Length 9.0–13.0 mm.

Description. Eye with long dense hairs. Face moderately to densely pruinose, dull yellow with dark shining median stripe.

Scutum black; in some specimens with brownish or bluish tinge, subshining, yellow or gray pruinose laterally. Scutellum dull yellow to dark brown, darkened anterolaterally or anteriorly. Ventral scutellar fringe long and dense, or sparse. Pleura black, with obscure or distinct grayish pruinosity on about upper half. Anterior anepisternum, meron, and metapleuron bare. Upper and lower katepisternal hair patches narrowly joined posteriorly, broadly separated or almost joined anteriorly. Metasternum bare. Wing entirely trichose or with bare areas on basal one-third. Hind coxa with or without posteromedial apical hair tuft.

Abdomen broadly oval to nearly parallel sided, unmargined or with indistinct to distinct margin from base of tergite 3 to apex of tergite 5. Tergite 1 black; tergite 2 almost entirely pale yellow to submetallic or with pair of large subquadrate pale yellow to submetallic spots; tergites 3 and 4 entirely black or each with pair of small pale spots, or tergite 3 narrowly pale anteriorly. Sternites partly pale, partly gray to black.

Biology. Larvae are probably aphidophagous (see under *L.* (*Ischyrosyrphus*) and *L.* (*Leucozona*)).

Key to subgenera of *Leucozona*

1. Wing unmarked. Abdomen with pair of yellowish or grayish subquadrate or subrectangular spots on each of tergites 2–4 (Fig. 140) .... ***Leucozona (Ischyrosyrphus)* Bigot**

   Wing with broad dark transverse band at about mid length. Abdomen with basal half mostly pale and apical half dark (Fig. 141) .... ***Leucozona (Leucozona)* Schiner**

Clé des sous-genres de *Leucozona*

1. Aile sans marque. Abdomen portant une paire de taches jaunâtres ou grisâtres, de forme presque carrée ou rectangulaire, sur chacun des tergites 2–4 (fig. 140) .... ***Leucozona (Ischyrosyrphus)* Bigot**

   Aile ornée d'une large bande transversale foncée, à peu près à mi-longueur. Abdomen largement pâle dans sa moitié basale et foncé dans la moitié apicale (fig. 141) .... ***Leucozona (Leucozona)* Schiner**

Subgenus *Leucozona* (*Ischyrosyrphus*) Bigot

Diagnosis. Species rather slender with pile of moderate length. Wing unmarked. Abdomen nearly parallel sided, with pale to submetallic spots on tergites 2–4. Length 9.0–12.0 mm.

Description. Face with dark median stripe about one-fifth as wide as face.

Scutellum dull yellow to brown, much darker on about anterior half. Ventral scutellar fringe sparse. Upper and lower sternopleural hair patches narrowly joined posteriorly, almost joined anteriorly. Wing unmarked except for dark brown stigma, entirely trichose or with bare areas on basal one-third. Hind coxa with or without posteromedial hair tuft.

Abdomen (Fig. 140) nearly parallel sided, unmargined, or with very weak margin on tergites 4 and 5. Tergite markings pale dull yellow to submetallic gray; tergite 2 with pair of large subquadrate basal spots; tergites 3 and 4 each with pair of subbasal subrectangular transverse spots.

Venter usually yellow on about basal half, entirely blackish in some specimens.

Distribution. Two Nearctic species, one of which not yet recorded from Canada; several Palaearctic species.

Biology. Dušek and Láska (1962) reared larvae of a Palaearctic species *L.* (*I.*) *glaucia* (Linnaeus), 1758, from eggs deposited in the laboratory; they used four species of aphids but no larvae matured. Rotheray and Gilbert (1989) reported rearing *L.* (*I.*) *laternaria* (Müller), 1776, from aphids of the genus *Cavariella*.

Key to New World species of *Leucozona* (*Ischyrosyrphus*)

1. Disc of scutum with many black hairs behind transverse suture. Wing membrane entirely trichose. Pale spots of tergite 2 at most three-fifths as long as tergite (Fig. 140). Western Canada .... ***velutina* (Williston)**

   Disc of scutum with only pale hairs. Base of cell c and most of anterior half of cell br without microtrichia. Pale spots of tergite 2 about five-sixths as long as tergite. Eastern United States, *not* Canada .... ***xylotoides* (Johnson)**

Clé des espèces du Nouveau Monde de *Leucozona* (*Ischyrosyrphus*)

1. Disque du scutum garni de nombreux poils noirs derrière la suture transversale. Membrane alaire entièrement velue. Taches pâles sur le tergite 2 d'une longueur ne dépassant pas les trois cinquièmes du tergite (fig. 140). Ouest du Canada .... ***velutina* (Williston)**

   Disque du scutum pourvu uniquement de poils pâles. Partie basilaire de la cellule c et majeure partie de la moitié antérieure de la cellule br sans microchètes. Taches pâles sur le tergite 2 d'une longueur équivalant à peu près aux cinq sixièmes du tergite. Est des États-Unis, *pas* au Canada .... ***xylotoides* (Johnson)**

Leucozona (Ischyrosyrphus) velutina (Williston)

Fig. 140; Map 33

Syrphus velutinus Williston, 1882:314.
Ischyrosyrphus tricolor Bigot, 1884:73.

Length. 9.0–12.0 mm.

Male and female. Characters as given in key. Abdomen as in Fig. 140.

Distribution. Alaska,* western Canada (Map 33), south to California* and Idaho.* B.C., VI–VIII.

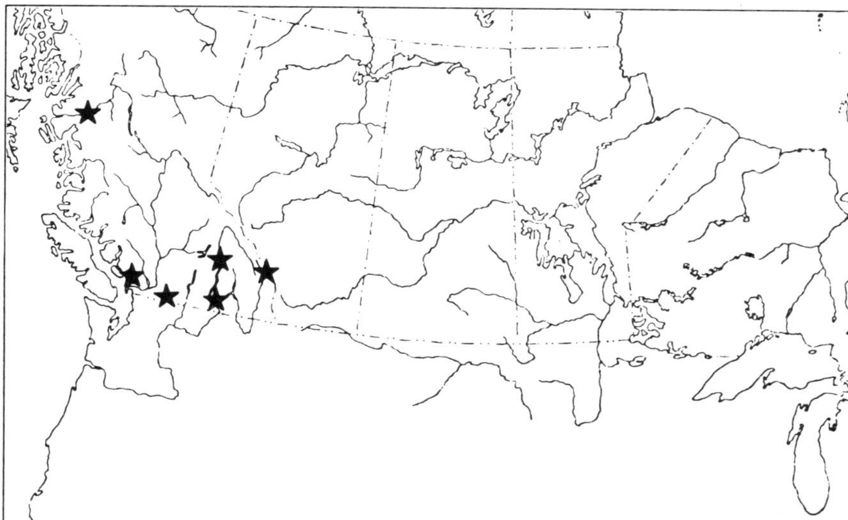

Map 33. Collection localities for *Leucozona velutina* (Williston).

Specimens identified. Canada, 9 ♂♂, 18 ♀♀.

Leucozona (Ischyrosyrphus) xylotoides (Johnson)

Syrphus xylotoides Johnson 1916:80.

Length. 10.0–11.8 mm.

Male and female. Characters as given in key.

Distribution. Vermont* and New Hampshire south to Virginia. Mass., Va., V, VI.

Specimens identified. United States, 1 ♂, 3 ♀♀.

Subgenus *Leucozona* (*Leucozona*) Schiner

Diagnosis. Species robust with rather long pile, with oblique brown anterior spot at mid length of wing, and with abdomen oval and whitish yellow to submetallic gray on basal half. Length 9.8–13.0 mm.

Description. Face with dark median stripe almost half as wide as face.

Scutellum dull yellow, darkened only anterolaterally. Ventral scutellar fringe dense. Upper and lower katepisternal hair patches narrowly joined posteriorly, otherwise broadly separated. Wing with moderate bare areas on basal one-third, with oblique brown anterior spot at mid length. Calypters, their fringes, and knob of halter dark brown. Hind coxa with posteromedial apical hair tuft.

Abdomen (Fig. 141) broadly oval, with strong margin from base of tergite 3 to apex of tergite 5; tergite 1 black; tergite 2 (except for narrow anterior black mid line) and usually narrow base of tergite 3 whitish yellow to submetallic gray; remaining tergites shining black. Venter pale yellow to submetallic gray on about basal half, otherwise black.

Distribution. One Holarctic species; second species only in Nepal.

Biology. Adults of *L. lucorum* Linnaeus assemble on hill tops, presumably for mating (D.M. Wood, personal communication). Dixon (1960) reported larvae of this species on *Rumex* sp.; they are presumably aphidophagous although this observation was not stated.

Leucozona (*Leucozona*) *lucorum* (Linnaeus)

Fig. 141; Map 34

Musca lucorum Linnaeus, 1758:592.
Leucozona lucorum var. *americana* Curran, 1923*b*:38.

Length. 9.8–13.0 mm.

Male and female. Characters as given for subgenus. Abdomen as in Fig. 141.

Distribution. Alaska, Canada (Map 34), south to Oregon,* Colorado, and New York; Europe; Asia. B.C., VI–IX; Ont., Que., VI–III.

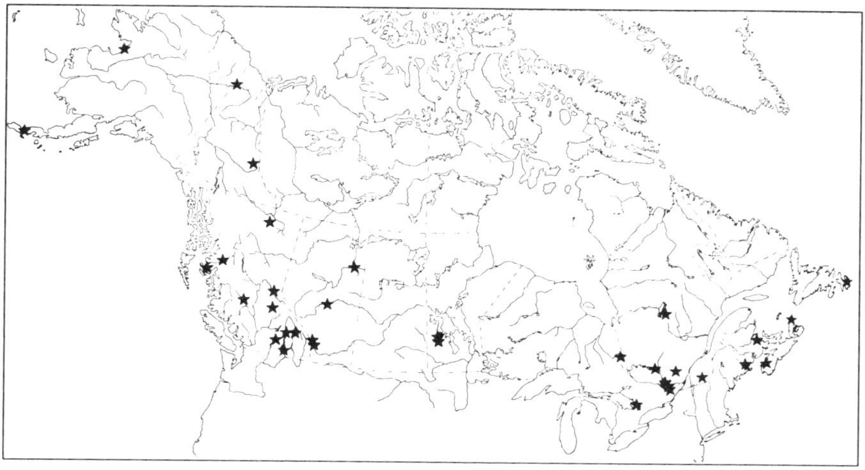

Map 34. Collection localities for *Leucozona lucorum* (Linnaeus).

Specimens identified. Alaska, 2 ♀♀; Canada, 35 ♂♂, 46 ♀♀; United States, 1 ♂, 3 ♀♀; Europe, 20 ♂♂, 25 ♀♀; Japan, 1 ♂.

Biology. Larvae are associated with arboreal and ground layer aphids in Great Britain (Rotheray and Gilbert 1989).

Genus *Melangyna* Verrall

Diagnosis. Species slender, usually dark, with yellow lateral scutal margin in some specimens, with yellow transverse or subquadrate abdominal spots, or with yellow to submetallic triangular yellow spots or entire bands. Length 6.0–12.0 mm.

Description. Eye bare or haired. Face entirely yellow or variously darkened, black with only membrane around antennal bases brownish in some specimens.

Scutum black, shining or subshining, pruinose laterally or with poorly or clearly defined yellowish lateral stripe in some specimens, rarely with pair of separated or confluent yellow prescutellar spots. Scutellum dull yellow to blackish brown, with black margin in some specimens. Ventral scutellar fringe complete. Pleura black, shining to densely gray pruinose, yellowish on upper half in some specimens. Upper and lower katepisternal hair patches separated or narrowly joined posteriorly. Anterior anepisternum, meron, metapleuron, and metasternum bare. Wing membrane entirely trichose or with small bare areas on basal one-third. Hind coxa with or without posteromedial apical hair tuft.

Abdomen nearly parallel sided, unmargined (Figs. 142–144). Tergite 2 usually with pair of yellow spots; tergites 3 and 4 usually with pair of yellow spots or entire yellow band; rarely markings submetallic or tergite 2 or all tergites black. Sternites yellow, obscurely darkened, with distinct dark bands or stripes, or entirely blackish.

Distribution. Two Nearctic subgenera entirely Holarctic; two other subgenera, perhaps a genus distinct from *Melangyna*, having about seven species in Australian region.

Biology. Larvae of several species of both subgenera are aphidophagous.

Key to subgenera of *Melangyna*

1. Spots of tergites 3 and 4 transverse, separated (Figs. 142, 143). Hind coxa with some hairs at posteromedial apical angle (as in Fig. 46) .... ***M. (Melangyna)* Verrall**

   Spots of tergites 3 and 4 slightly to strongly oblique on posterior margin, separated (Fig. 144) or joined medially. Hind coxa without hairs at posteromedial apical angle .... ***M. (Meligramma)* Frey**

Clé des sous-genres de *Melangyna*

1. Tergites 3 et 4 maculés de taches transversales et séparées (fig. 142, 143). Hanche postérieure pourvue de quelques poils à l'angle apical postéro-médian (comme dans la fig. 46) .... ***M. (Melangyna)* Verrall**

   Tergites 3 et 4 maculés de taches de légèrement à fortement obliques sur le bord postérieur (fig. 144), séparées ou réunies au milieu. Hanche postérieure glabre à l'angle apical postéro-médian .... ***M. (Meligramma)* Frey**

Subgenus *Melangyna* (*Melangyna*) Verrall

Diagnosis. As in genus *Melangyna*. Length 6.0–12.0 mm.

Description. Eye bare or haired. Face at least slightly widened below; color of face as described for genus.

Scutum black, shining or subshining, slightly or densely pruinose laterally in some specimens, never distinctly yellowish laterally. Pleura black, shining or pruinose, at most with obscure yellowish area on posterodorsal part of anepisternum. Katepisternal hairs as described for genus. Hind coxa with tuft of hairs at posteroventral apical angle.

Tergites 2–4 with transverse yellow spots always well-separated, usually subrectangular, if slightly broadened laterally then not oblique; spots of tergite 2 usually shorter than those of tergite 3, rarely much larger and subquadrate; very rarely spots of tergite 2, or all pale abdominal markings, completely absent. Sternites with black bands or entirely darkened.

Distribution. Seven species in Nearctic region; five of these in Holarctic region. About 21 species recorded from Palaearctic region. In North America group predominantly boreal, occurring in low Arctic and extending into southern United States only in mountainous areas.

Biology. Dixon (1960) and Goeldlin (1974) described the larvae of *M. umbellatarum* and *M. lasiophthalma*, respectively, but neither author recorded the larval host. Goeldlin reported that the latter species has a lengthy larval diapause. Rotheray and Gilbert (1989) give aphid host records for several European species of *Melangyna*. My attempts, as well as those of Láska (written communication), to have females oviposit in the presence of aphids were unsuccessful.

Discussion. Most of the differences in the color patterns of tergites and sternites used by previous authors (e.g., Fluke 1935) to distinguish apparent species are, in my opinion, variable within most species and of little taxonomic value at the species level. Therefore I refer in the following descriptions to normal tergite markings (Fig. 142). Among Nearctic species, only *M. fisherii*, with very large subquadrate spots on tergite 2 (Fig. 143), is clearly distinguished by the markings on the tergites.

Key to New World species of *Melangyna* (*Melangyna*)

1. Male (eyes meeting on frons) 2
   Female (eyes widely separated on frons) 7

2. Wing with basal one-sixth to two-thirds of cell c, and at least anterobasal one-fifth of cell bm, bare. Widespread .... ***umbellatarum*** **(Fabricius)**

Wing membrane entirely trichose or at most with indistinct bare line along middle of cell bm 3

3. Face very dark, gray to grayish black in some specimens; at most sides obscurely yellowish on about upper half and not sharply distinguished from broad blackish median stripe; face entirely densely gray pruinose. Scutellum with posterior margin darkened; disc dull yellow-brown or yellow-gray. Eye very nearly bare. Boreal and western Canada .... ***arctica*** **(Zetterstedt)**

Face yellow with well-defined median dark stripe (or rarely entirely yellow), and with at least upper three-fifths of sides bright to dull yellow; face moderately pruinose or subshining. Scutellum usually with posterior margin darkened only laterally, narrowly to broadly black posteriorly in some specimens. Eye bare to densely haired 4

4. Entire posterior margin of scutellum narrowly to broadly darkened, at least slightly darker than disc. Many hairs of upper part of anepisternum, and at least some hairs of upper posterior part of katepisternum, black. Eye with short but moderately abundant hairs. Widespread .... ***lasiophthalma*** **(Zetterstedt)**

Posterior margin of scutellum darkened only at lateral angles, concolorous with disc over most of its width. Hairs of anepisternum and katepisternum white to brown, but none black. Eye either nearly bare or with abundant rather long hairs 5

5. Eye with dense long hairs separated from one another by less than own length. Western .... ***coei*** **Nielsen**

Eye nearly bare, at most with some very short hairs separated from one another by more than own length 6

6. Median black facial stripe broad, usually tapering strongly from oral margin upward. Yellow spots of tergites 3 and 4 not extending to lateral margins in some specimens. Widespread .... ***labiatarum*** **(Verrall)**

Median black facial stripe usually absent, if present narrow, parallel-sided, commonly not extending to oral margin. Yellow spots of tergites 3 and 4 extending broadly to lateral margins. Western Canada .... ***subfasciata*** **(Curran)**

7. Eye distinctly although rather sparsely haired in some specimens 8

Eye bare or nearly so, scattered hairs if present separated from one another by much more than own length 9

8. Eye hairs rather long and dense. Pruinose band of frons large, undivided, extending about one-quarter length of frons. Cell bm entirely trichose. Posterior margin of scutellum yellow except at extreme lateral angles. Western Canada .... ***coei* Nielsen**

Eye hairs variable, moderately dense to very sparse. Pruinose band of frons smaller, entire or medially divided, covering about one-sixth to one-eighth length of frons. Cell bm slightly to extensively bare anterobasally in some specimens. Posterior margin of scutellum narrowly to broadly black in some specimens. Widespread .... ***lasiophthalma* (Zetterstedt)** (in part)

9. Each yellow spot of tergite 2 large, subquadrate, little wider than greatest length, nearly reaching anterior margin of tergite over entire width and lateral margin over entire length (Fig. 143). Widespread .... ***fisherii* (Walton)**

Each yellow spot of tergite 2 usually narrow, at least 1.5 times as wide as greatest length, well-removed from anterior margin of tergite or reaching it only laterally (Fig. 142), if spots subcircular or subquadrate then small and far-removed from lateral margin of tergite 10

10. Wing membrane partly bare; cell bm slightly to extensively bare anterobasally or almost entirely bare; cell c usually bare at least at extreme base, bare up to basal four-fifths in some specimen 11

Wing membrane entirely trichose 12

11. Notopleuron densely grayish white pruinose, not contrasting with densely pruinose anepisternum. Frons with pruinose band large, entire, its length just lateral to mid line subequal to that of black area in front of it. Facial hairs white. Posterior margin of scutellum pale yellow except at lateral angles. Cell c bare on as much as basal four-fifths in some specimens. Widespread .... ***umbellatarum* (Fabricius)**

Notopleuron scarcely pruinose, almost shining, contrasting with moderately pruinose upper posterior part of anepisternum. Frons with pruinose band small, entire or medially divided, its length just lateral to mid line about half that of black area in front of it. Facial hairs partly black in some specimens. Posterior margin of scutellum commonly narrowly to broadly black. Cell c bare only at extreme base. Widespread .... ***lasiophthalma* (Zetterstedt)** (in part)

12. Black median facial stripe broad, at level of tubercle from one-third to half width of face; lateral margins of stripe commonly obscured by dense grayish pruinosity covering all face except tubercle. Boreal and western Canada .... ***arctica* (Zetterstedt)**

Black or brown median facial stripe narrower or even absent, at level of tubercle at most one-third as wide as face; face laterally subshining or with yellowish white pruinosity not obscuring margins of stripe 13

13. Black median facial stripe present, distinct, broadly confluent with oral margin. Widespread ...... ***labiatarum* (Verrall)**

Face entirely yellow or with narrow brown median stripe commonly not extending to oral margin. Western Canada .... ***subfasciata* (Curran)**

Clé des espèces du Nouveau Monde de *Melangyna* (*Melangyna*)

1. Mâle (yeux contigus sur le front) 2

Femelle (yeux largement séparés sur le front) 7

2. Aile dont un sixième aux deux tiers de la portion basilaire de la cellule c et au moins un cinquième de la portion antéro-basilaire de la cellule bm sont glabres. Espèce répandue .... ***umbellatarum* (Fabricius)**

Aile dont la membrane est entièrement velue ou tout au plus ornée d'un liséré glabre peu visible le long de la partie médiane de la cellule bm 3

3. Face très foncée, gris à gris noir chez certains spécimens; côtés tout au plus de couleur jaunâtre floue sur environ la moitié supérieure, se distinguant peu de la large rayure médiane noirâtre; pruinosité grise sur toute la figure. Scutellum orné d'un liséré postérieur foncé; disque brun jaune terne ou gris jaune. Oeil presque glabre. Régions boréales et ouest du Canada .... ***arctica* (Zetterstedt)**

Face jaune soulignée d'une rayure médiane foncée et bien définie (ou rarement entièrement jaune), jaune clair à terne sur au moins les trois cinquièmes supérieurs des côtés. Face modérément pruineuse ou presque brillante. Scutellum orné habituellement d'un liséré postérieur foncé uniquement sur les côtés pourvu, chez certains spécimens, d'une zone noire étroite à large sur la face postérieure. Oeil de glabre à très poilu 4

4. Liséré postérieur du scutellum en son entier légèrement à très foncé, tout au moins légèrement plus foncé que le disque. Il y a de nombreux poils noirs sur la partie supérieure de l'anépisterne et tout au moins de quelques poils noirs sur la partie postérieure supérieure du katépisterne. Oeil orné de poils courts assez abondants. Espèce répandue .... ***lasiophthalma*** **(Zetterstedt)**

Liséré postérieur du scutellum foncé uniquement aux angles latéraux, de même couleur que le disque sur presque toute sa largeur. Poils de l'anépisterne et du katépisterne de blancs à bruns; aucun poil noir. Oeil presque glabre ou pourvu de poils abondants et assez longs 5

5. Oeil garni de poils longs et denses, séparés d'une distance inférieure à leur longueur. Ouest du Canada .... ***coei*** **Nielsen**

Oeil presque glabre, orné tout au plus de quelques poils très courts séparés d'une distance supérieure à leur longueur 6

6. Face pourvue d'une large rayure médiane noire, devenant en général de plus en plus fuselée à partir de la marge buccale. Taches jaunes sur les tergites 3 et 4 n'atteignant pas les bords latéraux chez certains spécimens. Espèce répandue .... ***labiatarum*** **(Verrall)**

Face dépourvue en général d'une rayure médiane noire; s'il y en a une, ses côtés parallèles sont étroits, n'atteignant habituellement pas le bord buccal. Taches jaunes sur les tergites 3 et 4 s'étendant largement jusqu'aux bords latéraux. Ouest du Canada .... ***subfasciata*** **(Curran)**

7. Oeil orné de poils distincts quoique assez clairsemés chez certains spécimens 8

Oeil glabre ou presque; lorsqu'il y a quelques poils, ils sont séparés par une distance bien supérieure à leur longueur 9

8. Oeil orné de poils assez longs et denses. Large bande frontale pruineuse et continue, qui s'étend sur environ un quart de la longueur du front. Cellule bm entièrement velue. Liséré postérieur du scutellum jaune, sauf à l'extrémité des angles latéraux. Ouest du Canada .... ***coei*** **Nielsen**

Oeil orné de poils de longueur et de densité variable, allant de modérément denses à très clairsemés. Bande pruineuse frontale plus petite, continue ou séparée au milieu, couvrant de un sixième à un huitième de la longueur du front. Chez certains spécimens, cellule bm de légèrement à fortement glabre sur la portion antéro-basilaire. Liséré postérieur du scutellum légèrement à largement noir chez certains spécimens. Espèce répandue .... ***lasiophthalma*** **(Zetterstedt)** (*partim*)

9. Chaque tache jaune du tergite 2 large, presque carrée, un peu plus large que la longueur maximale, atteignant presque le bord antérieur du tergite sur toute la largeur et le bord latéral sur toute la longueur (fig. 143). Espèce répandue .... ***fisherii*** **(Walton)**

Chaque tache jaune du tergite 2 habituellement étroite, au moins 1,5 fois plus large que la longueur maximale, bien séparée du bord antérieur du tergite ou ne l'atteignant que sur les côtés (fig. 142). Si les taches sont presque circulaires ou carrées, elles sont alors petites et bien éloignées du bord latéral du tergite 10

10. Membrane alaire en partie glabre; cellule bm légèrement à fortement glabre sur la portion antéro-basilaire ou presque entièrement glabre; cellule c généralement glabre, tout au moins à l'extrémité basilaire, et glabre jusqu'aux quatre cinquièmes de la partie basilaire chez certains spécimens 11

Membrane alaire entièrement velue 12

11. Notopleurite pourvu d'une forte pruinosité blanc grisâtre, qui ne fait pas contraste avec l'anépisterne lui aussi très pruineux. Front orné d'une large bande pruineuse complète, d'une longueur, juste à côté de la ligne médiane, presque égale à celle de la zone noire située à l'avant. Pilosité faciale blanche. Liséré postérieur du scutellum jaune pâle, sauf aux angles latéraux. Chez certains spécimens, cellule c glabre jusqu'aux quatre cinquièmes de la partie basale. Espèce répandue .... ***umbellatarum*** **(Fabricius)**

Notopleurite rarement pruineux, presque lustré, faisant contraste avec la partie postérieure supérieure modérément pruineuse de l'anépisterne. Front orné d'une petite bande pruineuse, complète ou séparée au milieu, d'une longueur, juste à côté de la ligne médiane, équivalant environ à la moitié de celle de la zone noire à l'avant. Chez certains spécimens, pilosité faciale en partie noire. Liséré postérieur du scutellum en général de légèrement à fortement noir. Cellule c glabre uniquement à l'extrémité de la partie basilaire. Espèce répandue .... ***lasiophthalma*** **(Zetterstedt)** (*partim*)

12. Rayure faciale médiane noire et large, au niveau du tubercule, couvrant de un tiers à la moitié de la largeur de la figure; bords latéraux de la bande généralement assombris par une dense

pruinosité grisâtre couvrant toute la face, sauf le tubercule. Régions boréales et ouest du Canada .... ***arctica* (Zetterstedt)**

Rayure faciale médiane noire ou brune plus étroite ou même absente, au niveau du tubercule, d'une largeur équivalant au plus au tiers de la figure; côtés de la face presque luisants ou présentant une pruinosité blanc jaunâtre qui n'assombrit pas les bords de la rayure 13

13. Face pourvue d'une rayure médiane noire, distincte, rejoignant largement le bord buccal. Espèce répandue .... ***labiatarum* (Verrall)**

Face entièrement jaune ou soulignée d'une étroite rayure médiane brune qui ne s'étend habituellement pas jusqu'au bord buccal. Ouest du Canada ...... ***subfasciata* (Curran)**

Melangyna (*Melangyna*) *arctica* (Zetterstedt)

Map 35

Scaeva arctica Zetterstedt, 1838:604.
Melanostoma glacialis Johnson, 1898:18.
Syrphus gracilis Coquillett, 1900:432 (preoccupied Meigen, 1822).
Melangyna coquilletti Sedman, 1965:566 (new name for *gracilis* Coquillett).

Length. 6.0–8.7 mm.

Male. Eye nearly bare. Frons black, densely gray pruinose; eye angle about 90–100°. Face distinctly broadened below, dark, with broad black median stripe, laterally yellowish on about upper half but yellow commonly obscured by dense gray pollen covering all face but tubercle; broad lower facial margin and cheek blackish.

Scutum with black hairs; upper part of anepisternum and anepimeron with black hairs; pleura commonly mostly black-haired. Notopleuron and pleura weakly pruinose. Scutellar hairs mostly black; scutellum with posterior margin black, disc dull yellow to grayish yellow. Wing entirely trichose. Legs dark brown to black, with bases of fore and mid tibiae reddish.

Tergites with normal markings; spots of tergite 2 usually small and rounded, not reaching margins. Sternites almost entirely black, with moderately broad paler incisures in some specimens, densely pale gray pruinose.

Female. Frons black with broad brown pruinose band, upper half of frons usually weakly but distinctly pruinose. Face usually paler than in male and less densely pruinose; yellow more apparent laterally and extending almost to oral margin; cheek usually yellow-orange on posterior half. Scutum with mostly black hairs; notopleuron and pleura densely pale gray pruinose, with white hairs. Scutellar hairs mixed black and white; scutellum usually brighter yellow than in male, with or without black posterior margin. Tergite 2 with yellow spots almost as large as those of tergite 3 in some specimens.

Distribution. Alaska, northern and western Canada (Map 35), south in mountains to Colorado, New York, and New Hampshire; Europe. B.C., VI–VIII; Que., VII, VIII.

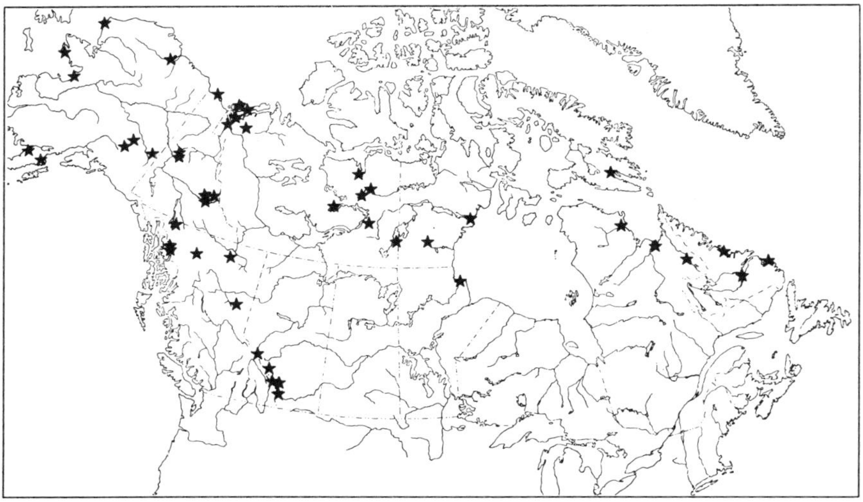

Map 35. Collection localities for *Melangyna arctica* (Zetterstedt).

Specimens identified. Alaska, 15 ♂♂, 35 ♀♀; Canada, 89 ♂♂, 139 ♀♀; United States, 13 ♂♂, 46 ♀♀; Europe, 1 ♂, 2 ♀♀.

Biology. Rotheray and Gilbert (1989) recorded the aphid *Pterocallis ulni* on *Alnus* as a host of *M. arctica* in Great Britain.

Melangyna (*Melangyna*) *coei* Nielsen

Map 36

Melangyna coei Nielsen, 1971:65.

Length. 7.5–10.2 mm.

Male. Eye with dense long hairs. Frons black, with rather dense brownish pruinosity; eye angle about 90–110°. Face bright to dull yellow, moderately to densely yellow pruinose, with broad black median stripe, broad black lower facial margin, blackish cheek.

Scutum with black hairs. Pleural hairs as in *M. arctica*. Notopleuron and pleura weakly pruinose. Scutellum dull pale yellow, narrowly black only laterally; scutellar hairs black. Wing entirely trichose. Leg color as in *M. arctica*.

Tergites with normal markings; yellow spots of tergite 2 rather large but subquadrate, not extending to lateral margins. Sternites dark, usually with very broad pale incisures, rather weakly pruinose.

Female. Eye hairs shorter and sparser than in male. Frons black, with well-defined moderately large entire yellow-brown pruinose band. Cheek mostly dull yellow. Scutum with yellow hairs, also with scattered black hairs in some specimens. Pleural hairs white. Pleura densely pale gray pruinose. Yellow spots of tergite 2 almost twice as wide as long.

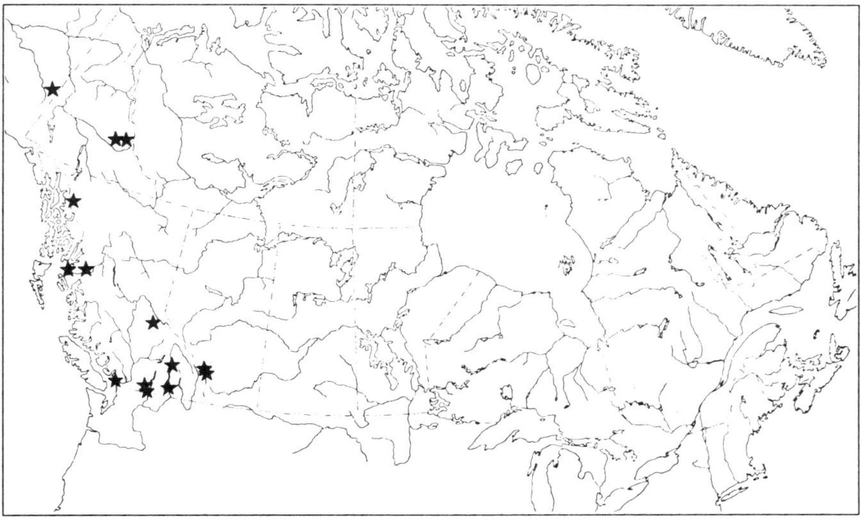

Map 36. Collection localities for *Melangyna coei* Nielsen.

Distribution. Alaska, western Canada (Map 36), south to Colorado; northern Europe. B.C., IV–X.

Specimens identified. Alaska, 1 ♀; Canada, 19 ♂♂, 18 ♀♀; United States, 1 ♂; Europe, 4 ♂♂, 6 ♀♀.

Discussion. *M. coei* differs consistently from *M. labiatarum* only in the presence of abundant and very distinct eye hairs. The absence of intermediate specimens indicates that the two forms are probably specifically distinct.

Melangyna (*Melangyna*) *fisherii* (Walton)

Fig. 143; Map 37

Syrphus fisherii Walton, 1911:319.

Length. 6.5–10.8 mm.

Male. Unknown.

Female. Frons black, with rather large arcuate pale gray or brown pruinose band at mid length. Face pale yellow with pale hairs, densely pale yellow pruinose, with narrow black median stripe; lower facial margin yellow or narrowly black; cheek black.

Scutum and pleura with white hairs. Notopleuron and upper two-thirds of pleura densely whitish gray pruinose. Scutellum yellow, very narrowly black laterally; scutellar hairs mostly black. Wing moderately bare; cell c usually bare on about basal one-sixth, rarely bare only at extreme base or on about basal half; cell bm usually about three-quarters bare, rarely with only some microtrichia near apex. Fore and mid femora dark brown to black on about basal one-third; rest of femora and tibiae yellow-orange; tarsi brownish; hind leg mostly dark brown to black.

Tergites (Fig. 143) with normal markings but spots of tergite 2 very large, slightly longer than wide, nearly reaching anterior margin in their full width and lateral margin in their full length; yellow spots of tergites 3 and 4 extending to lateral margins in their full length. Sternites pale yellow; sternite 1 with central black spot; sternite 2 with narrow black band, sternites 3–5 usually each with broad black band extending forward laterally to anterior margin; black areas shining or very weakly pruinose, considerably reduced in some specimens.

Distribution. Canada (Map 37), south in eastern North America to North Carolina. B.C., VI–VIII; Ont., VII–IX.

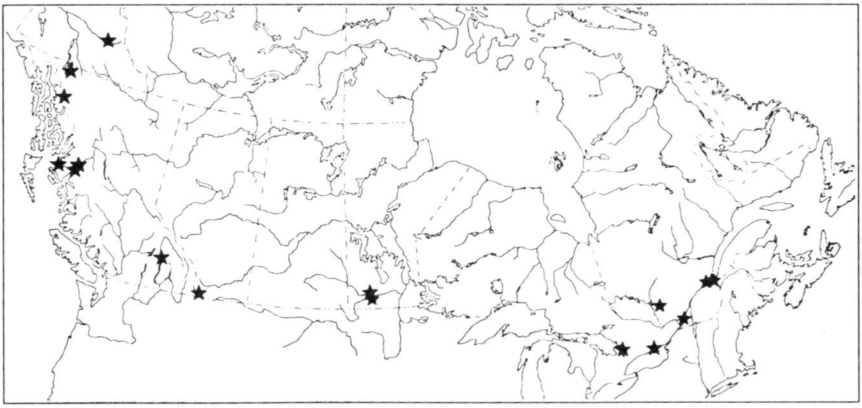

Map 37. Collection localities for *Melangyna fisherii* (Walton).

Specimens identified. Canada, 20 ♀♀; United States, 3 ♀♀.

Discussion. Although Fluke (1935) referred to males of *M. fisherii*, I have seen no males that I would refer here. It is possible either that they have not been distinguished from those of *M. umbellatarum*, or that *M. fisherii* is based on females of that species with large subquadrate yellow spots on tergite 2. However, I have seen no specimens with an intermediate condition of this character.

Melangyna (*Melangyna*) *labiatarum* (Verrall)

Map 38

Syrphus labiatarum Verrall, 1901:415.

Length. 7.0–10.6 mm.

Male. Eye nearly bare; eye angle about 90°. Frons black, mostly brown pruinose. Face black, densely yellow-gray pruinose, with broad black median stripe; lower facial margin broadly black; cheek black, reddish posteriorly in some specimens.

Scutum and pleura with hairs yellow-brown to dark brown; notopleuron and pleura weakly pruinose. Scutellum dull yellow, blackish only laterally, with black hairs. Wing entirely trichose. Legs mostly dark brown to black; basal half of fore and mid tibiae usually slightly paler.

Tergites with normal markings; spots of tergites 3 and 4 extending broadly to lateral margins in some specimens. Sternites dull yellow, weakly pruinose, with obscure to distinct broad black bands.

Female. Frons black, with large and distinct to small and obscure brown pruinose band. Face weakly pruinose in some specimens; oral margin and cheek commonly yellow. Scutal hairs mostly yellow, some black hairs present laterally and postsuturally in some specimens; pleural hairs white; notopleuron and pleuron weakly pruinose as in male or, especially in western specimens, moderately to densely whitish gray pruinose. Scutellum with black hairs posteriorly or on most of surface.

Distribution. Alaska, northern and western Canada (Map 38), south to Colorado; Europe. B.C., VI–VIII; Que., VII–VIII.

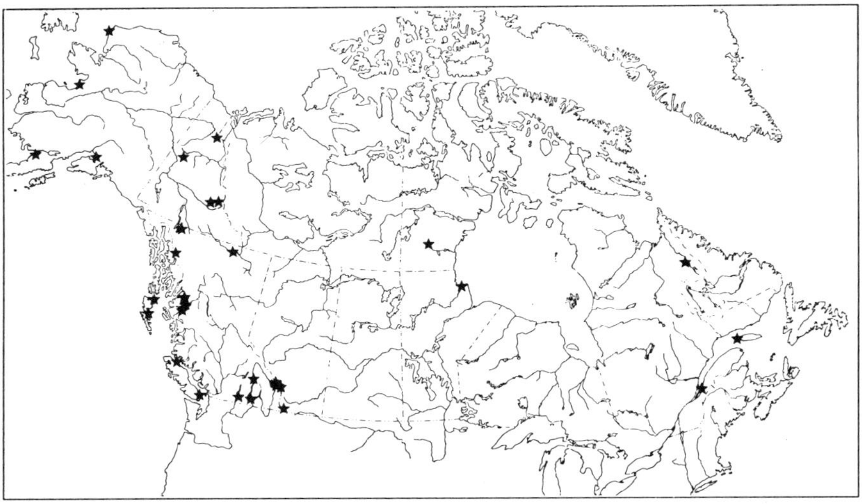

Map 38. Collection localities for *Melangyna labiatarum* (Verrall).

Specimens identified. Alaska, 8 ♂♂, 50 ♀♀; Canada, 30 ♂♂, 65 ♀♀; United States, 1 ♂, 1 ♀; Europe, 12 ♂♂, 17 ♀♀.

Discussion. Specimens from the west side of Hudson Bay and northern Quebec differ from most western specimens in several ways: the face is slightly more produced below; the pale spots of tergites 3 and 4 always extend to the lateral margins; and, in the female, the pruinose band of the frons is much smaller and less distinct and the notopleuron and pleura are much less pruinose. The eastern specimens are also smaller than most of the western specimens. However, a few of the western specimens show some of these characters so it is not apparent that two species are involved.

Melangyna (*Melangyna*) *lasiophthalma* (Zetterstedt)

Fig. 142; Map 39

Scaeva lasiophthalma Zetterstedt, 1843:735.
Syrphus sexquadratus Walker, 1849:586.
Syrphus mentalis Williston, 1887:72.
Stenosyrphus vittifacies Curran, 1923*a*:66.
Epistrophe abruptus Curran, 1924*c*:80.
Stenosyrphus garretti Curran, 1925:109.
Stenosyprhus columbiae Curran, 1925:110.
Syrphus flavosignatus Hull, 1930:139.

Length. 6.1–10.9 mm.

Male. Eye with dense moderately long hairs little shorter than those of *M. coei*; eye angle about 90–100°. Frons black, subshining to moderately gray pruinose. Face dark yellow, subshining, with moderately to very broad black median stripe; facial hairs white or partly black; lower facial margin very broadly black; cheek black.

Scutal hairs mostly or entirely black; pleural hairs black at least above, mostly black in some specimens; notopleuron and pleura scarcely pruinose. Scutellum with posterior margin narrowly to broadly black, disc yellow; scutellar hairs black. Wing entirely trichose or with narrow bare area along centre of cell bm. Legs mostly dark brown to black, most of fore and mid tibiae slightly paler.

Tergites (Fig. 142) with normal markings; yellow spots rarely reaching lateral margins; spots of tergite 2 in some early spring specimens tiny or absent. Sternites subshining, black with narrow to broad yellow incisures, with paired anterior yellow spots confluent medially in some specimens.

Female. Eye with short but usually moderately abundant and distinct hairs, in some eastern specimens hairs rather sparse. Frons black with pruinose band at mid length small, arcuate, usually divided, gray or in some specimens obscure brown. Face strongly shining to distinctly pruinose, with median stripe usually narrower than in male; cheek usually mostly yellow. Scutal hairs pale yellow or partly black; pleural hairs white; notopleuron and pleura scarcely more pruinose than in male. Scutellar hairs black posteriorly or mostly black. Wing as in male or, especially in eastern specimens, with cell c narrowly bare basally and with cell bm up to two-thirds bare.

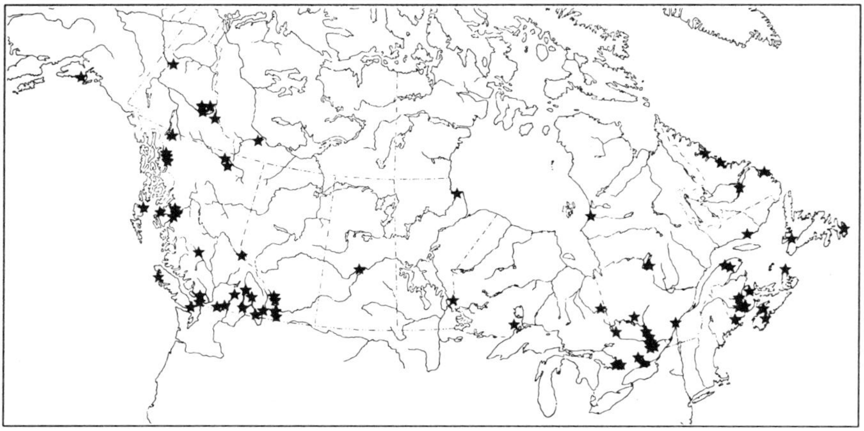

Map 39. Collection localities for *Melangyna lasiophthalma* (Zetterstedt).

Distribution. Alaska, Canada (Map 39), south to Colorado and Maryland; Europe; Asia. B.C., II–VIII; Ont., Que., IV–VII.

Specimens identified. Alaska, 1 ♂, 2 ♀♀; Canada, 77 ♂♂, 221 ♀♀; United States, 11 ♀♀; Europe, 10 ♂♂, 8 ♀♀; Siberia, 1 ♂, 1 ♀.

Melangyna (*Melangyna*) *subfasciata* (Curran)

Map 40

Stenosyrphus subfasciatus Curran, 1925:111.

Length. 7.0–10.2 mm.

Male. Eye bare; eye angle about 90°. Frons black, weakly pruinose on upper two-thirds. Face bright yellow, shining or weakly pruinose laterally, usually without darker median stripe, rarely with narrow median brown stripe incomplete or narrowed ventrally; lower facial margin yellow or narrowly dark brown; cheek yellow to blackish.

Scutal hairs mostly yellow, some black hairs present laterally and postsuturally; pleural hairs yellow, commonly brownish above. Notopleuron scarcely pruinose; pleura weakly pruinose. Wing entirely trichose. Fore and hind femora irregularly brownish on basal one-quarter to half; hind femur mostly dark or irregularly darkened; hind tibia with brownish areas in some specimens; other legs yellowish brown.

Tergites with normal markings; yellow spots rather large, usually extending to lateral margins. Sternites scarcely pruinose, yellow with broad black bands having anterior margins convex.

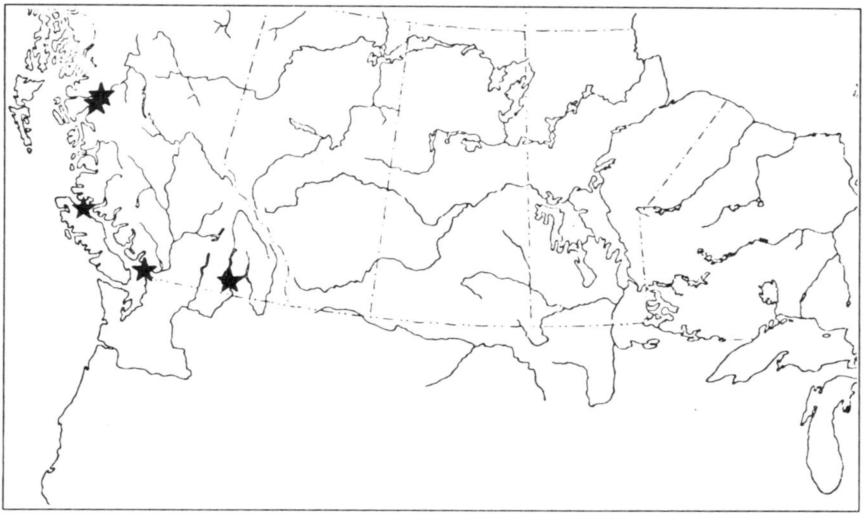

Map 40. Collection localities for *Melangyna subfasciata* (Curran).

Female. Frons black, with small to large, commonly obscure, pruinose band. Scutal hairs yellow; pleural hairs white. Upper part of pleura slightly more pruinose than in male. Legs usually paler than in male; femora entirely yellowish in some specimens.

Distribution. British Columbia (Map 40), Alberta,* south to Oregon.* B.C., V–VII.

Specimens identified. Canada, 18 ♂♂, 62 ♀♀; United States, 1 ♂, 2 ♀♀.

Discussion. This form differs significantly from *M. labiatarum* only in the reduction or absence of a median facial stripe. It may be a color variant of that species with a markedly restricted geographic distribution.

Melangyna (*Melangyna*) *umbellatarum* (Fabricius)

Map 41

Syrphus umbellatarum Fabricius, 1794:307.
Syrphus pullulus Snow, 1895:237.
Melanostoma cherokeenensis Jones, 1917:219.
Stenosyrphus albipunctatus Curran, 1925:104.
Stenosyrphus diversipunctatus Curran, 1925:106.
Stenosyrphus remotus Curran, 1925:108.

Length. 7.2–12.0 mm.

Male. Eye bare; eye angle 90°. Frons black, weakly to strongly pruinose on upper two-thirds. Face yellow with black hairs, weakly to rather densely yellow-gray pruinose, with rather narrow median brown to black stripe broadening slightly ventrally; lower facial margin narrowly to broadly blackish; cheek yellow to brownish black.

Scutal hairs brown to brownish black; pleural hairs pale to dark brown above, yellow-brown below. Notopleuron slightly to moderately pruinose; pleura more densely pruinose. Scutellum dull dark yellow to yellow-brown, narrowly blackish only laterally; scutellar hairs black. Wing slightly to extensively bare; cell c bare on basal one-sixth to two-thirds; cell bm bare on at least basal one-fifth, with microtrichia only on apical one-eighth in some specimens. Fore and mid femora brown to black on basal one-third to half; hind femur brown to black on basal four-fifths; hind tibiae and all tarsi mostly brown; legs otherwise dull yellow to yellow-brown.

Tergites with normal markings; yellow spots rarely extending narrowly and obscurely to lateral margins. Sternites subshining to moderately pruinose, almost entirely yellow to mostly black.

Female. Frons black with large arcuate brown pruinose band. Facial hairs yellow; lower facial margin narrowly brown in some specimens; cheek yellow. Scutal hairs mostly yellow; some black hairs present postsuturally; pleural hairs white. Notopleuron and almost all pleura densely white pruinose. Scutellum pale yellow, narrowly black only laterally; scutellar hairs black posteriorly or on most of surface. Legs slightly paler than in male. Tergites with yellow spots commonly extending broadly to lateral margins.

Distribution. Alaska, Canada (Map 41), south to Arizona and North Carolina; Europe. B.C., VI–IX; Ont., Que., VI–VIII.

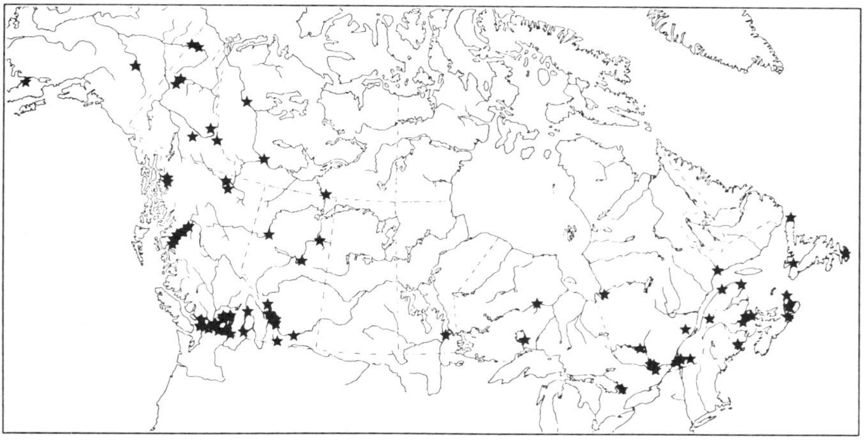

Map 41. Collection localities for *Melangyna umbellatarum* (Fabricius).

Specimens identified. Alaska, 2 ♂♂; Canada, 91 ♂♂, 170 ♀♀; United States, 20 ♂♂, 42 ♀♀; Europe 2 ♂♂, 5 ♀♀.

Subgenus *Melangyna* (*Meligramma*) Frey

Diagnosis. As for genus *Melangyna*. Length 7.0–10.0 mm.

Description. Eye bare. Face narrow, parallel-sided below, yellow to dark grayish yellow, never with darker median stripe.

Scutum shining black, with obscure or distinct broad lateral yellow margin, with single or divided yellow prescutellar spot in some specimens. Pleura black, distinctly pruinose above, with extensive but obscure yellowish areas on upper two-thirds. Upper and lower katepisternal hair patches broadly separated throughout. Hind coxa without hairs at posteromedial apical angle.

Abdominal markings yellow to submetallic. Tergite 2 with pair of small spots; tergites 3 and 4 each with pair of spots broader laterally, with oblique posterior margin; in some specimens spots confluent medially forming an entire nearly basal band. Sternites with black bands or with obscure lateral dark marks.

Distribution. Three Nearctic and three Palaearctic species (two of these Holarctic).

Biology. Heiss (1938) recorded *M. triangulifera* (as *Epistrophe cinctus*) as probably univoltine and as having a larval diapause. Larvae of two European species are recorded as aphidophagous.

Key to New World species of *Melangyna* (*Meligramma*)

1. Male (eyes meeting on frons) 2

   Female (eyes widely separated on frons) 5

2. Scutum with one or two rounded yellow spots immediately in front of scutellum 3

   Scutum with entire disc black 4

3. Tergites 3 and 4 each with pair of well-separated yellow spots (Fig. 144). Widespread .... ***guttata* (Fallén)** (in part)

   Tergites 3 and 4 each with entire yellow band. British Columbia .... ***vespertina* Vockeroth**

4. Frons entirely yellow or darkened on at most upper one-third. Face entirely yellow. Tergites 3 and 4 each with pair of well-separated yellow spots (Fig. 144). Widespread .... ***guttata* (Fallén)** (in part)

   Frons usually dark along most of mid line, mostly or entirely dark in some specimens. Face darkened ventrolaterally, in some specimens mostly dark. Tergites 3 and 4 each with entire or emarginate yellow to submetallic band or with pair of slightly or well-separated yellow to submetallic spots. Widespread .... ***triangulifera* (Zetterstedt)**

5. Frons, at level of ocellar triangle, with narrow pruinose stripe along eye margin. Scutum with two small yellow spots or single large yellow spot in front of scutellum. Scutellum with some short erect black hairs on disk. Yellow spots on tergites 2–4 small, separated; spots on tergite 2 well-separated from anterior margin of tergite; only spots on tergite 4 in some specimens reaching lateral margin of tergite (as Fig. 144). Widespread .... ***guttata* (Fallén)**

   Frons, at level of ocellar triangle, shining over entire width. Scutum usually black in front of scutellum, rarely with yellow spot or spots. Scutellum usually entirely yellow-haired; black hairs if present usually long and fine. Tergite 2 with pair of yellow to submetallic spots usually extending forward laterally to, or almost to, anterior margin of tergite; tergites 3 and 4 each with entire yellow to submetallic band or pair of spots commonly reaching lateral margins of tergites. Widespread .... ***triangulifera* (Zetterstedt)**

Clé des espèces du Nouveau Monde de *Melangyna* (*Meligramma*)

1. Mâle (yeux contigus sur le front) 2

   Femelle (yeux largement séparés sur le front) 5

2. Scutum maculé d'une ou de deux taches jaunes arrondies, immédiatement à l'avant du scutellum 3

   Scutum pourvu d'un disque tout noir 4

3. Tergites 3 et 4 ornés tous deux d'une paire de taches jaunes bien séparées (fig. 144). Espèce répandue .... ***guttata* (Fallén)** *(partim)*

   Tergites 3 et 4 ornés tous deux d'une bande jaune complète. Colombie-Britannique .... ***vespertina* Vockeroth**

4. Front tout jaune, ou foncé au plus sur le tiers supérieur. Face entièrement jaune. Tergites 3 et 4 ornés tous deux d'une paire de taches jaunes bien séparées (fig. 144). Espèce répandue .... ***guttata* (Fallén)** *(partim)*

   Front habituellement foncé le long de la ligne médiane, entièrement ou largement foncé chez certains spécimens. Face foncée ventro-latéralement et largement foncée chez certains spécimens. Tergites 3 et 4 portant tous deux une bande complète ou échancrée, de couleur jaune à presque métallique, ou une paire de taches jaunes à presque métalliques, légèrement ou bien séparées. Espèce répandue .... ***triangulifera* (Zetterstedt)**

5. Front pourvu d'une étroite rayure pruineuse le long du bord de l'œil, au niveau du triangle ocellaire. Scutum orné de deux petites taches jaunes ou d'une grosse tache jaune à l'avant du scutellum. Scutellum pourvu de quelques poils noirs courts et raides sur le disque. Petites taches jaunes séparées sur les tergites 2–4; sur le tergite 2, les taches sont bien séparées du bord antérieur du tergite; seules les taches sur le tergite 4 atteignent, chez certains spécimens, le bord latéral (comme dans la fig. 144). Espèce répandue .... ***guttata* (Fallén)**

   Front lustré sur toute sa largeur au niveau du triangle ocellaire. Scutum généralement noir à l'avant du scutellum, rarement orné d'une ou de plusieurs taches jaunes. Pilosité du scutellum habituellement tout jaune; lorsqu'il y a des poils noirs, ceux-ci sont généralement longs et fins. Tergite 2 ayant une paire de taches jaunes à presque métalliques, qui s'étendent habituellement vers l'avant pour atteindre ou presque le bord antérieur du tergite. Tergites 3 et 4 portant chacun une bande complète jaune ou légèrement métallique, ou une paire de taches atteignant généralement le bord latéral. Espèce répandue .... ***triangulifera* (Zetterstedt)**

Melangyna (*Meligramma*) *guttata* (Fallén)

Fig. 144; Map 42

Scaeva guttata Fallén, 1817:44.
Xanthogramma habilis Snow, 1895:238.

Length. 7.6–9.6 mm.

Male. Frons pale or bright yellow, blackish on upper one-third in some specimens. Face pale yellow.

Scutum subshining black, with broad obscure or distinct yellow lateral margin; some specimens with pair of small submedian yellow spots in front of scutellum.

Tergite 2 with pair of small subtriangular yellow spots not reaching margins; tergites 3 and 4 each with pair of slightly larger yellow spots not distinctly reaching lateral margins (Fig. 144).

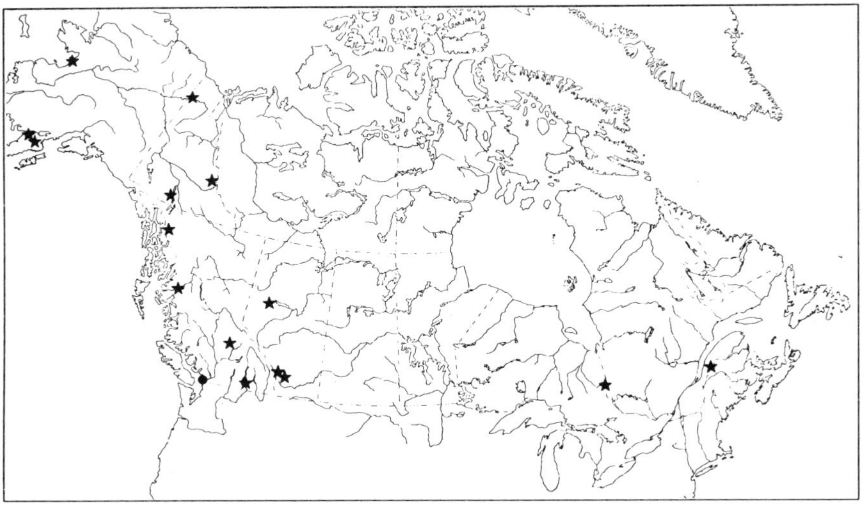

Map 42. Collection localities for *Melangyna guttata* (Fallén) (★) and for *M. vespertina* Vockeroth (●).

Female. Frons black on upper one-third, otherwise yellow with narrow black median stripe extending in some specimens to anterior margin; frons pruinose laterally; pruinose stripe extending narrowly to level of posterior ocelli. Scutum with two small yellow spots or single large yellow spot in front of scutellum. Scutellum with some short black hairs among longer yellow hairs on disc. Yellow spots on tergites 2–4 small, well-separated; spots on tergite 2 far from anterior margin of tergite; spots on tergite 4 reaching lateral margin of tergite in some specimens.

Distribution. Alaska, Canada (Map 42), south to Arizona* and New Mexico; Europe. B.C., VII–IX.

Specimens identified. Alaska, 2 ♂♂, 5 ♀♀; Canada, 5 ♂♂, 11 ♀♀; United States, 1 ♂, 2 ♀♀; Europe, 2 ♂♂, 7 ♀♀.

Melangyna (*Meligramma*) *triangulifera* (Zetterstedt)

Map 43

Scaeva triangulifera Zetterstedt, 1843:747.
Xanthogramma tenuis Osburn, 1908:8.
Syrphus oronoensis Metcalf, 1917:162.
Scaeva cincta of American authors (not *S. cincta* Fallén, 1817).

Length. 7.0–10.0 mm.

Male. Frons dull yellow, usually darkened on mid line or mostly or entirely dark. Face yellow, darkened ventrolaterally; some specimens mostly dark.

Scutum shining black, obscurely yellow laterally, without yellow prescutellar markings.

Tergite 2 with pair of small to large subtriangular yellow or submetallic spots not distinctly reaching lateral margins; tergites 3 and 4 each with entire or posteriorly emarginate yellow to submetallic band or with pair of slightly or well-separated yellow to submetallic spots; these markings rarely reaching tergite margins.

Female. Frons dark on upper one-third to half and on narrow to broad median stripe, yellow anterolaterally, pruinose along eye margin on lower two-thirds, shining over entire width at level of ocellar triangle. Face pale yellow, with lower margin blackish in some specimens. Scutum with well-defined lateral yellow margin, with or without pair of small yellow prescutellar spots. Scutellum usually only with yellow hairs; black hairs if present usually long and fine. Tergite 2 with pair of yellow to submetallic spots usually extending forward laterally to, or almost to, anterior margin of tergites; tergites 3 and 4 each with entire yellow to submetallic band or pair of spots commonly reaching lateral margins of tergites.

Distribution. Alaska, Canada (Map 43), south to Colorado, Virginia, and North Carolina;* Europe. B.C., IV–VIII; Ont., Que., IV–IX.

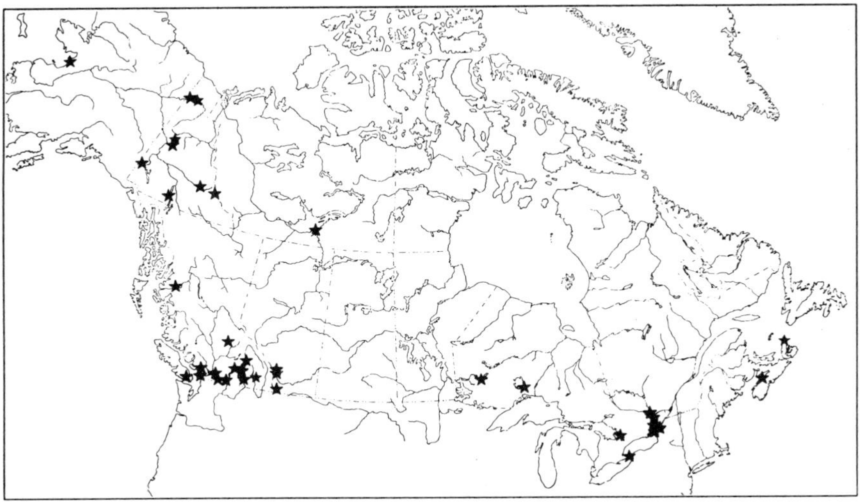

Map 43. Collection localities for *Melangyna triangulifera* (Zetterstedt).

Specimens identified. Alaska, 3 ♀♀; Canada, 41 ♂♂, 71 ♀♀; United States, 7 ♂♂, 13 ♀♀; Europe, 4 ♂♂, 3 ♀♀.

Biology. Adults have been reared from larvae feeding on several species of aphids on deciduous shrubs and trees (Heiss 1938; as *Epistrophe cincta* and *E. triangulifera*) and on *Cinara carolina* (record from CNC). Láska and Starý (1980) listed four species of aphids as larval hosts in Czechoslovakia.

Melangyna (Meligramma) vespertina Vockeroth

Map 42

Melangyna (Meligramma) vespertina Vockeroth, 1980:777.

Length. 8.2 mm.

Male. Frons and face pale yellow, translucent.

Scutum shining black, with broad well-defined yellow lateral margin and with large yellow prescutellar spot.

Tergite 2 with pair of well-separated subtriangular yellow spots reaching anterolateral corners of tergite; tergites 3 and 4 each with entire basal or nearly basal yellow band slightly narrowed posteromedially and extending broadly to lateral margins of tergite.

Female. Unknown.

Distribution. British Columbia (Map 42); VII.

Specimens identified. Canada, 1 ♂.

Larval host. *Macrosiphum* sp. on *Pteridium aquilinum*.

Genus *Melanostoma* Schiner

Diagnosis. Species small, slender or very slender, with entirely black head and thorax; usually with distinct pairs of yellow to orange spots on tergites 2–4 but with spots reduced or absent in some specimens. Length 4.8–10.0 mm.

Description. Eye bare. Frons of male broadly pruinose along eye margins, otherwise shining to subshining; frons of female with pair of distinct to obscure pruinose spots just below mid length; spots variable in size and separated medially or confluent. Face (Fig. 5) slightly receding below, with low broad pruinose to shining tubercle, otherwise moderately densely pruinose to almost shining. Antenna varying from entirely black to yellow with first flagellomere brownish above.

Thoracic hairs short, usually entirely yellow, rarely dark on dorsum. Scutum mostly shining, slightly pruinose anteriorly and laterally. Scutellum shining. Ventral scutellar fringe complete. Pleura slightly pruinose. Anterior anepisternum, meron, metapleuron, and metasternum bare. Upper and lower katepisternal hair patches widely separated. Metasternum (Fig. 44) with deep posterior incision on each side so only narrow anterior band and median stripe sclerotized. Wing membrane entirely trichose or with small bare areas near base, at most extreme base of cell c and about basal half of cell bm partly bare. Hind coxa without posteromedial apical hair tuft. Legs slender, with first tarsomere of hind leg scarcely swollen, without outstanding hairs or bristles, varying from almost entirely black to entirely yellow except for black coxae.

Abdomen unmargined, variable in proportions and markings. Male with abdomen (Fig. 145) nearly parallel sided, from two to five times as long as greatest width. Tergites 2–4 usually with distinct yellow to yellow-orange spots, but with spots in some specimens darkened and pruinose or submetallic, or, in some arctic specimens, indicated only by dark brown slightly shining areas; spots of tergite 2 well-separated from anterior and posterior margins, usually extending broadly to lateral margins but in some specimens reduced in size, indistinct, or absent; tergites 3 and 4 usually with distinct subquadrate or subrectangular basal yellow spots extending to lateral margins on at least anterior half of their length; spots in some specimens reduced in size and not reaching lateral margins, in many arctic specimens spots scarcely distinguishable.

Female with abdomen (Fig. 146) varying from nearly parallel sided to oval, from 1.7 to 2.5 times as long as greatest width; tergites 2–5 usually with yellow spots distinct but in some specimens reduced in size or absent, especially in arctic specimens; spots of tergite 2 elongate to rounded, well-separated from anterior and posterior margins but anteriorly in some specimens extending narrowly to lateral margins; tergites 3 and 4 usually with yellow basal spots of characteristic shape, strongly narrowed posterolaterally and extending only narrowly to lateral margins; tergite 5 usually with large to small anterolateral spots. Sternites with variable markings, ranging from entirely black to yellow with narrow complete or partial brown median line.

Distribution. Only one very widespread and varied New World species, extending from arctic Canada to Mexico recognized (see discussion under *M. mellinum*). Three species currently recognized in Europe (Speight 1978) and more than 20 in Afrotropical, Oriental, and Australian regions.

Biology. Although adults of *Melanostoma* are possibly as abundant as those of any other genus of Syrphinae in the Holarctic region, almost nothing is known of their biology. Old records (Coe 1953) of larvae feeding on adults of *Musca* or of Anthomyiidae, or on larvae of *Tortrix*, require verification. Metcalf (1916) found eggs, larvae, and pupae on the underside of leaves of rape sparsely infested with *Myzus persicae*. He noted that they preferred very damp places. He reared these larvae, and others from eggs deposited by females in the laboratory, on *Myzus persicae* and *Aphis cornifolia*. He offered them other foods (unspecified) but they would eat only aphids. Neither Goeldlin (1974) nor Láska and Starý (1980) found, or were able to rear in the laboratory, any larvae of the genus.

Discussion. The genus *Melanostoma* is similar to *Platycheirus* and was first clearly distinguished from the latter by Andersson (1970). Until recently several Nearctic species of *Platycheirus* were referred to *Melanostoma* (Vockeroth 1990). The most reliable distinguishing feature is the difference in the structure of the metasternum, as indicated in the key to genera. Unfortunately in some specimens the metasternum is not easy to see, but supplementary characters are useful. The face of *Melanostoma* is slightly receding with a low, rounded tubercle and with uniformly distributed pruinosity (Fig. 5); that of *Platycheirus* (Figs. 6–11) may be similar but is commonly produced forward below and may have a larger or more abrupt tubercle or rippled or punctate pruinosity laterally. The pale abdominal markings of *Melanostoma* usually differ from those of *Platycheirus*. Most males of *Melanostoma* have the spots of tergites 3 and 4 subquadrate or subrectangular, separated medially and narrowed only near the lateral margin (Fig. 145). Some species of *Platycheirus* with broadened fore tarsi have similar abdominal markings, but those with slender fore tarsi have distinctly different abdominal markings (e.g., Figs. 171, 172) and also have the face distinctly swollen or produced forward below. Most females of *Melanostoma* have very characteristic abdominal markings (Fig. 146), with the pale spots of tergites 3 and 4 strongly narrowed posterolaterally and reaching the margins only very narrowly; no species of *Platycheirus* have abdominal spots of this form. Only specimens of *Melanostoma* with the spots on the tergites considerably reduced in size or absent are likely to be difficult to distinguish from specimens of *Platycheirus*.

Melanostoma mellinum (Linnaeus)

Figs. 5, 44, 145, 146; Map 44

Musca mellina Linnaeus, 1758:594.
Melanostoma ? pictipes Bigot, 1884:78.
Melanostoma ? pruinosa Bigot, 1884:79.
Melanostoma ? pachytarse Bigot, 1884:80.
Melanostoma montivagum Johnson, 1916:78.
Melanostoma fallax Curran, 1923*e*:271.
Melanostoma pallitarse Curran, 1926:83.
Melanostoma melanderi Curran, 1930*b*:64.

Length. 4.8–10.0 mm.

Male and female. Characters as given for genus. Head as in Fig. 5. Abdomen of male as in Fig. 145, of female as in Fig. 146.

Distribution. Alaska, Canada (including Arctic) (Map 44), south to California, Mexico (Mexico), and Georgia; Europe; Asia.

Specimens identified. Alaska, 50 ♂♂, 216 ♀♀; Canada, 641 ♂♂, 3818 ♀♀; United States, 194 ♂♂, 309 ♀♀; Mexico, 5 ♀♀; Europe, 122 ♂♂, 154 ♀♀; Asia Minor, 28 ♂♂; Siberia, 2 ♂♂; Japan, 1 ♂, 2 ♀♀.

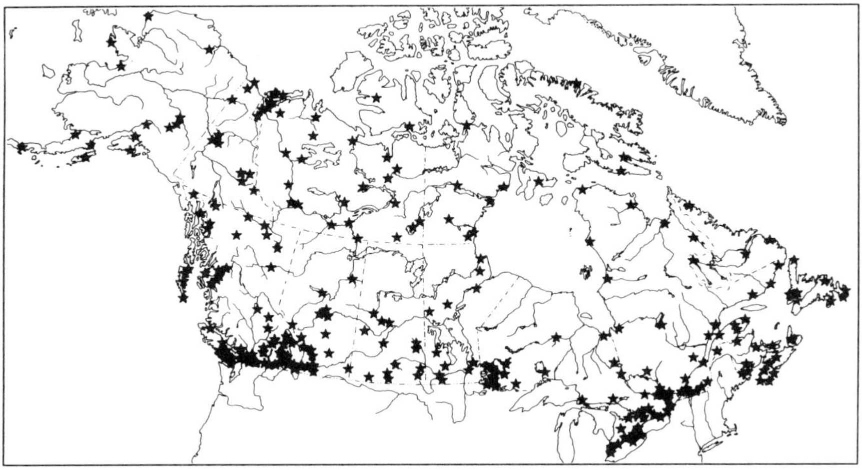

Map 44. Collection localities for *Melanostoma mellinum* (Linnaeus).

Discussion. The taxonomy of this genus in the Holarctic region is notoriously difficult. Many European authors have recognized two or three species, two of which, *M. mellinum* and *M. dubium* (Zetterstedt), 1838, have been recognized by various North American authors. Speight (1978) has recognized three species in Great Britain. British material that I have examined seemed to consist of specimens of two of these [*M. mellinum* and *M. scalare* (Fabricius)]; the third species, *M. dubium*, is rare in Britain. However, many specimens from western Europe in the CNC, from northern Sweden to Spain, do not always conform to his diagnoses; there seem to be a number of rather varied intermediates.

In North America the situation is even more confused. Curran (1930*b*) included males of four species in a key to eastern species; two others, which he had described earlier, were not mentioned, even though one was described from Wisconsin. Examination of several thousand specimens from many parts of North America convinces me that almost all characters used by earlier workers to distinguish species are either highly variable, sometimes even in what may be a series of a single species from one locality, or so little different between supposed species that they are useless. These characters include pruinosity of frons and of face; color of antenna and shape of first flagellomere; length of aristal hairs; color of thoracic hairs; color of femora; shape and proportions of abdomen and of individual tergites and sternites; color of abdominal hairs; presence or absence, as well as color, extent, and proportions, of pale markings of tergites; and color of sternites. To these may be added differences in distribution of wing microtrichia, a character of considerable taxonomic value in many genera but here not noticeably correlated with other differences. I am unable to recognize more than one Nearctic species; because many specimens agree well with British specimens that agree with Speight's diagnosis of *M. mellinum*, I use that name.

Genus *Meliscaeva* Frey

Diagnosis. Species slender, medium-sized, with moderately pruinose scutum and broad entire slightly arcuate yellow bands on abdomen. Length 8.0–10.8 mm.

Description. Eye bare. Frons mostly black, yellowish only anterolaterally, densely yellow-gray to brown pruinose except for shining semicircular black area above antennae. Face yellow, rather densely yellow pruinose except for tubercle; tubercle slightly to strongly darkened in some specimens.

Scutum black, slightly pruinose, notopleuron densely pale pruinose. Scutellum yellow. Ventral scutellar fringe long, dense. Pleura black, rather densely yellow pruinose on most of upper two-thirds, otherwise subshining. Anterior anepisternum extensively haired. Upper and lower katepisternal hair patches broadly separated. Metapleuron and metasternum bare. Wing with minute black dots along entire posterior margin; membrane with restricted bare areas near base; cell bm bare on about anterobasal one-third. Hind coxa without posteromedial apical hair tuft.

Abdomen (Fig. 147) slender, slightly oval to parallel-sided, unmargined. Tergite 1 yellow anterolaterally; tergite 2 with pair of large yellow lateral spots; tergites 3 and 4 each with broad entire yellow band reaching lateral margins in nearly full width, bands with nearly straight anterior margin and usually with slightly arcuate posterior margin; tergite 5 yellow with posterior black band or triangle.

Distribution. One Nearctic species (also in Europe and Asia); one other Palaearctic species; about 12 Oriental species; one unidentified Australian species.

Biology. Láska and Starý (1980) listed three species of aphids as larval hosts of *M. cinctella* in Czechoslovakia. Rotheray and Gilbert (1989) listed several additional hosts. Goeldlin (1974) suggested that in Switzerland females of *M. auricollis* (Meigen), 1822, hibernate.

Discussion. Nearctic species with the distribution and relationships of *M. cinctella* (Zetterstedt) have, in many cases, been introduced into North America by human agency since 1492 but the Nearctic distribution of *M. cinctella*, as well as lack of any apparent association with humans, suggest that it was not so introduced. The only other predominantly Palaeotropical genera or subgenera of Syrphinae occurring in North America are *Melanostoma* Schiner and *Paragus* (*Pandasyopthalmus*) Stuckenberg.

Meliscaeva cinctella (Zetterstedt)

Fig. 147; Map 45

Scaeva cinctella Zetterstedt, 1843:742.
Syrphus diversipes Macquart, 1850:459.

Length. 8.0–10.8 mm.

Male and female. Characters as given for genus. Abdomen as in Fig. 147.

Distribution. Alaska, Canada (Map 45), south to California, Colorado, and Georgia; Europe; Asia. B.C., V–IX; Ont., Que., VI–X.

Specimens identified. Alaska, 22 ♀♀; Canada, 233 ♂♂, 403 ♀♀; United States, 21 ♂♂, 58 ♀♀; Europe, 11 ♂♂, 24 ♀♀; Siberia, 1 ♂; Japan, 2 ♂♂, 1 ♀.

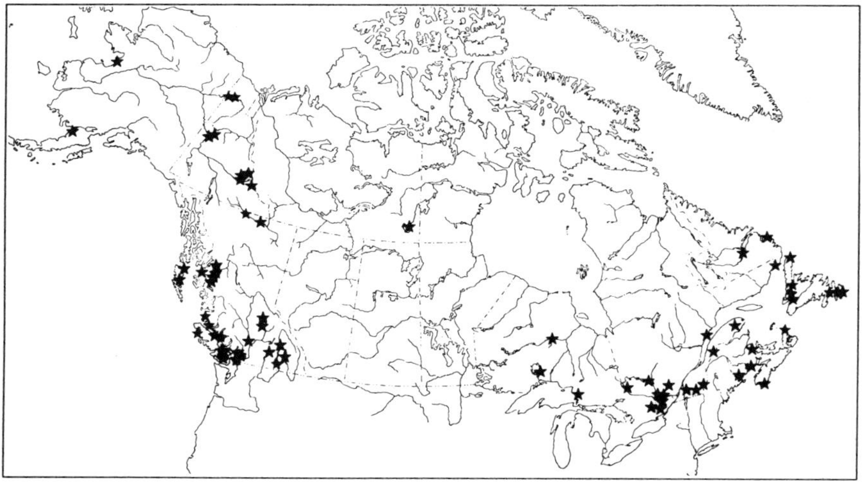

Map 45. Collection localities for *Meliscaeva cinctella* (Zetterstedt).

Genus *Ocyptamus* Macquart

Diagnosis. Species moderately slender to very slender, very diverse in appearance. In Canadian species abdomen usually very slender, petiolate, and mostly dark, more rarely parallel-sided and with two distinct dark yellow bands. Wing usually with dark markings, rarely hyaline. Length 6.8–13.3 mm.

Description. Eye bare. Frons of female broad or strongly narrowed posteriorly. Face rather broad and parallel-sided or narrow and distinctly narrowed below, entirely yellow, darkened laterally or with brown to black median stripe.

Scutum brown to black, faintly striped in some species, shining to moderately pruinose. Pleura brown to black, subshining to distinctly pruinose, with yellow markings in some species. Scutellum dull yellow to brown or blackish brown. Scutum with one to two transverse rows of long hairs anteriorly in some species. Ventral scutellar fringe usually dense, rarely very sparse or absent. Anterior anepisternum haired dorsally or on entire surface; meron bare; metepisternum haired or bare; metepimeron and metasternum bare. Wing usually brown on anterior one-third or more or with transverse brown band (Figs. 32–35), rarely hyaline; membrane with microtrichia very variable in extent. Hind coxa without posteromedial apical hair tuft.

Abdomen usually very slender, petiolate, 10–20 times as long as least width of tergite 2, dark with subbasal yellow to reddish bands or lateral spots on tergites 3 and 4; rarely moderately broad, parallel-sided, about 3.5 times as long as width of tergite 2, dark with an entire dark yellow band on each of tergites 3 and 4.

Distribution. Fourteen Nearctic species (five from southern Canada southward, nine more from Texas, Georgia, and Florida); about 250 Neotropical species.

Biology. Larvae of Canadian species have been reported as predators of Coccidae and of Aphididae. Larvae of Neotropical species have been reported to feed also on Aleyrodidae or on mites.

Key to Canadian species of *Ocyptamus*

1. Abdomen moderately broad, nearly parallel sided, about 3.5 times as long as width of tergite 2. Wing hyaline, with microtrichia only near apex and along posterior margin. Metepisternum bare. British Columbia .... ***diversifasciatus* (Knab)**

Abdomen very slender, petiolate, at least 10 times as long as least width of tergite 2 (as in Fig. 2). Wing brown on at least anterior one-third or with dark transverse band across crossveins, with dense microtrichia on at least apical one-third. Metepisternum with some long hairs 2

2. Wing brown on at least anterior one-third (Figs. 32, 34); cells c, br, and bm entirely trichose 3

Wing with transverse brown band across crossveins, otherwise almost entirely hyaline (Figs. 33, 35); cells c, br, and bm mostly bare 4

3. Face entirely yellow. Tibiae entirely dark brown. Wing with cell dm brown on at least apical one-third; crossvein r-m beyond level of Sc apex (Fig. 34). Ventral scutellar fringe at most six hairs; hairs commonly entirely absent. Southern Ontario .... ***fuscipennis* (Say)**

Face yellow with broad median black stripe. Fore and mid tibiae pale orange; hind tibiae with basal one-third yellow, apical two-thirds dark brown. Wing with cell dm hyaline; crossvein r-m slightly before or at level of Sc apex (Fig. 32). Ventral scutellar fringe dense, long. Southern Ontario .... ***costatus* (Say)**

4. Wing band large, covering about apical two-fifths of cell c and basal half of cell dm (Fig. 33); cell r_{4+5} entirely trichose. Manitoba eastward .... ***fascipennis* (Wiedemann)**

Wing band small, not extending into cell c and covering at most basal one-third of cell dm (Fig. 35); cell r_{4+5} bare on most of basal one-quarter to half. Saskatchewan westward .... ***lemur* (Osten Sacken)**

Clé des espèces canadiennes de *Ocyptamus*

1. Abdomen modérément large, à côtés presque parallèles, environ 3,5 fois plus long que la largeur du tergite 2. Aile transparente, pourvue de microchètes uniquement près de l'apex et le long de la marge postérieure. Métépisterne glabre. Colombie-Britannique .... ***diversifasciatus* (Knab)**

Abdomen très mince et pétiolé, au moins 10 fois plus long que la largeur minimale du tergite 2 (comme dans la fig. 2). Aile brune sur au moins le tiers antérieur ou ornée d'une bande transversale foncée d'un côté à l'autre des nervures transversales, pourvue de denses microchètes sur au moins le tiers apical. Métépisterne garni de quelques longs poils 2

2. Aile brune sur au moins le tiers inférieur (fig. 32, 34); cellules c, br et bm entièrement velues 3

Aile presque toute transparente, à l'exception d'une bande brune transversale d'un côté à l'autre des nervures transversales (fig. 33, 35); cellules c, br et bm largement glabres 4

3. Face totalement jaune. Tibias entièrement brun foncé. Aile pourvue d'une cellule dm brune, au moins sur le tiers apical; nervure transversale r-m au-delà du niveau de l'apex de Sc (fig. 34). Frange ventrale du scutellum ayant au plus six poils; souvent, il n'y a aucun poil. Sud de l'Ontario .... ***fuscipennis* (Say)**

Face jaune ornée d'une large rayure médiane noire. Tibias antérieur et médian orange pâle; tibias postérieurs jaunes sur le tiers inférieur et brun foncé sur les deux tiers apicaux. Aile pourvue d'une cellule dm transparente; nervure transversale r-m légèrement en avant ou au niveau de l'apex de Sc (fig. 32). Frange ventrale du scutellum longue et dense. Sud de l'Ontario .... ***costatus* (Say)**

4. Bande alaire large couvrant environ les deux cinquièmes apicaux de la cellule c et la moitié basilaire de la cellule dm (fig. 33); cellule r_{4+5} entièrement velue. Manitoba et plus à l'est .... ***fascipennis* (Wiedemann)**

Bande alaire étroite, ne rejoignant pas la cellule c et couvrant au plus le tiers basilaire de la cellule dm (fig. 35); cellule r_{4+5} glabre sur au plus le quart à la moitié de la partie basale. Saskatchewan et plus à l'ouest .... ***lemur* (Osten Sacken)**

Ocyptamus costatus (Say)

Fig. 32; Map 46

Baccha costata Say, 1829:161.
Baccha costalis Wiedemann, 1830:97.
Baccha tarchetius Walker, 1849:549.

Length. 8.7–11.5 mm.

Male. Frons mostly black, with narrow yellow area only anterolaterally. Face narrow, narrowed below, yellow with broad black median stripe not reaching lower facial margin.

Scutum shining on disc, subshining laterally; scutellum dark brown; pleura subshining, posterior part of anepisternum and upper part of katepisternum yellow. Scutum without longer hairs anteriorly; ventral scutellar fringe long, dense; anterior anepisternum haired posterodorsally; metepisternum with some hairs. Wing (Fig. 32) brown anteriorly, in front of M to just beyond crossvein r-m, in front of R_{4+5} beyond this point; membrane mostly trichose, with only most of alula and small area at base of anal lobe bare; crossvein r-m slightly before or at level of apex of Sc. Fore and mid legs pale orange, with tarsi slightly brownish; hind femur with basal half and extreme apex pale orange, otherwise dark brown; hind tibia with basal one-third yellow, apical two-thirds dark brown; hind tarsus yellow.

Abdomen slender, petiolate, about 10 times as long as least width of tergite 2. Tergites mostly opaque black; tergite 1 and margins of several other tergites shining blue-black; tergites 3 and 4 each with pair of small triangular lateral yellow spots; tergite 5 with trace of such spots.

Female. Frons strongly narrowed posteriorly; lateral ocellus about its own diameter from eye margin. Tergite 6 and sternite 6 fused laterally, depressed anteriorly, slightly compressed posteriorly; tergite 7 and sternite 7 fused laterally, strongly compressed.

Distribution. Eastern Canada (Map 46), south to Louisiana,* Mississippi, Florida, and Cuba. Ont., VII.

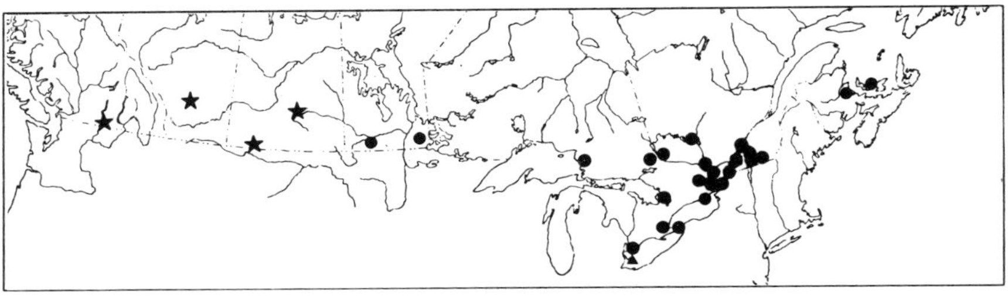

Map 46. Collection localities for *Ocyptamus costatus* (Say) (▲), for *O. fascipennis* (Wiedemann) (●), and for *O. lemur* (Osten Sacken) (★).

Specimens identified. Canada, 1 ♀; United States, 22 ♂♂, 35 ♀♀; Cuba, 1 ♀.

Biology. Larvae have been reported feeding on the coccid *Toumeyella liriodendri* in the eastern United States (Burns and Donley 1970).

Ocyptamus diversifasciatus (Knab)

Map 47

Syrphus diversifasciatus Knab, 1914:151.
Syrphus rubripleuralis Curran, 1921*b*:172.

Length. 8.4–10.1 mm.

Male. Frons black posteriorly, orange-brown anteriorly, with black arc just above antennae. Face broad, parallel-sided, dull yellow with brown to dark brown median stripe; lower margin of face and cheeks black.

Scutum shining, moderately pruinose laterally. Scutellum yellow with blackish margin. Pleura distinctly pruinose. Scutum without longer hairs anteriorly; ventral scutellar fringe long, dense; anterior anepisternum haired over entire surface; metepisternum bare. Wing hyaline; membrane almost bare, with distinct microtrichia only near apex and along narrow posterior margin; crossvein r-m distinctly before level of apex of Sc. Femora mostly gray to black, distinctly pruinose; tibiae mostly brown to blackish, paler basally; tarsi brown.

Abdomen moderately broad, nearly parallel sided, about 3.5 times as long as width of tergite 2. Tergites black, opaque on disc, shining on margins; tergites 3 and 4 each with entire slightly arcuate yellow to orange (or rarely grayish) band not reaching lateral margins. Sternites black; sternites 3 and 4 each with basal yellow-orange band.

Female. Frons broad, only slightly narrowed posteriorly, black on upper two-thirds, yellow on lower one-third, with brown to black arc above antennae; lateral ocellus about twice its own diameter from eye margin. Tergite and sternite 6 and tergite and sternite 7 not fused laterally, slightly convex.

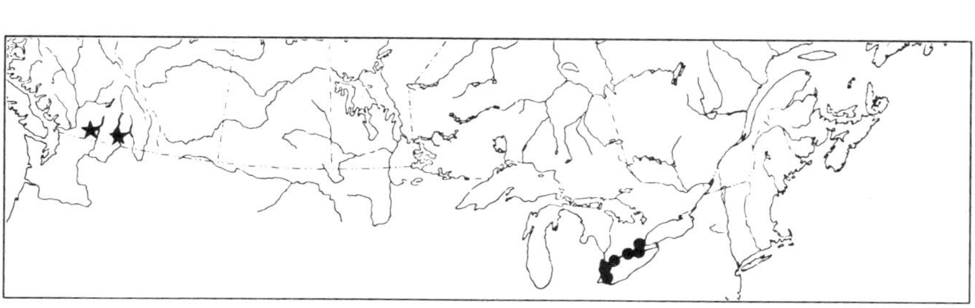

Map 47. Collection localities for *Ocyptamus diversifasciatus* (Knab) (★) and for *O. fuscipennis* (Say) (●).

Distribution. Western Canada (Map 47), south to California, Arizona, and Mexico.* B.C., VI–X.

Specimens identified. Canada, 3 ♀♀; United States, 20 ♂♂, 23 ♀♀.

Oyptamus fascipennis (Wiedemann)

Fig. 33; Map 46

Baccha fascipennis Wiedemann, 1830:96.
Baccha aurinota Walker, 1849:548.

Length. 9.0–13.3 mm.

Male. Frons brownish black with yellow M-shaped mark above antenna. Face narrow, narrowed below, yellow, narrowly to broadly pale to dark gray laterally.

Scutum subshining; scutellum yellow-brown to brown; pleura subshining, without yellow markings. Scutum without longer hairs anteriorly; ventral scutellar fringe long, dense; anterior anepisternum haired posterodorsally; metepisternum with some hairs. Wing (Fig. 33) brown to level of crossvein h, with broad brown band across crossveins; band narrowing posteriorly, ending just before posterior margin; wing otherwise hyaline; wing bare in hyaline portions of cells c, r, br, and bm, anterobasal part of cell *cup*, base of anal lobe, and basal three-quarters of alula; rest of wing trichose, including entire area beyond brown band; crossvein r-m under or very slightly beyond apex of Sc. Legs yellow with apical one-third of hind tibia and all tarsi brown.

Abdomen very slender, petiolate, about 20 times as long as least width of tergite 2, dark brown, with pair of small subbasal lateral yellow or reddish spots on each of tergites 3 and 4.

Female. Frons narrowly yellow laterally on anterior half, strongly narrowed posteriorly; lateral ocellus about its own diameter from eye margin. Face at most slightly grayish laterally. Alula trichose only at extreme apex. Tergite 2 commonly obscurely reddish at lateral margin; tergite 5 with pair of subbasal pale spots similar to those of tergites 3 and 4; tergite and sternite 6 and tergite and sternite 7 not fused laterally, depressed.

Distribution. Eastern Canada (Map 46), south to Texas and Florida. Ont., Que., VI–IX.

Specimens identified. Canada, 20 ♂♂, 70 ♀♀; United States, 8 ♂♂, 18 ♀♀.

Biology. Larvae have been reported as predators of the coccid *Pseudococcus citri* on *Chrysanthemum* sp. in Minnesota (Harris and Hamrun 1968). Larvae, especially early instars, were almost hidden among the coccids. The period from oviposition to adult emergence was 22–33 days at outdoor temperatures. Many specimens were collected in Malaise traps in a large sphagnum bog near Ottawa, Ont.

Ocyptamus fuscipennis (Say)

Fig. 34; Map 47

Baccha fuscipennis Say, 1823:100.
Ocyptamus fascipennis Macquart, 1834:554 (preoccupied Wiedemann 1830).
Syrphus amissas Walker, 1849:589.
Syrphus peas Walker, 1849:590.
Syrphus radaca Walker, 1849:590.
Baccha lugens Loew, 1863:14.
Ocyptamus longiventris Loew, 1866:38.
Baccha fuscipennis var. *fenestrata* Hull, 1949:280, fig. 381.

Length. 6.8–11.3 mm.

Male. Frons blackish brown, commonly reddish above antenna. Face narrow, narrowed below, yellow to reddish, slightly translucent laterally.

Scutum weakly pruinose; scutellum yellow brown to dark brown; pleura weakly pruinose with posterior part of anepisternum obscurely yellowish. Scutum anteriorly with irregular double row of long stiff white hairs about twice as long as other scutal hairs; ventral scutellar fringe absent or very sparse, with at most six long hairs; anterior anepisternum haired posterodorsally; metepisternum with some hairs. Wing (Fig. 34) usually mostly brown with only alula, anal area, and posteroapical part of wing hyaline; some specimens also with most of cells dm, cua_1, and cu*p* hyaline or nearly so but always with apex of cell dm broadly brown; membrane with only alula and extreme base of anal lobe bare; crossvein r-m slightly beyond apex of Sc. Legs mostly dark yellow to orange or orange-brown; femoral apices and much of tibiae and tarsi brownish.

Abdomen very slender, petiolate, about 12 times as long as least width of tergite 2. Tergites dark brown; most of tergites 2 and 3 narrowly and obscurely reddish laterally.

Female. Frons broadly reddish yellow anterolaterally, strongly narrowed posteriorly; lateral ocellus about its own diameter from eye margin. Katepisternum usually obscurely yellowish dorsally. Tergites commonly partly obscurely reddish; reddish areas in some specimens forming distinct broad bands at base of tergites 3–5; tergite and sternite 6 and tergite and sternite 7 not fused laterally, depressed.

Distribution. Eastern Canada (Map 47), south to Colorado,* Texas, and Florida. Ont., VI–VII, IX.

Specimens identified. Canada, 2 ♂♂, 14 ♀♀; United States, 17 ♂♂, 74 ♀♀.

Biology. Folsom (1909) reported *O. fuscipennis* as a possible predator of the aphid *Macrosiphum pisi*, but the evidence was very weak. Miller (1929) reported it (as *O. lugens*) as a predator of the aphid *Aphis spiraecola* in Florida. The period from oviposition to adult emergence was 14–19 days.

Ocyptamus lemur (Osten Sacken)

Fig. 35; Map 46

Baccha lemur Osten Sacken, 1877:331.

Length. 7.2–11.2 mm.

Male and female. Very similar in both sexes to *O. fascipennis*, differing as follows: face commonly darker, with only tubercle distinctly yellow in some specimens.

Scutellum bright yellow in some specimens. Wing (Fig. 35) with transverse brown band narrower, not extending into cell c and covering at most basal one-third of cell dm; membrane with moderate bare areas in cells r_{2+3} and r_{4+5} beyond brown band.

Tergites 3 and 4 with pale spots larger, subrectangular, nearly joined medially or joined to form entire band; female with pale spots of tergite 5 tiny or indistinct.

Distribution. Western Canada (Map 46), south to California, Arizona, Texas,* and Wisconsin.* B.C., VII.

Specimens identified. Canada, 2 ♂♂, 3 ♀♀; United States, 48 ♂♂, 62 ♀♀.

Biology. Larvae were recorded as predators of a coccid, *Pseudococcus* sp., in California by Clausen (1915).

Genus *Paragus* Latreille

Diagnosis. Species small, slender to moderately robust. Thorax black or with apex of scutellum pale. Abdomen usually extensively red-orange to dark red, rarely entirely black. Length 4.3–7.2 mm.

Description. Eye with dense short hairs. Antenna with first flagellomere three times as long as wide, slightly pointed apically. Face produced slightly from base of antenna to very obscure tubercle, then receding at most very slightly to lower facial margin (Fig. 14), yellow with narrow black lower margin and commonly black median stripe.

Scutum black, usually with grayish pruinose submedian stripes anteriorly. Scutellum black or with posterior margin yellow to reddish. Ventral scutellar fringe complete, very short. Pleura black. Anterior anepisternum, meron, metapleuron, and metasternum bare. Wing membrane slightly or extensively bare on about basal half. Hind coxa without posteromedial apical hair tuft. Legs with coxae, trochanters, and about basal two-thirds of femora black; tibiae pale yellow basally to yellow-orange apically with dark ring on hind pair in some specimens; tarsi yellow-orange with hind pair dark above in some specimens.

Abdomen (Fig. 148) nearly parallel sided, unmargined, with tergites minutely punctured and with slightly depressed transverse areas. Tergite 1 well-developed especially on disc, at least half as long as tergite 2, always extending well beyond scutellum, sublaterally about three-quarters as long as tergite 2; tergite 1 usually entirely black; other tergites partly or entirely red-orange to dark red or entirely black (color pattern not reliable for specific identification). Sternites usually mostly orange, darker when tergites extensively blackish. Male terminalia very different in two subgenera; aedeagal guide distinct, very small or very large; paramere articulated with hypandrium; aedeagus unsegmented, with very small to very large compressed lateral lobes.

Distribution. Subgenus *Paragus*: seven species in New World (none south of United States); about 20 species in Palaearctic, Afrotropical, and Oriental regions. Subgenus *Pandasyopthalmus*: one Holarctic species in New World (as far south as Costa Rica); some species in Palaearctic and Oriental regions; probably 15 or more species in Afrotropical region.

Biology. All recorded larval hosts are aphids. Heiss (1938) summarized Nearctic host records, but her records for *P. bicolor* (Fabricius), as well as later records by Cole (1969), are unreliable because *P. bicolor* of North American authors may be any of seven species. Heiss listed seven species of aphids as hosts of *P. haemorrhous* Meigen [as *P. tibialis* (Fallén)]. Goeldlin (1976) and Láska and Starý

(1980) gave host records from Europe for *P. haemorrhous* and for several species of *Paragus* s.s.

Goeldlin (1976) has given a brief account of the biology. Larvae and adults occur mostly near the ground in low vegetation; they prefer hot, dry areas. Larvae are diurnal and feed fully exposed among the host aphids; they leave the colony only when ready to pupate. Most of the European species, at least, are polyvoltine. The extensive range of collecting dates of most Nearctic species suggests that they too are polyvoltine. I have several times taken adults by sweeping low vegetation where *Fragaria* is in bloom, but I find no definite evidence of association with this plant.

Key to subgenera of *Paragus*

1. Scutellum entirely black. Eye hairs of uniform length and color. Male with distinct black facial stripe extending well above tubercle. Surstylus subrectangular, nearly truncate apically; paramere stout, straight, directed posteriorly; lateral lobe of aedeagus very small, hidden by paramere in lateral view .... ***Paragus (Pandasyopthalmus)* Stuckenberg**

   Scutellum black with apex narrowly yellow or reddish. Eye with vertical median band of shorter and usually darker hairs on about upper three-quarters. Male with face yellow except along lower margin or with at most very narrow dark median stripe not extending above tubercle (Fig. 14). Surstylus strongly narrowed to slender apex; paramere slender, slightly or strongly curved, directed upward; lateral lobe of aedeagus large, exposed in lateral view (Figs. 229–235) .... ***Paragus (Paragus)* Meigen**

Clé des sous-genres de *Paragus*

1. Scutellum tout noir. Poils de l'œil de longueur et de couleur uniformes. Mâle portant une rayure faciale noire distincte, qui s'étend bien au-dessus du tubercule. Surstylus quasi rectangulaire, presque tronqué dans sa partie apicale; paramère épais, droit et dirigé vers l'arrière; lobe latéral de l'édéage très petit, caché par le paramère vu de côté .... ***Paragus (Pandasyopthalmus)* Stuckenberg**

   Scutellum noir, faiblement jaune ou rougeâtre à l'apex. Oeil orné d'une bande médiane verticale constituée de poils plus courts et habituellement plus foncés, sur environ les trois quarts supérieurs. Chez le mâle, la face est jaune sauf le long du bord inférieur, ou elle est ornée tout au plus d'une rayure médiane foncée très étroite qui ne s'étend pas au-dessus du tubercule (fig. 14). Apex du surstylus fortement rétréci à mince; paramère mince, de légèrement à fortement recourbé et dirigé vers le haut; lobe latéral de l'édéage large, exposé de profil (fig. 229–235) .... ***Paragus (Paragus)* Meigen**

Subgenus *Paragus* (*Pandasyopthalmus*) Stuckenberg

Diagnosis. As for genus *Paragus*. Length 4.3–5.9 mm.

Description. Eye hairs uniform in length and in color. Male briefly holoptic or with eyes separated by as much as diameter of one facet. Face of both sexes with distinct black median stripe extending well above tubercle usually to antennal bases.

Scutellum entirely black. Wing membrane extensively bare on basal half; cell c bare or with some microtrichia anteriorly; cell bm with some scattered microtrichia on apical one-third.

Surstylus subrectangular in dorsal view, slightly narrowed beyond base, broadly rounded apically. Hypandrium with very shallow lateral emargination without marginal spines; aedeagal guide tiny, at most one-quarter as long as midventral length of hypandrium. Paramere broad, subrectangular or subtriangular, directed posteroventrally, straight, without marginal serrations. Lateral lobe of aedeagus tiny, entirely hidden by paramere in lateral view, without marginal teeth. Ejaculatory apodeme large.

Paragus (*Pandasyopthalmus*) *haemorrhous* Meigen

Fig. 148; Map 48

Paragus haemorrhous Meigen, 1822:182.
Paragus dimidiatus Loew, 1863:308.
Paragus auricaudatus Bigot, 1883:540.
Paragus tibialis of American authors (not *P. tibialis* Fallén, 1817).
Paragus obscurus: Davidson 1916:457 (not *P. obscurus* Meigen, 1822).

Length. 4.3–5.9 mm.

Male. Hind tibia usually with apical half yellow-orange or with obscure brownish ring, in some Mexican specimens with black ring.

Tergites black; tergite 2 black or with posterior margin red; tergites 3–5 usually entirely red-orange to dark red, in some specimens with red areas partly darkened (Fig. 148). Paramere variable in shape and size, with apex nearly truncate to strongly oblique, and with length from one to two times that of surstylus.

Female. Similar to male; usually with wider black facial stripe; abdomen entirely black in some specimens.

Distribution. Canada (Map 48), south to Baja California, Costa Rica, and Florida; Europe; Asia; Africa.* B.C. V–IX; Ont., Que., V–IX.

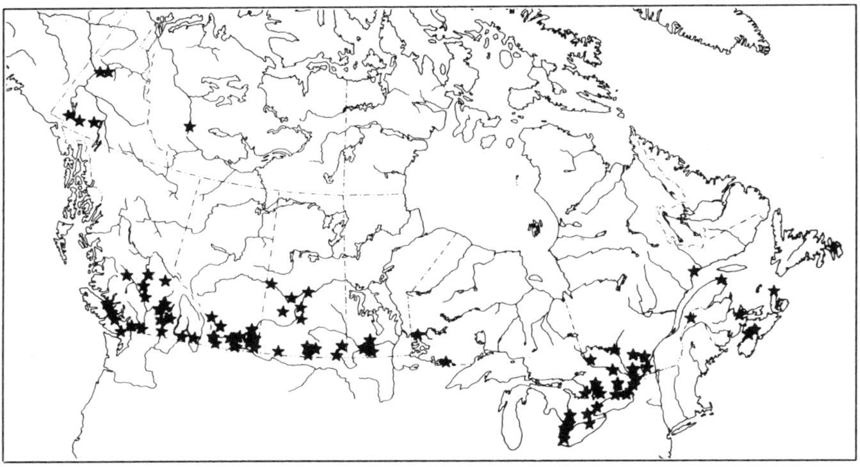

Map 48. Collection localities for *Paragus haemorrhous* Meigen.

Specimens identified. Canada, 201 ♂♂, 204 ♀♀; United States, 1493 ♂♂, 995 ♀♀; Mexico, 24 ♂♂, 31 ♀♀; Central America, 3 ♂♂; Europe, 6 ♂♂; Siberia, 2 ♂♂.

Discussion. This species has almost consistently been known as *P. tibialis* (Fallén) in North America. Goeldlin (1976) showed that almost all the European specimens previously known as *P. tibialis* are either of *P. tibialis* (with very large paramere) or of *P. haemorrhous* (with small paramere). Nearctic specimens show considerable variation in shape and size of the paramere (it varies in length from that of European *P. haemorrhous* to about two-thirds that of *P. tibialis*). The two extremes taken alone would suggest that at least two species occur in North America, but a division into two or more discrete groups appears impossible so all New World specimens are here referred to one species.

Subgenus *Paragus* (*Paragus*) Latreille

Diagnosis. As for genus *Paragus*. Length 4.7–7.2 mm.

Description. Eye with median vertical band of shorter and usually darker hairs on about upper three-quarters. Male holoptic. Face of male (Fig. 14) yellow with black lower margin or rarely with very narrow darkened median stripe not reaching above tubercle; face of female with distinct narrow to broad median black stripe.

Scutellum with apex narrowly to rather broadly yellow or reddish. Wing membrane mostly trichose; cells c and bm both with abundant microtrichia on at least apical half.

Male terminalia in Figs. 229–235. Surstylus tapering strongly from mid length or before to slender apex. Hypandrium with deep lateral emargination commonly bearing one or more marginal spines or spinose processes; aedeagal guide very large, more than half as long as midventral length of hypandrium. Paramere slender, moderately or strongly curved or bent, directed upward, with some short hairs on lateral surface and usually with margin and surface in some specimens beyond bend distinctly serrate or denticulate. Lateral lobe of aedeagus large, scarcely hidden by paramere in lateral view; margin usually with one or two strong teeth. Ejaculatory apodeme small.

Discussion. The only characters found to distinguish reliably the New World species of *Paragus* s.s. are those of the male terminalia. Females are therefore not described separately. It is possible that collecting females in association with males, especially in areas with only one or two species, may resolve this problem, but I think it unlikely.

Key to New World species of *Paragus* (*Paragus*)

1. Margin of hypandrium, between base of aedeagal guide and base of paramere, with one or two distinct spines or spine-bearing processes (Figs. 230*a*–232*a*, 235*a*) 2

   Margin of hypandrium without spines (Figs. 229*a*, 233*a*, 234*a*) 5

2. Margin of hypandrium, near base of aedeagal guide, with one large spine or with process bearing several spines and, near base of paramere, with one long strong spine (Figs. 232*a*,*b*); surstylus, in lateral view, with distinct strong triangular anteroventral lobe (Fig. 232*a*). Widespread . . . . ***bispinosus*** **Vockeroth**

   Margin of hypandrium, between base of aedeagal guide and base of paramere, with single spine or spinose process;

surstylus, in lateral view, with at most weak rounded anteroventral lobe (Figs. 230, 231, 235) 3

3. Aedeagal guide at most 1.5 times as long as its width at mid length (Fig. 235*b*). Widespread . . . . . ***variabilis*** **Vockeroth**

Aedeagal guide at least 2.5 times as long as its width at mid length (Figs. 230*b*, 231*b*) 4

4. Margin of hypandrium, well below base of paramere, with strong lateral process bearing several short spines on its posterior margin and with additional spines on adjacent posterior surface of arm of hypandrium; aedeagal guide about four times as long as its width at mid length (Figs. 231*a*,*b*); surstylus moderately broad on basal one-third, with weak but distinct anteroventral lobe (Figs. 231*a*,*c*). Southwestern United States . . . . ***arizonensis*** **Vockeroth**

Margin of hypandrium, well below base of paramere, with single short spine (Figs. 230*a*,*b*); aedeagal guide 2.5 times as long as its width at mid length; surstylus strongly narrowed from base to apex, without trace of anteroventral lobe (Fig. 230*a*). Eastern Canada . . . . ***angustistylus*** **Vockeroth**

5. Paramere curved through about 45°, with apex enlarged and smooth (Fig. 233*a*); lateral lobe of aedeagus with posteroventral margin only very slightly emarginate and with only one or two minute teeth (Fig. 233*a*); aedeagal guide with broad shallow apical emargination (Fig. 233*b*). Widespread . . . . ***cooverti*** **Vockeroth**

Paramere curved through about 90°, with apex enlarged and spiculate or tapering and smooth (Figs. 229*a*, 234*a*); lateral lobe of aedeagus with ventral margin deeply emarginate and with two strong teeth (Figs. 229*a*, 234*a*); aedeagal guide with narrow deep apical emargination (Figs. 229*b*, 234*b*) 6

6. Surstylus in dorsal view broad and nearly parallel sided on basal three-fifths, abruptly narrowed on apical two-fifths (Fig. 229*c*); paramere with apex slightly enlarged and strongly denticulate (Fig. 229*a*); posterior end of aedeagal apodeme with very short ventral tooth (Fig. 229*a*). Widespread . . . . ***angustifrons*** **Loew**

Surstylus in dorsal view tapering from just beyond base (Fig. 234*c*); paramere with apex tapering and smooth (Fig. 234*a*); posterior end of aedeagal apodeme with long curved ventral spine (Fig. 234*a*). California, Nevada, Idaho . . . . ***longistylus*** **Vockeroth**

Clé des espèces du Nouveau Monde de *Paragus* (*Paragus*)

1. Marge de l'hypandrium, entre la base du guide de l'édéage et la base du paramère, garnie d'une ou de deux épines distinctes ou d'appendices épineux (fig. 230*a*–232*a*, 235*a*) 2

   Marge de l'hypandrium dépourvue d'épines (fig. 229*a*, 233*a*, 234*a*) 5

2. Marge de l'hypandrium pourvue, près de la base du guide de l'édéage, d'une grosse épine ou d'un appendice portant plusieurs épines et, près de la base du paramère, d'une épine longue et robuste (fig. 232*a*,*b*); vu de profil, surstylus pourvu d'un lobe antéro-ventral triangulaire épais et distinct (fig. 232*a*). Espèce répandue .... ***bispinosus* Vockeroth**

   Marge de l'hypandrium pourvue, entre la base du guide de l'édéage et la base du paramère, d'une seule épine ou d'un seul appendice épineux; de profil, surstylus présentant tout au plus un faible lobe antéro-ventral arrondi (fig. 230, 231, 235) 3

3. Guide de l'édéage au plus 1,5 fois plus long que sa largeur à mi-longueur (fig. 235*b*). Espèce répandue .... ***variabilis* Vockeroth**

   Guide de l'édéage au moins 2,5 fois plus long que sa largeur à mi-longueur (fig. 230*b*, 231*b*) 4

4. Marge de l'hypandrium pourvue, bien en dessous de la base du paramère, d'un grand appendice latéral portant plusieurs courtes épines sur le bord postérieur et d'autres épines sur la face postérieure adjacente du bras de l'hypandrium; guide de l'édéage environ quatre fois plus long que sa largeur à mi-longueur (fig. 231*a*,*b*); surstylus modérément large sur le tiers basilaire, avec un lobe antéro-ventral faible, mais distinct (fig. 231*a*,c). Sud-ouest des États-Unis .... ***arizonensis* Vockeroth**

   Marge de l'hypandrium pourvue, bien en dessous de la base du paramère, d'une seule épine courte (fig. 230*a*,*b*); guide de l'édéage d'une longueur de 2,5 fois sa largeur à mi-longueur; surstylus fortement effilé de la base à l'apex; aucune trace de lobe antéro-ventral (fig. 230*a*). Est du Canada .... ***angustistylus* Vockeroth**

5. Paramère incurvé à environ 45° pourvu d'un apex élargi et lisse (fig. 233*a*); lobe latéral de l'édéage orné d'un bord postéro-ventral très légèrement échancré et pourvu seulement d'une à deux minuscules dents (fig. 233*a*); guide de l'édéage

orné d'une large échancrure apicale peu profonde (fig. 233*b*). Espèce répandue .... ***cooverti*** **Vockeroth**

Paramère incurvé à environ 90° pourvu d'un apex élargi et muni de petites épines, ou effilé et lisse (fig. 229*a*, 234*a*); lobe latéral de l'édéage fortement échancré sur le bord ventral et garni de deux fortes dents (fig. 229*a*, 234*a*); guide de l'édéage portant une échancrure apicale étroite et profonde (fig. 229*b*, 234*b*) 6

6. En vue dorsale, surstylus large et à côtés presque parallèles sur les trois cinquièmes de la portion basilaire, brusquement rétréci sur les deux cinquièmes apicaux (fig. 229*c*); apex du paramère légèrement élargi et fortement denticulé (fig. 229*a*); extrémité postérieure de l'apodème de l'édéage pourvue d'une dent ventrale très courte (fig. 229*a*). Espèce répandue .... ***angustifrons*** **Loew**

En vue dorsale, surstylus devenant fuselé à partir de la base (fig. 234*c*); apex du paramère effilé et lisse (fig. 234*a*); extrémité postérieure de l'apodème de l'édéage munie d'une longue épine ventrale incurvée (fig. 234*a*). Californie, Nevada, Idaho .... ***longistylus*** **Vockeroth**

Paragus (*Paragus*) *angustifrons* Loew

Figs. 14, 229; Map 49

Paragus angustifrons Loew, 1863:309.

Length. 4.8–6.1 mm.

Male. Hind tibia usually with brown-to-black band on apical half but band obscure in some specimens; hind leg with first tarsomere orange-to-brown above.

Tergites 1 and 2 black; tergites 3–5 usually dark red to red-orange; tergite 3 blackish on disc in some specimens. Terminalia in Fig. 229; surstylus broad, in dorsal view almost parallel sided on basal half, then narrowed abruptly to blunt apex, in lateral view with barely perceptible anteroventral broadening. Hypandrium moderately excavated laterally, without marginal spines; aedeagal guide as long as its width at mid length, slightly narrowed from base to apex, with posterolateral angles weakly denticulate. Paramere bent through about 90°, slightly broadened beyond bend, with apical part strongly denticulate. Lateral lobe of aedeagus with deep rounded emargination on ventral surface (rather than on posteroventral surface as in other species), with strong tooth at each of posteroventral and anteroventral angles.

Female. Unrecognized.

Distribution. Southern Canada (Map 49), south to Colorado and Georgia. Ont., Que., V–IX.

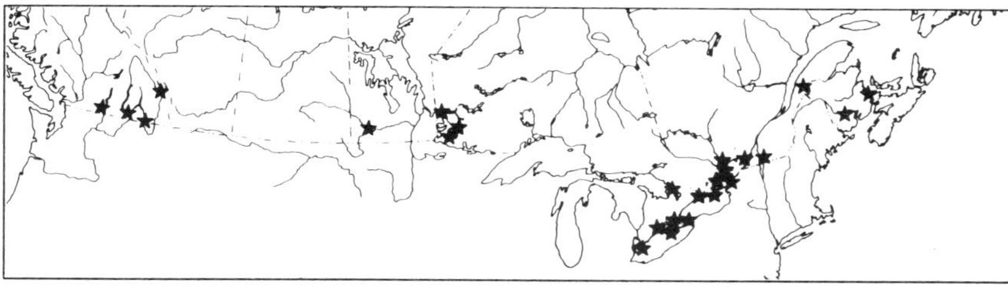

Map 49. Collection localities for *Paragus angustifrons* Loew.

Specimens identified. Canada, 136 ♂♂; United States, 238 ♂♂.

Paragus (*Paragus*) *angustistylus* Vockeroth

Fig. 230; Map 50

Paragus (*Paragus*) *angustistylus* Vockeroth, 1986*a*:190.

Length. 4.7–6.9 mm.

Male. Hind tibia with broad black band on apical half; hind leg with first tarsomere pale brown to dark brown above.

Tergite 2 black or orange-red to dark red posteriorly; tergites 3–5 usually orange-red to dark red but tergites 3 and 4 with blackish band in some specimens. Terminalia in Fig. 230; surstylus slender, in dorsal view strongly curved and almost uniformly narrowed from base to apex, in lateral view without trace of anteroventral lobe. Hypandrium deeply excavated laterally, with short and in some specimens blunt marginal spine just below base of paramere; aedeagal guide 2.5 times as long as its width at mid length, slightly narrowed, with lateral margins infolded beyond mid length, and with surface smooth. Paramere bent through about 90°, with irregular projection at bend, with portion beyond bend short, slightly narrowed, and strongly denticulate. Lateral lobe of aedeagus with very shallow posteroventral emargination; posterodorsal angle broadly rounded but usually with small tooth; posteroventral angle with small but distinct tooth.

Female. Unrecognized.

Distribution. Eastern Canada (Map 50), south to Illinois and New Jersey. Ont., Que., V–IX.

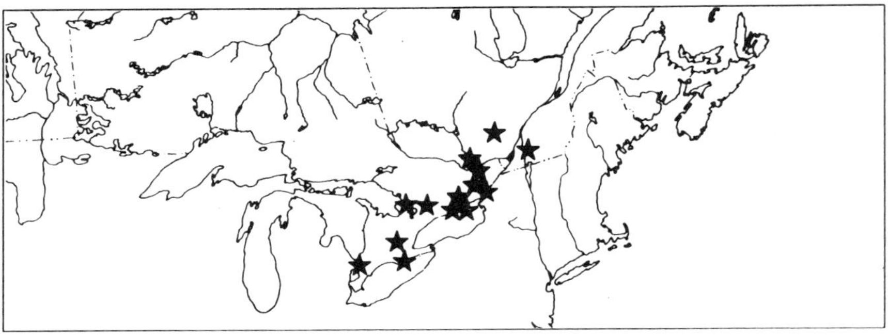

Map 50. Collection localities for *Paragus angustistylus* Vockeroth.

Specimens identified. Canada, 57 ♂♂; United States, 74 ♂♂.

Paragus (*Paragus*) *arizonensis* Vockeroth

Fig. 231

Paragus (*Paragus*) *arizonensis* Vockeroth, 1986*a*:191.

Length. 5.5–6.6 mm.

Male. Hind tibia yellow to orange or with faint brownish rings; hind tarsus yellow-orange.

Abdomen with tergites mostly red-orange, only lateral one-third or one-quarter of tergite 1 and anterolateral angles of tergite 2 black. Tergites 2–5 in some specimens each with narrow distinct or obscure arcuate gray pruinose band. Terminalia in Fig. 231; surstylus in dorsal view moderately broad on basal one-third, in lateral view with low rounded anteroventral lobe. Hypandrium deeply excavated laterally, below base of paramere with strong lateral lobe having irregularly serrate margin and several short teeth on posterior surface; lateral arm of hypandrium, near base of lobe, also with short teeth; aedeagal guide about four times as long as its width at mid length, slightly narrowed from base to apex, with lateral portions distinctly denticulate. Paramere bent through about 90°, slightly narrowed preapically, with apex minutely serrate. Lateral lobe of aedeagus strongly narrowed basally, with small tooth at posterodorsal angle, with rather deep broad posteroventral emargination, with strong acute tooth at each of posteroventral and anteroventral angles. Posterior end of aedeagal apodeme with two short acute ventral teeth.

Female. Unrecognized.

Distribution. Utah, New Mexico, and Arizona. Ariz., IV–VI, VIII.

Specimens identified. United States, 11 ♂♂.

Paragus (*Paragus*) *bispinosus* Vockeroth

Fig. 232; Map 51

Paragus bicolor: Metcalf 1921:pl. X [not *P. bicolor* (Fabricius), 1794].
Paragus (*Paragus*) *bispinosus* Vockeroth, 1986*a*:192.

Length. 5.5–6.6 mm.

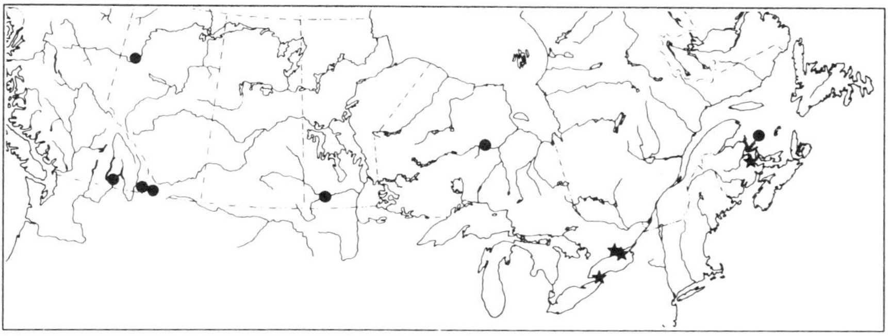

Map 51. Collection localities for *Paragus bispinosus* Vockeroth (★) and for *P. cooverti* Vockeroth (●).

Male. Hind tibia with brown to black band on apical half; hind leg with first tarsomere brown to black above.

Abdomen usually with tergite 2 red-orange medially or posteriorly and with tergites 3–5 entirely red-orange; in some specimens tergites 3–5 with dark areas or black with only posterior margin of tergite 4 and all tergite 5 dark red. Terminalia in Fig. 232; surstylus in posterior view strongly curved, evenly narrowed from base to apex, in lateral view with strong subtriangular anteroventral lobe. Hypandrium shallowly excavated laterally, margin of excavation near base of aedeagal guide with single large spine or process bearing several spines of various sizes; margin near base of paramere with long strong spine extending well beyond base of paramere; aedeagal guide as wide as long, with posterolateral angles denticulate. Paramere slender, sharply bent through about 90°, with strong posterolateral tooth at bend, and with dorsal margin beyond bend coarsely and irregularly serrate. Lateral lobe of aedeagus with distinct broad posteroventral emargination and with strong tooth at each posterior angle.

Female. Unrecognized.

Distribution. Eastern Canada (Map 51), south to Colorado, Missouri, and New Hampshire. Ont., VII.

Specimens identified. Canada, 10 ♂♂; United States, 31 ♂♂.

Paragus (*Paragus*) *cooverti* Vockeroth

Fig. 233; Map 51

Paragus (*Paragus*) *cooverti* Vockeroth, 1986*a*:192.

Length. 5.0–6.4 mm.

Male. Hind tibia with broad black ring on most of apical half to three-fifths; hind leg with first tarsomere and usually also next two or three tarsomeres black above.

Tergites usually entirely black, rarely with narrow lateral and posterior margins of tergite 5 dark reddish. Terminalia in Fig. 233; surstylus in dorsal view narrowed from about one-third its length but moderately stout and scarcely curved, in lateral view without trace of anteroventral lobe. Hypandrium moderately excavated laterally, without marginal spines; aedeagal guide twice as long as its width at mid length (in dry specimens often with margins inrolled and appearing much narrower), slightly narrowed from base to mid length and then broadened to apex, with surface smooth. Paramere curved through about 45°, in some specimens slightly narrowed at mid length, irregularly and variably broadened apically, without marginal serrations. Lateral lobe of aedeagus with barely perceptible posteroventral emargination and with only one or two minute teeth; posterodorsal and posteroventral angles broadly rounded and without teeth.

Female. Unrecognized.

Distribution. Canada (Map 51), south to Oregon, Utah, and New York. B.C. VII; Ont., VII.

Specimens identified. Canada, 12 ♂♂; United States, 21 ♂♂.

Paragus (Paragus) longistylus Vockeroth

Fig. 234

Paragus (*Paragus*) *longistylus* Vockeroth, 1986*a*:193.

Length. 6.2–7.2 mm.

Male. Hind tibia yellow to orange or with faint brownish ring; hind tarsus yellow orange.

Abdomen with tergites mostly red-orange to dark red, with tergite 1 and anterolateral angles of tergite 2 black. Terminalia in Fig. 234; surstylus long, in dorsal view moderately curved, regularly narrowed from just beyond base to mid length, then slightly narrowed to apex, in lateral view without trace of anteroventral lobe. Hypandrium moderately excavated laterally, without marginal spines but with small obscure projection well below base of paramere; aedeagal guide as long as its width at mid length, strongly narrowed only near apex, the posterolateral angles strongly denticulate. Paramere bent through about 90°, slightly narrowed beyond bend, with surface and margins smooth. Lateral lobe of aedeagus with deep posteroventral emargination, with very strong tooth at each of posterodorsal and posteroventral angles. Posterior end of aedeagal apodeme with long slender acute curved ventral tooth.

Female. Unrecognized.

Distribution. Idaho, Nevada, and California. Calif., VI, VIII, IX.

Specimens identified. United States, 14 ♂♂.

Paragus (Paragus) variabilis Vockeroth

Fig. 235; Map 52

Paragus quadrifasciatus: Metcalf 1921:pl. X (not *P. quadrifasciatus* Meigen, 1822).

Paragus (*Paragus*) *variabilis* Vockeroth, 1986*a*:193.

Length. 4.8–6.5 mm.

Male. Hind tibia orange or with brown to black ring on apical one-third to three-fifths; hind leg with first tarsomere and in some specimens also next two tarsomeres brown to black above.

Abdomen with tergites very variable in color. Tergite 2 rarely entirely black, usually with broad reddish band at mid length or extensively reddish posteromedially; tergites 3–5 usually red orange; in some specimens tergites 4 and 5 and less often 3 extensively darkened. Terminalia in Fig. 235; surstylus variable in shape, in dorsal view parallel-sided on about basal one-third then narrowed toward apex or gradually narrowed from just beyond base, in lateral view usually with distinct but short rounded anteroventral lobe but in some specimens only with obscure lobe. Hypandrium deeply excavated laterally, with single short to long marginal spine; spine in some specimens slightly serrated and extending beyond base of paramere; aedeagal guide 1–1.5 times as long as broad with lateral portions strongly denticulate. Paramere bent through about 90°, of equal width throughout or very slightly narrowed or broadened beyond the bend, with fine dorsal and apical serrations. Lateral lobe of aedeagus with distinct broad posteroventral emargination; posterodorsal angle with large tooth; posteroventral angle usually with strong rather blunt tooth but in some specimens rounded, with small tooth or untoothed. Posterior end of aedeagal apodeme with short tooth or strong curved hook.

Female. Unrecognized.

Distribution. Canada (Map 52), south to California and Colorado (up to 3538 m). B.C., V–VIII; Que., VIII.

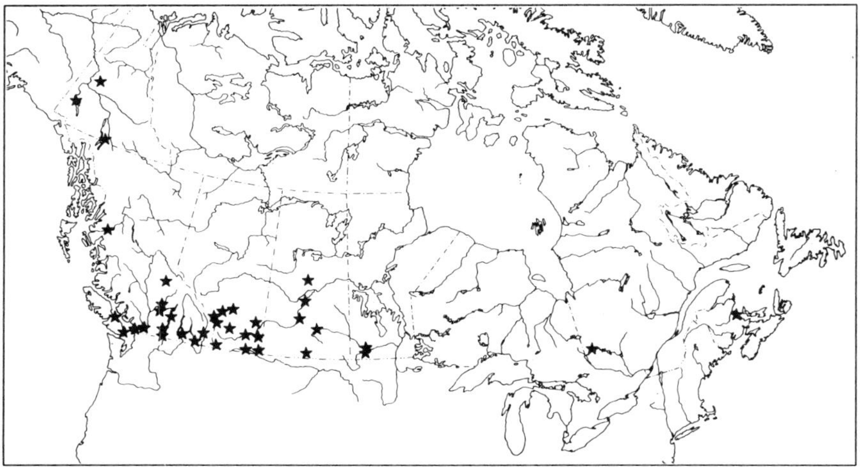

Map 52. Collection localities for *Paragus variabilis* Vockeroth.

Specimens identified. Canada, 88 ♂♂; United States, 362 ♂♂.

Discussion. This species shows great variation in abdominal color, in the shape of the surstylus and lateral lobe of the aedeagus, and in the length and strength of the marginal spine of the hypandrium. I have been unable to find any correlation between color or structural characters that would indicate distinct species within a complex.

Genus *Parasyrphus* Matsumura

Diagnosis. Species small to medium sized, usually robust, usually with posteriorly rounded yellow spots or straight or posteriorly emarginate yellow bands on tergites, rarely with yellow abdominal markings reduced or absent. Length 5.6–11.3 mm.

Description. Eye with minute scattered hairs, with short but distinct hairs, or with long dense hairs. Frons in some specimens swollen and very broad. Face with tubercle usually distinct and rather prominent, rarely with tubercle low and broad; face usually yellow or with distinct black median stripe, in some specimens generally darkened or entirely dark brown to black.

Scutum black, subshining. Scutellum dull yellow, commonly darkened laterally. Ventral scutellar fringe complete. Pleura black, weakly pruinose. Anterior anepisternum with many long fine erect hairs. Upper and lower katepisternal hair patches separated or narrowly joined posteriorly. Meron, metapleuron, and metasternum bare. Wing membrane entirely trichose or with small bare areas near base. Hind coxa with posteromedial apical tuft of several strong hairs (Fig. 46).

Abdomen oval, usually unmargined, with very weak margin only in *P. nigritarsis*. Tergite 2 usually with pair of spots; tergites 3 and 4 usually each with pair of yellow spots or an entire or posteriorly deeply emarginate yellow band; yellow spots or bands in some specimens reduced or absent; tergite 4 and usually tergite 5 with posterior margin narrowly yellow. Sternites yellow, usually with faint or distinct dark median spots, stripes, or broad triangles or with black subapical bands.

Distribution. Only in Holarctic region. In general, most northerly genus of Syrphinae. In North America two species north of 80°, none as far south as Gulf States or Mexico; those extending southward in eastern or western United States almost entirely at high altitudes. Ten species in North America and about 10 in Palaearctic region.

Biology. *P. nigritarsis* (Zetterstedt) is a predator of eggs and larvae of Chrysomelidae (see account under that species). Goeldlin (1974) reared larvae of the European species *P. lineolus* (Zetterstedt) and *P. vittiger* (Zetterstedt) on *Aphis fabae* from eggs laid in the laboratory on the aphid colonies. He also found larvae of the latter species in nature among aphids on cassis. Rotheray and Gilbert (1989) gave several aphid host records (mostly on conifers) for *P. vittiger*. Goeldlin found that most larvae of *P. lineolus* pupated about a week after reaching their full size but that two larvae pupated after 51 days; he concluded that the species was oligovoltine with a facultative larval diapause.

Although Goeldlin found larvae of one species among aphids and reared two species on aphids, it seems unlikely that aphids, or at least those in exposed colonies, are the normal larval hosts. Several species of the genus are moderately common in western Europe and these or others are found in eastern or western Canada. Moreover, *P. tarsatus* (Zetterstedt) is one of the most abundant boreal Syrphinae in North America. If larvae of these species feed on exposed aphids, records would likely be more numerous. Láska and Starý (1980) gave more than 220 syrphine–aphid host records from Czechoslovakia but none of these were for *Parasyrphus*.

Discussion. Larger specimens of *Parasyrphus* much resemble those of several species of *Syrphus* but may be easily distinguished by the haired anterior anepisternum, the lack of hairs on the upper surface of the lower squama, and the unmargined or only very weakly margined abdomen. All the species recognized here, except *P. tarsatus* and *P. groenlandicus*, show distinct differences in the male terminalia; but these are mentioned and illustrated only when the species are otherwise difficult to distinguish.

Key to New World species of *Parasyrphus*

1. Tergites entirely black or with pairs of yellow spots (Figs. 149, 151, 153) 2

   Tergite 4 and usually tergite 3 with entire yellow band (Figs. 150, 152) 10

2. Tergite 2 with pair of yellow spots; tergites 3 and 4 black . . . . melanic females of several species

   Tergites entirely black or tergites 3 and 4 with yellow to orange spots 3

3. Male (eyes meeting on frons or nearly so) 4

   Female (eyes widely separated on frons) 7

4. Frons large, swollen, very broad, with long dense hairs; eye angle about 130° (Fig. 15). Tergites 2–4 with small orange spots (Fig. 151) or some or all of tergites 2–4 entirely black. Arctic, including Greenland . . . . . ***groenlandicus* (Nielsen)**

Frons smaller, not swollen, narrower, with shorter and less dense hairs; eye angle at most 100° (Fig. 16). Tergites 2–4 usually with large yellow to orange spots (Fig. 153); in some arctic specimens spots very reduced or absent 5

5. Face yellow medially, or with brownish tubercle or generally darkened, without distinct black median stripe. Widespread, including Arctic and Greenland ... ***tarsatus* (Zetterstedt)**

Face with distinct black median stripe extending from lower margin to well above tubercle 6

6. Eye with short sparse hairs separated from one another by more than their length. Scutum entirely yellow haired or with some black hairs; pleura with at most some black hairs. Paramere (Fig. 248) dorsoapically with one strong tooth and rounded lobe. Western Canada .... ***currani* (Fluke)**

Eye with rather long dense hairs separated by less than their length. Scutum and pleura with many black hairs. Paramere (Fig. 251) dorsoapically with two strong teeth, without rounded lobe. Western Canada .... ***macularis* (Zetterstedt)**

7. Face yellow medially or with brownish tubercle or generally darkened, without distinct black median stripe 8

Face with distinct black median stripe extending from lower margin to well above tubercle 9

8. Tergites 2–4 with moderately large distinct yellow orange spots (as in Fig. 153). Northern limit of trees southward, Greenland .... ***tarsatus* (Zetterstedt)** (in part)

Tergites 2–4 with small obscure orange spots or some or all these tergites entirely black. Northern limit of trees northward .... ***groenlandicus* (Nielsen)**
.... ***tarsatus* (Zetterstedt)** (in part)

9. Eye with short sparse hairs separated from one another by more than their length. Scutum entirely yellow haired. Western Canada .... ***currani* (Fluke)**

Eye with rather long dense hairs separated from one another by less than their length. Scutum in some specimens with some black hairs. Western Canada .... ***macularis* (Zetterstedt)**

10. Face yellow, without black median stripe, with at most lower margin black. Boreal regions .... ***nigritarsis* (Zetterstedt)**

Face with black median stripe or mostly or entirely black 11

11. Male (eyes meeting on frons) 12

Female (eyes widely separated on frons) 16

12. Wing membrane partly bare, cell bm with at least narrow bare median stripe on most of basal one-third. Median black facial stripe distinct, less than one-third as wide as face. Widespread .... ***genualis* (Williston)**

Wing membrane entirely trichose. Median black facial stripe variable, indistinct laterally in some specimens or more than one-third as wide as face 13

13. Facial stripe at least four-sevenths as wide as face, usually indistinct laterally so face is entirely dark brown to black. Fore and mid tarsi yellow. Frons very lightly pruinose above; border between upper pruinose and lower shining areas indistinct. Boreal and western Canada .... ***insolitus* (Osburn)**

Facial stripe usually less than half as wide as face; face distinctly yellow laterally; if stripe wider or face obscurely darkened laterally then fore and mid tarsi dark above. Frons usually densely pruinose above, with border between upper pruinose and lower shining areas distinct 14

14. Frons narrowly but distinctly pruinose along eye margins; about anteromedian four-fifths of frons shining. Yellow bands of tergites 3 and 4 deeply emarginate posteriorly or one band narrowly divided medially in some specimens. Eastern Canada .... ***semiinterruptus* (Fluke)**

Frons broadly although in some specimens rather obscurely pruinose along eye margins; about anteromedian one-third of frons shining. Yellow bands of tergites 3 and 4 shallowly or deeply emarginate posteriorly 15

15. Sternites without distinct markings or with obscure or distinct longitudinal or strongly triangular dark markings. Hind femur in some specimens broadly yellow basally. Paramere (Fig. 252*c*) dorsoapically with one strong tooth and rounded lobe; aedeagal base (Fig. 252*d*) with two acute teeth but without strong ventrolateral prong. Boreal and western Canada .... ***relictus* (Zetterstedt)**

Sternites with at least moderately distinct black bands at most slightly produced forward medially. Hind femur black basally. Paramere (Fig. 247*a*) dorsoapically with two strong teeth, without rounded lobe; aedeagal base (Fig. 247*b*) with strong ventrolateral prong. Widespread ....
.... new species (**Thompson** *in litt.*)

16. Wing with cell bm broadly bare anteriorly on at least basal half. Sternites with distinct subapical black bands. Widespread .... ***genualis* (Williston)**

Wing membrane entirely trichose. Sternites with or without distinct black bands 17

17. Sternites with distinct subapical black bands .... ***semiinterruptus* (Fluke)** .... new species **(Thompson *in litt.*)**

Sternites with faint or distinct broad dark triangles, faint longitudinal median dark marks, or entirely yellow (although in some specimens darkened by postmortem changes) ... 18

18. Fore and mid tarsi entirely yellow. Fore and mid femora yellow; hind femur entirely yellow or with dark preapical ring. Black facial stripe from one-third to three-fifths as wide as face, extending to upper facial margin. Sternites yellow. Boreal and western Canada .... ***insolitus* Osburn**

Fore and mid tarsi brown to blackish above, at least distinctly brownish above toward apex. Fore and mid femur commonly black at base; hind femur mostly black to entirely yellow. Black facial stripe not more than one-quarter as wide as face, not extending to its upper margin. Sternites usually with dark median longitudinal spots or median triangles, entirely yellow in some specimens. Boreal and western Canada .... ***relictus* (Zetterstedt)**

Clé des espèces du Nouveau Monde de *Parasyrphus*

1. Tergites tout noirs ou maculés de paires de taches jaunes (fig. 149, 151, 153) 2

Tergite 4 et habituellement tergite 3 ornés d'une bande jaune complète (fig. 150, 152) 10

2. Tergite 2 portant une paire de taches jaunes; tergites 3 et 4 noirs .... femelles mélaniques de plusieurs espèces

Tergites tout noirs ou tergites 3 et 4 ornés de taches jaune à orange 3

3. Mâle (yeux contigus ou presque sur le front) 4

Femelle (yeux largement séparés sur le front) 7

4. Front renflé et très large, garni de poils longs et denses; angle oculaire d'environ 130° (fig. 15). Tergites 2–4 marqués de petites taches orange (fig. 151), ou tergites 2–4 entièrement ou

partiellement noirs. Arctique, y compris le Groënland .... ***groenlandicus* (Nielsen)**

Front plus petit et plus étroit, non renflé et portant des poils plus courts et moins denses; angle oculaire d'au plus 100° (fig. 16). Tergites 2–4 ornés habituellement de grosses taches jaune à orange (fig. 153); chez certains spécimens arctiques, il n'y a pas de taches ou elles sont très réduites 5

5. Face jaune dans la région médiane, ou ornée d'un tubercule brunâtre ou généralement foncée, sans bande médiane noire distincte. Espèce répandue, y compris dans l'Arctique et au Groënland .... ***tarsatus* (Zetterstedt)**

Face ornée d'une rayure médiane noire distincte, qui s'étend du bord inférieur jusqu'à bien au-delà du tubercule 6

6. Oeil orné de poils courts et clairsemés, séparés les uns des autres d'une distance supérieure à leur propre longueur. Poils du scutum entièrement jaunes ou quelques poils noirs; pleurites pourvus au plus de quelques poils noirs. Paramère (fig. 248) pourvu, dans la région dorso-apicale, d'une forte dent et d'un lobe arrondi. Ouest du Canada .... ***currani* (Fluke)**

Oeil orné de poils denses et assez longs, séparés les uns des autres d'une distance inférieure à leur propre longueur. Scutum et pleurites pourvus de nombreux poils noirs. Paramère (fig. 251) pourvu, dans la région dorso-apicale, de deux fortes dents; sans lobe arrondi. Ouest du Canada .... ***macularis* (Zetterstedt)**

7. Face jaune au milieu, ou pourvue d'un tubercule brunâtre ou généralement foncée; sans bande médiane noire distincte 8

Face soulignée d'une rayure médiane noire distincte qui s'étend du bord inférieur jusqu'à bien au-delà du tubercule 9

8. Tergites 2–4 maculés de taches jaune orangé assez grosses et distinctes (comme dans la fig. 153). De la limite forestière septentrionale vers le sud, Groënland .... ***tarsatus* (Zetterstedt)** (partim)

Tergites 2–4 ornés de petites taches orange floues ou quelques-uns ou la totalité de ces tergites tout noirs. De la limite forestière septentrionale vers le nord .... ***groenlandicus* (Nielsen)**
.... ***tarsatus* (Zetterstedt)** (partim)

9. Oeil orné de poils courts et clairsemés, séparés les uns des autres d'une distance supérieure à leur longueur. Poils du scutum tout jaunes. Ouest du Canada .... ***currani* (Fluke)**

Oeil orné de poils denses et assez longs, séparés les uns des autres d'une distance inférieure à leur longueur. Chez certains spécimens, scutum garni de quelques poils noirs. Ouest du Canada .... ***macularis* (Zetterstedt)**

10. Face jaune, sans rayure médiane noire, pourvue tout au plus d'un liséré inférieur noir. Régions boréales .... ***nigritarsis* (Zetterstedt)**

Face ornée d'une rayure médiane noire, ou largement ou entièrement noire 11

11. Mâle (yeux contigus sur le front) 12

Femelle (yeux largement séparés sur le front) 16

12. Membrane alaire partiellement glabre; cellule bm portant tout au moins une étroite bande médiane glabre sur la majeure partie du tiers basilaire. Présence d'une rayure faciale médiane noire, couvrant au plus le tiers de la largeur de la face. Espèce répandue .... ***genualis* (Williston)**

Membrane alaire entièrement velue. Rayure faciale médiane noire variable, peu visible sur les côtés chez certains spécimens ou d'une largeur supérieure au tiers de la figure 13

13. Rayure faciale équivalant au moins aux quatre septièmes de la largeur de la face, habituellement peu visible sur les côtés, de sorte que la face apparaît entièrement brun foncé ou noire. Tarses antérieur et médian jaunes. Front très légèrement pruineux sur le dessus; aucune séparation nette entre la région pruineuse supérieure et la zone lustrée inférieure. Régions boréales et ouest du Canada .... ***insolitus* (Osburn)**

Rayure faciale habituellement deux fois moins large que la face; face soulignée sur les côtés de zones jaunes distinctes. Si la rayure est plus large ou si la face est plus foncée sur les côtés, les tarses antérieur et médian sont alors foncés sur le dessus. Front habituellement très pruineux sur le dessus, une marge distincte séparant la zone supérieure pruineuse de la zone inférieure lustrée 14

14. Front pourvu d'une bande pruineuse étroite mais apparente bordant l'œil; front lustré sur environ les quatre cinquièmes de la portion antéro-médiane. Bandes jaunes des tergites 3 et 4 fortement échancrées vers la partie postérieure ou, chez certains spécimens, bande étroitement séparée au milieu. Est du Canada .... ***semiinterruptus* (Fluke)**

Front largement pruineux autour de l'œil quoique, chez certains spécimens, la pruinosité soit assez floue; front lustré sur environ le tiers de la portion antéro-médiane. Bandes jaunes des tergites 3 et 4 légèrement ou fortement échancrées vers la partie postérieure 15

15. Sternites dépourvus de marques distinctes ou portant des marques foncées floues ou nettement longitudinales ou triangulaires. Chez certains spécimens, le fémur postérieur est largement jaune dans sa partie basilaire. Paramère (fig. 252*c*) pourvu, dans sa portion dorso-apicale, d'une forte dent et d'un lobe arrondi; base de l'édéage (fig. 252*d*) portant deux dents acérées, mais dépourvue d'une forte pointe sur la face ventro-latérale. Régions boréales et ouest du Canada .... ***relictus* (Zetterstedt)**

Sternites ornés au moins de bandes noires assez distinctes, qui s'avancent tout au plus légèrement vers l'avant au milieu. Portion basilaire du fémur postérieur noire. Portion dorso-apicale du paramère (fig. 247*a*) portant deux fortes dents; pas de lobe arrondi; base de l'édéage (fig. 247*b*) pourvue d'une forte pointe sur la portion ventro-latérale. Espèce répandue .... nouvelle espèce **(Thompson *in litt.*)**

16. Aile pourvue d'une cellule bm largement glabre sur la portion antérieure, au moins dans la moitié basilaire. Sternites ornés de bandes noires subapicales distinctes. Espèce répandue ... . ***genualis* (Williston)**

Aile pourvue d'une membrane entièrement velue. Sternites ornés ou non de bandes noires distinctes 17

17. Sternites ornés de bandes noires subapicales distinctes .... ***semiinterruptus* (Fluke)**
.... nouvelle espèce **(Thompson *in litt.*)**

Sternites ornés de larges triangles foncés, flous ou distincts, de faibles marques longitudinales médianes foncées, ou entièrement jaunes (quoique, chez certains spécimens, noircis par les changements en post-mortem) 18

18. Tarses antérieur et médian tout jaunes. Fémurs antérieur et médian jaunes; fémur postérieur tout jaune ou souligné d'un anneau pré-apical foncé. Rayure faciale noire, couvrant du tiers aux trois cinquièmes de la largeur de la face et s'étendant jusqu'au bord facial supérieur. Sternites jaunes. Régions boréales et ouest du Canada .... ***insolitus* Osburn**

Tarses antérieur et médian bruns à noirâtres sur le dessus, tout au moins nettement bruns vers l'apex. Fémurs antérieur et médian généralement noirs à la base; fémur postérieur variant de largement noir à tout jaune. Bande faciale noire couvrant au plus le quart de la largeur de la face et n'atteignant pas le bord supérieur. Sternites ornés habituellement de taches médianes foncées, longitudinales ou triangulaires, tout jaunes chez certains spécimens. Régions boréales et ouest du Canada .... ***relictus* (Zetterstedt)**

Parasyrphus currani (Fluke)

Figs. 149, 248; Map 53

Epistrophe currani Fluke, 1935:29.

Length. 7.7–11.3 mm.

Male. Eye with short sparse hairs separated from one another by more than their length. Frons black, mostly densely pale pruinose, shining above antennae, yellowish immediately above antennae. Face dull yellow, densely pruinose, with ventral margin narrowly black, with rather well-defined black median stripe from one-quarter to one-third width of face at level of tubercle. Antenna black.

Scutum subshining, with fine slightly olivaceous pruinosity on disc and denser yellowish pruinosity laterally, with only yellow hairs or with mixed black and yellow hairs on disc. Scutellum dull yellow with only black hairs or with some yellow hairs laterally. Pleura weakly gray pruinose, with only yellow hairs or with some black hairs on upper half. Wing membrane entirely trichose. Legs dark brown to black with broad apices of fore and mid femora, of hind femur in some specimens, of fore and mid tibiae, and of hind tibia in some specimens dull yellow.

Abdomen in Fig. 149. Tergite 2 with pair of large obliquely subquadrate yellow spots not reaching margins; tergites 3 and 4 each with pair of large subbasal yellow spots with straight anterior margins and curved posterior margins; spots in some specimens extending to lateral margins; tergite 5 yellow with large black central triangle. Sternites entirely yellow or with large obscure black median triangles. Paramere (Fig. 248) with one stout tooth and rounded lobe at dorsoapical angle.

Female. Frons black, yellow immediately above antenna, mostly olivaceous pruinose, with darker pruinosity above. Face with well-defined black median stripe not more than one-quarter as wide as face. Hairs of scutum and pleura entirely yellow; hairs of scutellum almost entirely black. Fore and mid femora yellow on apical half or slightly more; hind femur almost entirely black; hind tibia dark, or yellowish on basal one-third. Yellow spots of tergites 2–4 extending broadly to margins; tergite 5 with anterolateral angles yellow. Sternites in some specimens with obscure or distinct dark median triangles.

Distribution. Alaska, British Columbia (Map 53). B.C., VI–VIII.

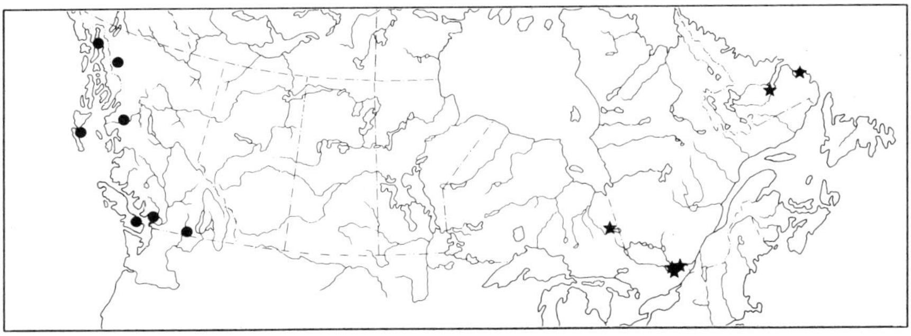

Map 53. Collection localities for *Parasyrphus currani* (Fluke) (●) and for *P. semiinterruptus* (Fluke) (★).

Specimens identified. Alaska, 1 ♂; Canada, 5 ♀♀.

Parasyrphus genualis (Williston)

Figs. 150, 249; Map 54

Syrphus genualis Williston, 1887:86.
Syrphus johnsoni Curran, 1924*c*:79.

Length. 5.9–8.9 mm.

Male. Eye with extremely short widely scattered hairs. Frons black, narrowly yellow above antennae, obscurely pruinose on about posterior half, shining on about anterior half. Face dull yellow, moderately pruinose, with black lower margin and well-defined dark brown to black median stripe about two-sevenths as wide as face. Antenna blackish brown above, dull orange below.

Scutum subshining, with yellow hairs. Scutellum dull yellow, narrowly darkened laterally in some specimens, with black hairs. Pleura weakly pruinose, with yellow or white hairs. Wing usually with cell c bare at extreme base, with distinct narrow to broad bare stripe on middle of at least basal one-third of cell bm. Legs with coxae and trochanters black; fore and mid femora black on basal one-third to half, otherwise yellow; hind femur black with extreme apex yellow; fore and mid tibiae yellow; hind tibia black with about basal one-third yellowish;

fore and mid tarsi yellowish, slightly brownish toward apex in some specimens; hind tarsus brown.

Abdomen in Fig. 150. Tergite 2 with pair of broad rounded yellow spots extending narrowly or broadly to lateral margin; tergites 3 and 4 each with subbasal yellow band about half as long as tergite, with each band slightly to strongly emarginate posteriorly and usually extending narrowly to lateral margins; tergite 5 black with anterolateral angles yellow. Sternites shining yellow, with narrow to broad, subapical black bands, obscure in some specimens. Paramere (Fig. 249*c*) with small closely appressed tooth lying beside rounded lobe at dorsoapical angle, and with basoventral hook. Distal segment of aedeagus (Fig. 249*e*) very broad, subcircular in posterior view, considerably wider than basal segment; latter without ventrolateral prong (Fig. 249*d*).

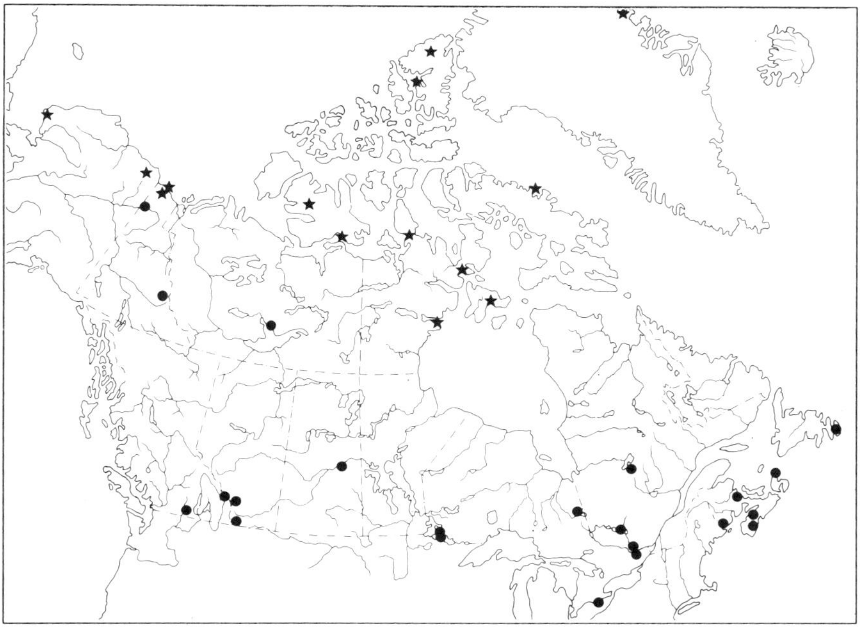

Map 54. Collection localities for *Parasyrphus genualis* (Williston) (●) and for *P. groenlandicus* (Nielsen) (★).

Female. Frons black, yellow immediately above antenna, lower one-third shining except on narrow lateral margins, middle one-third with two large pruinose lateral spots narrowly separated medially, upper one-third subshining. Wing usually with membrane more extensively bare, cell bm bare anteriorly on up to basal one-third. Fore and mid femora yellow; hind femur yellow on basal one-quarter to one-third. Yellow bands of tergites 3 and 4 shorter, reaching lateral margins in almost their full width. Sternites with black bands distinct.

Distribution. Canada (Map 54), south to Colorado and Connecticut.* B.C., VIII; Ont., Que., V–VIII.

Specimens identified. Canada, 28 ♂♂, 52 ♀♀; United States, 8 ♂♂, 9 ♀♀.

Parasyrphus groenlandicus (Nielsen)

Figs. 15, 151; Map 54

Catabomba groenlandica Nielsen, 1910:61.
Stenosyrphus bulbosus Fluke, 1954:8.

Length. 7.2–7.9 mm.

Male. Eye with long dense hairs. Frons very broad, rather swollen, black with narrow yellow anterior margin, with dense black hairs; eye angle about 130°; eyes contiguous dorsally or separated by at most width of one ommatidium (Fig. 15). Face very broad, with rather low tubercle, dull yellow with broad black ventral margin, darkened laterally in some specimens, with long dense black hairs. Antenna black.

Scutum subshining; scutellum dull yellow to dark brown with blackish posterior margin; pleura subshining; thoracic hairs entirely black or partly yellow. Wing membrane entirely trichose. Legs black with narrow apices of femora, most of fore and mid tibiae, and about basal one-third and apex of hind tibia in some specimens dull orange.

Abdomen in Fig. 151. Tergite 2 black or with pair of small broadly oval dull yellow-orange spots; tergites 3 and 4 each with pair of dull yellow-orange spots with straight anterior margins and rounded posterior margins; spots usually reaching margins anterolaterally; tergite 5 black. Sternites black.

Female. Frons black, narrowly yellow immediately above antennae, with slight to dense grayish or slightly olivaceous pruinosity weaker along lower margin and on about upper one-third. Face very wide, at most recessive point between tubercle and oral margin about three-fifths as wide as head, commonly extensively brownish or blackish. Yellow-orange spots of tergites 2 absent, spots of tergites 3 and 4 present in some specimens but very short and broad, usually absent.

Distribution. Arctic Alaska, Canada, and Greenland (Map 54). N.W.T., VI, VII.

Specimens identified. Alaska, 3 ♂♂; Canada, 52 ♂♂, 115 ♀♀; Greenland, 2 ♂♂, 2 ♀♀.

Discussion. Males of *P. groenlandicus* are definitely separable from *P. tarsatus* (Zetterstedt), but distinguishing between the females is, I believe, not always possible (see discussion following *P. tarsatus*).

Parasyrphus insolitus (Osburn)

Fig. 250; Map 55

Syrphus insolitus Osburn, 1908:5.

Length. 5.6–8.1 mm.

Male. Similar to male of *P. genualis*, differing as follows: face with tubercle lower and broader; median facial stripe at least four-sevenths as wide as face; face usually entirely dark brown to black.

Scutum and pleura in some specimens with some black hairs. Wing membrane entirely trichose. Fore and mid tarsi yellow.

Sternites with dark marks very obscure; mark on sternite 2 usually subtriangular. Paramere (Fig. 250*c*) with larger basoventral hook; distal segment of aedeagus (Fig. 250*e*) with margins less-rounded, only slightly broader than basal segment.

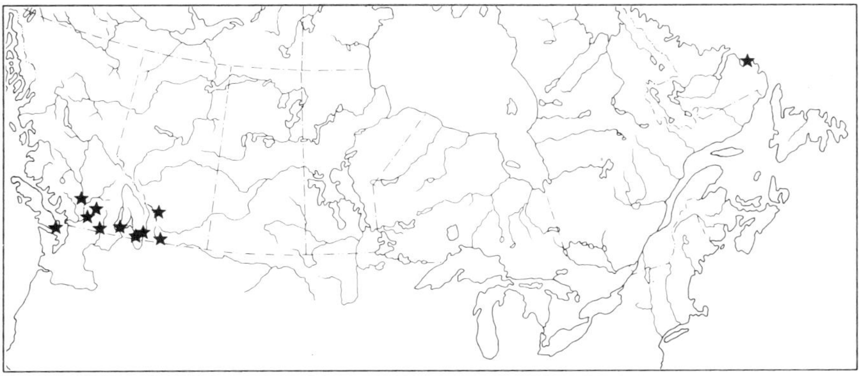

Map 55. Collection localities for *Parasyrphus insolitus* (Osburn).

Female. Frons with pruinose spots rather indistinct, separated medially. Lower margin of face broadly black; facial stripe from one-third to three-fifths as wide as face, face clearly yellow laterally, not entirely darkened. Scutum and pleura with only yellow hairs. Fore and mid femora yellow; hind femur yellow, usually with dark preapical ring. Fore and mid tarsi yellow. Sternites yellow (darkened in some specimens by postmortem changes).

Distribution. Alaska,* western and northern Canada (Map 55), south to Oregon* and Montana.* B.C., V; Labrador, VI.

Specimens identified. Canada, 24 ♂♂, 44 ♀♀.

Parasyrphus macularis (Zetterstedt)

Fig. 251; Map 56

Scaeva macularis Zetterstedt, 1843:730.
Stenosyrphus nigrifacies Curran, 1923*c*:62.

Length. 7.9–10.2 mm.

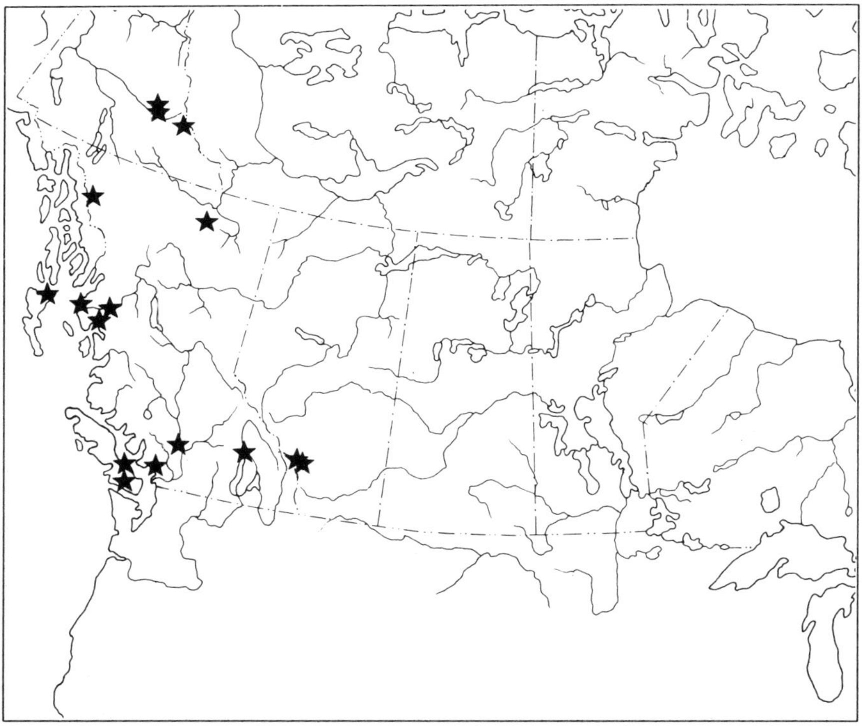

Map 56. Collection localities for *Parasyrphus macularis* (Zetterstedt).

Male. Similar to male of *P. currani*, differing as follows: eye with long dense hairs separated from one another by less than their length. Face darker, broadly black below, with broad median stripe at least half as wide as face, commonly not clearly delimited from dark sides of face. Antenna in some specimens with first flagellomere obscurely orange below.

Scutum without yellow pruinosity laterally, mostly black-haired, with some yellow hairs laterally but with some black hairs on notopleuron and postcallus. Scutellum narrowly darkened laterally, entirely black-haired. Pleura with many black hairs near upper margin. Fore and mid tibia brownish on apical half; hind tibia dark with base narrowly yellowish.

Tergite 5 with anterolateral angles yellow. Sternites with large obscure or distinct blackish triangles or mostly dark. Paramere (Fig. 251) with one stout tooth and one slender tooth at dorsoapical angle.

Female. Frons as in *P. currani* but darker anteriorly. Scutum usually with some black hairs sublaterally. Abdominal markings as in *P. currani*.

Distribution. Alaska,* western Canada (Map 56), south to Oregon;* Europe. B.C., VI–VIII.

Specimens identified. Canada, 44 ♂♂, 53 ♀♀; United States, 2 ♂♂; Europe, 5 ♂♂, 4 ♀♀.

Parasyrphus nigritarsis (Zetterstedt)

Map 57

Scaeva nigritarsis Zetterstedt, 1843:710.
Stenosyrphus imperialis Curran, 1925:100.

Length. 9.0–11.3 mm.

Male. Eye with extremely short sparse hairs. Frons blackish with dense pale gray pruinosity on upper one- to two-thirds, shining yellow below. Face yellow, with lower margin narrowly black except medially in some specimens. Antenna orange with first flagellomere brownish above.

Scutum weakly olivaceous pruinose, with yellow hairs. Scutellum pale yellow, with mixed black and white hairs. Pleura with weak grayish pruinosity, with yellow hairs. Wing membrane entirely trichose. Legs with coxae and trochanters black; fore and mid femora black on basal one-fifth to half, otherwise yellow; hind femur black on basal one-fifth to two-thirds, otherwise yellow; tibiae yellow; fore and mid tarsi varying from yellow to black above; hind tarsus varying from brown to black above.

Abdomen usually with weak lateral margin on tergite 4; tergite 2 with large oblique posteriorly rounded yellow spots extending broadly anterolaterally to lateral margin; tergites 3 and 4 each with broad entire subbasal yellow band strongly narrowed laterally and strongly emarginate posteriorly or rarely narrowly divided medially; tergite 5 yellow, usually with small to moderately large black central triangle or obscure spot. Sternites yellow, with obscure dark central spot in some specimens.

Female. Frons black, narrowly shining yellow anteriorly, with entire band of olivaceous pruinosity on about three-quarters of its length. Hind coxa yellow in some specimens; femora yellow or narrowly black basally; first tarsomere of hind leg yellow above in some specimens. Tergite 5 with large black triangle reaching lateral margins in some specimens. Abdomen entirely narrowly yellow laterally in some specimens.

Distribution. Alaska, Canada (Map 57), south to Colorado; Europe; Asia. B.C., V–VII; Ont., Que., V–VII.

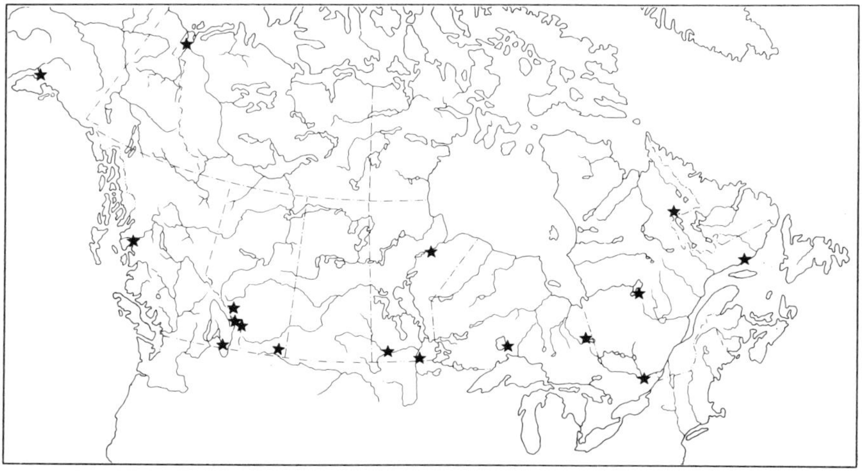

Map 57. Collection localities for *Parasyrphus nigritarsis* (Zetterstedt).

Specimens identified. Alaska, 1 ♂, 3 ♀♀; Canada, 18 ♂♂, 13 ♀♀; United States, 1 ♀; Europe, 1 ♂, 3 ♀♀; Siberia, 2 ♂♂.

Biology. Schneider (1953) in Switzerland found eggs of this species among egg masses of the Chrysomelid beetle *Melasoma vigintipunctata*. Young larvae fed on the beetle eggs and then on beetle larvae. Full-grown larvae stopped feeding in late May, overwintered outdoors, pupated in March, and emerged as adults 6 weeks later. Schneider also established that specimens reared on larvae of *Melasoma aenea* in Finland and identified as *Syrphus ribesii* (Linnaeus) by Kanervo (1946) were actually of *P. nigritarsis*. Both authors tried feeding larvae on larvae of other Chrysomelidae; they found that the larvae would feed on *Melasoma* spp. and three other genera of Chrysomelinae but not on one other genus of Chrysomelinae or on five species of three genera of Galerucinae.

Parasyrphus relictus (Zetterstedt)

Fig. 252; Map 58

Scaeva relicta Zetterstedt, 1838:603.
Syrphus quinquelimbatus Bigot, 1884:91.
Syrphus conjunctus Osburn, 1908:7 (preoccupied Wiedemann, 1830).
Syrphus bimaculatus Lovett, 1919:244 (preoccupied Roser, 1840).
Syrphus rectoides Curran, 1921*a*:159.
Stenosyrphus melanderi Curran, 1925:103.
Syrphus incisus Goot, 1964:219 (new name for *S. conjunctus* Osburn).
Syrphus lovetti Goot, 1964:218 (new name for *S. bimaculatus* Lovett).

Length. 5.7–11.1 mm.

Male. Similar to male of *P. genualis*, differing as follows: frons strongly and distinctly pruinose on posterior half or more. Facial stripe narrow, at most one-quarter as wide as face.

Scutellum not darkened laterally. Wing membrane entirely trichose. Fore and mid femora entirely yellow or black on up to basal two-fifths; hind femur black except at apex, broadly yellow basally, or entirely yellow except for apical one-third of dorsal surface. Fore and mid tarsi often entirely dark brown above, at least slightly brownish above toward apex.

Sternites entirely yellow, with median longitudinal dark markings or with obscure or distinct dark triangles. Paramere (Fig. 252*c*) with distinct strong tooth lying behind rounded lobe at dorsoapical angle and with bifid basoventral hook. Aedeagal base (Fig. 252*d*) without ventrolateral prong.

Female. Frons with large lateral pruinose spots on middle third; spots confluent or nearly so medially, extending broadly forward along eye margins to anterior end of frons. Facial stripe up to one-third as wide as face.

Distribution. Alaska, western and northern Canada (Map 58), south to California and Colorado; Europe.* B.C., V–VIII, X; Que., VII.

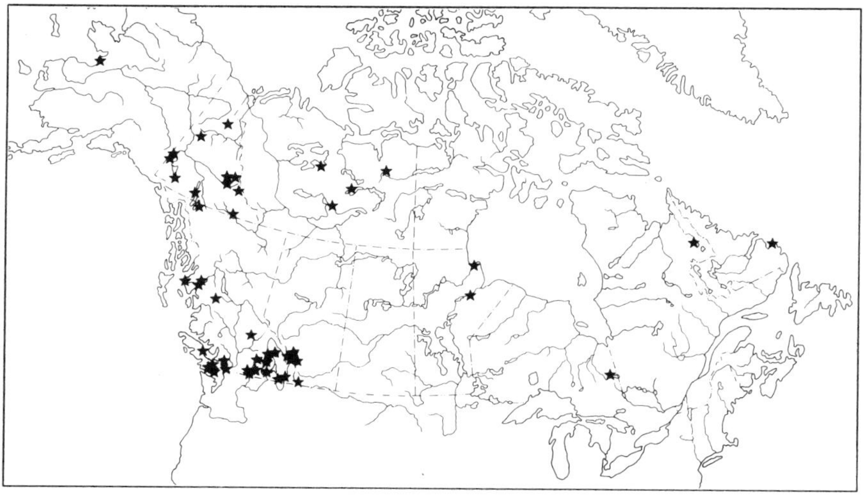

Map 58. Collection localities for *Parasyrphus relictus* (Zetterstedt).

Specimens identified. Alaska, 8 ♀♀; Canada, 141 ♂♂, 204 ♀♀; United States, 24 ♂♂, 77 ♀♀.

Discussion. It is possible that two Nearctic species are included under the name *P. relictus*. Most specimens either have the hind femur black with the apex narrowly yellow or have the hind femur broadly yellow basally and black or at least darkened above preapically. However, the extent and degree of the preapical darkening is variable and in some specimens the basal half of the femur is paler than the apical half yet is still darkened. Specimens of the first group usually have the fore and mid femora black on about the basal one-third and the fore and mid tarsi entirely darkened above; those of the second group usually have the femora only narrowly darkened basally or entirely yellow and have the tarsi darkened above only toward the apex. However, these characters vary in both groups and are not completely correlated with each other. No differences appear between the two forms in male terminalia or in other characters. I think it likely that all the specimens are of one species. If two species are present the name *P. melanderi* is available for that with the paler hind femora.

I use the name *P. relictus* (Zetterstedt) on the advice of Dr. F.C. Thompson who has compared European and North American specimens of the species.

Parasyrphus semiinterruptus (Fluke)

Figs. 152, 253; Map 53

Epistrophe semiinterruptus Fluke, 1935:21.

Length. 6.2–8.4 mm.

Male. Similar to male of *P. genualis*, differing as follows: frons narrowly but distinctly gray pruinose along eye margins, about anteromedian four-fifths of frons shining black. Facial stripe from three- to four-sevenths as wide as face.

Scutum in some specimens with some black hairs sublaterally. Scutellum darkened laterally. Wing membrane entirely trichose. Fore and mid tarsi brown above.

Yellow bands of tergites 3 and 4 deeply emarginate posteriorly; tergite 3 in some specimens narrowly divided (Fig. 152). Paramere (Fig. 253) dorsoapically with two small teeth at dorsoapical angle. Distal segment of aedeagus narrower than basal segment.

Female. Not definitely distinguishable from female of new species (Thompson *in litt.*).

Distribution. Eastern Canada (Map 53). Ont., Que., V, VI.

Specimens identified. Canada, 6 ♂♂.

Parasyrphus tarsatus (Zetterstedt)

Fig. 153; Map 59

Scaeva tarsata Zetterstedt, 1838:601.
Syrphus adolescens Walker, 1849:584.
Syrphus contumax Osten Sacken, 1875*a*:147.
Syrphus sodalis Williston, 1887:74.
Syrphus bryantii Johnson, 1898:17.
Syrphus sodalis var. *interruptus* Malloch, 1919:55 (preoccupied Philippi, 1865).

Syrphus mallochi Curran, 1923*d*:74 (new name for *interruptus* Malloch)

Syrphus nigropilosa Curran, 1927*a*:12.

Syrphus monachus Hull, 1930:140.

Petersina lanata Enderlein, 1938:206.

Petersina lanata var. *evanescens* Enderlein, 1938:207.

Petersina lanata var. *extrema* Enderlein, 1938:207.

Petersina lanata var. *flavifacies* Enderlein, 1938:207.

Petersina lanata var. *violaceiventris* Enderlein, 1938:207.

Length. 6.9–11.0 mm.

Male. Similar to male of *P. groenlandicus*, differing as follows: frons less swollen, with shorter and less dense hairs, black in most arctic specimens, grayish in more southern specimens; eye angle about 100°; eyes contiguous dorsally or, in some arctic specimens, separated by about width of two ommatidia. Face less strongly protruding below.

Thoracic hairs at least partly yellow, in southern specimens mostly or entirely yellow. Scutellum paler, mostly dull pale yellow. Legs usually paler, with femora yellow on up to apical half and fore and mid tibiae entirely yellow.

Tergites 2–4 usually with large yellow semicircular spots (Fig. 153); in northern specimens spots on tergite 2 commonly not reaching lateral margins; spots on tergites 3 and 4 smaller; in north Greenland specimens tergite 2 with spots tiny and obscure or absent but tergite 3 always with at least small obscure spots.

Female. Not always distinguishable from female of *P. groenlandicus* at or above northern limit of trees. Face usually only little more than half as wide as head but slightly variable; color of face, thoracic hairs, and legs varying as in male. Abdomen of specimens from below treeline with large distinct yellow spots, similar to those of male but with straighter posterior margins; specimens from near treeline or southern arctic tundra commonly with spots greatly reduced; some specimens from southern Greenland and all from northern Greenland with pale abdominal spots absent.

Distribution. Alaska, Canada, Greenland (Map 59), south in mountains to Colorado and New Hampshire; Europe; Asia. B.C., VI–VIII; Ont., Que., VI–VIII.

Specimens identified. Alaska, 37 ♂♂, 120 ♀♀; Canada, 517 ♂♂, 1147 ♀♀; Greenland, 146 ♂♂, 53 ♀♀; United States, 10 ♂♂, 33 ♀♀; Europe, 9 ♂♂, 9 ♀♀; Siberia, 1 ♀.

Discussion. The marked variation in this species has been studied by F.C. Thompson (personal communication). He distinguished four morphs, based on color and on characters of the male head and hind leg. He designated these as high arctic, low arctic, Rocky Mountain, and typic and has found intermediates between all but the first two (which may be because too few specimens have been collected). He concluded, and I agree, that specimens of these forms are conspecific.

Males of *P. tarsatus* can be readily distinguished from those of the arctic species *P. groenlandicus* by the differences in the frons indicated in the key and descriptions. Thompson concluded that females of the two species could be distinguished by the wider face of *P. groenlandicus* relative to head width, but, in my opinion, the slight variation in this character and the difficulty of measuring it make it unreliable. Females from south of the area of distribution of *P. groenlandicus* have moderately large yellow or orange spots on the tergites. Females from localities where males of both species occur (Maps 54, 59) have the spots either very variable in size or absent; they cannot be definitely referred to one species or the other. Females from high arctic localities in Canada, where males of only *P. groenlandicus* have been taken, either have very small obscure spots or, more commonly, lack spots. Finally, the females of a long series of specimens from southern Greenland, of which the males have the characters of *P. tarsatus*, have the abdomen either with small orange spots or entirely black. Females of an even longer series of 172 specimens, from northern Greenland, with the males referrable to *P. tarsatus*, have the abdominal spots absent.

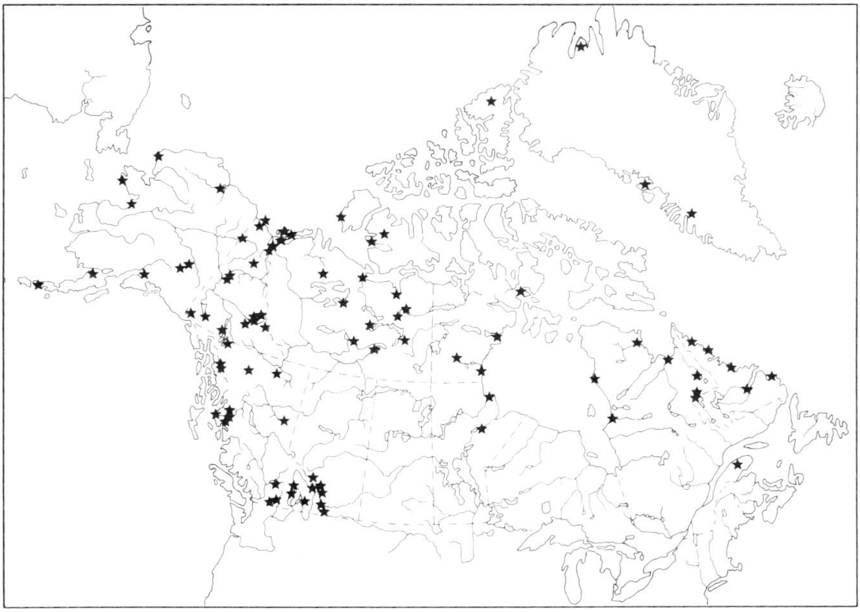

Map 59. Collection localities for *Parasyrphus tarsatus* (Zetterstedt).

Parasyrphus new species (Thompson *in litt.*)

Fig. 247; Map 60

Parasyrphus new species (Thompson *in litt.*)

Length. 6.3–8.2 mm.

Male. Similar to male of *P. genualis*, differing as follows: frons with pruinose area obscure or distinct. Facial stripe up to three-sevenths width of face.

Scutum in some specimens with some brown to blackish hairs sublaterally. Wing membrane entirely trichose. Fore and mid tarsi usually entirely brown above.

Paramere (Fig. 247*a*) with two strong teeth and with strong basoventral hook at dorsoapical angle. Distal segment of aedeagus narrower than basal segment; latter with strong ventrolateral prong (Fig. 247*b*).

Female. Not definitely distinguishable from female of *P. semiinterruptus*.

Distribution. Canada (Map 60). B.C., VII; Ont., Que., IV–VI.

Specimens identified. Canada, 74 ♂♂.

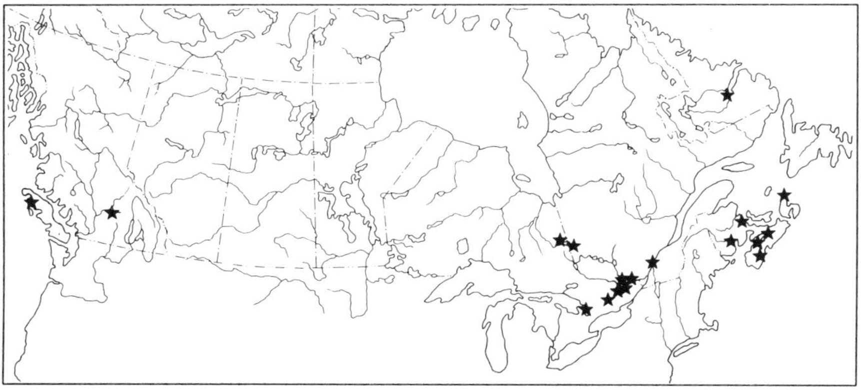

Map 60. Collection localities for *Parasyrphus* new species (Thompson *in litt.*).

Genus *Platycheirus* Lepeletier & Serville

Diagnosis. Species slender to rather robust with completely black head and thorax; abdomen usually black with distinct paired yellow to orange or silver to gray pruinose or bluish submetallic spots; abdomen in some species almost completely black or mostly yellow or orange. Length 4.7–10.5 mm.

Description. Eye bare. Face black, varying from slightly receding below with obscure tubercle to rather strongly produced below with prominent tubercle, usually uniformly pruinose or with only tubercle shining, with pruinosity in some species obliquely rippled or with oblique rows of distinct punctures (Figs. 6–11).

Scutum black, usually shining or subshining, with notopleuron in some species strongly pruinose, rarely entirely moderately pruinose. Scutellum and pleura black, at most weakly pruinose. Ventral scutellar fringe complete. Upper and lower katepisternal hair patches broadly separated; former reduced or absent in some species. Anterior anepisternum, meron, metapleuron, and metasternum bare. Metasternum well-developed, with two halves not deeply excavated posteriorly (Fig. 45). Wing membrane usually entirely trichose; some species with bare areas near base or bare on most of basal half. Hind coxa without posteromedial apical hair tuft. Legs in Figs. 47–99. Fore trochanter commonly with short stiff setulae below; fore femur commonly with distinctive hair tufts, modified hairs, or distinct bristles on posterior or ventral surfaces; fore tibia usually flattened and broadened toward apex or with row of posterior bristles; fore leg usually with first one, two, or three tarsomeres flattened and broadened. Mid coxa with slender ventral process in *P. scutatus* (Meigen); mid femur usually with distinctive hairs or bristles, with anterior excavation just beyond mid length in some species; mid tibia in some species slightly flattened or broadened, commonly with distinctive hairs or bristles; mid tarsus only rarely modified. Hind leg usually simple, with distinctive hairs or bristles in some species; first tarsomere of hind leg commonly swollen.

Abdomen nearly parallel sided in males, narrowly oval in females, unmargined, with varied markings (Figs. 155–172). Sternites yellow, orange-brown or black; color correlated with predominant color of tergites. Male terminalia in Figs. 236–246, 269–271. Surstylus usually slender, flattened, tapering slightly to blunt or broadly rounded apex, with large thumblike basomedian lobe; rarely elongate oval, or slightly narrowed near mid length, more rarely with basal lobe very small or absent. Epandrium (sternite 9) truncate apically. Paramere (superior lobe) usually with slender base and slender tapering hooklike process, rarely with stout base and stout toothlike process or with two toothlike processes. Aedeagus unsegmented, usually strongly swollen basally, then abruptly constricted with short nearly tubular apical section.

Distribution. Genus predominantly Holarctic and markedly boreal; of 70 species from north of Mexico 65 in Canada and Alaska; 45 in Alaska, Yukon Territory, and Northwest Territories; 23 or more in Palaearctic region. Additional species in temperate parts of Mexico, Central America, and South America (especially in Chile), and in Palaearctic region. In Oriental region several species occur at high altitudes in Taiwan, the Philippines, and Nepal. Absent from Indonesia, New Guinea, and Australia but well-developed in New Zealand. No species known from Afrotropical region.

Biology. Larvae of 10 Nearctic species have been reported to be aphid predators either in Europe or in North America. Curran (quoted by Davidson 1922) found that larvae of *P. obscurus* (Say) ate decomposing chickweed as well as aphids and were more successful on the former (this remark is ambiguous but I think it unlikely larvae developed successfully on chickweed alone). Goeldlin (1974) found that larvae of *P. immarginatus* (Zetterstedt) lived for 10 days by feeding on bean cotyledons but grew much more slowly than those fed on aphids and failed to produce adults even though aphids were added to their diet. He concluded that the larvae of the genus are aphidophagous but are facultatively phytophagous.

Both authors reported that larvae of various species did most of their feeding at night and usually hid in dark places during the day. Heiss (1938) observed the same and mentioned that larvae preferred moist situations. Goeldlin found that larvae of several species lived only on low-growing plants; only those of *P. ambiguus* (Fallén), a close relative of *P. coerulescens* (Williston), were taken on aphid colonies at higher levels in orchards. Rotheray and Gilbert (1989) have suggested that many species may be predators in leaf litter.

Goeldlin reported that several species are polyvoltine or oligovoltine but that two, of which *P. parmatus* Rondani (as *P. ovalis* Becker) occurs in North America, are univoltine with an obligatory larval diapause of 8–10 months. Davidson (1922) reported that two species in California developed without diapause.

Goeldlin observed that females of seven species deposited their eggs in rather regular rows in clusters of 2–10. Davidson (1922) showed a photograph of eggs of *P. stegnus* (Say) deposited in this way. Chandler (1968) gave a key to eggs of six British species (five of which are Holarctic) and illustrated some striking differences in chorionic sculpture. Dixon (1960) gave a key to larvae of four British species (all Holarctic).

Goot and Grabandt (1970), using specimens from the Netherlands, and Leereveld (1982), using considerably more specimens from several European countries and Canada, analyzed pollen in the gut of a considerable number of specimens of *Platycheirus* spp. The results agreed in large part. They determined that specimens of *P. angustatus* (Zetterstedt), *P. clypeatus* (Meigen), *P. fulviventris* Macquart,

P. immarginatus (Zetterstedt), and *P. scambus* (Staeger), and some specimens of *P. granditarsus* (Forster), fed entirely or largely on pollen of Gramineae, Cyperaceae, and *Plantago lanceolata* (or species of the genus with similar pollen). These plants are generally considered to be wind-pollinated (anemophilous) and unattractive to insects. Specimens of the other four species fed on pollen of a variety of other, mostly showy, flowers of many families. Many of the specimens of both groups used in the first study were taken in the same small area at the same time, so availability of plants was almost certainly not a factor. Stelleman and Meeuse (1976) studied *P. clypeatus*, *P. fulviventris*, *P. scambus*, and *Melanostoma mellinum* (Linnaeus) in the field and found that some or all of the four species were effective pollinators of *Plantago lanceolata*.

Many species, especially those with broadened male fore tarsi, are most abundant in moist areas with low vegetation, e.g., marshes and lake shores. Other species, e.g., *P. obscurus* (Say), are common in mesophytic woodland. Several high boreal or arctic species, e.g., *P. carinatus* (Curran), occur on barren hilltops.

Discussion. Males of most species can be readily distinguished on the basis of characters other than those of the terminalia. The terminalia (Figs. 236–246, 269–271) usually show only insignificant differences in closely related species, but in some groups, especially those with simple legs, differences are rather striking. Differences are mentioned only when other characters do not allow ready identification. Females of most species cannot be definitely identified at the present time; those that are distinct are included in the key below. Some others, provisionally identified by association with males or on the basis of geographic distribution, are not included in the key and usually not in the descriptions. As there have been many misidentifications of species in this genus, few published distribution records have been cited.

The genus *Platycheirus* as defined here includes the genera *Platycheirus*, *Carposcalis* Enderlein, *Pyrophaena* Schiner, and part of the genus *Melanostoma* Schiner of Wirth et al. (1965). Of the five Nearctic species not known from Canada or Alaska three occur in Colorado, one in Oregon and California, and one in southern California; as all but the last of these may occur in Canada they are treated here.

Key to Nearctic species of *Platycheirus*

1. Male (eyes meeting on frons) 2

   Female (eyes widely separated on frons) 72

2. Fore tibia slightly to distinctly broadened toward apex, or first tarsomere of fore leg distinctly broadened, or both (Figs. 47*a*, 49*a*–57*a*, 65*a*–69*a*) 3

Fore tibia and fore tarsus slender throughout (Figs. 91*a*, 92*a*, 97*a*, 99*b*) 45

3. Fore tibia at most slightly broadened near apex and distinctly narrower than first tarsomere of fore leg (Figs. 47*a*, 49*a*–55*a*) 4

Fore tibia usually broadened on apical half or more and at least as wide as first tarsomere (Figs. 59*a*, 65*a*–83*a*, 85*a*–88*a*, 90*a*), strongly broadened in some species only near apex and slightly narrower than very large first tarsomere (Figs. 56*a*, 57*b*, 60*b*) 11

4. Fore tarsus entirely black, with first tarsomere bearing long triangular anterior process (Fig. 47*a*). Mid tarsus with first three tarsomeres much wider than apex of mid tibia (Fig. 47*b*). Posterior half of tergite 2, tergite 3, and anterior half of tergite 4 mostly or entirely orange (Fig. 155). Widespread .... ***granditarsis* (Forster)**

Fore tarsus with first tarsomere whitish, without anterior process, nearly symmetrical in outline (Figs. 49*a*–55*a*). Mid tarsus slender. Tergites 2–4 black with silvery or yellowish sublateral spots; some spots broadly confluent medially (Fig. 157) 5

5. Mid tibia with hairs on basal one-third of posterior to posteroventral surface dense, fine and wavy, at least twice as long as tibial diameter and as long or almost as long as hairs on apical half of posterior surface; hairs on basal one-third of anteroventral surface usually dense, fine, and black and at least three times as long as tibial diameter (Figs. 49*j*, 51*j*, 53*k*, 54*k*, 55*j*) 6

Mid tibia with hairs on basal one-third of posterior surface short and nearly straight; hairs subequal in length to tibial diameter and less than half as long as long fine or bristlelike hairs on apical half of posterior surface; hairs on anteroventral surface only very short, straight, inconspicuous (Figs. 50*k*, 52*k*) 10

6. Basal one-third of hind tibia with dense wavy black anterodorsal hairs up to three times as long as tibial diameter. First tarsomere of fore leg about three times as wide as apex of fore tibia; apex of tarsomere very oblique (Fig. 53*a*). Colorado .... ***oreadis* Vockeroth**

Hind tibia with only short hairs or rarely with some longer anterodorsal hairs at or beyond mid length. First tarsomere of fore tarsus at most 2.2 times as wide as apex of fore tibia; apex of tarsomere slightly oblique or transverse (Figs. 49*a*, 51*a*, 54*a*, 55*a*) 7

7. First tarsomere of foreleg at least twice as long as wide (Fig. 54*a*). Northwestern Canada .... ***subordinatus*** **Becker**

First tarsomere of fore leg at most 1.6 times as long as wide (Figs. 49*a*, 51*a*, 55*a*) 8

8. Mid tibia with anteroventral hairs short, at most slightly longer than tibial diameter (Fig. 55*j*). Widespread .... ***thylax*** **Hull**

Mid tibia with tuft of long wavy hairs on about basal one-third of anteroventral surface; hairs at least three times as long as tibial diameter and much longer than hairs elsewhere on anteroventral surface (Figs. 49*j*, 51*j*) 9

9. Mid leg with first two tarsomeres yellow; first tarsomere strongly compressed; second tarsomere slightly compressed; last three tarsomeres black (Fig. 49*j*). Widespread .... ***discimanus*** **Loew**

Mid leg entirely black; tarsomeres not compressed (Fig. 51*j*). Boreal and arctic Canada .... ***groenlandicus*** **Curran**

10. Ventral surface of fore trochanter with many stiff black setulae. Ventral surface of fore femur with only long fine hairs. Apical half of posterior surface of fore and mid tibiae with many fine irregularly arranged black hairs of various lengths (Figs. 52*a*,*k*). Apical one-fifth of mid femur with short stiff black anteroventral hairs. First tarsomere of fore leg at least 2.3 times as wide as apex of fore tibia (Fig. 52*a*). Alaska .... ***manicatus*** **(Meigen)**

Ventral surface of fore trochanter with some fine pale setae. Basal half of ventral surface of fore femur with row of three or four stiff black or yellow setae about two-thirds as long as femoral diameter. Apical half of posterior surface of fore and mid tibiae with single row of straight black bristlelike hairs up to 2.5 times tibial diameter in length (Figs. 50*a*,*k*). Apical one-fifth of mid femur with long fine tangled black anteroventral hairs. First tarsomere of fore leg from 1.3–2.0 times as wide as apex of fore tibia (Fig. 50*a*). Boreal regions .... ***flabellus*** **Hull**

11(3). Posterior surface of fore femur densely covered with rather strong nearly uniform slightly flattened black hairs, without outstanding hair tufts near base, and without outstanding longer hairs or bristles along femoral length (Fig. 56*a*). Anterior surface of mid femur just beyond mid length with distinct concavity bordered below by very short curved black setulae (Fig. 56*e*) 12

Posterior surface of fore femur without flattened black hairs, either with uniform weak hairs, or with black or white hair tufts near base or with longer outstanding strong hairs or bristles along its length (Figs. 59*a*, 65*a*–83*a*, 85*a*–88*a*, 90*a*). Anterior surface of mid femur either without concavity or with at most shallow concavity 19

12. First tarsomere of fore leg without dorsal keel; second tarsomere as wide as widest part of fore tibia and about nine-tenths as wide as first tarsomere (Fig. 60*b*). Western Canada .... ***latitarsis* Vockeroth**

First tarsomere of fore leg with weak dorsal keel on apical half or with strong dorsal keel on most of its length; second tarsomere at most four-fifths as wide as fore tibia and first tarsomere (Figs. 56*a*, 57*b*) 13

13. First tarsomere of fore leg with moderately strong dorsal keel on apical half but without distinct keel on basal half (Fig. 56*a*). Anepimeron in some species with some hairs on lower half 14

First tarsomere of fore leg with distinct dorsal keel over its entire length although keel in some species becoming a little lower toward base of tarsomere (Fig. 57*b*). Anepimeron with hairs only on upper half 17

14. Apical half of ventral surface of mid tibia with suberect tangled dark hairs on basal part of strong apical tibial swelling about as long as least subapical tibial diameter (Fig. 62*L*). Anepimeron with hairs only on upper half; hairs forming compact tuft. Wing membrane entirely trichose. Boreal and western Nearctic; Palaearctic .... ***nielseni* Vockeroth**

Apical half of ventral surface of mid tibia with very short scarcely discernible hairs; apical tibial swelling weak (Fig. 61*L*). Anepimeron in some species with some hairs on lower half; hairs on upper half not forming compact tuft. Wing membrane entirely trichose or with bare areas near base 15

15. Basal half of anteroventral surface of mid tibia with tuft of short nearly straight subappressed hairs; longest hairs at most 1.5 times as long as least tibial diameter (Fig. 61*k*). Anepimeron with hairs on upper half only. Scutellar hairs mostly or entirely yellow. Widespread ....
.... ***nearcticus* Vockeroth**

Basal half of anteroventral surface of mid tibia with tuft of long erect wavy black hairs; longest hairs more than twice as long as tibial diameter (Fig. 56*e*). Anepimeron in some species with hairs on lower half. Scutellar hairs entirely yellow or about half black 16

16. First tarsomere of hind leg strongly and abruptly constricted at mid length (Figs. 51*m*,*p*). Scutellar hairs yellow. Anepimeron usually with hairs only on upper half, rarely with hairs on lower half. Bare area of cell bm usually at least half as long as cell and reaching or nearly reaching its anterior margin, rarely less extensive. Eastern Canada . . . . ***inversus* Ide**

First tarsomere of hind leg gradually narrowed over apical three-quarters of its length (Figs. 56*m*,*p*). Scutellar hairs about half black. Anepimeron with at least some hairs on lower half. Bare area of cell bm at most one-third as long as cell and well separated from its anterior margin. Boreal and western Canada . . . . ***amplus* Curran**

17. Lower katepisternal hairs long, at least two-thirds as long as arista. Hind tibia with setae of anterior surface sparse, those of basal one-third much shorter than those of apical two-thirds (Fig. 64*n*). First tarsomere of fore leg with keel becoming distinctly lower toward base of tarsomere. Wing membrane entirely trichose or with tiny and indistinct bare areas at base of cells c and bm. Western Canada . . . . ***peltatoides* Curran**

Lower katepisternal hairs short, at most half as long as arista. Hind tibia with setae of anterior surface very dense; setae of basal one-third almost as long as those of apical two-thirds (Figs. 57*m*, 63*n*). First tarsomere of fore leg with keel of nearly uniform height throughout (Fig. 57*b*). Wing membrane entirely trichose or with moderate bare areas at base of cells c and bm 18

18. Wing membrane entirely trichose. Basal half of anteroventral surface of mid tibia with tuft of long wavy erect black or yellow hairs; longest hairs about three times as long as tibial diameter (as in Fig. 56*e*). Anepimeron with very dense tuft of hairs, their bases visible only at lower edge of tuft. Boreal and western Canada . . . . ***holarcticus* Vockeroth**

Wing with about basal one-sixth of cell c and basal one-quarter of cell bm bare. Basal half of anteroventral surface of mid tibia with tuft of shorter straighter usually pale subappressed hairs; longest hairs at most 1.5 times as long as tibial diameter (as in Fig. 61*k*). Anepimeron with moderately dense tuft of hairs, with their bases visible throughout. Western Canada . . . . ***octavus* Vockeroth**

19(11). Posterior surface of fore femur with two large dense or rather diffuse tufts of long wavy coarse black hairs preceded by similar tuft of some wavy white hairs (Figs. 65*a*–69*a*) 20

Posterior surface of fore femur without two large subbasal tufts of black hairs, either with uniform fine hairs, or with one or two tufts of long hairs with flattened and broadened apices, or with subbasal tuft of several long white hairs followed in some species by single weak tuft of several black hairs, or with uniform row of widely spaced strong hairs or bristles over most of its length (Figs. 59*a*, 70*a*–83*a*, 85*a*–90*a*) 24

20. Mid coxa with slender fingerlike ventral process (Fig. 31*h*). Fore femur beyond subbasal tufts with many long strong black hairs; second tarsomere of fore leg about one-sixth as long as first tarsomere (Fig. 68*a*). Widespread .... ***scutatus* (Meigen)**

Mid coxa without ventral process. Fore femur beyond subbasal tufts with three well-spaced long black bristles among shorter hairs; second tarsomere of fore leg about two-fifths as long as first tarsomere (Figs. 65*a*–67*a*, 69*a*) 21

21. Fore tibia strongly broadened on apical one-third and with posteroapical angle broadly rounded; first tarsomere of fore leg strongly broadened posteriorly on about basal two-thirds, sharply angulate at this point, then with margin straight to apex (Fig. 69*a*). First four tarsomeres of mid leg yellow; fifth tarsomere brown above. Wing with cell bm entirely trichose. Alaska, British Columbia, Quebec .... ***urakawensis* (Matsumura)**

Fore tibia less strongly broadened and with posteroapical angle subacute or narrowly rounded; first tarsomere only gradually broadened posteriorly, without distinct angle (Figs. 65*a*–67*a*). At least last four tarsomeres of mid leg brown to dark brown above. Cell bm with at least small bare area near base, narrowly bare anteriorly over most of its length in some species 22

22. Fore tibia uniformly broadened from base to apex; first tarsomere of fore leg gradually broadened from base to apex, its margins slightly divergent throughout and its apex distinctly arcuate; second tarsomere with both basal and apical margins distinctly arcuate (Fig. 67*a*). Cell bm with only tiny bare area near base. High boreal regions .... ***nigrofemoratus* Kanervo**

Fore tibia uniformly broadened on basal three-quarters and then slightly more strongly broadened posteriorly; first tarsomere gradually broadened on basal half, parallel-sided on apical half and with apex transversely truncate; second tarsomere subrectangular with apex transversely truncate or nearly so (Figs. 65*a*, 66*a*). Cell bm with small to rather large bare area. Boreal and western Canada 23

23. Posteroapical angle of fore tibia distinctly rounded and extending slightly beyond level of posterior margin of first tarsomere (Fig. 66*a*). Mid leg with first tarsomere dark, at most slightly paler than fifth tarsomere. Pacific coastal region .... ***ciliatus*** **Bigot**

Posteroapical angle of fore tibia subacute, not extending beyond level of first tarsomere (Fig. 65*a*). Mid leg with first tarsomere yellow to brown, usually much paler than fifth tarsomere. Boreal and western Canada .... ***albimanus*** **(Fabricius)**

24. Fore tibia with distinct longitudinal dorsal keel becoming higher toward apex of tibia; first tarsomere of fore leg rather slender, slightly to strongly constricted at mid length (Fig. 85*a*). Tergites 3 and 4 yellow-orange with black posterior margin, without black median line (Fig. 159). Widespread .... ***normae*** **Fluke**

Fore tibia without dorsal keel; first tarsomere of fore leg usually gradually widened beyond base, not constricted (Figs. 59*a*, 70*a*–83*a*, 86*a*–88*a*, 90*a*). Tergites 3 and 4 with or without black median line, mostly black in some species (Figs. 160–170) 25

25. Mid femur near base of posteroventral surface with dense brush of stiff orange and black setae (Fig. 81*i*). Mid tibia distinctly and nearly uniformly broadened on about apical three-quarters, with dense anteroventral yellow hairs; first tarsomere of mid leg distinctly broadened and depressed (Fig. 81*f*). Tergites extensively yellow. Widespread .... ***quadratus*** **(Say)**

Mid femur without subbasal brush of setae, with at most seven well-spaced yellow or black posteroventral bristles on basal half (Figs. 78*i*, 79*g*, 82*i*, 86*i*, 88*i*, 89*i*). Mid tibia slender or slightly and irregularly broadened on at most apical half, with or without dense ventral hairs; first tarsomere of mid leg neither broadened nor depressed (Figs. 70*e*, 71*e*, 73*k*, 76*k*, 87*f*). Tergites variable in color, mostly yellow to mostly black 26

26. Posterior surface of fore femur just beyond base with compact tuft of long hairs each bearing spearhead-like broadening at apex (Figs. 70*a*–72*a*) 27

Posterior surface of fore femur with or without subbasal hair tuft; if tuft present then individual hairs not broadened apically (Figs. 57*a*, 73*a*–83*a*, 86*a*–88*a*) 29

27. Fore femur with one subbasal tuft of long hairs with slightly broadened apices; this tuft in some specimens preceded by

single long slender pale hair (Fig. 72*a*). Mid tibia without long appressed or erect anteroventral hairs. Fore and mid tarsi entirely yellow. Tergite 5 with large anterolateral yellow spots (Fig. 160). Eastern Canada .... ***thompsoni* Vockeroth**

Fore femur with two subbasal tufts of long hairs with slightly to strongly broadened apices (Figs. 70*a*, 71*a*). Basal half of mid tibia with long appressed or erect anteroventral hairs (Figs. 70*e*, 71*e*). Fore and mid leg usually with one or more tarsomeres distinctly darker above than yellow first tarsomere. Tergite 5 with anterolateral yellow spots or entirely black 28

28. Tergite 5 black (Fig. 161). Hairs of basal tuft of fore femur pale throughout; fore tibia and fore tarsus rather narrow (Fig. 71*a*). Mid femur with many posteroventral hairs longer than femoral diameter; mid tibia with long erect black anteroventral hairs on basal one-third (Fig. 71*e*). Mid leg usually with third and fourth tarsomeres darker above than first tarsomere. Boreal regions .... ***pilatus* Vockeroth**

Tergite 5 with pair of large anterolateral yellow spots. Hairs of basal tuft of fore femur with apices brown to black; fore tibia and tarsus wider (Fig. 70*a*). Mid femur with posteroventral hairs much shorter than femoral diameter; mid tibia with long strongly appressed black anteroventral hairs on basal half (Fig. 70*e*). Mid leg with first four tarsomeres yellow; fifth tarsomere usually brown to black above. Boreal and western Canada .... ***nodosus* Curran**

29. Posterior surface of fore femur with subbasal tuft of two or three closely appressed long white or yellowish hairs with wavy apices (Figs. 73*a*–80*a*, 82*a*, 83*a*) 30

Posterior surface of fore femur without subbasal tuft of white hairs, either with nearly uniform fine hairs or with well-spaced longer hairs or long slender bristles over most of its length (Figs. 59*a*, 86*a*–88*a*, 90*a*) 39

30. Apical half of anteroventral surface of mid femur with nearly regular row of 7–16 short stout black setae, with one or two strong curved hairs near or beyond end of row in some species (Figs. 78*i*, 82*i*). Posterior surface of fore femur with three to five long, moderately strong, slightly wavy, black (or in some species white) evenly spaced bristles on apical three-quarters (Figs. 78*a*, 82*a*) 31

Anteroventral surface of mid femur without row of strong setae, in some species either with cluster of weak setae at about mid length or with fine preapical hairs (Fig. 79*g*). Posterior surface of fore femur with uniform fine hairs or with at most one row of four or five long weak black bristles (Figs. 73*a*–77*a*, 79*a*, 80*a*, 83*a*) 32

31. Tergites 3 and 4 each with pair of large yellow spots. Mid femur and tibia entirely yellow. Hind femur and tibia entirely yellow or each with dark ring. Fore tibia nearly uniformly broadened from base to apex (Fig. 78*a*). Widespread .... ***immarginatus* (Zetterstedt)**

Tergites 2–4 each with pair of silvery spots on entirely dark background (Fig. 162). Mid and hind femora and tibiae mostly black. Fore tibia abruptly broadened at about three-quarters its length (Fig. 82*a*). British Columbia .... ***setipes* Vockeroth**

32. Fore tibia strongly and abruptly broadened on apical two-fifths, slightly narrower apically than preapically (Figs. 80*a*, 83*a*) 33

Fore tibia less strongly broadened, uniformly broadened from base to apex (Figs. 73*a*–77*a*, 79*a*) 34

33. Second and third tarsomeres of fore leg each at most three-quarters as long as wide (Fig. 80*a*). Basal half of anteroventral surface of mid tibia with erect or subappressed black hairs at least three times as long as tibial diameter. Hairs of anepisternum and anepimeron mostly yellow-brown. Boreal and western Canada .... ***podagratus* (Zetterstedt)**

Second and third tarsomeres of fore leg each at least as long as wide (Fig. 83*a*). Anteroventral surface of mid tibia with only very short inconspicuous hairs. Hairs of anepisternum and anepimeron usually entirely black. Western Canada .... ***tenebrosus* Coquillett**

34. Tergite 5 mostly or entirely yellow, with at most median stripe and narrow posterior margin black; tergites 3 and 4 with large lateral yellow spots at least nine-tenths as long as tergites and commonly confluent medially (Fig. 164). Apical one-third of anteroventral surface of mid femur with weak wavy hairs at least as long as femoral diameter (Fig. 79*g*). Widespread .... ***perpallidus* Verrall**

Tergite 5 entirely black or with at most small anterolateral obscure yellow spots; tergites 3 and 4 with lateral spots either silvery or submetallic, or, if distinct subquadrate yellow then spots at most four-fifths as long as tergite and not confluent medially (Figs. 165, 166). Apical one-third of anteroventral surface of mid femur bare or with some hairs much shorter than femoral diameter 35

35. Fore and mid femora black with apex narrowly yellow. First tarsomere of fore leg parallel-sided except at extreme base (Figs. 73*a*, 76*a*) 36

Fore and mid femora entirely or mostly yellow, at most with dark stripes on part or all of their lengths; first tarsomere of fore leg parallel-sided only on apical half (Figs. 74*a*, 75*a*) or very slightly narrowed toward apex (Fig. 77*a*) 37

36. Face slightly but distinctly produced forward below (Fig. 8). Cells bm and *cup* bare anteriorly on most of basal half. Posterior surface of fore femur with regular row of about five long strong black hairs on most of its length; fore tibia rather broad, with posterior margin irregular (Fig. 76*a*). Mid tibia with anteroventral hairs straight, rather sparse and of uniform length throughout and with three long strong black hairs on apical half of posteroventral surface (Fig. 76*k*). Spots of tergites 3 and 4 distinctly yellowish with overlay of silvery pruinosity (Fig. 165) British Columbia .... ***hispidipes*** **Vockeroth**

Face nearly vertical (Fig. 11). Wing membrane entirely trichose. Posterior surface of fore femur with uniform fine hairs decreasing in length toward apex; fore tibia narrower, with regular margins (Fig. 73*a*). Mid tibia with short dense wavy anteroventral hairs on about middle three-quarters and without long posteroventral hairs (Fig. 73*k*). Spots of tergites 3 and 4 entirely silvery or with yellow background. Boreal and western Canada .... ***aeratus*** **Coquillett**

37. Apex of fore tibia with posterior angle at most very slightly produced so that tibial apex nearly obliquely truncate; first tarsomere of fore leg with posterior margin slightly curved on apical three-quarters, slightly narrower at apex than at mid length (Fig. 77*a*). Spots of tergites 3 and 4 entirely dark with strong silvery sheen (most northern specimens) or slightly to extensively yellow with an overlay of dense silvery pruinosity best seen in posterodorsal view (some northern and all southern specimens). Widespread ....
.... ***hyperboreus*** **(Staeger)**

Apex of fore tibia with posterior angle distinctly produced so that tibial apex notched; first tarsomere of fore leg with apical half parallel-sided (Fig. 74*a*, 75*a*). Spots of tergites 3 and 4 entirely yellow or slightly metallic, with at most very sparse silvery pruinosity 38

38. Wing membrane with cell bc mostly bare and with tiny bare area at base of cell c and at base of cell bm. Fore femur beyond subbasal white hair tuft with fine mostly pale hairs at most little longer than femoral diameter (Fig. 74*a*). Tergite 2 slightly

longer than wide; tergites 3 and 4 subquadrate. Widespread .... ***angustatus* (Zetterstedt)**

Wing membrane entirely trichose. Posterior surface of fore femur beyond subbasal white hair tuft with longer stronger usually black hairs some at least 1.5 times as long as tibial diameter (Fig. 75*a*). Tergite 2 at least slightly wider than long; tergites 3 and 4 about 1.3 times as wide as long. Widespread .... ***clypeatus* (Meigen)**

39(29). Tergites black with silvery spots, without trace of yellow markings (Fig. 167). First tarsomere of fore leg as wide as long and strongly narrowed on basal half; following four tarsomeres each progressively slightly narrower (Fig. 80*a*). Mid and hind tibiae black with bases and some apices narrowly yellow. Boreal and western Canada, Greenland .... ***varipes* Curran**

Tergites 3 and 4 with large yellow spots or almost entirely yellow (Figs. 168–170). First tarsomere of fore leg at least 1.5 times as long as wide and strongly narrowed only very near base; following four tarsomeres each progressively slightly narrower or second or third tarsomere much narrower than preceding tarsomere (Figs. 59*a*, 86*a*–88*a*). Mid and hind tibiae black with bases narrowly yellow, or mostly or entirely yellow 40

40. Face distinctly produced forward below (Fig. 10). Yellow spots of tergites 3 and 4 slightly wider than long (Fig. 168). Second tarsomere (*P. jaerensis*) or third tarsomere (*P. parmatus*) of fore leg much narrower than preceding tarsomere (Figs. 59*a*, 87*a*). Hind femur black with apex narrowly yellow; hind tarsus entirely dark brown to black above 41

Face vertical or slightly receding below (Fig. 10). Yellow spots of tergites 3 and 4 distinctly longer than wide, confluent medially in some species (Figs. 169, 170). Second to fifth tarsomeres of fore leg each slightly narrower than preceding tarsomere (Figs. 86*a*, 88*a*). Hind femur entirely yellow or with at most black ring on apical half; hind leg with second and third tarsomeres partly or entirely dull to bright yellow above 42

41. Fore tibia with many posterior hairs longer than tibial width; fore leg with second tarsomere much wider than long and only slightly narrower than first tarsomere (Fig. 87*a*). Mid femur anteriorly without concavity, anteroventrally with only long fine hairs; mid tibia with many long fine anteroventral hairs much longer than tibial diameter (Fig. 87*f*). Hind tibia on apical half with several anterodorsal hairs about twice as long as tibial diameter. Boreal regions ..... ***parmatus* Rondani**

Fore tibia with posterior hairs much shorter than tibial width; fore leg with second tarsomere slightly longer than wide and much narrower than first tarsomere (Fig. 59*a*). Mid femur anteriorly beyond mid length with shallow concavity bordered below by short dense slightly curved black setulae, otherwise with only short anteroventral hairs; mid tibia with anteroventral hairs of basal half slightly longer than tibial diameter, otherwise with very short hairs. Hind tibia with only very short hairs. Eastern Canada, probably widespread .... ***jaerensis* Nielsen**

42. Tergites 3–5 with black median line broadly broken or absent or represented by faint brownish line (Fig. 169). Lower katepisternal hairs less than half as long as first flagellomere. Mid tibia anteroventrally on basal two-thirds with dense wavy black hairs about three times as long as tibial diameter, otherwise with short appressed hairs (Fig. 84*g*). Widespread .... ***modestus* Ide**

Tergites 3–5 with distinct black median line (Fig. 170). Lower katepisternal hairs about as long as first flagellomere. Mid tibia ventrally with only very short appressed hairs or with dense fine wavy mostly pale erect hairs subequal in length to tibial diameter on apical three-quarters 43

43. Mid tibia on apical three-quarters of ventral surface with dense wavy erect mostly pale hairs at least as long as tibial diameter; mid femur anteroventrally with at most some scattered black setae (Fig. 86*i*). Eastern Canada, coastal regions .... ***orarius* Vockeroth**

Mid tibia ventrally with short pale or partly dark appressed hairs, without erect hairs; mid femur anteroventrally usually with nearly regular row of 3–15 short stiff black or yellow setae (Figs. 88*i*, 89*i*) 44

44. Mid tibia with appressed anteroventral and posteroventral hairs mostly black; mid femur anteroventrally usually with row of 3–15 short stiff black setae (rarely with only yellow setae), and with four to six strong black posteroventral bristles on basal half (Fig. 89*i*). Widespread .... ***scambus* (Staeger)**

Mid tibia with appressed anteroventral and posteroventral hairs entirely yellow; mid femur anteroventrally with short stiff yellow setae or also with one black seta, and with three or four very weak black posteroventral bristles (Fig. 88*i*). Eastern Canada .... ***scamboides* Curran**

45(2). Apical half or more of posterior surface of fore femur with regular row of five or more rather long stiff slightly flattened black setae; row ending with longer and more slender seta with strongly curved apex (Fig. 91*a*) 46

Posterior surface of fore femur with uniform or nearly uniform fine hairs, with at most some near apex slightly longer and with curved apices (Figs. 92*a*, 97*a*) 47

46. Fore femur entirely orange or with blackish brown posterior stripe, subbasally with posterior row of three to five rather long strong yellow setae. Mid femur subbasally with similar row of slightly longer yellow setae. Wing with cell bm entirely bare on at least basal half, usually with some microtrichia near apex. Widespread .... ***coerulescens*** **(Williston)**

Fore femur blackish brown with only narrow apex yellow-orange, subbasally usually with black setae but with some pale setae in some specimens. Mid femur with subbasal setae black. Wing with cell bm usually entirely trichose, bare on at most anterobasal one-third. Subarctic, including Greenland .... ***lundbecki*** **(Collin)**

47. Fore (and mid) tibiae posteriorly with nearly regular row of weak to strong black bristles; longest bristles at least one-fifth as long as tibia (Figs. 93*c*–96*c*, 97*a*). Face with pruinosity uniform or with weak ripples or weak to strong punctures (Figs. 6, 267, 268) 48

Fore (and mid) tibiae posteriorly with only short weak hairs at most twice as long as tibial diameter (Figs. 92*a*, 99*a*). Face with pruinosity uniform, neither rippled nor punctate 59

48. First tarsomere of fore leg with about six long weak posterior bristles, last one or two bristles with curved apices and about two-thirds as long as tarsomere (Fig. 97*a*). Arctic and alpine, including Greenland 49

First tarsomere of fore leg with only very short posterior hairs 50

49. Mid femur, at about one-third its length, with compact cluster of two to four moderately strong black bristles 1.50–1.75 times as long as femoral diameter and without distinct anteroventral bristles near base. Fore femur, at about one-third its length, with loose to compact cluster of three to five strong ventral bristles some or all about twice as long as much weaker posteroventral bristles on basal one-third of femur. Silver-gray spots on tergite 2 as wide as long, on tergites 3 and 4 about 1.6 times as wide as long. Surstylus with shorter lobe at its mid length from 1.25 to 2.0 times as wide as longer lobe at its mid length (as in Fig. 236). Length 7.9–8.9 mm. Northwestern

Canada, high boreal and alpine regions .... *yukonensis* **Vockeroth**

Mid femur in some specimens without distinct ventral bristles, in some with up to nine very weak anteroventral bristles near base; at one-third its length with one slender bristle about 1.5 times as long as femoral diameter or two moderately strong well-separated bristles at most 1.25 times as long as femoral diameter. Fore femur with two to nine irregularly spaced ventral bristles only slightly longer and stronger than posteroventral bristles on basal one-third of femur. Silver-gray spots on tergite 2 about four-fifths as wide as long, on tergites 3 and 4 subquadrate. Surstylus with shorter lobe, at its mid length, from 0.80 to 1.17 times as wide as longer lobe at its mid length. Length 4.8–7.6 mm. High boreal and arctic regions, including Greenland .... ***carinatus* (Curran)**

50. Pruinosity of face with oblique rows of punctures (Fig. 6) or with faint lateral ripples (Figs. 267, 268). Wing membrane in some species slightly or extensively bare basally 51

Pruinosity of face uniformly distributed with only tubercle or median stripe bare. Wing membrane entirely trichose ... 58

51. Face with oblique rows of rounded punctures over most of its surface (Fig. 6). Fore tibia with strong black posterior bristles (Figs. 94*c*, 96*c*). Abdominal markings metallic bluish or bronze, never with orange background. Western Canada 52

Face with faint oblique lateral ripples (Figs. 267, 268). Fore tibia with weaker posterior bristles (Figs. 93*c*, 95*c*). Abdominal markings with orange background in some species. Widespread 55

52. Wing with cells c and bm bare or with some scattered microtrichia near apex 53

Wing with cells c and bm mostly densely trichose, bare only basally or anterobasally 54

53. Face with only tubercle shining black; face slightly broader, with coarser punctures (Fig. 6). Scutellum usually with only white hairs. Western Canada .... ***stegnus* (Say)**

Face with shining median black stripe more extensive, usually reaching lower margin of face and extending above upper limit of tubercle; face slightly narrower, with finer punctures. Scutellum with black hairs at least posteriorly. Western United States .... *spinipes* **Vockeroth**

54. Wing with cells c and bm entirely trichose. Upper pleural hairs black. Face with only tubercle shining black; area between tubercle and lower facial margin pruinose (as in Fig. 6). Fore tibia posteriorly with rather slender bristles on entire length; bristles decreasing in length toward base and very short on about basal one-third (Fig. 96*c*). Western Canada .... ***stegnoides* Vockeroth**

Wing with cell c bare on about basal one-sixth and cell bm bare on about anterobasal one-third. Upper pleural hairs usually entirely pale. Face with shining median black stripe extending from lower facial margin to well above tubercle; fore tibia posteriorly with five to seven strong bristles of nearly equal length, without bristles on about basal one-quarter (Fig. 94*c*). Oregon, California .... ***hesperius* Vockeroth**

55. Face moderately produced below, with anterior oral margin produced at least as far and usually slightly farther forward than facial tubercle (Fig. 268). Wing with cell bm bare or nearly so on at least basal half of anterior margin. Surstylus with dorsobasal lobe large (Fig. 270). Widespread .... ***obscurus* (Say)**

Face only slightly produced below, with anterior oral margin not extending as far forward as facial tubercle (Fig. 267). Wing with microtrichia variable in extent but commonly present along most or all of anterior margin of cell bm. Surstylus with dorsobasal lobe smaller (Figs. 269, 271) 56

56. Wing membrane extensively bare; cell c trichose on at most apical one-fifth; cell bm with only slender patch of microtrichia near apex; cell cua_1 bare on entire width at base. Anepisternum usually entirely white haired, rarely with some black hairs. Surstylus with dorsobasal lobe very small; process beyond lobe curved and nearly as broad as base of surstylus (Fig. 271). Northwestern Canada .... ***sabulicola* Vockeroth**

Wing membrane more trichose; cell c trichose on at least apical half; cell bm trichose along entire posterior margin; cell cua_1 entirely trichose. Anepisternum with at least some black hairs near upper margin. Surstylus with dorsobasal lobe larger; process beyond lobe nearly straight and more slender than base of surstylus before lobe (Fig. 269) 57

57. Lateral hairs of tergites almost all black, strong and long; longest hairs on tergite 3 at least half as long as apical width of tergite. Lower katepisternal hairs (near mid-ventral line) black. Wing membrane entirely trichose. Posterior bristles of fore and mid tibiae longer and stronger; longest bristles about half as long as tibia; first bristle on fore tibia near base and

preceded by some short hairs (Fig. 95). Western Canada .... ***squamulae* (Curran)**

Lateral hairs of tergites mostly pale, fine, and short; longest hairs on tergite 3 about one-third as long as apical width of tergite. Lower katepisternal hairs white. Wing of eastern specimens with membrane mostly trichose; wing of western specimens with cell c with at least small bare area at base and cell bm with at least bare median stripe near base and commonly with much of anterior margin bare. Posterior bristles of fore and mid tibia shorter and weaker; longest bristles about one-third as long as tibia; first distinct bristle on fore tibia at one-third or more tibial length and preceded by several shorter and weaker hairlike bristles (Fig. 93). Widespread .... ***confusus* (Curran)**

58(50). Face with weak but distinct keel between tubercle and upper end of face, with rather small tubercle. First tarsomere of mid leg with three strong black anteroventral setae on apical two-thirds of its length; all setae at least as long as diameter of tarsomere (Fig. 98*j*). Boreal regions .... ***setitarsis* Vockeroth**

Face without keel above antenna, with large tubercle. First tarsomere of mid leg without distinct anteroventral setae. Colorado .... ***protrusus* Vockeroth**

59(47). Abdomen with extensive orange markings 60

Abdomen without orange markings, entirely dark 64

60. Legs almost entirely black, at most extreme apices of femora and bases of tibiae yellowish. Tergite 5 mostly or entirely orange (Fig. 171). Northwestern Canada .... ***rufigaster* Vockeroth**

Fore and mid femora yellow on at least apical half of anterior surface; fore and mid tibiae mostly or entirely yellow. Tergite 5 black or with pair of obscure orange basal spots (Fig. 172) 61

61. Fore femur posteriorly near apex with two or three long black hairs with curled apices; these hairs contrasting with preceding shorter dense pale hairs; fore tibia posteriorly with dense fine pale hairs some longer than tibial diameter (Fig. 92*a*). Face about five-eighths as wide as head. Wing with cell c bare at least at base; cell bm bare except at apex. Tergite 2 with pair of orange spots; tergites 3 and 4 with orange spots widely separated from lateral margins. Western Canada .... ***kelloggi* (Snow)**

Fore femur with sparser and shorter hairs decreasing in length toward apex; hairs without curled apices; fore tibia with only very short hairs. Face not more than half as wide as head. Wing membrane entirely trichose or extensively bare. Tergite 2 black or mostly red-orange; tergites 3 and 4 either with orange spots reaching lateral margins in some species or entirely red-orange 62

62. Wing membrane entirely trichose. Spots of tergites 3 and 4 yellow-orange, distinct, narrowed laterally (Fig. 172). Hairs of scutum and scutellum longer; longest scutellar hairs about two-thirds as long as arista; lower part of katepisternum, between upper and lower patches of hair, weakly but distinctly pruinose. Widespread . . . . ***rosarum*** **(Fabricius)**

Wing with base of cell c and most of cell bm bare. Tergites 3 and 4 either with spots red-orange, usually obscure, not narrowed laterally, or entirely red-orange. Hairs of scutum and scutellum very short; longest scutellar hairs less than half as long as arista. Lower part of katepisternum, between upper and lower patches of hair, strongly shining or weakly pruinose 63

63. Tergite 2 black with submetallic bluish spots; tergites 3 and 4 with basal red-orange spots; tergite 5 black or with obscure basal orange spots. Lower part of katepisternum, between upper and lower patches of hair, strongly shining with only narrow posterior margin pruinose. Western Canada . . . . ***rufimaculatus*** **Vockeroth**

Tergite 2 red-orange with anterior margin and lateral margins narrowly black; tergites 3–5 entirely red-orange. Lower part of katepisternum, between upper and lower patches of hair, entirely or almost entirely moderately pruinose. Southern California . . . . ***russatus*** **Vockeroth**

64. Hairs of notopleural area partly or entirely black; scutellar hairs black 65

Hairs of notopleural area white to yellow brown; scutellar hairs partly to entirely pale 67

65. Face above tubercle with two or three weak vertical ridges bordering weak median keel or shallow groove. Lower part of katepisternum, between upper and lower patches of hair, weakly but distinctly pruinose. Western Canada, arctic and alpine regions . . . . ***pullatus*** **Vockeroth** (in part)

Face above tubercle smoothly rounded, without trace of ridges or median keel. Lower part of katepisternum, between upper and lower patches of hair, mostly or entirely shining black 66

66. Wing membrane entirely trichose. Shining area of katepisternum, above lower patch of hairs, extending to

posterior margin of sclerite. Surstylus elongate-oval, broadest at mid length (Fig. 243*b*). Paramere (Fig. 243*c*) with two spines. Yukon Territory .... ***coracinus* Vockeroth**

Wing with extreme base of cell c and posterobasal part of cell bm bare. Shining area of katepisternum, above lower patch of hairs, bordered posteriorly by narrow but distinct band of minute pile. Surstylus straplike, narrowest at mid length (Fig. 244*b*). Paramere (Fig. 244*c*) with one spine. Western Canada .... ***latus* (Curran)**

67. Legs almost entirely black; only extreme apices of femora and extreme bases of tibiae obscurely yellowish. Antenna black. Wing membrane entirely trichose 68

Legs with at least broad apices of fore and mid femora and most of fore and mid tibiae orange to red-orange. Antenna with first flagellomere broadly orange below. Wing membrane entirely trichose or with cells c and bm slightly or extensively bare 69

68. Fore tibia posteriorly with dense fine mostly pale erect hairs almost twice as long as tibial diameter; first tarsomere of fore leg with similar but shorter hairs (Fig. 99*b*). Mid femur anteroventrally without black setae, with long fine white hairs on basal half. Yukon Territory .... ***woodi* Vockeroth**

Fore tibia posteriorly with short black appressed hairs; first tarsomere of fore leg with only very short hairs. Mid femur anteroventrally with irregular row of short weak stiff black setae on most of its length, without pale hairs. Western Canada, arctic and alpine regions ..... ***pullatus* Vockeroth** (in part)

69. Femora entirely red-orange. Face near upper margin usually with faint indication of median groove. Cell bm bare or nearly so on at least basal one-third. Western Canada .... ***willistoni* (Goot)**

Hind femur dark brown to black on at least basal half, fore and mid femora usually black at least at base. Face near upper margin either smoothly rounded medially, or with distinct ridges and median groove, or with weak but distinct median keel. Wing microtrichia variable; membrane almost entirely trichose to extensively bare 70

70. Upper part of face evenly rounded medially, without ridges, grooves, or median keel. Paramere stout, with short curved spine closely appressed to base (Fig. 245*c*). Widespread .... ***concinnus* (Snow)**

Upper part of face either with weak but distinct median keel or with weak submedian ridges and shallow median groove. Paramere slender, with long spine well-separated from base (Fig. 246*c*) 71

71. Face sparsely pruinose, laterally with minute oblique striations in integument. Upper part of face with moderately strong median keel. Wing with cell bm at least one-third bare. Paramere with spine moderately separated from base (Fig. 246*c*). Central Nearctic . . . . ***luteipennis*** **(Curran)**

Face more densely pruinose, without striations in integument. Upper part of face either with very weak median keel or with very shallow median groove. Wing with cell bm entirely trichose or with at most narrow bare median stripe near base. Paramere with spine more widely separated from base. Boreal and western Canada . . . . ***striatus*** **Vockeroth**

72(1). First tarsomere of fore leg distinctly and uniformly broadened toward apex; second tarsomere slightly wider than long; third tarsomere almost twice as wide as long (Fig. 48); tarsus mostly or entirely dark brown to black above. Yellow spots of tergite 2 confluent medially (Fig. 156). Widespread . . . . ***granditarsis*** **(Forster)**

First tarsomere of fore leg usually parallel-sided, if very slightly broadened toward apex then tarsus entirely yellow-orange; second tarsomere slightly longer than wide; third tarsomere only slightly wider than long. Yellow spots of tergite 2, if present, separated by black median line 73

73. Face with oblique rows of rounded punctures (as in Fig. 6) or with faint oblique lateral ripples in pruinosity (as in Figs. 267, 269) 74

Face with pruinosity uniformly distributed except on tubercle, with neither punctures nor ripples 80

74. Face with distinct oblique rows of punctures over most of surface (as in Fig. 6). Western Canada 75

Face with faint lateral ripples in pruinosity (as in Figs. 267, 269). Widespread 78

75. Wing with cells c and bm extensively trichose; cell bm with microtrichia on most or all of posteroapical half 76

Wing with cells c and bm bare or with at most very few microtrichia near apex 77

76. Wing with cells c and bm entirely trichose. Face with only tubercle shining; area between tubercle and lower facial margin

pruinose (as in Fig. 6). Western Canada .... ***stegnoides* Vockeroth**

Wing with cell c bare on about basal one- to two-thirds; cell bm bare on about anterobasal half. Face with shining median black stripe extending from lower facial margin to well above tubercle. Oregon, California .... ***hesperius* Vockeroth**

77. Face with only tubercle shining black; face slightly broader with coarser punctures (as in Fig. 6). Western Canada .... ***stegnus* (Say)**

Face with shining median black stripe more extensive, usually reaching lower margin of face and extending above upper limit of tubercle; face slightly narrower with finer punctures. Western United States .... ***spinipes* Vockeroth**

78. Face moderately produced below, with anterior oral margin produced at least as far and usually slightly farther forward than facial tubercle (as in Fig. 269). Widespread .... ***obscurus* (Say)**

Face only slightly produced below, with anterior oral margin not extending as far forward as facial tubercle (as in Fig. 267) 79

79. Wing with membrane entirely trichose or with at most minute bare area at base of cell c. Western Canada .... ***squamulae* (Curran)**

In eastern specimens wing with at least small bare area near base of cell bm; in western specimens wing with base of cells c and bm both extensively bare .... ***confusus* (Curran)**
.... ***sabulicola* Vockeroth**

80. Tergites 3–5 entirely orange; tergite 2 black. Northwestern Canada .... ***rufigaster* Vockeroth**

Tergites 3–5 with at least posterior margins narrowly black; if these tergites mostly orange then tergite 2 also mostly orange 81

81. Tergite 3 with orange spots at least half as long as tergite and almost twice as long as those of tergite 4; spots of both tergites extending to lateral margins; tergites 2 and 5 black or with very obscure reddish markings. Widespread .... ***rosarum* (Fabricius)**

Tergites 3 and 4 with or without orange spots; if with, spots about same size on both tergites, not reaching lateral margins in some species, with similar although less distinct orange spots in some species on tergite 2 82

82. Tergites 3 and 4 each with pair of large orange spots widened laterally; tergite 5 black. Cells c and bm entirely bare. Face without trace of median keel above tubercle. Western Canada .... ***kelloggi* (Snow)**

Tergites very variable in pattern, commonly with dark metallic or gray or silver pruinose spots; if tergites 3 and 4 each with pair of large yellow or orange spots then tergite 5 at least partly yellow or cells c and bm extensively trichose. Face in some species with distinct median keel or ridge on upper part .... other spp.

Clé des espèces néarctiques de *Platycheirus*

1. Mâle (yeux contigus sur le front) 2

Femelle (yeux largement séparés sur le front) 72

2. Tibia antérieur s'élargissant faiblement à fortement vers l'apex, ou premier article du tarse de la patte antérieure distinctement élargi, ou les deux (fig. 47*a*, 49*a*–57*a*, 65*a*–69*a*) 3

Tibia et tarse antérieurs minces sur toute leur longueur (fig. 91*a*, 92*a*, 97*a*, 99*b*) 45

3. Tibia antérieur tout au plus légèrement élargi vers l'apex et nettement plus étroit que le premier article du tarse de la patte antérieure (fig. 47*a*, 49*a*–55*a*) 4

Tibia antérieur habituellement élargi sur au moins la moitié apicale, au moins aussi large que le premier article du tarse (fig. 59*a*, 65*a*–83*a*, 85*a*–88*a*, 90*a*); chez certaines espèces, très élargi seulement près de l'apex et légèrement plus mince que le premier article très large du tarse (fig. 56*a*, 57*b*, 60*b*) 11

4. Tarse antérieur tout noir et pourvu, sur le premier article, d'un long appendice antérieur triangulaire (fig. 47*a*). Trois premiers articles du tarse médian beaucoup plus larges que l'apex du tibia médian (fig. 47*b*). Moitié postérieure du tergite 2, tergite 3 et moitié antérieure du tergite 4 largement ou entièrement orange (fig. 155). Espèce répandue .... ***granditarsis* (Forster)**

Tarse antérieur ayant un premier article blanchâtre et dépourvu d'appendice antérieur, de profil presque symétrique (fig. 49*a*–55*a*). Tarse médian plus mince. Tergites 2–4 noirs, soulignés de taches sublatérales argentées ou jaunâtres; quelques taches largement réunies au milieu (fig. 157) ... 5

5. Tibia médian pourvu, sur le tiers basilaire de la face postérieure à postéro-ventrale, de poils denses, fins et ondulés, d'une longueur d'au moins deux fois le diamètre du tibia et aussi longs, ou presque, que les poils ornant la moitié apicale de la face postérieure; sur le tiers basilaire de la face antéro-ventrale, les poils sont habituellement denses, fins et noirs et au moins trois fois plus longs que le diamètre du tibia (fig. 49*j*, 51*j*, 53*k*, 54*k*, 55*j*) 6

Tibia médian pourvu, sur le tiers basilaire de la face postérieure, de poils courts et presque droits, d'une longueur presque égale au diamètre du tibia et deux fois moins longs, au moins, que les poils longs et fins comme des soies qui ornent la moitié apicale de la face postérieure; sur la face antéro-ventrale, les poils sont très courts, droits et peu visibles (fig. 50*k*, 52*k*) 10

6. Tibia postérieur dont le tiers de la base est orné sur la face antéro-dorsale de poils noirs, denses et ondulés, jusqu'à trois fois plus longs que le diamètre du tibia. Premier article du tarse de la patte antérieure environ trois fois plus large que l'apex du tibia antérieur; apex du premier article très oblique (fig. 53*a*). Colorado .... ***oreadis* Vockeroth**

Tibia postérieur pourvu seulement de quelques poils courts ou rarement, sur la face antéro-dorsale, de quelques poils plus longs à mi-longueur ou au-delà. Premier article du tarse antérieur tout au plus 2,2 fois plus large que l'apex du tibia antérieur; apex de l'article légèrement oblique ou transversal (fig. 49*a*, 51*a*, 54*a*, 55*a*) 7

7. Premier article du tarse de la patte antérieure au moins deux fois plus long que large (fig. 54*a*). Nord-ouest du Canada .... ***subordinatus* Becker**

Premier article du tarse de la patte antérieure au moins 1,6 fois plus long que large (fig. 49*a*, 51*a*, 55*a*) 8

8. Tibia médian pourvu, sur la face antéro-ventrale, de poils courts, tout au plus légèrement plus longs que le diamètre du tibia (fig. 55*j*). Espèce répandue .... ***thylax* Hull**

Tibia médian orné d'une touffe de longs poils ondulés sur environ le tiers basilaire de la face antéro-ventrale; poils au moins trois fois plus longs que le diamètre du tibia et beaucoup plus longs que les autres poils ornant le reste de la face antéro-ventrale (fig. 49*j*, 51*j*) 9

9. Patte médiane dont les deux premiers articles du tarse sont jaunes; premier article fortement comprimé; deuxième article

légèrement comprimé; trois derniers articles noirs (fig. 49*j*). Espèce répandue .... ***discimanus*** **Loew**

Patte médiane entièrement noire; articles du tarse non comprimés (fig. 51*j*). Régions boréales et arctiques du Canada .... ***groenlandicus*** **Curran**

10. Face ventrale du trochanter antérieur ornée de nombreuses petites soies noires et raides. Face ventrale du fémur antérieur pourvue seulement de poils longs et fins. Tibias antérieur et médian pourvus, sur la moitié apicale de la face postérieure, de nombreux poils noirs et fins, disposés de façon irrégulière et de longueurs variées (fig. 52*a*,*k*). Fémur médian garni, sur un cinquième apical, de poils courts, raides et noirs sur la face antéro-ventrale. Premier article du tarse de la patte antérieure au moins 2,3 fois plus large que l'apex du tibia antérieur (fig. 52*a*). Alaska .... ***manicatus*** **(Meigen)**

Face ventrale du trochanter antérieur pourvue de quelques soies pâles et fines. Fémur antérieur orné, sur la moitié basilaire de la face ventrale, d'une rangée de trois à quatre soies raides, jaunes ou noires, mesurant environ les deux tiers du diamètre du fémur. Portion apicale de la face postérieure des tibias antérieur et médian ornée d'une seule rangée de poils droits et noirs, mesurant jusqu'à 2,5 fois le diamètre du tibia (fig. 50*a*,*k*). Un cinquième de l'apex du fémur médian pourvu de longs poils noirs, fins et enchevêtrés, sur la face antéro-ventrale. Premier article du tarse de la patte antérieure de 1,3 à 2 fois plus large que l'apex du tibia antérieur (fig. 50*a*). Régions boréales .... ***flabellus*** **Hull**

11(3). Face postérieure du fémur antérieur recouvert de poils noirs et denses, assez forts, presque uniformes et légèrement aplatis; aucune touffe de poils en saillie près de la base ainsi que de poils ou de soies plus longs le long du fémur (fig. 56*a*). Face antérieure du fémur médian présentant une concavité distincte juste au-delà de la longueur médiane et ornée en dessous de soies noires incurvées et très courtes (fig. 56*e*) 12

Face postérieure du fémur antérieur dépourvue de poils noirs aplatis; faibles poils uniformes, touffes de poils blancs ou noirs près de la base ou de poils ou soies forts, plus longs et en saillie sur toute la longueur (fig. 59*a*, 65*a*–83*a*, 85*a*–88*a*, 90*a*). Face antérieure du fémur médian non concave ou tout au plus légèrement concave 19

12. Premier article du tarse de la patte antérieure dépourvu de carène dorsale; deuxième article aussi large que la portion la plus large du tibia antérieur et d'une largeur équivalant

environ aux neuf dixièmes du premier article (fig. 60*b*). Ouest du Canada .... ***latitarsis* Vockeroth**

Premier article du tarse de la patte antérieure présentant une faible carène dorsale sur la moitié apicale ou une carène dorsale prononcée sur presque toute sa longueur; deuxième article d'une largeur tout au plus égale aux quatre cinquièmes de celle du tibia antérieur et du premier article (fig. 56*a*, 57*b*) 13

13. Premier article du tarse de la patte antérieure présentant une carène dorsale assez prononcée dans la moitié apicale, mais dépourvu de carène nette dans la moitié basilaire (fig. 56*a*). Chez certaines espèces, l'anépimère est orné de quelques poils dans la moitié inférieure 14

Premier article du tarse de la patte antérieure pourvu d'une carène dorsale distincte sur toute sa longueur, bien que, chez certaines espèces, la carène devienne un peu plus petite près de la base de l'article (fig. 57*b*). Anépimère poilu uniquement dans la partie supérieure 17

14. Moitié apicale de la face ventrale du tibia médian ornée de poils foncés, enchevêtrés et légèrement dressés, sur la partie basilaire du renflement apical marqué du tibia, à peu près aussi longs que le diamètre subapical minimal du tibia (fig. 62*L*). Anépimère poilu uniquement dans sa portion supérieure; poils formant une touffe épaisse. Membrane alaire entièrement velue. Régions boréales et de l'ouest, néarctiques; paléoarctiques .... ***nielseni* Vockeroth**

Moitié apicale de la face ventrale du tibia médian ornée de poils très courts à peine visibles; faible renflement apical du tibia (fig. 61*L*). Chez certaines espèces, l'anépimère porte quelques poils dans la partie inférieure; poils ne formant pas une touffe dense dans la partie supérieure. Membrane alaire entièrement velue ou portant des zones glabres près de la base 15

15. Moitié basale de la face antéro-ventrale du tibia médian portant une touffe de poils légèrement affaissés, courts et presque droits; les poils les plus longs font au plus 1,5 fois le diamètre minimal du tibia (fig. 61*k*). Anépimère poilu dans la portion supérieure seulement. Poils du scutellum largement ou entièrement jaunes. Espèce répandue .... ***nearcticus* Vockeroth**

Moitié basale de la face antéro-ventrale du tibia médian ornée d'une touffe de longs poils noirs ondulés et dressés; les poils les plus longs font plus de deux fois le diamètre du tibia (fig. 56*e*). Chez certaines espèces, il y a des poils dans la moitié inférieure de l'anépimère. Poils du scutellum entièrement jaunes ou à moitié noirs 16

16. Premier article du tarse de la patte postérieure se rétrécissant d'une façon brusque et marquée, à mi-longueur (fig. 51*m*,*p*). Poils du scutellum jaunes. Anépimère habituellement poilu dans la moitié supérieure seulement; il y a rarement des poils dans la portion inférieure. Zone glabre sur la cellule bm, couvrant habituellement au moins la moitié de la cellule et atteignant ou presque le bord antérieur; rarement moins large. Est du Canada .... ***inversus* Ide**

Premier article du tarse de la patte postérieure se rétrécissant graduellement sur les trois quarts apicaux de sa longueur (fig. 56*m*,*p*). Poils du scutellum à moitié noirs, ou presque. Anépimère pourvu tout au moins de quelques poils dans la portion inférieure. Zone glabre sur la cellule bm couvrant au plus le tiers de la cellule et bien séparée du bord antérieur. Régions boréales et ouest du Canada ...... ***amplus* Curran**

17. Portion inférieure du katépisterne ornée de longs poils, d'une longueur égale au moins aux deux tiers de l'arista. Face antérieure du tibia postérieur pourvue de soies clairsemées, celles du tiers basilaire étant beaucoup plus courtes que les autres sur les deux tiers apicaux (fig. 64*n*). Premier article du tarse de la patte antérieure souligné d'une carène devenant de moins en moins profonde à proximité de la base de l'article. Membrane alaire entièrement velue ou marquée de minuscules zones glabres peu visibles à la base des cellules c et bm. Ouest du Canada .... ***peltatoides* Curran**

Portion inférieure du katépisterne ornée de poils courts, d'une longueur équivalant tout au plus à la moitié de l'arista. Tibia postérieur pourvu de soies très denses sur la face antérieure; les soies sur le tiers basilaire sont presque aussi longues que celles des deux tiers apicaux (fig. 57*m*, 63*n*). Premier article du tarse de la patte antérieure souligné d'une carène de hauteur presque uniforme (fig. 57*b*). Membrane alaire entièrement velue ou pourvue de moyennes zones glabres à la base des cellules c et bm 18

18. Aile pourvue d'une membrane entièrement velue. Moitié basale de la face antéro-ventrale du tibia médian portant une touffe de longs poils ondulés et dressés, noirs ou jaunes; les poils les plus longs sont environ trois fois plus longs que le diamètre du tibia (comme dans la fig. 56*e*). Anépimère orné d'une touffe de poils très denses, dont la base n'est visible que dans la partie inférieure. Régions boréales et ouest du Canada .... ***holarcticus* Vockeroth**

Aile, environ un sixième basilaire de la cellule c et un quart basilaire de la cellule bm, glabres. Moitié basilaire de la face antéro-ventrale du tibia médian ornée d'une touffe de poils légèrement affaissés, plus courts et plus droits et habituellement pâles; les poils les plus longs font au plus 1,5 fois le diamètre du tibia (comme dans la fig. 61*k*). Anépimère pourvu d'une touffe assez dense de poils, dont la base est visible partout. Ouest du Canada .... ***octavus* Vockeroth**

19(11). Face postérieure du fémur antérieur pourvue de deux larges touffes denses ou assez clairsemées de longs poils noirs, ondulés et grossiers, précédées d'une touffe similaire faite de quelques poils blancs ondulés (fig. 65*a*–69*a*) 20

Face postérieure du fémur antérieur dépourvue des deux larges touffes de poils noirs, mais pourvue de poils fins et uniformes, d'une ou de deux touffes de poils longs ayant un apex aplati et élargi, d'une touffe subbasale constituée de plusieurs poils longs et blancs et suivie, chez certaines espèces, d'une faible touffe unique faite de plusieurs poils noirs, ou d'une rangée uniforme de soies ou de poils forts et très espacés sur presque toute sa longueur (fig. 59*a*, 70*a*–83*a*, 85*a*–90*a*) 24

20. Hanche médiane pourvue d'un mince appendice ventral digitiforme (fig. 31*h*). Fémur antérieur, au-delà des touffes subbasilaires, avec de nombreux poils noirs, longs et forts; deuxième article du tarse de la patte antérieure d'une longueur équivalant environ à un sixième du premier article (fig. 68*a*). Espèce répandue .... ***scutatus* (Meigen)**

Hanche médiane dépourvue d'appendice ventral. Fémur antérieur, au-delà des touffes subbasilaires, avec trois longues soies noires bien espacées parmi des poils plus courts; deuxième article du tarse de la patte antérieure d'une longueur à peu près égale aux deux cinquièmes du premier article (fig. 65*a*–67*a*, 69*a*) 21

21. Tibia antérieur fortement élargi sur le tiers apical, avec un angle postéro-apical très arrondi; premier article du tarse de la patte antérieure très élargi vers l'arrière sur environ les deux tiers de la base, formant brusquement un angle à cet endroit puis continuant en ligne droite jusqu'à l'apex (fig. 69*a*). Quatre premiers articles du tarse de la patte médiane jaunes; cinquième article brun. Aile pourvue d'une cellule bm entièrement velue. Alaska, Colombie-Britannique, Québec .... ***urakawensis* (Matsumura)**

Tibia antérieur moins élargi, avec un angle postéro-apical subaigu ou légèrement arrondi; élargissement graduel du premier article du tarse postérieurement, sans angle distinct (fig. 65*a*–67*a*). Au moins les quatre derniers articles du tarse de la patte médiane sont de couleur brune à brun foncé sur le dessus. Cellule bm présentant tout au moins une petite zone glabre près de la base et pourvue antérieurement, chez certaines espèces, d'une étroite zone glabre sur presque toute sa longueur 22

22. Tibia antérieur s'élargissant uniformément, de la base à l'apex; élargissement graduel du premier article du tarse de la patte antérieure de la base à l'apex, ses bords étant légèrement divergents partout et son apex nettement arqué; marges basilaire et apicale du deuxième article nettement arquées (fig. 67*a*). Cellule bm ne présentant qu'une minuscule zone glabre près de la base. Régions boréales .... ***nigrofemoratus* Kanervo**

Tibia antérieur s'élargissant uniformément sur les trois quarts de la moitié basilaire, suivi d'un élargissement un peu plus marqué; élargissement graduel du premier article du tarse sur la moitié basilaire, lequel présente des côtés parallèles dans la moitié apicale et un apex transversalement tronqué; deuxième article presque rectangulaire pourvu d'un apex transversalement tronqué ou presque (fig. 65*a*, 66*a*). Cellule bm pourvue d'une zone glabre de petite à assez large. Régions boréales et ouest du Canada 23

23. Angle postéro-apical du tibia antérieur nettement arrondi et s'avançant légèrement au-delà du bord postérieur du premier article du tarse (fig. 66*a*). Premier article du tarse de la patte médiane foncé, tout au plus légèrement plus pâle que le cinquième article. Côte du Pacifique .... ***ciliatus* Bigot**

Angle postéro-apical du tibia antérieur subaigu, n'avançant pas au-delà du premier article du tarse (fig. 65*a*). Premier article du tarse de la patte médiane de jaune à brun, habituellement beaucoup plus pâle que le cinquième article. Régions boréales et ouest du Canada .... ***albimanus* (Fabricius)**

24. Tibia antérieur présentant une carène dorsale longitudinale bien marquée, qui s'élève à proximité de l'apex du tibia; premier article du tarse de la patte antérieure plutôt mince, de légèrement à fortement comprimé à mi-longueur (fig. 85*a*). Tergites 3 et 4 jaune orangé avec un liseré postérieur noir, sans bande médiane noire (fig. 159). Espèce répandue .... ***normae* Fluke**

Tibia antérieur sans carène dorsale; élargissement habituellement graduel du premier article du tarse de la patte antérieure, à partir de la base; aucun rétrécissement (fig. 59*a*, 70*a*–83*a*, 86*a*–88*a*, 90*a*). Tergites 3 et 4 avec ou sans bande médiane noire, surtout noirs chez certaines espèces (fig. 160–170) 25

25. Fémur médian pourvu d'une épaisse brosse de soies raides oranges et noires près de la base de sa face postéro-ventrale (fig. 81*i*). Élargissement marqué et presque uniforme du tibia médian sur environ les trois quarts apicaux; poils denses et jaunes sur la face antéro-ventrale. Premier article du tarse de la patte médiane nettement élargi et affaissé (fig. 81*f*). Tergites largement jaunes. Espèce répandue ...... ***quadratus* (Say)**

Fémur médian sans brosse de soies, pourvu tout au plus de sept soies jaunes ou noires bien espacées sur la face postéro-ventrale de la moitié basilaire (fig. 78*i*, 79*g*, 82*i*, 86*i*, 88*i*, 89*i*). Tibia médian mince ou élargi légèrement et de façon irrégulière sur au plus la moitié apicale, avec ou sans poils denses sur la face ventrale. Premier article du tarse de la patte médiane ni élargi, ni affaissé (fig. 70*e*, 71*e*, 73*k*, 76*k*, 87*f*). Tergites de couleur variable, pour la plupart jaunes ou noirs 26

26. Face postérieure du fémur antérieur pourvu d'une touffe compacte de poils longs tous terminés par un renflement lancéolé, juste au-delà de la base (fig. 70*a*–72*a*) 27

Face postérieure du fémur antérieur avec ou sans touffe subbasilaire; s'il y a une touffe, les poils ne sont pas renflés à l'apex (fig. 57*a*, 73*a*–83*a*, 86*a*–88*a*) 29

27. Fémur antérieur pourvu d'une touffe subbasilaire faite de longs poils ayant un apex légèrement élargi; chez certains spécimens, cette touffe est précédée d'un unique poil pâle, long et mince (fig. 72*a*). Tibia médian dépourvu, sur la face antéro-ventrale, de longs poils affaissés ou dressés. Tarses antérieur et médian entièrement jaunes. Tergite 5 maculé de grosses taches jaunes sur la face antéro-latérale (fig. 160). Est du Canada .... ***thompsoni* Vockeroth**

Fémur antérieur pourvu de deux touffes subbasilaires, faites de longs poils ayant un apex de légèrement à fortement élargi (fig. 70*a*, 71*a*). Moitié basilaire du tibia médian orné, sur la face antéro-ventrale, de poils longs, affaissés ou dressés (fig. 70*e*, 71*e*). Généralement, un ou plusieurs articles du tarse des pattes antérieure et médiane sont nettement plus foncés sur le dessus que le premier article jaune. Tergite 5 souligné de taches jaunes antéro-latérales ou entièrement noir 28

28. Tergite 5 noir (fig. 161). Touffe basilaire sur le fémur antérieur composée uniquement de poils pâles; tibia et tarse antérieurs assez étroits (fig. 71*a*). Fémur médian pourvu sur la face postéro-ventrale de nombreux poils plus longs que le diamètre du fémur; tibia médian orné, sur la face antéro-ventrale du tiers basilaire, de longs poils noirs dressés (fig. 71*e*). Sur la patte médiane, le troisième et le quatrième articles du tarse

sont habituellement plus foncés que le premier. Régions boréales .... ***pilatus* Vockeroth**

Tergite 5 orné d'une paire de grosses taches jaunes antéro-latérales. Sur le fémur antérieur, touffe basilaire de poils ayant un apex de brun à noir; tibia et tarse antérieurs plus larges (fig. 70*a*). Fémur médian pourvu, sur la face postéro-ventrale, de poils beaucoup plus courts que le diamètre du fémur; tibia médian orné de longs poils noirs fortement affaissés sur la moitié basilaire de la face antéro-ventrale (fig. 70*e*). Quatre premiers articles du tarse de la patte médiane jaunes; cinquième article généralement brun ou noir sur le dessus. Régions boréales et ouest du Canada .... ***nodosus* Curran**

29. Face postérieure du fémur antérieur pourvue d'une touffe subbasilaire de deux à trois longs poils affaissés, blancs ou jaunâtres, ayant un apex ondulé (fig. 73*a*–80*a*, 82*a*, 83*a*) 30

Face postérieure du fémur antérieur dépourvue de touffe subbasilaire de poils blancs, mais ornée de poils fins presque uniformes, de poils plus longs et bien espacés ou de longues soies plus minces sur presque toute sa longueur (fig. 59*a*, 86*a*–88*a*, 90*a*) 39

30. Face antéro-ventrale du fémur médian pourvue, dans la moitié apicale, d'une rangée presque uniforme faite de 7 à 16 soies noires, courtes et fortes avec, chez certaines espèces, un ou deux poils incurvés près ou au-delà de l'extrémité de la rangée (fig. 78*i*, 82*i*). Face postérieure du fémur antérieur ayant de 3 à 5 soies noires (ou blanches chez certaines espèces), longues, assez robustes, légèrement ondulées et également espacées, sur les trois quarts apicaux (fig. 78*a*, 82*a*) 31

Face antéro-ventrale du fémur médian dépourvue d'une rangée de soies solides; chez certaines espèces, il y a un groupe de soies minces à mi-longueur environ ou de poils fins dans la région pré-apicale (fig. 79*g*). Face postérieure du fémur antérieur ornée de poils fins et uniformes ou, tout au plus, d'une rangée de quatre à cinq soies noires, longues et minces (fig. 73*a*–77*a*, 79*a*, 80*a*, 83*a*) 32

31. Tergites 3 et 4 ornés tous deux d'une paire de grosses taches jaunes. Fémur et tibia médians entièrement jaunes. Fémur et tibia postérieurs entièrement jaunes ou soulignés chacun d'un anneau foncé. Élargissement presque uniforme du tibia antérieur, de la base à l'apex (fig. 78*a*). Espèce répandue .... ***immarginatus* (Zetterstedt)**

Tergites 2–4 tous pourvus d'une paire de taches argentées sur fond foncé (fig. 162). Fémurs et tibias médian et postérieur largement noirs. Élargissement soudain du tibia antérieur, aux trois quarts environ de sa longueur (fig. 82*a*). Colombie-Britannique .... ***setipes* Vockeroth**

32. Tibia antérieur fortement et brusquement élargi sur les deux cinquièmes apicaux, légèrement plus étroit dans la portion apicale que pré-apicale (fig. 80*a*, 83*a*) 33

Tibia antérieur s'élargissant uniformément et d'une façon moins prononcée, de la base à l'apex (fig. 73*a*–77*a*, 79*a*) 34

33. Deuxième et troisième articles du tarse de la patte antérieure tous deux d'une longueur équivalant au plus aux trois quarts de leur largeur (fig. 80*a*). Portion basilaire de la face antéro-ventrale du tibia médian ornée de poils noirs dressés ou affaissés, au moins trois fois plus longs que le diamètre du tibia. Poils de l'anépisterne et de l'anépimère surtout jaune brun. Régions boréales et ouest du Canada .... ***podagratus* (Zetterstedt)**

Deuxième et troisième articles du tarse de la patte antérieure tous deux au moins aussi longs que larges (fig. 83*a*). Face antéro-ventrale du tibia médian pourvue seulement de poils très courts et peu visibles. Poils de l'anépisterne et de l'anépimère habituellement tout noirs. Ouest du Canada ... . ***tenebrosus* Coquillett**

34. Tergite 5 largement ou entièrement jaune, orné tout au plus d'une rayure médiane noire et d'un étroit liséré postérieur noir; tergites 3 et 4 maculés de larges taches latérales jaunes faisant au moins les neuf dixièmes de la longueur du tergite et souvent réunies sur la face médiane (fig. 164). Tiers apical de la face antéro-ventrale du fémur médian pourvu de faibles poils ondulés, au moins aussi longs que le diamètre du fémur (fig. 79*g*). Espèce répandue .... ***perpallidus* Verrall**

Tergite 5 entièrement noir ou orné tout au plus de petites taches antéro-latérales jaunes floues; tergites 3 et 4 présentant des taches latérales argentées ou presque métalliques; lorsqu'il y a des taches jaunes, presque carrées et distinctes, celles-ci mesurent au plus les quatre cinquièmes de la longueur du tergite et elles ne convergent pas sur le plan médian (fig. 165, 166). Tiers apical de la face antéro-ventrale du fémur médian glabre ou pourvu de quelques poils beaucoup plus courts que le diamètre du fémur 35

35. Fémurs antérieur et médian noirs ayant un apex faiblement jaune. Premier article du tarse de la patte antérieure à côtés parallèles, sauf à l'extrémité basilaire (fig. 73*a*, 76*a*) 36

Fémurs antérieur et médian entièrement ou largement jaunes, ornés tout au plus de rayures foncées sur une partie ou la totalité de leur longueur; premier article du tarse de la patte antérieure à côtés parallèles, uniquement dans la moitié apicale (fig. 74*a*, 75*a*) ou très légèrement rétréci vers l'apex (fig. 77*a*) 37

36. Face légèrement mais nettement avancée vers l'avant, en dessous (fig. 8). Cellules bm et *cup* glabres sur la face antérieure, sur presque toute la moitié basilaire. Face postérieure du fémur antérieur garnie sur presque toute sa longueur d'une rangée régulière comptant environ cinq poils noirs longs et forts; tibia antérieur assez large, à bord postérieur irrégulier (fig. 76*a*). Tibia médian pourvu, sur toute la face antéro-ventrale, de poils droits, assez clairsemés et de longueur uniforme et de trois poils noirs longs et épais sur la moitié apicale de la face postéro-ventrale (fig. 76*k*). Tergites 3 et 4 marqués de taches jaunâtres distinctes, recoupées d'une pruinosité argentée (fig. 165). Colombie-Britannique .... ***hispidipes*** **Vockeroth**

Face presque verticale (fig. 11). Membrane alaire entièrement velue. Face postérieure du fémur antérieur ornée de poils fins et uniformes, de longueur décroissante vers l'apex; tibia antérieur plus étroit, à bords réguliers (fig. 73*a*). Tibia médian orné de poils courts, denses et ondulés sur environ les trois quarts médians de la face antéro-ventrale, mais dépourvu de poils longs sur la face postéro-ventrale (fig. 73*k*). Taches sur les tergites 3 et 4 entièrement argentées ou sur fond jaune. Régions boréales et ouest du Canada .... ***aeratus*** **Coquillett**

37. Apex du tibia antérieur à angle postérieur tout au plus très légèrement en saillie, de sorte que l'apex est presque tronqué; premier article du tarse de la patte antérieure à bord postérieur légèrement incurvé sur les trois quarts apicaux, légèrement plus étroit à l'apex qu'à mi-longueur (fig. 77*a*). Taches sur les tergites 3 et 4 entièrement foncées, ayant un reflet argenté prononcé (la plupart des spécimens du nord) ou de légèrement à largement jaunes et recoupées d'une pruinosité argentée dense surtout visible en face postéro-dorsale (certains spécimens du nord et tous ceux du sud). Espèce répandue .... ***hyperboreus*** **(Staeger)**

Apex du tibia antérieur à angle postérieur nettement en saillie, de sorte que l'apex est échancré; premier article du tarse de la patte antérieure à côtés parallèles sur la moitié apicale (fig. 74*a*, 75*a*). Taches sur les tergites 3 et 4 tout jaunes ou légèrement métalliques, avec tout au plus une pruinosité argentée très clairsemée 38

38. Membrane alaire pourvue d'une cellule bc largement glabre et présence d'une minuscule zone glabre à la base des cellules c et bm. Au-delà de la touffe subbasilaire de poils blancs, fémur antérieur orné de poils fins et largement pâles, au plus légèrement plus longs que le diamètre du fémur (fig. 74*a*). Tergite 2 un peu plus long que large; tergites 3 et 4 presque carrés. Espèce répandue .... ***angustatus* (Zetterstedt)**

Membrane alaire entièrement velue. Au-delà de la touffe subbasilaire de poils blancs, face postérieure du fémur antérieur pourvue de poils plus longs et plus épais, habituellement noirs, dont certains sont au moins 1,5 fois plus longs que le diamètre du tibia (fig. 75*a*). Tergite 2 au moins un peu plus large que long; tergites 3 et 4 à peu près 1,3 fois plus larges que longs. Espèce répandue .... ***clypeatus* (Meigen)**

39(29). Tergites noirs ornés de taches argentées; aucune trace de marques jaunes (fig. 167). Premier article du tarse de la patte antérieure aussi large que long et nettement plus étroit dans la portion basilaire; quatre articles suivants devenant progressivement un peu plus étroit (fig. 80*a*). Tibias médian et postérieur noirs, soulignés d'un peu de jaune à la base et aussi à l'apex, chez certains. Régions boréales et ouest du Canada, Groënland .... ***varipes* Curran**

Tergites 3 et 4 ornés de larges taches jaunes, ou presque entièrement jaunes (fig. 168–170). Premier article du tarse de la patte antérieure au moins 1,5 fois plus long que large et fortement rétréci, uniquement très près de la base; quatre articles suivants devenant progressivement plus étroits, ou deuxième ou troisième article beaucoup plus étroit que le précédent (fig. 59*a*, 86*a*–88*a*). Tibias médian et postérieur noirs avec un peu de jaune à la base, ou largement ou entièrement jaunes 40

40. Face s'avançant nettement vers l'avant, en dessous (fig. 10). Taches jaunes sur les tergites 3 et 4 un peu plus larges que longues (fig. 168). Deuxième article du tarse (*P. jaerensis*) ou troisième article (*P. parmatus*) de la patte antérieure beaucoup plus étroit que le précédent (fig. 59*a*, 87*a*). Fémur postérieur noir ayant un peu de jaune à l'apex; tarse postérieur entièrement brun foncé à noir sur le dessus 41

Face verticale ou légèrement rentrée en dessous (fig. 10). Taches jaunes sur les tergites 3 et 4 nettement plus longues que larges, convergeant au milieu chez certaines espèces (fig. 169, 170). Rétrécissement graduel du tarse de la patte antérieure, du deuxième au cinquième article (fig. 86*a*, 88*a*). Fémur postérieur tout jaune ou marqué tout au plus d'un anneau noir sur la portion apicale; deuxième et troisième articles de la patte postérieure partiellement ou entièrement jaune terne à brillant sur le dessus 42

41. Tibia antérieur pourvu de nombreux poils postérieurs, plus longs que la largeur du tibia; deuxième article du tarse de la patte antérieure beaucoup plus large que long et seulement un peu plus étroit que le premier (fig. 87*a*). Face antérieure du fémur médian non concave, pourvue sur la portion antéro-ventrale de poils longs et fins; tibia médian garni sur la face antéro-ventrale de nombreux poils longs et fins, beaucoup plus longs que le diamètre du tibia (fig. 87*f*). Portion apicale du tibia postérieur pourvue, sur la face antéro-dorsale, de plusieurs poils environ deux fois plus longs que le diamètre du tibia. Régions boréales .... ***parmatus* Rondani**

Tibia antérieur orné de poils postérieurs beaucoup plus courts que la largeur du tibia; deuxième article du tarse de la patte antérieure un peu plus long que large et nettement plus mince que le premier (fig. 59*a*). À partir de la mi-longueur, face antérieure du fémur médian légèrement concave et bordée en dessous de courtes soies noires et denses, légèrement recourbées; ailleurs, poils courts uniquement sur la face antéro-ventrale; dans la portion basilaire de la face antéro-ventrale du tibia médian, les poils sont un peu plus longs que le diamètre du tibia; ailleurs, ils sont très courts. Tibia postérieur orné seulement de poils très courts. Est du Canada, espèce probablement répandue .... ***jaerensis* Nielsen**

42. Tergites 3–5 avec une ligne médiane noire largement coupée, absente ou représentée par une ligne brunâtre peu visible (fig. 169). Poils du katépisterne inférieur d'une longueur inférieure à la moitié du premier article du flagelle. Face antéro-ventrale du tibia médian pourvue, sur les deux tiers basilaires, de poils noirs, denses et ondulés, environ trois fois plus longs que le diamètre du tibia; ailleurs, les poils sont courts et affaissés (fig. 84*g*). Espèce répandue .... ***modestus* Ide**

Tergites 3–5 ornés d'une ligne médiane noire distincte (fig. 170). Poils du katépisterne inférieur presque aussi longs que le premier article du flagelle. Face ventrale du tibia médian ornée seulement de poils affaissés très courts ou de poils denses, fins, ondulés, dressés et assez pâles, d'une longueur presque égale au diamètre du tibia, sur les trois quarts apicaux .. 43

43. Face ventrale du tibia médian ornée, sur les trois quarts apicaux, de poils denses, ondulés, dressés et assez pâles, au moins aussi longs que le diamètre du tibia; il y a au plus quelques soies noires clairsemées sur la face antéro-ventrale du fémur médian (fig. 86*i*). Est du Canada, régions côtières .... ***orarius*** **Vockeroth**

Face ventrale du tibia médian garnie de poils courts et affaissés, pâles ou partiellement foncés; aucun poil dressé; face antéro-ventrale du fémur médian ornée habituellement d'une rangée presque régulière de 3 à 15 soies courtes et raides, de couleur noire ou jaune (fig. 88*i*, 89*i*) 44

44. Faces antéro-ventrale et postéro-ventrale du tibia médian garnies de poils affaissés, pour la plupart noirs; fémur médian portant généralement une rangée de 3 à 15 soies antéro-ventrales noires, courtes et raides (rarement entièrement jaunes) et de quatre à six soies postéro-ventrales noires et épaisses dans la portion basilaire (fig. 89*i*). Espèce répandue .... ***scambus*** **(Staeger)**

Faces antéro-ventrale et postéro-ventrale du tibia médian ornées de poils affaissés entièrement jaunes; fémur médian pourvu antéro-ventralement de soies jaunes, courtes et fortes ou d'une seule soie noire et de trois à quatre soies noires très faibles sur la face postéro-ventrale (fig. 88*i*). Est du Canada .... ***scamboides*** **Curran**

45(2). Face postérieure, tout au moins sur la moitié apicale de la face postérieure du fémur antérieur, pourvue d'une rangée régulière constituée d'au moins cinq soies noires assez longues, raides et légèrement aplaties; rangée se terminant par une soie plus longue et plus mince, à apex fortement incurvé (fig. 91*a*) 46

Face postérieure du fémur antérieur garnie de poils fins, uniformes ou presque avec, tout au plus, quelques poils un peu plus longs près de l'apex et à apex recourbé (fig. 92*a*, 97*a*) 47

46. Fémur antérieur entièrement orange ou orné d'une rayure postérieure brun noirâtre avec, dans la région subbasilaire, une rangée postérieure de trois à cinq soies assez longues, jaunes et fortes. Portion subbasilaire du fémur médian portant une rangée similaire de soies jaunes un peu plus longues. Aile pourvue d'une cellule bm entièrement glabre sur au moins la moitié basilaire, généralement avec quelques microchètes près de l'apex. Espèce répandue ...... ***coerulescens*** **(Williston)**

Fémur antérieur brun noirâtre ayant seulement un peu de jaune orangé au niveau de l'apex; portion subbasilaire garnie habituellement de soies noires, quoique certains spécimens portent quelques soies pâles. Soies noires dans la portion subbasilaire du fémur médian. Aile pourvue d'une cellule bm habituellement entièrement velue, glabre au plus sur le tiers antéro-basilaire. Régions subarctiques, y compris le Groënland .... ***lundbecki* (Collin)**

47. Tibias antérieur (et médian) ornés à la partie postérieure d'une rangée presque régulière de soies noires faibles à fortes; les soies les plus longues font au moins le cinquième de la longueur du tibia (fig. 93*c*–96*c*, 97*a*). Pruinosité faciale uniforme ou face marquée de faibles rides ou de ponctuations de faibles à prononcées (fig. 6, 267, 268) 48

Tibias antérieur (et médian) garnis à la partie postérieure de poils courts et faibles, mesurant au plus deux fois le diamètre du tibia (fig. 92*a*, 99*a*). Pruinosité faciale uniforme, sans rides ni ponctuations 59

48. Premier article du tarse de la patte antérieure orné d'environ six soies longues et faibles, la dernière ou les deux dernières ayant un apex recourbé et mesurant environ les deux tiers de la longueur de l'article (fig. 97*a*). Régions arctiques et de haute altitude, y compris le Groënland 49

Premier article du tarse de la patte antérieure orné seulement de poils postérieurs très courts 50

49. Fémur médian ayant, à environ un tiers de sa longueur, un groupe compact de deux à quatre soies noires assez fortes, de 1,50 à 1,75 fois plus longues que le diamètre du fémur; aucune soie antéro-ventrale distincte près de la base. À environ un tiers de sa longueur, le fémur antérieur présente un groupe de compact à clairsemé, constitué de trois à cinq soies ventrales fortes dont quelques-unes ou la totalité sont environ deux fois plus longues que les soies postéro-ventrales beaucoup plus délicates sur le tiers basilaire du fémur. Sur le tergite 2, taches gris argenté aussi larges que longues et environ 1,6 fois plus larges que longues sur les tergites 3 et 4. Surstylus pourvu d'un lobe plus court à mi-longueur, de 1,25 à 2 fois plus large que le lobe le plus long à sa mi-longueur (comme dans la fig. 236). Longueur variant de 7,9 à 8,9 mm. Nord-ouest du Canada, régions boréales et de haute altitude .... ***yukonensis* Vockeroth**

Fémur médian dépourvu, chez certains spécimens, de soies ventrales distinctes; d'autres possédant jusqu'à neuf soies antéro-ventrales très faibles près de la base; à un tiers de sa longueur, garni d'une soie mince environ 1,5 fois plus longue que le diamètre du fémur ou de deux soies assez fortes et bien séparées, mesurant au plus 1,25 fois le diamètre du fémur. Fémur antérieur pourvu de deux à neuf soies ventrales espacées de façon irrégulière, à peine plus longues et plus fortes que les soies postéro-ventrales sur le tiers basilaire du fémur. Taches gris argenté sur le tergite 2 dont la largeur équivaut à peu près aux quatre cinquièmes de la longueur; taches presque carrées sur les tergites 3 et 4. Surstylus pourvu d'un lobe plus court à mi-longueur, de 0,80 à 1,17 plus large que le lobe le plus long à mi-longueur. Longueur variant de 4,8 à 7,6 mm. Régions boréales et arctiques de haute altitude, y compris le Groënland . . . . ***carinatus* (Curran)**

50. Pruinosité faciale marquée de rangées obliques de ponctuations (fig. 6) ou de rides latérales peu prononcées (fig. 267, 268). Chez certaines espèces, membrane alaire légèrement à fortement glabre dans sa portion basilaire 51

Pruinosité faciale répartie uniformément, seul le tubercule ou la rayure médiane étant glabre. Membrane alaire entièrement velue 58

51. Face marquée sur presque toute sa surface de rangées obliques de ponctuations rondes (fig. 6). Tibia antérieur orné de fortes soies postérieures noires (fig. 94*c*, 96*c*). Marques abdominales de couleur bleu ou bronze métallique, jamais sur fond orange. Ouest du Canada 52

Face marquée de rides latérales obliques peu prononcées (fig. 267, 268). Tibia antérieur pourvu de poils postérieurs plus faibles (fig. 93*c*, 95*c*). Chez certaines espèces, marques abdominales sur fond orange. Espèce répandue 55

52. Aile pourvue de cellules c et bm glabres ou pourvue de quelques microchètes clairsemées près de l'apex 53

Aile pourvue de cellules c et bm largement velues, glabres uniquement dans la portion basilaire ou antéro-basilaire 54

53. Face dont seul le tubercule est noir et lustré; face légèrement plus large, marquée de ponctuations plus grossières (fig. 6). Scutellum en général orné uniquement de poils blancs. Ouest du Canada . . . . ***stegnus* (Say)**

Face ornée d'une rayure médiane noire et lustrée plus large, qui atteint habituellement le bord inférieur de la face et qui s'étend au-dessus de la limite supérieure du tubercule; face légèrement plus étroite, marquée de ponctuations plus fines. Scutellum garni de poils noirs, du moins à la partie arrière. Ouest des États-Unis . . . . ***spinipes* Vockeroth**

54. Aile pourvue de cellules c et bm entièrement velue. Poils des pleurites supérieurs noirs. Face dont seul le tubercule est noir et lustré; zone pruineuse séparant le tubercule et le bord

inférieur de la face (comme dans la fig. 6). Tibia antérieur garni postérieurement de soies assez minces sur toute sa longueur; soies de longueur décroissante vers la base, très courtes sur environ un tiers basilaire (fig. 96*c*). Ouest du Canada .... ***stegnoides*** **Vockeroth**

Aile pourvue d'une cellule c glabre sur environ un sixième basilaire et cellule bm glabre sur le tiers environ de sa portion antéro-basilaire. Poils des pleurites supérieurs en général entièrement pâles. Face ornée d'une rayure médiane noire et lustrée, qui s'étend du bord inférieur jusqu'à bien au-delà du tubercule; face postérieure du tibia antérieur ornée de cinq à sept soies fortes de longueur presque égale; aucune soie sur environ un quart de la base (fig. 94*c*). Orégon, Californie .... ***hesperius*** **Vockeroth**

55. Face s'avançant modérément en dessous; marge buccale antérieure faisant saillie vers l'avant, au moins aussi loin et habituellement un peu plus loin que le tubercule facial (fig. 268). Aile pourvue d'une cellule bm glabre ou presque, sur au moins la moitié basilaire de la marge antérieure. Surstylus avec large lobe dorsobasilaire (fig. 270). Espèce répandue .... ***obscurus*** **(Say)**

Face s'avançant légèrement en dessous; marge buccale antérieure ne s'étendant pas aussi loin vers l'avant que le tubercule facial (fig. 267). Aile pourvue de microchètes en quantités variables, mais qui se trouvent habituellement sur la majeure partie ou la totalité de la marge antérieure de la cellule bm. Surstylus pourvu d'un lobe dorsobasilaire plus petit (fig. 269, 271) 56

56. Membrane alaire largement glabre; cellule c velue au plus sur un cinquième apical; cellule bm garnie seulement d'une mince bande de microchètes près de l'apex; cellule cua_1 glabre sur toute sa largeur à la base. Pilosité de l'anépisterne en général entièrement blanche, rarement avec quelques poils noirs. Surstylus pourvu d'un lobe dorsobasilaire très petit; appendice recourbé au-delà du lobe, presque aussi large que la base du surstylus (fig. 271). Nord-ouest du Canada .... ***sabulicola*** **Vockeroth**

Membrane alaire plus velue; cellule c velue sur au moins la moitié apicale; cellule bm velue tout le long de la marge postérieure; cellule cua_1 entièrement velue. Anépisterne pourvu au moins de quelques poils noirs près du bord supérieur. Lobe dorsobasilaire du surstylus plus large; au-delà du lobe, appendice presque droit, plus mince que la base du surstylus (fig. 269) 57

57. Poils latéraux des tergites presque entièrement noirs, forts et longs; sur le tergite 3, les poils les plus longs font au moins la moitié de la largeur apicale du tergite. Poils inférieurs du katépisterne (près de la ligne ventrale médiane) noirs. Membrane alaire entièrement velue. Soies postérieures des tibias antérieur et médian plus longues et plus épaisses; les soies les plus longues font environ la moitié du tibia; première soie sur le tibia antérieur située près de la base et précédée de quelques poils courts (fig. 95). Ouest du Canada .... ***squamulae*** **(Curran)**

Poils latéraux des tergites surtout pâles, fins et courts; les poils les plus longs sur le tergite 3 font environ le tiers de la largeur apicale du tergite. Poils inférieurs du katépisterne blancs. Chez les spécimens de l'est, membrane alaire largement velue; chez les spécimens de l'ouest, aile pourvue d'une cellule c ornée tout au moins d'une petite zone glabre à la base et cellule bm portant au moins une rayure médiane glabre près de la base et souvent glabre sur presque toute la marge antérieure. Soies postérieures des tibias antérieur et médian plus courtes et plus faibles; les soies les plus longues font environ le tiers du tibia; première soie distincte sur le tibia antérieur située au moins au tiers de la longueur du tibia et précédée de plusieurs soies plus courtes et plus faibles ressemblant à des cheveux (fig. 93). Espèce répandue .... ***confusus*** **(Curran)**

58(50). Face ornée d'une carène faible mais distincte, entre le tubercule (assez petit) et l'extrémité supérieure de la face. Premier article du tarse de la patte médiane portant trois soies antéro-ventrales noires et fortes sur les deux tiers apicaux de sa longueur; toutes les soies sont d'une longueur au moins égale au diamètre de l'article du tarse (fig. 98*j*). Régions boréales .... ***setitarsis*** **Vockeroth**

Face sans carène au-dessus de l'antenne, pourvue d'un large tubercule. Premier article du tarse de la patte médiane dépourvu de soies antéro-ventrales distinctes. Colorado .... ***protrusus*** **Vockeroth**

59(47). Abdomen avec de larges marques orange 60

Abdomen tout noir, sans marques orange 64

60. Pattes presque entièrement noires, soulignées tout au plus de zones jaunâtres à l'extrémité apicale des fémurs et à la base des tibias. Tergite 5 largement ou entièrement orange (fig. 171). Nord-ouest du Canada .... ***rufigaster*** **Vockeroth**

Fémurs antérieur et médian jaunes, tout au moins sur la moitié apicale de la face antérieure; tibias antérieur et médian largement ou entièrement jaunes. Tergite 5 noir ou orné d'une paire de taches basilaires floues, de couleur orange (fig. 172) 61

61. Fémur antérieur dont la face postérieure est pourvue, près de l'apex, de deux à trois longs poils noirs, ayant un apex recourbé; ces poils font contraste avec les précédents, qui sont pâles, denses et plus courts. Face postérieure du tibia antérieur garnie de poils denses, fins et pâles, dont certains sont plus longs que le diamètre du tibia (fig. 92*a*). Face d'une largeur équivalant environ aux cinq huitièmes de la tête. Aile pourvue d'une cellule c glabre, au moins à la base; cellule bm glabre, sauf à l'apex. Tergite 2 maculé d'une paire de taches orange; tergites 3 et 4 avec des taches orange nettement séparées des bords latéraux. Ouest du Canada .... ***kelloggi* (Snow)**

Fémur antérieur pourvu de poils plus courts et plus clairsemés, de longueur décroissante vers l'apex; poils ayant un apex non recourbé; tibia antérieur orné seulement de poils très courts. Largeur de la face ne dépassant pas la moitié de celle de la tête. Membrane alaire entièrement velue ou très glabre. Tergite 2 noir ou surtout rouge orangé; tergites 3 et 4 entièrement rouge orangé ou soulignés de taches orange qui atteignent les bords latéraux chez certaines espèces 62

62. Aile dont la membrane est entièrement velue. Tergites 3 et 4 maculés de taches jaune orangé distinctes, plus étroites sur les côtés (fig. 172). Poils du scutum et du scutellum plus longs; les poils les plus longs sur le scutellum font à peu près les deux tiers de l'arista; partie inférieure du katépisterne, entre les plaques supérieure et inférieure de poils, faiblement mais nettement pruineuse. Espèce répandue .... ***rosarum* (Fabricius)**

Aile pourvue d'une base de la cellule c et de presque toute la cellule bm glabres. Tergites 3 et 4 ornés de taches rouge orangé, habituellement floues et non rétrécies sur les côtés, ou entièrement rouge orangé. Poils du scutum et du scutellum très courts; les poils les plus longs sur le scutellum font moins de la moitié de l'arista. Partie inférieure du katépisterne, entre les plaques supérieure et inférieure de poils, très lustrée ou faiblement pruineuse 63

63. Tergite 2 noir ayant des taches bleu métallique; tergites 3 et 4 ornés de taches basilaires rouge orangé; tergite 5 noir ou marqué de taches basilaires floues, de couleur orange. Partie inférieure du katépisterne, entre les plaques supérieure et inférieure de poils, très lustrée avec seulement une étroite bande postérieure pruineuse. Ouest du Canada .... ***rufimaculatus* Vockeroth**

Tergite 2 rouge orangé et souligné d'une étroite bande noire sur la marge antérieure et les bords latéraux; tergites 3–5 entièrement rouge orangé. Partie inférieure du katépisterne, entre les plaques supérieure et inférieure de poils, modérément pruineuse sur toute sa surface ou presque. Sud de la Californie .... ***russatus* Vockeroth**

64. Poils du notopleurite partiellement ou totalement noirs; poils du scutellum noirs 65

Poils du notopleurite de blanche à jaune brun; poils du scutellum de partiellement à entièrement pâles 67

65. Face, au-dessus du tubercule, ornée de deux ou trois délicates crêtes verticales bordant une faible carène médiane ou un sillon peu profond. Partie inférieure du katépisterne, entre les plaques inférieure et supérieure de poils, faiblement mais nettement pruineuse. Ouest du Canada, régions arctiques et de haute altitude .... ***pullatus* Vockeroth** (partim)

Face bien arrondie au-dessus du tubercule, sans trace de crêtes ou de carène médiane. Partie inférieure du katépisterne, entre les plaques inférieure et supérieure de poils, largement ou totalement noire et lustrée 66

66. Aile ayant une membrane entièrement velue. Zone lustrée sur le katépisterne, au-dessus de la plaque inférieure de poils et s'étendant jusqu'au bord postérieur du sclérite. Surstylus ovale à oblong, de largeur maximale à mi-longueur (fig. 243*b*). Paramère (fig. 243*c*) garni de deux épines. Territoire du Yukon .... ***coracinus* Vockeroth**

Aile pourvue d'une extrémité basilaire de la cellule c et d'une portion postéro-basilaire de la cellule bm glabres. Zone lustrée sur le katépisterne, au-dessus de la plaque inférieure de poils, bordée à la partie arrière par une bande étroite mais distincte de minuscules poils. Surstylus en forme de lanière, de largeur minimale à mi-longueur (fig. 244*b*). Paramère (fig. 244*c*) garni d'une seule épine. Ouest du Canada .... ***latus* (Curran)**

67. Pattes presque entièrement noires; seules l'extrémité apicale des fémurs et l'extrémité basilaire des tibias sont jaunâtres d'une façon floue. Antenne noire. Membrane alaire entièrement velue 68

Pattes orange à rouge orangé sur au moins une large portion de l'apex des fémurs antérieur et médian et sur la majeure partie des tibias antérieur et médian. Premier article du flagelle largement orange en-dessous. Membrane alaire entièrement velue ou ayant des cellules c et bm, de légèrement à très glabres 69

68. Tibia antérieur orné à la partie arrière de poils dressés, denses et fins, surtout pâles, d'une longueur équivalant presque au double du diamètre du tibia; premier article du tarse de la patte antérieure garni de poils similaires, mais plus courts (fig. 99*b*). Portion antéro-ventrale du fémur médian dépourvue de soies noires mais garnie de longs et fins poils blancs dans la moitié basilaire. Territoire du Yukon .... ***woodi* Vockeroth**

Tibia antérieur orné postérieurement de poils noirs, courts et affaissés; premier article du tarse de la patte antérieure ayant seulement des poils très courts. Face antéro-ventrale du fémur médian ornée sur presque toute sa longueur d'une rangée irrégulière de soies noires, courtes, raides et faibles; aucun poil pâle. Ouest du Canada, régions arctiques et de haute altitude .... ***pullatus* Vockeroth** (partim)

69. Fémur entièrement rouge orangé. Près du bord supérieur, face présentant habituellement un sillon médian flou. Cellule bm glabre ou presque, sur au moins un tiers basilaire. Ouest du Canada .... ***willistoni* (Goot)**

Fémur postérieur brun foncé à noir, tout au moins sur la moitié basilaire; fémurs antérieur et médian habituellement noirs, du moins à la base. Près du bord supérieur, face bien arrondie sur le plan médian, ornée de crêtes distinctes et d'un sillon médian, ou portant une carène médiane peu profonde mais distincte. Microchètes variables sur l'aile; membrane variant de presque entièrement velue à très glabre 70

70. Portion supérieure de la face arrondie uniformément sur la partie médiane; aucune crête, sillon ou carène médiane. Paramère fort, garni d'une courte épine incurvée étroitement appuyée contre la base (fig. 245*c*). Espèce répandue .... ***concinnus* (Snow)**

Portion supérieure de la face ornée, soit d'une carène médiane peu profonde mais distincte, soit de faibles crêtes submédianes et d'un sillon médian peu profond. Paramère mince, garni d'une longue épine bien séparée de la base (fig. 246*c*) 71

71. Face un peu pruineuse, ornée sur les côtés de minuscules stries obliques dans le tégument. Portion supérieure soulignée d'une carène médiane assez prononcée. Aile pourvue d'une cellule bm glabre sur au moins le tiers. Paramère pourvu d'une épine assez séparée de la base (fig. 246*c*). Régions néarctiques centrales .... ***luteipennis* (Curran)**

Face plus pruineuse, dépourvue de stries dans le tégument. Portion supérieure de la face ornée d'une carène médiane très faible ou d'un sillon médian très peu profond. Aile pourvue d'une cellule bm entièrement velue ou ayant, tout au plus, une étroite rayure médiane glabre près de la base. Paramère pourvu d'une épine plus séparée de la base. Régions boréales et ouest du Canada .... ***striatus* Vockeroth**

72(1). Premier article du tarse de la patte antérieure nettement et uniformément élargi vers l'apex; deuxième article légèrement plus large que long; troisième article presque deux fois plus large que long (fig. 48); tarse largement ou totalement brun foncé à noir sur le dessus. Sur le tergite 2, taches jaunes réunies sur le plan médian (fig. 156). Espèce répandue .... ***granditarsis* (Forster)**

Premier article du tarse de la patte antérieure ayant d'habitude des côtés parallèles; lorsqu'il y a un très léger élargissement vers l'apex, le tarse est alors entièrement jaune orangé; deuxième article légèrement plus large que long; troisième article à peine plus large que long. Sur le tergite 2, lorsqu'il y a des taches jaunes, elles sont séparées par une ligne médiane noire 73

73. Face marquée de rangées obliques de ponctuations arrondies (comme dans la fig. 6) ou de légères rides latérales obliques dans la pruinosité (comme dans les fig. 267, 269) 74

Face uniformément pruineuse, sauf sur le tubercule; aucune ponctuation ou ride 80

74. Face marquée, sur presque toute sa surface, de rangées obliques distinctes de ponctuations (comme dans la fig. 6). Ouest du Canada 75

Face pruineuse marquée de rides latérales légères (comme dans les fig. 267, 269). Espèce répandue 78

75. Aile pourvue des cellules c et bm très velues; cellule bm pourvue de microchètes sur toute la moitié postéro-apicale, ou presque 76

Aile pourvue des cellules c et bm glabres ou ayant, tout au plus, très peu de microchètes près de l'apex 77

76. Aile pourvue des cellules c et bm entièrement velues. Seul le tubercule de la face est lustré; zone pruineuse séparant le tubercule et le bord inférieur de la face (comme dans la fig. 6). Ouest du Canada .... ***stegnoides* Vockeroth**

Aile pourvue de la cellule c glabre sur environ un tiers à deux tiers de la portion basilaire; cellule bm glabre sur la moitié antéro-basale. Face ornée d'une rayure médiane noire et lustrée, qui s'étend du bord inférieur jusqu'à bien au-delà du tubercule. Orégon, Californie .... ***hesperius* Vockeroth**

77. Face dont seul le tubercule est noir et lustré; face légèrement plus large et garnie de ponctuations plus grossières (comme dans la fig. 6). Ouest du Canada .... ***stegnus* (Say)**

Face ornée d'une rayure médiane noire lustrée plus large, qui atteint habituellement le bord inférieur de la face et qui s'étend au-dessus de la limite du tubercule; face un peu plus étroite et garnie de ponctuations plus fines. Ouest des États-Unis .... ***spinipes* Vockeroth**

78. Face s'avançant modérément en dessous, le bord antérieur buccal faisant saillie vers l'avant, au moins aussi loin et habituellement un peu plus loin que le tubercule facial (comme dans la fig. 269). Espèce répandue .... ***obscurus* (Say)**

Face légèrement avancée en dessous, le bord antérieur buccal ne s'avançant pas aussi loin que le tubercule facial (comme dans la fig. 267) 79

79. Aile ayant une membrane entièrement velue ou ornée tout au plus d'une minuscule zone glabre à la base de la cellule c. Ouest du Canada .... ***squamulae* (Curran)**

Aile ayant au moins une petite zone glabre chez les spécimens des régions de l'est près de la base de la cellule bm; chez les spécimens de l'ouest, la base des cellules c et bm est largement glabre .... ***confusus* (Curran)**
.... ***sabulicola* Vockeroth**

80. Tergites 3–5 entièrement orange; tergite 2 noir. Nord-ouest du Canada .... ***rufigaster* Vockeroth**

Tergites 3–5 pourvus tout au moins d'un étroit liséré postérieur noir; si ces tergites sont largement orange, le tergite 2 est alors lui aussi surtout orange 81

81. Tergite 3 orné de taches orange d'une longueur équivalant au moins à la moitié du tergite et près de deux fois plus longues que celles sur le tergite 4; sur les deux tergites, les taches s'étendent jusqu'aux bords latéraux; tergites 2 et 5 noirs ou avec des marques rougeâtres très floues. Espèce répandue .... ***rosarum* (Fabricius)**

Tergites 3 et 4 avec ou sans taches orange; lorsqu'il y a des taches, elles sont à peu près de la même taille sur les deux tergites et elles n'atteignent pas les bords latéraux chez certaines espèces; taches orange similaires, quoique moins apparentes, sur le tergite 2 de certaines espèces 82

82. Tergites 3 et 4 maculés tous deux d'une paire de larges taches orange élargies sur les côtés; tergite 5 noir. Cellules c et bm entièrement glabres. Face sans trace de carène médiane

au-dessus du tubercule. Ouest du Canada .... ***kelloggi* (Snow)**

Tergites à motif très variable, souvent ornés de taches métalliques foncées, grises ou pruineuses et argentées; si les tergites 3 et 4 portent chacun une paire de larges taches jaunes ou orange, le tergite 5 est alors partiellement jaune ou les cellules c et bm sont très velues. Chez certaines espèces, la face porte une carène ou une crête médiane distincte dans la partie supérieure .... autres esp.

Platycheirus aeratus Coquillett

Figs. 11, 73; Map 61

Platycheirus aeratus Coquillett, 1900:430.
Platycheirus pauper Hull, 1944:77.

Length. 5.3–7.3 mm.

Male. Face (Fig. 11) nearly vertical, blackish, with very low slightly shining tubercle, otherwise with sparse to very sparse gray pruinosity.

Scutum shining with disc and lateral margins slightly pruinose. Thoracic hairs yellow to yellow-brown; posterior part of anepisternum with hairs uniform, moderately dense, crinkly. Wing membrane entirely trichose. Knob of halter yellow. Legs mostly black with narrow apices of femora, fore tibia except posterior streak, basal half and apex of mid tibia, broad base and narrow apex of hind tibia, and fore and mid tarsi yellow to yellow-brown. Fore leg in Fig. 73; trochanter with short pale or black setulae on ventral surface; femur with posterior subbasal tuft of two or three long wavy closely appressed white hairs, otherwise with only fine black hairs; hairs on posterior and posteroventral surfaces up to twice as long as femoral diameter; tibia slightly and nearly uniformly broadened from base to apex, with posteroapical angle only slightly produced; first tarsomere only slightly broadened beyond base, almost parallel sided, about 1.75 times as long as wide; second tarsomere subquadrate or slightly wider than long. Mid femur usually with some short rather weak black setae near middle of anteroventral surface, otherwise with fine black hairs; some hairs of posteroventral surface usually slightly stronger and about twice as long as femoral diameter; mid tibia scarcely broadened or depressed, on middle half of anteroventral surface with dense subappressed wavy black hairs about as long as tibial diameter, on second quarter of posteroventral surface with similar but shorter hairs. First tarsomere of hind leg scarcely swollen, about 4.5 times as long as its greatest depth.

Abdomen slender, with pale markings densely silvery gray pruinose, rarely with obscure orange-brown ground color. Tergite 2 mostly dull black, submetallic along most of lateral margins, rarely with trace of silvery spots; tergites 3 and 4 each with pair of subquadrate spots on anterior half reaching anterior but not lateral margin; tergite 5 with anterolateral angles submetallic, not distinctly pruinose.

Female. Not distinguishable from females of several similar species.

Distribution. Alaska, northern and western Canada (Map 61), south to California and Colorado (2900–3900 m). B.C., VII, VIII; Que., VII.

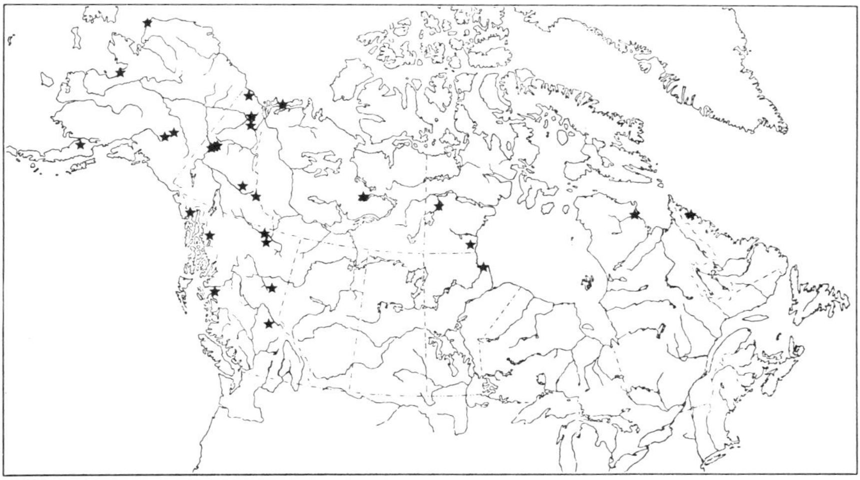

Map 61. Collection localities for *Platycheirus aeratus* Coquillett.

Specimens identified. Alaska, 23 ♂♂; Canada, 45 ♂♂; United States, 28 ♂♂.

Platycheirus albimanus (Fabricius)

Fig. 65; Map 62

Syrphus albimanus Fabricius, 1781:434.

Length. 6.2–9.6 mm.

Male. Face slightly protruding below (as in Fig. 8), with low shining tubercle, otherwise weakly to moderately gray pruinose. Antenna black; first flagellomere in some specimens obscurely orange below.

Scutum and scutellum weakly pruinose, with mixed black and white or yellow hairs. Pleura little more densely pruinose, with white or partly black hairs. Hairs of lower part of katepisternum fine, all white or partly black; posterior hairs much longer than anterior hairs. Wing with tiny to moderately large bare areas at base of cells c and bm; cell c bare on basal half in some specimens; cell bm narrowly bare anteriorly on basal five-sixths in some specimens; knob of halter brown to blackish brown. Legs mostly dark brown to black with narrow apices of femora, narrow bases of tibiae, broad apex of mid tibia, and first tarsomere of mid leg yellow to brown; dorsal surface of fore tibia and all fore tarsus (except brownish fifth tarsomere in some specimens) whitish yellow. Fore leg in Fig. 65; ventral surface of trochanter with short strong white to dark brown setae; femur posteriorly with subbasal tuft of three or four closely appressed long crinkly white hairs followed by two loose tufts of about five and ten coarse crinkly black hairs, respectively, followed by three or four widely spaced longer slender black bristles; tibia uniformly broadened from base to about three-quarters its length, then slightly more strongly broadened posteriorly and with posteroapical angle triangularly produced and subacute, with some strong black or white posterior hairs up to three times as long as tibial width; first tarsomere slightly widened posteriorly from base to about mid length, then parallel-sided to transverse apex, about twice as long as wide and about as wide as tibia; second tarsomere subrectangular, nearly parallel sided with transverse or nearly transverse apex, 1.25–1.75 times as wide as long and slightly narrower than first; third tarsomere similar but slightly smaller; last two tarsomeres progressively slightly narrower. Mid femur with nearly regular anteroventral row of 6–12 short stiff black setae on apical half to two-thirds followed by one long recurved black hair; mid tibia simple, usually with one to three long black posterior hairs just beyond mid length; longest hair as much as 3.5 times as long as tibial diameter. First tarsomere of hind leg scarcely swollen, about five times as long as its greatest depth.

Abdomen moderately slender. Spots of tergites strongly metallic, silvery gray pruinose; spots of tergites 3 and 4 obscurely orange-brown in some specimens; spots of tergite 2 obscure, restricted to middle one-third or extending forward laterally to its anterior margin; spots of tergites 3 and 4 on anterior half of tergite, extending only obscurely to anterior and lateral margins; tergite 5 with at most obscure anterolateral metallic spots.

Female. Not distinguishable from females of several similar species.

Distribution. Alaska, northern and western Canada (Map 62), south to California (2700 m), New Mexico, and South Dakota; Europe; Siberia. B.C., IV–VIII, X; Ont., Que., VI.

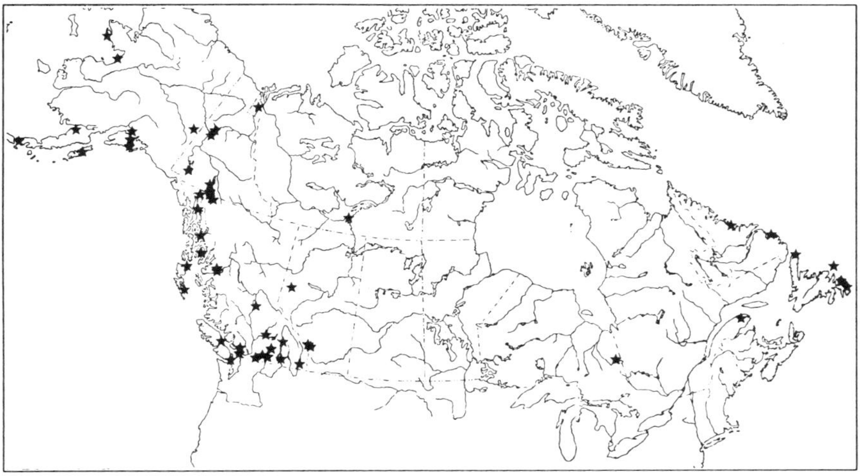

Map 62. Collection localities for *Platycheirus albimanus* (Fabricius).

Specimens identified. Alaska, 27 ♂♂; Canada, 119 ♂♂; United States, 40 ♂♂; Europe, 71 ♂♂; Siberia, 3 ♂♂.

Platycheirus amplus Curran

Figs. 7, 56; Map 63

Platycheirus amplus Curran, 1927*a*:4.

Length. 7.2–8.7 mm.

Male. Face produced forward below (Fig. 7), with rather dense olivaceous pruinosity and shining tubercle. Scape and pedicel dark brown, obscurely paler below in some specimens; first flagellomere dark brown above, yellow-orange on lower half or slightly less.

Scutum subshining, mostly white-haired with some black hairs sublaterally. Scutellar hairs varying from half black to mostly black. Pleural hairs varying from white to brown, mostly with crinkly apices. Upper half of anepimeron with hairs abundant but not forming dense tuft; lower half with some hairs. Hairs of lower part of katepisternum all pale, fine, with longest at most one-third as long as arista. Wing membrane usually with small bare areas near base of cells c and bm, and behind basal half of vein CuP. Knob of halter brown. Legs dull yellow; posterior surface or more of fore femur, basal half or more of mid femur, about apical one-third of mid tibia, and upper surface of first tarsomere of mid leg usually brown to dark brown; first tarsomere of fore leg yellowish white; coxae and hind leg dark brown to black. Fore leg in Fig. 56*a*; trochanter with weak black ventral setae about one-third as long as basal diameter of femur; entire length of posterior surface of fore femur with many long slightly flattened dark brown to black hairs; tibia gradually broadened from base, more abruptly broadened preapically and narrowed at apex, with moderately long posterior hairs on apical half; first tarsomere about twice as long as wide, slightly wider than fore tibia, with apex obliquely truncate and with weak dorsal keel on about apical half; second tarsomere about as long as wide, just over half as wide as first, without dorsal keel; last three tarsomeres simple. Mid leg in Fig. 56*e*; femur with shallow bare excavation on about third-quarter of anterior surface; excavation preceded by some slender black anteroventral setae, bordered below by very short stout curved black setae, followed by cluster of fine crinkly upward-directed black hairs almost as long as femoral diameter; tibia slightly arcuate, slightly constricted subbasally, only very slightly swollen near apex, on basal half with dense anteroventral tuft of erect black crinkly hairs up to three times as long as tibial diameter, with dense short black hairs on third-quarter of posterior surface, and with scarcely discernible hairs on apical one-third of ventral surface. Anterior surface of hind tibia (Fig. 56*m*) with short stout upcurved black setae on basal one-third, with similar but slightly longer setae on apical two-thirds; first tarsomere of hind leg (Fig. 56*p*) swollen on basal half, tapering almost uniformly to slender apex.

Abdomen (as in Fig. 158) robust, black, with dull yellow or slightly silvery markings in some specimens not reaching lateral margins. Tergite 2 with pair of subquadrate to subrectangular spots of variable size; tergites 3 and 4 each with large pair of subquadrate spots; tergite 5 with pair of small subtriangular spots of variable size. Sternites orange-brown to black.

Female. Not distinguishable from females of several related species.

Distribution. Alaska, northern and western Canada (Map 63), south to Washington and Colorado; Iceland; Europe.

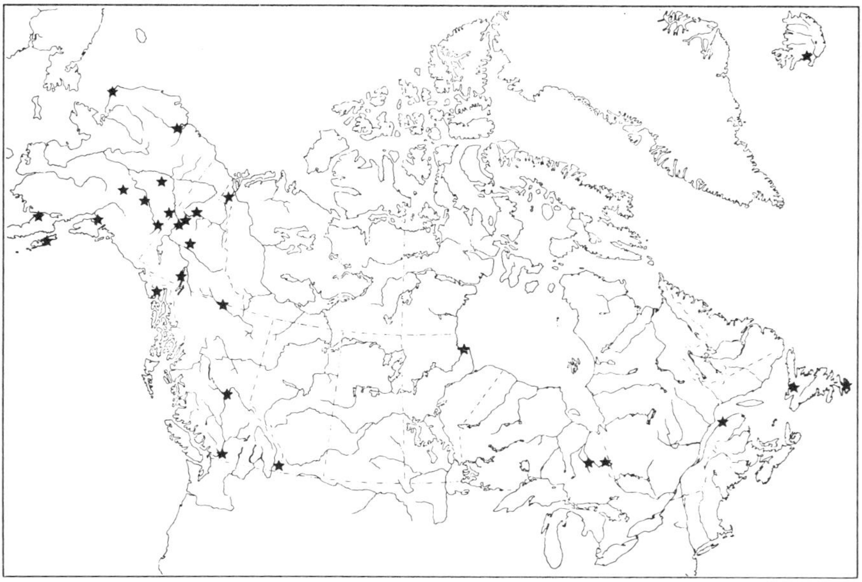

Map 63. Collection localities for *Platycheirus amplus* Curran.

Specimens identified. Alaska, 33 ♂♂; Canada, 18 ♂♂; United States, 3 ♂♂; Iceland, 1 ♂; Europe, 1 ♂.

Platycheirus angustatus (Zetterstedt)

Figs. 74, 166; Map 64

Scaeva angustata Zetterstedt, 1843:762.

Length. 5.7–7.9 mm.

Male. Face nearly vertical, with weak subshining tubercle (as in Fig. 11), otherwise densely gray pruinose. Antenna black, first flagellomere in some specimens obscurely orange below at extreme base.

Scutum mostly shining with weak pruinosity anteromedially and on notopleuron, with pale yellow hairs. Scutellum shining, with pale yellow hairs. Pleura weakly pruinose, with pale hairs; katepisterum with anterior hairs of lower patch stiff, black. Wing with cell bc mostly bare and with small bare areas at base of cells c and bm. Knob of halter yellow. Legs mostly yellow; coxae and trochanters black; fore and mid femora in some specimens with dark brown posterior stripe; hind femur with apex yellow and with broad black ring leaving only narrow base in some specimens; hind tibia with black ring on basal one-third to half; hind tarsus mostly brown to black above; second and third tarsomeres dark yellow in some specimens. Fore leg in Fig. 74; trochanter with very short stiff yellow setulae below; femur with subbasal posterior tuft of two closely appressed long crinkly white hairs, with other hairs fine, rather sparse, mostly pale, at most little longer than femoral diameter; tibia uniformly broadened from base to apex, with posteroapical angle distinctly produced and extending about one-third length of first tarsomere, without distinctive hairs; first tarsomere slightly narrower than apex of tibia, about three-fifths as wide as long, narrowed posteriorly on basal one-third, parallel-sided on apical two-thirds; last four tarsomeres slightly narrower than first. Mid femur usually with irregular group of up to 13 short weak black anteroventral setae on basal half; setae in some specimens yellow and less conspicuous; some specimens also with two or three long weak black ventral bristlelike hairs on basal half; mid tibia usually with fine wavy erect dark anteroventral hairs about twice as long as tibial diameter on basal half, in some specimens with hairs shorter pale and appressed and with short appressed crinkly pale hairs on most of ventral surface. First tarsomere of hind leg only slightly swollen, about three times as long as its greatest depth.

Abdomen (Fig. 166) slender. Tergite 2 slightly or distinctly longer than wide; tergites 3 and 4 subquadrate or slightly longer than wide; pale markings of tergites usually yellow to yellow-orange, rarely with brassy tint, never with distinct silvery pruinosity; tergite 2 with pair of spots of variable size separated from margins and with rounded inner margin; tergites 3 and 4 each with pair of subquadrate spots reaching anterior and lateral margins; on tergite 3 spots from four-sevenths to four-fifths length of tergite; on tergite 4 spots from half to two-thirds length of tergite; tergite 5 black or with small obscure submetallic or orange anterolateral spots.

Female. Not distinguishable from females of several similar species.

Distribution. Alaska, Canada (Map 64), south to Colorado, Wisconsin, and Massachusetts; Europe; Siberia; Japan.* B.C., V–VII; Ont., Que., IV–VII, IX.

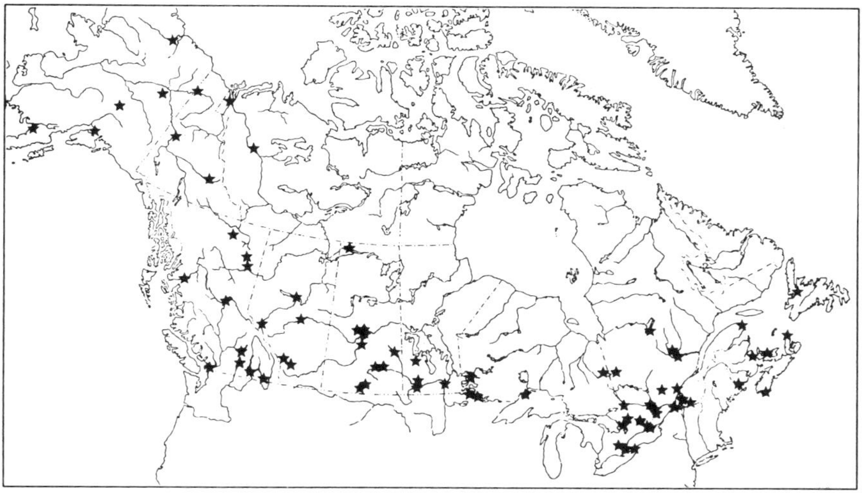

Map 64. Collection localities for *Platycheirus angustatus* (Zetterstedt).

Specimens identified. Alaska, 15 ♂♂; Canada, 233 ♂♂; United States, 50 ♂♂; Europe, 20 ♂♂; Siberia, 5 ♂♂.

Platycheirus carinatus (Curran)

Fig. 97; Map 65

Melanostoma carinata Curran, 1927*a*:11.

Length. 4.8–7.6 mm.

Male. Face very weakly grayish pruinose, without punctures or ripples, produced slightly forward below, with weak but distinct median keel extending from tubercle to upper end of face, with tubercle shining, variable in size, as in Fig. 8 or slightly more or less abrupt. Antenna black, with first flagellomere slightly longer than deep.

Scutum and scutellum subshining, with hairs varying from entirely white to mostly black. Pleura weakly pruinose, with hairs entirely white or mostly black on upper half. Wing membrane entirely trichose. Knob of halter pale brown. Legs black, with extreme apices of fore and mid femora and, in some specimens, of hind femur and narrow bases of fore and mid tibiae and, in some specimens, of hind tibia yellowish. Fore femur with rather long strong black posterior hairs, some near apex having curled apices, posteroventrally on basal one-third with 6–10 short stiff weak black bristles, ventrally near basal one-third with two to nine loosely and irregularly spaced, slightly longer and stronger black setae about as long as femoral diameter; fore tibia with single series of up to 17 weak black posterior bristles, short on basal one-third of tibia and up to one-third tibial length near apex of tibia; last three or four bristles very close together and almost forming fascicle; fore tarsus simple, with first tarsomere posteriorly having about six long weak curved black bristles; longest bristles about five-sixths as long as tarsomere (Fig. 97). Mid femur in some specimens with up to nine short weak black anteroventral bristles on basal one-third; in some specimens with one slender black bristle about 1.5 times as long as femoral diameter or with two moderately strong well-separated bristles at most 1.25 times as long as femoral diameter; mid tibia with posterior bristles similar to those of fore tibia but slightly weaker and strongly appressed on at least basal half of tibia; mid tarsus simple. First tarsomere of hind leg moderately swollen, about four times as long as its greatest depth.

Abdomen opaque black with shining lateral margins and with large rectangular or subquadrate densely silvery gray pruinose spots; spots on tergite 2 about four-fifths as wide as long; spots on tergites 3 and 4 subquadrate, from some angles showing brown-orange background. Surstylus with shorter lobe, at its mid length, from 0.80 to 1.17 times as wide as longer lobe at its mid length.

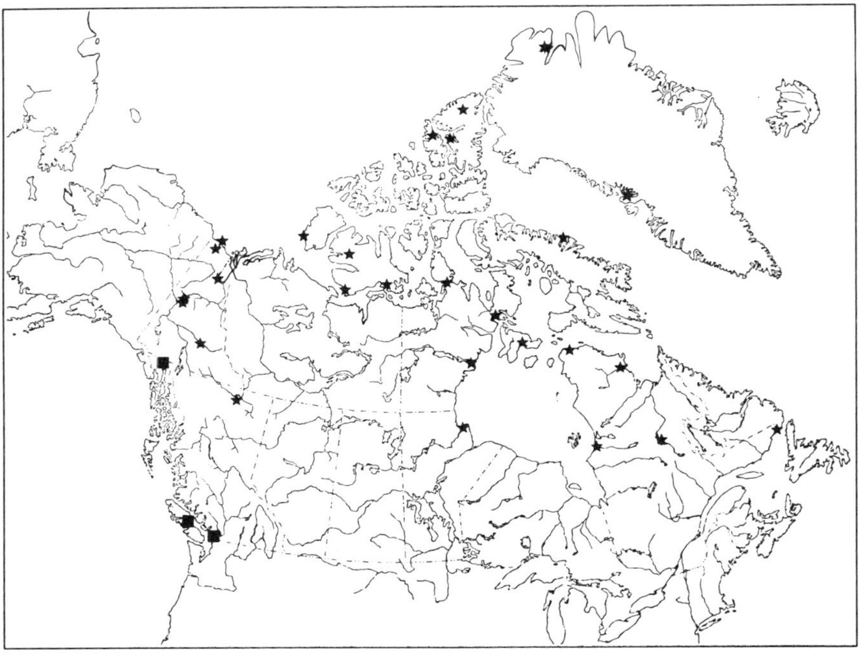

Map 65. Collection localities for *Platycheirus carinatus* (Curran) (★) and for *P. ciliatus* Bigot (■).

Female. Not distinguishable from females of several similar species.

Distribution. Arctic and high boreal Canada, Greenland (Map 65); Siberia. B.C., VI, VII; Que., VI, VII.

Specimens identified. Canada, 137 ♂♂; Greenland, 56 ♂♂; Siberia, 2 ♂♂.

Biology. *P. carinata* lives in the higher and more barren parts of the mountain slopes in northern Yukon Territory (D.M. Wood, personal communication). This accords with its wide distribution in the Arctic and its abundance in northern Ellesmere Island (63 ♂♂ taken at Lake Hazen) and northern Greenland. At the two southernmost localities at which *P. carinatus* has been taken in western Canada (Summit Lake, B.C.; Dickson Lake, Mt. Mye, Y.T.) it occurred at 1700 m.

Platycheirus ciliatus Bigot

Fig. 66; Map 65

Platycheirus ciliatus Bigot, 1884:74.
Platycheirus frontosus Lovett, 1919:247.

Length. 7.7–10.0 mm.

Male. Extremely similar to male of *P. albimanus*, differing as follows: fore leg with last tarsomere brown to dark brown above; mid tarsus brown to dark brown throughout, scarcely paler near base than at apex; fore tibia (Fig. 66) with posteroapical angle distinctly rounded and projecting slightly beyond level of posterior margin of first tarsomere.

Female. Not distinguishable from females of several related species.

Distribution. Western British Columbia, south to central California (always near Pacific coast) (Map 65). B.C., VI.

Specimens identified. Canada, 3 ♂♂; United States, 14 ♂♂.

Discussion. The differences between this form and *P. albimanus* are very slight and the two may be conspecific, as they were considered to be by Verrall (1901). The difference in shape of the fore tibia is so constant that I feel it better to treat the forms as distinct, at least until more material is available. As *P. albimanus* is known to occur in California only at one locality at 2700 m and as the species represented by the type of *P. frontosus* occurs along the coast, the type of *P. ciliatus* (from California) was likely of the latter species rather than of *P. albimanus*.

Platycheirus clypeatus (Meigen)

Fig. 75; Map 66

Syrphus clypeatus Meigen, 1822:335.

Length. 6.0–8.8 mm.

Male. Very similar to male of *P. angustatus*, differing as follows: hairs of thorax slightly longer and more brassy in color. Wing membrane entirely trichose. Legs slightly darker; mid femur usually dark brown on about basal half. Fore femur (Fig. 75) posteriorly beyond subbasal white tuft with longer stronger usually black hairs at least 1.5 times as long as femoral diameter. Mid femur in some specimens with short black anteroventral setae on entire length, with long strong black ventral hairs more numerous; mid tibia with anteroventral hairs on basal half and ventral hairs erect or suberect.

Abdomen more robust. Tergite 2 at least slightly wider than long, 1.3 times as wide as long in some specimens; tergites 3 and 4 about 1.3 times as wide as long; pale spots of tergites in some specimens with sparse silvery pruinosity apparent in oblique posterior view, more variable in size; spots of tergite 2 usually small and wider than long; spots of tergites 3 and 4 commonly not reaching lateral margin; tergite 5 usually without pale spots, very rarely with orange lateral spots.

Female. Not distinguishable from females of several similar species.

Distribution. Alaska, Canada (Map 66), south to Oregon, Colorado, and New York; Europe; Siberia; Japan. B.C., V, VIII; Que., VI, VII.

Specimens identified. Alaska, 46 ♂♂; Canada, 164 ♂♂; United States, 9 ♂♂; Europe, 57 ♂♂; Siberia, 5 ♂♂; Japan, 5 ♂♂.

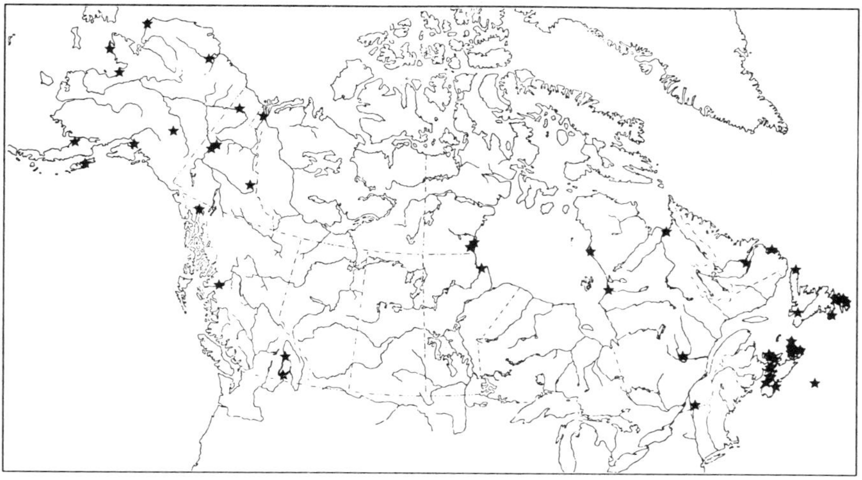

Map 66. Collection localities for *Platycheirus clypeatus* (Meigen).

Biology. Specimens were taken in Nova Scotia in damp, open sphagnum fen and in *Carex* marsh.

Platycheirus coerulescens (Williston)

Fig. 91; Map 67

Melanostoma coerulescens Williston, 1887:49.
Scaeva ambigua of American authors (not *S. ambigua* Fallén, 1817).

Length. 5.7–9.1 mm.

Male. Face very slightly protruding below, with uniform moderately dense grayish pruinosity and rather large shining tubercle (profile as in Fig. 268). Antenna black with first flagellomere at least narrowly orange below at base.

Scutum and scutellum nearly shining, pleura slightly whitish pruinose; thoracic hairs all white. Wing extensively bare on basal half; cell c bare on most of basal half; cell r almost entirely bare; cell bm bare on at least basal four-fifths; cell *cup* narrowly bare anteriorly. Knob of halter yellow. Legs mostly dark brown to black, with following areas dull yellow-orange: fore femur except posterior streak, apical half of mid and apex of posterior femur, basal half of fore and mid tibiae, basal one-quarter of hind tibia, and last tarsomere of fore leg. Fore leg in Fig. 91; femur posteriorly on apical seven-ninths with regular row of 9–12 strong straight black bristles increasing in strength towards femoral apex and up to twice as long as femoral diameter; this row followed by single longer bristle with slender recurved apex, ventrally on basal half with row of three to five short strong acute yellow bristles; longest bristle slightly shorter than femoral diameter; tibia slender, not at all depressed, posteriorly on apical half with row of long straight subappressed white hairs; longest hairs about 2.5 times as long as tibial diameter; tarsus slender, neither broadened nor depressed. Mid femur with row of two to five short strong acute yellow or rarely black bristles anteroventrally on basal one-third; longest bristle about twice as long as femoral diameter; mid tibia cylindrical, posteriorly near apex with some long pale hairs similar to those of fore tibia. First tarsomere of hind leg slightly swollen, about four times as long as its greatest depth.

Abdomen with dull yellow or orange strongly silvery gray pruinose spots. Tergite 2 with spots small, just beyond mid length of tergite, rounded anteriorly and with posterior margin oblique, greatly reduced in some specimens; tergites 3 and 4 with spots on anterior half but spots obliquely produced posterolaterally.

Female. Not distinguishable from females of several similar species.

Distribution. Alaska, Canada (Map 67), south in western mountains to California and New Mexico. B.C., IV, VII, VIII.

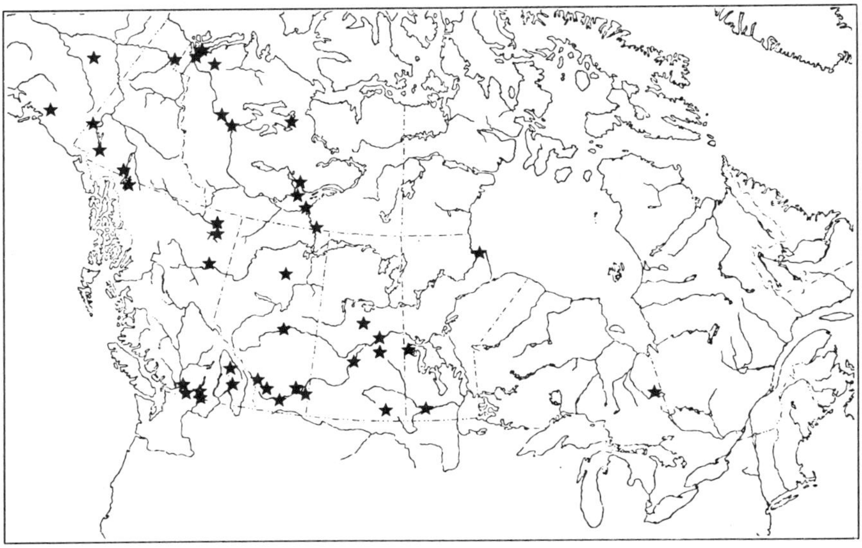

Map 67. Collection localities for *Platycheirus coerulescens* (Williston).

Specimens identified. Alaska, 6 ♂♂; Canada, 112 ♂♂; United States, 110 ♂♂.

Platycheirus concinnus (Snow)

Fig. 245; Map 68

Melanostoma concinnum Snow, 1895:229.

Length. 7.7–10.5 mm.

Male. Face nearly vertical, very slightly protruding below, with low slightly compressed tubercle, moderately gray pruinose, with tubercle shining but otherwise without median shining stripe, with upper part smoothly rounded and without median keel or grooves. Antenna black with first flagellomere extensively orange below.

Thorax usually with slight bluish tint, not brassy, with white hairs rather variable in length; longest scutellar hairs varying from slightly shorter to slightly longer than arista. Pleura subshining to shining; katepisternum, between upper and lower patches of hair, strongly shining except along narrow posterior margin. Wing with microtrichia variable; cell c bare at least at base, entirely bare on basal one-third in some specimens and narrowly bare posteriorly on middle one-third; cell bm varying from almost entirely trichose to bare except narrowly posteriorly and at apex, usually bare on most of basal one-third to half. Knob of halter orange. Legs variable in color; coxae and trochanters dark brown to black; fore and mid femora in some specimens entirely orange or with brown areas posteriorly, more often black at base or over most of length; hind femur usually black with apex narrowly orange, rarely narrowly orange at base; tibiae orange with posterior surface partly brown and with hind tibia extensively darkened in some specimens; fore and mid tarsi orange to brown above; hind tarsus dark brown above. Legs unmodified, without outstanding hairs or bristles; femoral hairs variable in length. First tarsomere of hind leg moderately swollen, about 3.5 times as long as its greatest depth.

Abdomen opaque black. Tergites with shining lateral margins and dark bluish spots with faint grayish pruinosity in some specimens; tergite 2 with small lateral spots; tergite 3 with large anterolateral spots narrowly separated medially and strongly broadened laterally; tergite 4 with similar spots confluent medially in some specimens and with shining posterior margin leaving an opaque black triangle on disc; tergite 5 shining. Surstylus (Fig. 245*b*) with slender thumblike basal lobe, with main arm slightly curved, nearly parallel sided. Paramere (Fig. 245*c*) with stout base and short curved closely appressed spine. Aedeagus as in Fig. 245*d*.

Female. Frons with pruinose spots extending narrowly inward to cover at least two-thirds width of frons. Entire antenna orange below. Thoracic hairs very short. Cell c bare on basal four-fifths in some specimens; cell bm bare on at least basal half. Abdomen with bluish spots commonly confluent medially on tergites 2–4 and less clearly defined than in male. (Description based on females taken with males.)

Distribution. Canada (Map 68), south to California, New Mexico, and Iowa. B.C., VI–VIII.

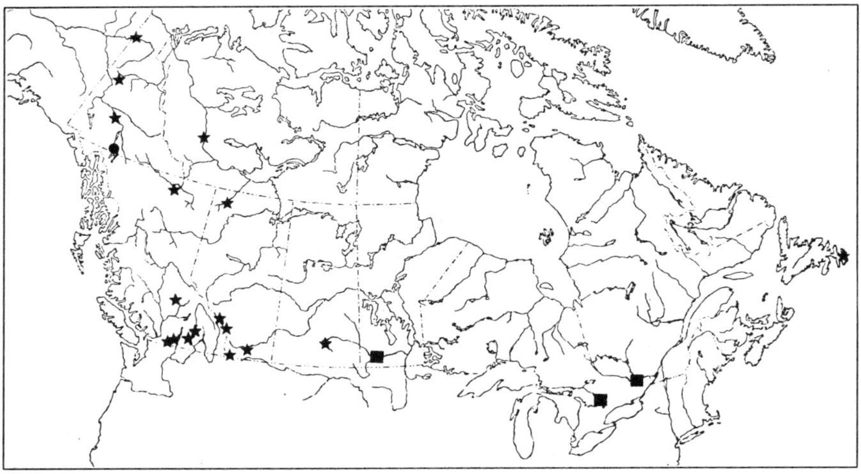

Map 68. Collection localities for *Platycheirus concinnus* (Snow) (★), for *P. coracinus* Vockeroth (●), and for *P. discimanus* Loew (■).

Specimens identified. Canada, 23 ♂♂; United States, 117 ♂♂, 12 ♀♀.

Discussion. The males show considerable variation in length of thoracic and femoral hairs and both sexes in wing microtrichia and in leg color. It is possible that more than one species is included under this name.

Platycheirus confusus (Curran)

Figs. 93, 267, 269; Map 69

Melanostoma confusum Curran, 1925:112.

Length. 6.0–8.2 mm.

Male. Face produced only slightly forward below, with anterior oral margin not extending as far forward as facial tubercle; with broad shining median stripe extending from lower margin to well above tubercle, laterally with faint oblique ripples in grayish pruinosity (Fig. 267). Antenna black with first flagellomere obscurely orange below. Angle of junction of eyes about 100°.

Scutum and scutellum nearly shining, with black hairs. Pleura very weakly pruinose; at least upper hairs, and in some specimens most hairs of anepisternum and anepimeron and in some specimens of upper part of katepisternum, black; other pleural hairs white. Wing membrane clear or very slightly brownish; eastern specimens with tiny bare area at base of cell c and narrow stripe along middle of basal one-fifth of cell bm bare; western specimens with from basal one-tenth to basal three-fifths of cell c bare or nearly so and usually with anterior strip bare, broad basally and narrowing apically on about basal three-quarters of cell bm; all specimens with cell cua_1 entirely trichose. Knob of halter blackish brown. Legs mostly dark brown to blackish, with narrow apices of femora and most of fore and mid tibiae obscurely orange-brown. Femora with only weak hairs; fore tibia with regular posterior row of weak curved black bristles; longest bristles about one-third as long as tibia, beginning at one-third or more tibial length (Fig. 93). Mid tibia with row of similar black bristles on about apical half to three-fifths; tarsi unmodified. First tarsomere of hind leg scarcely swollen, about five times as long as its greatest depth.

Abdomen moderately slender. Lateral hairs of tergites mostly pale, fine, and short; longest hairs of tergite 3 about one-third as long as apical width of tergite; lateral spots on tergite 2 elongate, metallic copper or bronzy, with inner margins slightly pruinose, and with widest points separated by about five-sevenths width of tergite; tergites 3 and 4 on anterior half with similar narrowly separated or obscurely confluent spots broadened posterolaterally where extending just beyond mid length of tergite; some spots with orange or orange-brown tint; tergite 5 with anterolateral angles obscurely metallic. Surstylus (Fig. 269) with shorter lobe of moderate size; longer lobe nearly straight and more slender than base of surstylus.

Female. Similar to male, differing as follows: frons mostly weakly brownish or grayish pruinose, shining only around ocelli and immediately above antennae. Thoracic hair short, white except for some black scutal and marginal scutellar hairs in some specimens. Bare areas of wing usually more extensive in western specimens; some with most of cell c and up to anterobasal three-quarters of cell bm bare. Legs usually paler; some specimens with much of fore and mid legs orange-brown. Fore and mid tibia in some specimens with some longer white or black posterior hairs shorter than tibial diameter. Abdomen more robust, with metallic spots of tergite 2 much larger and separated by about one-third width of tergite.

Distribution. Canada (Map 69), south to California, Colorado, and North Carolina. B.C., V–VIII; Ont., Que., IV–VII.

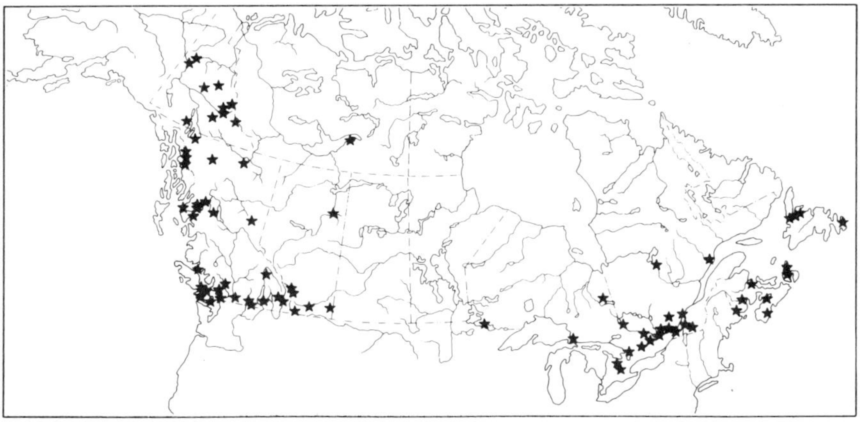

Map 69. Collection localities for *Platycheirus confusus* (Curran).

Specimens identified. Canada, 254 ♂♂, 301 ♀♀; United States, 52 ♂♂, 29 ♀♀.

Biology. Specimens taken in the Ottawa area were mostly from a large marsh with many boreal Diptera (in May) and from a large sphagnum bog (in June, with one in July). Many specimens were taken on Cape Breton Island, N.S., in early June, mostly from upland bogs and adjacent coniferous forests. Two specimens from New Brunswick (in CNC) were reared from *Adelges piceae*.

Discussion. The apparent wide geographic separation of the two populations of this species (Map 9), as well as the almost consistent difference in extent of wing microtrichia between the eastern and western specimens, suggests that two species may be involved. However, the male terminalia are nearly identical in specimens from widely separated localities. *Platycheirus squamulae* (Curran) may be a variant of *P. confusus*, but the wing microtrichia are even more extensive than in eastern specimens of *P. confusus*, and the leg bristles and abdominal hairs differ in the males, so that it seems likely that *P. squamulae* is specifically distinct. *P. sabulicola* Vockeroth, with a different surstylus, reduced wing microtrichia, and a very different habitat, is undoubtedly distinct from *P. confusus*, although females of the two species may not always be separable.

Platycheirus coracinus Vockeroth

Fig. 243; Map 68

Platycheirus coracinus Vockeroth, 1990:691.

Length. 7.3 mm.

Male. Face vertical, with large almost round not prominent shining tubercle, otherwise very slightly gray pruinose. Antenna black with first flagellomere obscurely orange below and as deep as long.

Thorax strongly shining, with hairs rather short and black, long and pale on lower part of katepisternum. Katepisternum shining between upper and lower hair patches, without posterior band of short pile. Wing membrane entirely trichose. Knob of halter yellow. Legs black with femora and tibiae shining and rather stout, without outstanding hairs and bristles, with posterior hairs of fore and mid femora shorter than femoral diameter. First tarsomere of hind leg moderately swollen, about four times as long as greatest depth.

Abdomen opaque black. Tergite 2 shining laterally; tergites 3 and 4 each with pair of large very obscure slightly grayish anterolateral spots. Surstylus (Fig. 243*b*) elongate oval, widest at mid length, without basal thumblike lobe. Paramere (Fig. 243*c*) with two strong spines. Aedeagus (Fig. 243*d*) with long slender subbasal lobe and with many ventral spicules on apical half.

Female. Unknown.

Distribution. Yukon Territory, 1830 m (Map 68). Y.T., VII.

Specimens identified. Canada, 1 ♂.

Platycheirus discimanus Loew

Fig. 49; Map 68

Platycheirus discimanus Loew, 1871:227.

Length. 6.2–6.8 mm.

Male. Face moderately protruding, strongly broadened below, with distinct median keel or submedian grooves above, or both, weakly silvery pruinose, with only tubercle and area below shining (as in Fig. 3 but with tubercle less abrupt above). Antenna black.

Scutum, scutellum, and pleura subshining with mixed white to dark brown hairs. Wing membrane entirely trichose. Knob of halter yellow. Legs dark brown, with first two tarsomeres of fore and mid legs pale to dark yellow. Fore leg in Fig. 49*a*; trochanter with many short, rather weak, ventral black setulae; femur without distinctive hairs or bristles; tibia very slightly broadened from base to apex, only very slightly depressed, with dense fine crinkly posterior hairs increasing in length toward apex; longest hairs about three times as long as tibial diameter; first tarsomere subtriangular, nearly transverse apically, widest just before apex, as long as wide and twice as wide as apex of fore tibia; second tarsomere subrectangular, twice as wide as long and about one-third as long as first tarsomere; last three tarsomeres normal. Mid leg in Fig. 49*g*; femur without distinctive hairs or bristles; tibia with dense anteroventral hairs on basal one-third; longest hairs at least four times tibial length, with similar but sparser posterior hairs on about middle one-third; first tarsomere strongly compressed, more than twice as deep as wide; second tarsomere very slightly compressed; last three tarsomeres normal; ventral setulae of tarsus yellow. Hind tibia with short hairs; first tarsomere of hind leg rather strongly swollen, thickest just before mid length, about three times as long as its greatest depth.

Abdomen moderately stout. Tergites 2–4 each with large subquadrate pair of silvery spots.

Female. Not distinguishable from females of several similar species.

Distribution. Central Canada (Map 68), Pennsylvania; Europe; Siberia. (Recorded by Wirth et al. [1965] from British Columbia to Quebec, south to Oregon and Maine, but these records not verified by me.) Manitoba, V.

Specimens identified. Canada, 13 ♂♂; United States, 1 ♂; Europe, 6 ♂♂; Siberia, 1 ♂.

Platycheirus flabellus Hull

Fig. 50; Map 70

Platycheirus flabella Hull, 1944:75.

Length. 6.8–9.1 mm.

Male. Similar to that of *P. discimanus*, differing as follows: face narrower, much more produced below, in profile as *P. amplus* (Fig. 7), densely yellow-gray pruinose, with only tubercle shining.

Thorax slightly more pruinose, with mixed black and white hairs. Knob of halter yellow to pale brown. Fore and mid tibiae in some specimens slightly to extensively yellowish basally; mid tarsus black. Fore leg in Fig. 50*a*; trochanter ventrally with pale fine hairs or minute black setae or both; femur on basal half with nearly regular ventral row of about six stiff black setae; longest setae slightly shorter than femoral diameter; femur on apical half with anteroventral row of very short stiff black setae; tibia posteriorly over most of its length with almost regular row of very weak black bristles; longest bristles almost three times as long as tibial diameter; first tarsomere broadened from base to apex, 1.5–2.0 times as long as broad; second tarsomere rather variable, subquadrate or up to 1.5 times as wide as long, from one-third to two-fifths as long as first. Mid femur ventrally with row of about seven very slender white or black bristles most slightly longer than femoral diameter, anteroventrally on apical one-quarter with rather dense long crinkly black hairs; tibia (Fig. 50*h*) with very short anterior hairs, posteroventrally on apical half with series of about five very slender black bristles; longest bristle about three times as long as tibial diameter; tarsus not at all compressed, with yellow ventral setulae. Hind tibia anteriorly on apical half with about six very slender bristles of variable length; longest bristle about twice as long as tibial diameter; first tarsomere of hind leg only slightly swollen, about five times as long as greatest depth.

Tergites with pruinose spots almost always with distinct orange-brown background, variable in size; spots on tergite 2 usually smaller than others.

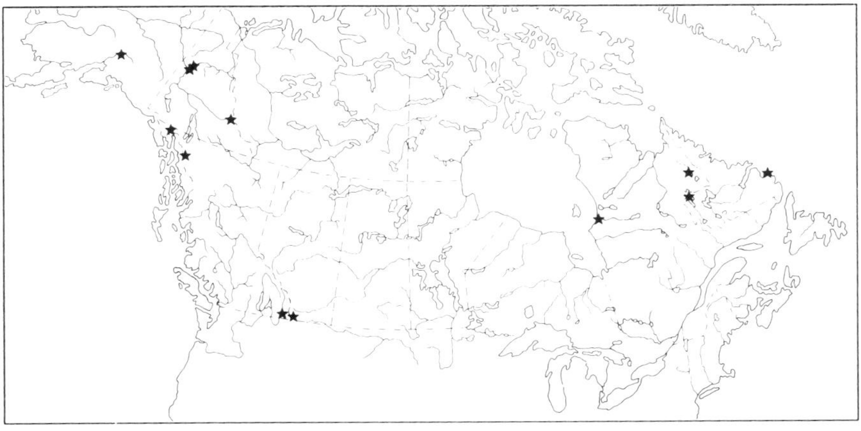

Map 70. Collection localities for *Platycheirus flabellus* Hull.

Female. Not distinguishable from females of several similar species.

Distribution. Alaska, northern and western Canada (Map 70), south to Washington, Idaho, and Maine. Y.T., VI–VII; Que., VII.

Specimens identified. Alaska, 1 ♂; Canada, 37 ♂♂; United States, 3 ♂♂.

Platycheirus granditarsis (Forster)

Figs. 47, 48, 155, 156, 237; Map 71

Musca granditarsa Forster, 1771:99.
Syrphus ocymi Fabricius, 1794:309.
Pyrophaena granditarsis var. *apicauda* Curran, 1925:115.
Pyrophaena digitalis Fluke, 1939:367.

Length. 7.7–10.5 mm.

Male. Face nearly vertical, with prominent tubercle, very slightly silver pruinose except on tubercle. Antenna black.

Thorax shining to subshining, with white to yellow hairs. Wing membrane entirely trichose. Knob of halter yellow. Legs black with basal one-third to one-fifth of fore and mid tibiae and basal one-fifth of hind tibia yellow. Fore leg in Fig. 47*a*; trochanter with some short fine hairs below; femur with very short stiff black setae in irregular anteroventral and irregular posteroventral row on basal three-fifths; tibia rather stout on apical three-fifths but not at all depressed; first tarsomere with large triangular anterior process having apical half strongly depressed; last four tarsomeres short and wide but only slightly depressed. Mid leg in Fig. 47*d*; femur posteroventrally with group of short stiff black setae near base; tibia similar to fore tibia; first four tarsomeres strongly depressed, each with broad apically rounded anterior process progressively shorter on first to third tarsomeres. First tarsomere of hind leg rather strongly swollen, about four times as long as greatest depth.

Abdomen (Fig. 155) rather broad, black with extensive yellow-orange areas as follows. Tergite 2 yellow-orange except anterior one-third and narrow lateral margins; anterior black area with triangular posteromedian extension extending as black median line to posterior margin in some specimens; tergite 3 yellow-orange except posterolateral angles; basal band on about one-third of tergite 4 yellow-orange; band commonly emarginate posteriorly or divided into pair of spots. Sternites black with posterolateral angles of 2, all of 3, and base of 4 yellow-orange. Surstylus (Fig. 237*b*) with poorly developed basal process. Paramere (Fig. 237*c*) subtriangular, with spine scarcely differentiated. Aedeagus as in Fig. 237*d*.

Female. Frons shining except for two tiny silvery pruinose lateral spots at two-thirds its length; face broadly shining medially. Legs mostly yellow with following areas black or rarely brown: coxae, trochanters, fore tarsus, last four tarsomeres of mid tarsus, small to large mark just beyond mid length of anterior surface of hind femur, broad ring on about apical two- to four-fifths of hind tibia, and hind tarsus; femora without stiff setae. Fore tarsus (Fig. 48) moderately depressed; first tarsomere slightly widened towards apex; next three tarsomeres each wider than long; fifth tarsomere subquadrate. Mid tibia and mid and hind tarsi rather slender. Abdomen (Fig. 156) with yellow-orange markings similar to those of male. Tergite 2 usually with narrow and obscure to broad and distinct posterior black band strongly produced forward laterally in some specimens; tergite 3 with similar posterior black band; tergite 5 with large triangular anterolateral yellow-orange spots. Sternites yellow-orange.

Distribution. Alaska, Canada (Map 71), south to California, New Mexico, and West Virginia; Europe; Siberia; Japan. B.C., VI–VIII; Ont., Que., V–VIII.

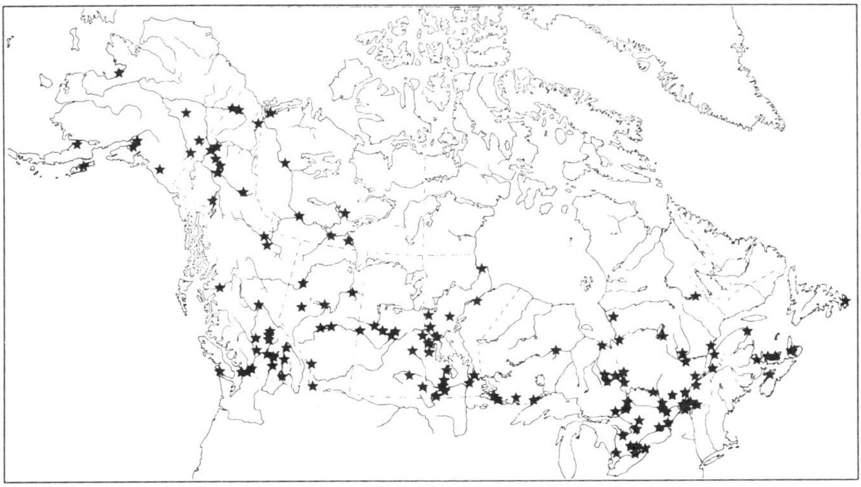

Map 71. Collection localities for *Platycheirus granditarsis* (Forster).

Specimens identified. Alaska, 52 ♂♂, 56 ♀♀; Canada, 400 ♂♂, 249 ♀♀; United States, 231 ♂♂, 224 ♀♀; Europe, 17 ♂♂, 18 ♀♀; Siberia, 2 ♂♂, 1 ♀; Japan, 2 ♂♂, 2 ♀♀.

Biology. Specimens were collected in Ontario and New Hampshire from *Carex* marsh.

Discussion. Some specimens from Utah and all those in a long series from the Mojave Desert, California, have the abdomen more extensively yellow-orange than do other Nearctic specimens. Some males from Utah are intermediate in color pattern, with tergite 5 with small to rather large sublateral pale spots. The male terminalia of all are as in the darker northern form. I consider these pale southwestern specimens (var. *apicauda* Curran) to be color variants of *P. granditarsis*.

Platycheirus groenlandicus Curran

Fig. 51; Map 72

Platycheirus groenlandicus Curran, 1927*a*:10.
Platycheirus monticolus Nielsen, 1972:91 (preoccupied Jones, 1917).
Platycheirus boreomontanus Nielsen, 1981:101 (new name for *P. monticolus* Nielson, 1972).

Length. 5.7–8.2 mm.

Male. Very similar to male of *P. discimanus*, differing as follows: face slightly more pruinose, with only tubercle shining. Thoracic hairs varying from almost entirely white to extensively black. Knob of halter dark brown. Fore and mid tibiae usually yellow on basal one-third; mid tarsus black; first tarsomere of fore leg from 1.25–1.5 times as long as wide (Fig. 51*a*). Mid tibia (Fig. 51*g*) with anteroventral hairs less abundant and posterior hairs rather variable in length and abundance; mid tarsus (Fig. 51*g*) not at all compressed, usually with black ventral setulae. Spots of tergites in some specimens with pruinosity grayish brown rather than silvery. Sternites black, rather densely gray pruinose.

Female. Not distinguishable from females of several similar species.

Distribution. Alaska, northern and western Canada, Greenland (arctic and alpine only) (Map 72); northern Europe; Siberia.* N.W.T., VI, VII.

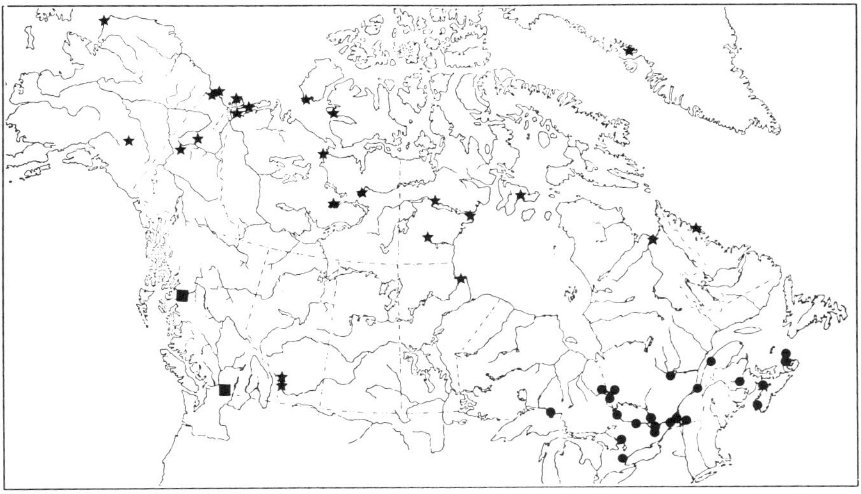

Map 72. Collection localities for *Platycheirus groenlandicus* Curran (★), for *P. hispidipes* Vockeroth (■), and for *P. inversus* Ide (●).

Specimens identified. Alaska, 4 ♂♂; Canada, 115 ♂♂; Greenland, 1 ♂; Europe, 3 ♂♂.

Platycheirus hesperius Vockeroth

Fig. 94

Melanostoma chaetopoda Davidson, 1922:35 (preoccupied Williston, 1887).

Platycheirus hesperius Vockeroth, 1990:696 (new name for *M. chaetopoda* Davidson, 1922).

Length. 5.7–7.9 mm.

Male. Face slightly produced forward below, broad, with low tubercle, with oblique rows of small indistinct punctures over most of surface (similar to punctures in Fig. 1 but smaller), and with narrow well-defined shining median stripe extending from lower facial margin to slightly above tubercle. Antenna black with first flagellomere obscurely orange below at base.

Scutum and scutellum nearly shining, with mostly black hairs; pleura nearly shining, with some black hairs near upper margin, otherwise with white hairs. Wing with moderate bare areas; cell c bare on about basal one-sixth; cell bm bare on most of anterobasal half to two-thirds. Knob of halter blackish. Legs black with narrow femoral apices and fore and mid tibiae orange, with fore and mid tarsi orange-brown and with hind tibia blackish brown; femora with only fine hairs. Fore tibia with regular posterior row of about seven strong black slightly curved bristles; longest bristles about one-third as long as tibia (Fig. 94). Mid tibia with about five very weak black posterior bristles; tarsi unmodified. Hind basitarsus moderately swollen, about 3.5 times as long as greatest depth.

Abdomen robust with shining metallic bluish or coppery spots. Spots on tergite 2 long and widely separated; spots on tergite 3 widened laterally, about four-fifths as long as tergite, and narrowly separated medially; spots on tergite 4 similar but little shorter and confluent medially in some specimens.

Female. Frons lightly gray pruinose below and brown pruinose above, shining only anteriorly and above antenna. Thoracic hairs mostly pale; some sublateral hairs on scutum and marginal hairs on scutellum black. Wing with cell c bare on at least basal half, in some specimens almost entirely bare and with cell bm bare on anterobasal half to two-thirds. Knob of halter yellow. Fore and mid tibiae with several weak posterior white or black hairs about as long as tibial diameter, otherwise without distinctive hairs. Metallic spots of tergite 2 separated by only about one-third width of tergite.

Distribution. Washington south to southern California. Oreg., Calif., II, IV, VI–X.

Specimens identified. United States, 30 ♂♂, 31 ♀♀.

Biology. Larvae, reared in the laboratory on three species of aphids, developed without diapause (Davidson 1922).

Platycheirus hispidipes Vockeroth

Figs. 8, 76, 165; Map 72

Platycheirus hispidipes Vockeroth, 1990:697.

Length. 8.7 mm.

Male. Face (Fig. 8) slightly protruding below, densely gray pruinose, with prominent slightly compressed shining tubercle. Antenna black with first flagellomere obscurely orange below at base.

Scutum and scutellum shining, with mixed white and black hairs. Pleura moderately whitish pruinose, with fine hairs all white except for some black hairs on upper posterior part of anepisternum. Wing rather extensively bare near base; cell bm bare on most of basal half. Knob of halter brown. Legs mostly brown to brownish black with much of fore and mid femora and tibiae, first, second, and fifth tarsomeres of fore leg, and first tarsomere of mid leg dull yellow. Fore leg in Fig. 76*a*; trochanter with fine white hairs on ventral surface; femur posteriorly with subbasal tuft of two long white closely appressed wavy hairs followed by loose tuft of about seven similar but shorter hairs and row of five long moderately strong black hairs; tibia slightly broadened on basal two-thirds, more strongly broadened posteriorly on apical one-third, with strongly produced posteroapical angle, with short white hairs and one long white hair posteriorly on basal two-thirds; first tarsomere uniformly broadened from base to apex, twice as long as broad; second and third tarsomeres slightly narrower, subquadrate. Mid femur with some weak black or pale anteroventral setae on apical half, posteriorly with long pale hairs and some slightly shorter black hairs; mid tibia (Fig. 76*k*) nearly cylindrical, with suberect hairs about half as long as tibial diameter on apical half of ventral surface, with three or four rather strong black hairs about four times as long as tibial diameter just beyond middle of dorsal surface. Hind femur with many long fine white hairs on anterior surface; longest hairs about four times as long as femoral diameter; hind tibia with fine white subappressed anterodorsal hairs about as long as tibial diameter on most of length; four hairs on apical half about three times as long as tibial diameter; first tarsomere of hind leg scarcely swollen, about 5.5 times as long as greatest depth.

Abdomen (Fig. 165) slender. Tergite 2 without pale spots but with anterior two-thirds of lateral margin submetallic; tergites 3 and 4 each with pair of subtriangular, dull orange, partly densely silvery pruinose sublateral spots on about anterior half; tergite 5 submetallic, dark gray.

Female. Unknown.

Distribution. British Columbia (Map 72). B.C., VI, VIII.

Specimens identified. Canada, 2 ♂♂.

Platycheirus holarcticus Vockeroth

Figs. 57, 158; Map 73

Scaeva rostrata Zetterstedt, 1838:607 (preoccupied Wiedemann, 1830).

Platycheirus holarcticus Vockeroth, 1990:698 (new name for *S. rostrata* Zetterstedt, 1838).

Length. 7.1–10.1 mm.

Male. Very similar to male of *P. amplus*, differing as follows: face produced slightly more strongly forward below.

Scutum and scutellum with hairs mostly black, with paler hairs yellow. Pleural hairs pale brown to black. Upper half of anepimeron with very dense upwardly directed tuft of dark brown to black crinkly hairs, with lower half bare. Longest hairs of lower part of katepisternum from one-quarter to half as long as arista. Wing membrane entirely densely trichose. Knob of halter pale to dark brown. Fore trochanter with ventral black setae about two-thirds as long as basal diameter of femur. First tarsomere of fore leg (Fig. 57*b*) with strong dorsal keel over whole length; second tarsomere with distinct broad dorsal keel. Mid tibia with anteroventral tuft on basal half usually less dense and slightly shorter and commonly with few to many pale hairs. Anterior surface of hind tibia (Fig. 57*m*) with very dense, rather slender, curved black setae; setae on basal one-third directed toward apex; setae on apical two-thirds curved upward; longest setae about two-thirds as long as tibial diameter; first tarsomere of hind leg (Fig. 57*m*) moderately swollen, tapering evenly to rather stout apex, about 4.5 times as long as greatest depth.

Abdomen as in Fig. 158.

Female. Not distinguishable from females of several related species.

Distribution. Alaska, boreal and western Canada (Map 73), south in mountains to Utah (3060 m), New Mexico (2950 m), Colorado (3000–3900 m), and New Hampshire (1650 m); northern Europe. B.C., VI–VIII; Que., VI–VIII.

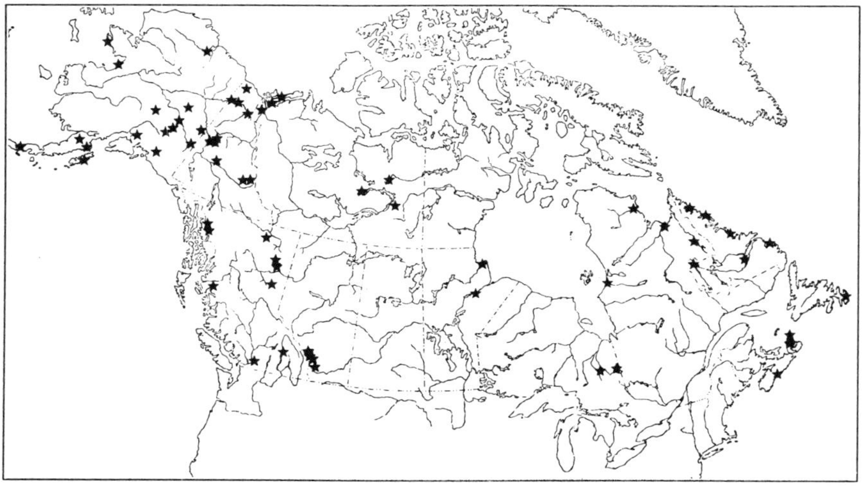

Map 73. Collection localities for *Platycheirus holarcticus* Vockeroth.

Specimens identified. Alaska, 34 ♂♂; Canada, 121 ♂♂; United States, 45 ♂♂; Europe, 11 ♂♂.

Biology. Specimens were taken in Nova Scotia in a damp open sphagnum fen at 300 m.

Platycheirus hyperboreus (Staeger)

Fig. 77; Map 74

Syrphus hyperboreus Staeger, 1845:362.
Platycheirus erraticus Curran, 1927*a*:7.
Platycheirus chirosphena Hull, 1944:76.

Length. 5.3–8.7 mm.

Male. Very similar to male of *P. angustatus*, differing as follows: first flagellomere extensively yellow below in southern specimens.

Wing membrane entirely trichose or with small bare area at base of cells c and bm. Fore leg in Fig. 77*a*; femur with posterior hairs in some specimens 1.5 times as long as femoral diameter; tibia scarcely produced posteroapically; apex obliquely truncate; first tarsomere with anterior margin straight and posterior margin slightly but evenly rounded so margins not parallel on apical half. Mid tibia with anteroventral hairs of basal half erect, with ventral hairs subappressed.

Abdomen moderately robust. Tergite 2 varying from slightly longer than wide to slightly wider than long; tergites 3 and 4 subquadrate or up to 1.3 times as wide as long; spots of tergites very variable in color; in most specimens taken at or near northern limit of trees spots gray or with only trace of yellow; spots of tergite 2 commonly very small or absent; in some northern specimens and in all those taken 100 km or more south of limit of trees or well below treeline in western mountains, spots grayish yellow or yellow-orange; in all specimens spots of tergite 4 almost entirely densely silvery pruinose; spots of tergite 3 densely silvery pruinose on about anteromedian half; spots of tergite 2 usually small, subcircular, restricted to posterior half of tergite, indistinct or absent in some specimens; spots of tergite 3 from half to two-thirds as long as tergite, extending broadly to anterior margin and usually to lateral margin, scarcely distinguishable in some northern specimens; spots of tergite 4 similar but scarcely more than half as long as tergite.

Female. Not distinguishable from females of several similar species.

Distribution. Alaska, Canada, Greenland (Map 74), south to California, Colorado, and North Carolina; northern Europe; Siberia. B.C., IV–VIII; Ont., Que., V–IX.

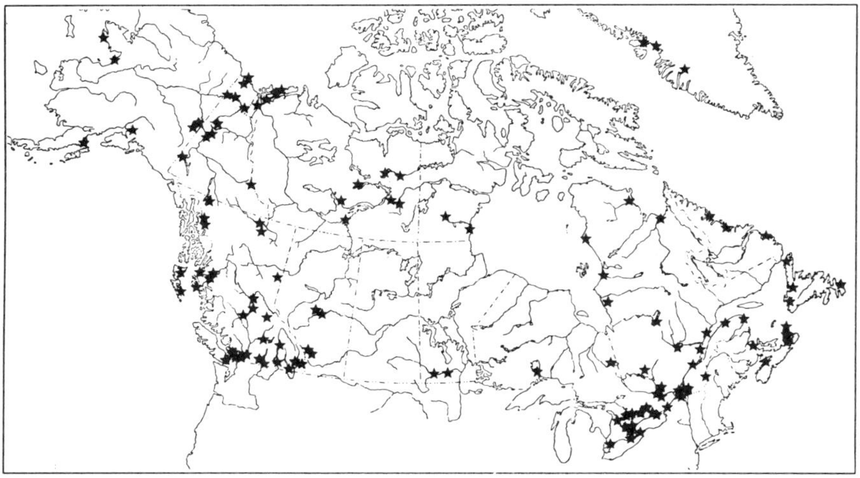

Map 74. Collection localities for *Platycheirus hyperboreus* (Staeger).

Specimens identified. Alaska, 21 ♂♂; Canada, 403 ♂♂; Greenland, 29 ♂♂; United States, 564 ♂♂; Europe, 7 ♂♂; Siberia, 1 ♂.

Biology. Fluke (1929) reported larvae as important predators of *Macrosiphum pisi* on pea and alfalfa in Wisconsin. Adults were collected in Nova Scotia in *Carex* marsh.

Platycheirus immarginatus (Zetterstedt)

Fig. 78; Map 75

Scaeva immarginata Zetterstedt, 1849:3149.
Platycheirus palmulosus Snow, 1895:231.
Platycheirus felix Curran, 1931*a*:94.

Length. 6.5–9.6 mm.

Male. Face vertical with small shining tubercle, otherwise moderately gray to grayish brown pruinose. Antenna black.

Scutum shining, slightly pruinose laterally, usually with only pale hairs but with some black hairs in some specimens. Scutellum shining, with pale hairs. Pleura with yellowish brown pruinosity above and grayish pruinosity below, with hairs usually entirely pale but partly brown to black in some specimens on upper half and stiff and black on lower part of katepisternum; upper part of katepisternum without hairs or with up to eight hairs; hairs of posterior part of anepisternum moderately dense and crinkly, with vertical submarginal row commonly stronger and almost bristlelike. Wing membrane entirely trichose. Knob of halter yellow. Legs usually mostly yellow with coxae and trochanters black; first, fourth, and fifth tarsomeres of hind leg brown to dark brown above; fore and mid femora in some specimens each with brown posterior stripe and hind femur and tibia each with brown ring on middle half; rarely hind femur and tibia dark brown except for narrow bases and apices. Fore leg in Fig. 78*a*; trochanter with many short stiff yellow setae on ventral surface; femur posteriorly with subbasal tuft of about three very closely appressed long wavy white hairs followed by nearly regular row of four or five long slender bristles with wavy tips, of which at least last three are black; longest bristle about twice as long as femoral diameter; tibia almost uniformly broadened from base almost to apex, with posteroapical angle subtriangular and bluntly rounded; first tarsomere with posterior margin oblique and anterior margin straight, about half as wide as long and slightly narrower than tibia; second tarsomere subrectangular, slightly wider than long and slightly narrower than first. Mid femur (Fig. 78*i*) anteroventrally on apical two-thirds with nearly regular single or in some specimens partly double row of 10–22 short stiff acute black setae and usually with one or two long slender strongly curved black hairs just before or just beyond end of row, ventrally on basal half with row of three to six variable but usually strong yellow or rarely black bristlelike hairs; longest hair about twice as long as femoral diameter; mid femur posteriorly and posteroventrally on apical half with few to many scattered short stiff yellow or black setae; mid tibia slightly broadened on apical three-fifths, anteroventrally usually with short but distinct crinkly yellow hairs on most of its length, posteroventrally usually with some similar yellow or dark hairs near mid length. First tarsomere of hind leg rather strongly swollen, about 3.3 times as long as its greatest depth.

Abdomen mostly yellow (as in Fig. 80 but with black areas slightly more extensive). Tergite 2 with narrow median line strongly broadened anteriorly, and with very narrow posterior margin, brownish black to black; tergites 3 and 4 similar to tergite 2 but with median dark line narrower and narrowed anteriorly or in some specimens indistinct; tergite 5 yellow with small posteromedian black triangle not reaching anterior margin in some specimens; rarely in northern specimens yellow spots reduced; spots of tergite 2 either not reaching to anterior or lateral margins, or small and rounded, or even absent; spots of tergites 3 and 4 not reaching anterior and lateral margins and separated by black median line about one-third as wide as tergite; spots of tergite 5 broadly separated or small and obscure.

Female. Not distinguishable from females of several similar species.

Distribution. Alaska, Canada (Map 75), south to southern California, Kansas, and New York; Europe; Siberia. B.C., V, VI; Ont., Que., V–VIII.

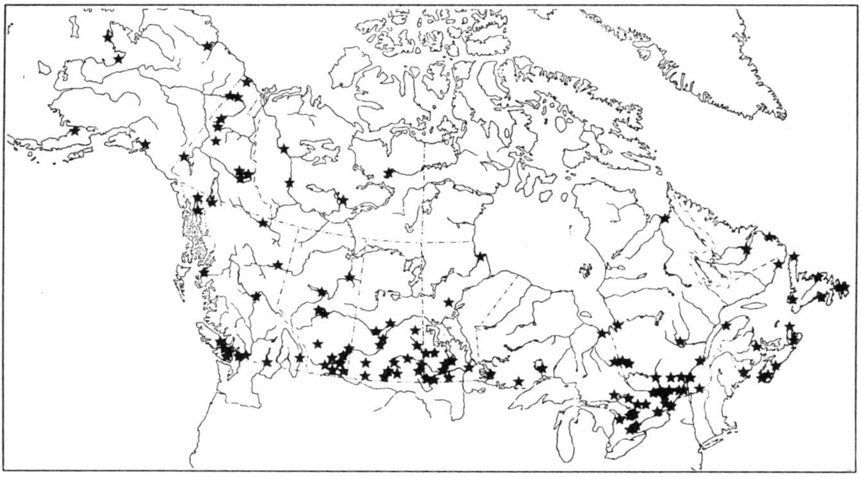

Map 75. Collection localities for *Platycheirus immarginatus* (Zetterstedt).

Specimens identified. Alaska, 42 ♂♂, Canada, 306 ♂♂; United States, 169 ♂♂; Europe, 4 ♂♂; Siberia, 1 ♂.

Biology. Goeldlin (1974) reared larvae on *Aphis fabae* in the laboratory in Switzerland and observed that they had a facultative diapause. Adults were collected in Nova Scotia and Quebec from *Carex* marsh.

Platycheirus inversus Ide

Fig. 58; Map 72

Platycheirus inversus Ide, 1926:156.

Length. 8.7–9.6 mm.

Male. Very similar to male of *P. amplus*, differing as follows: anepimeron with hairs restricted to upper half to two-thirds in some specimens. Wing membrane more extensively bare; cell bm with bare area usually at least half as long as cell, reaching both anterior and posterior margins; cell *cup* with bare area almost as long as anterior margin of cell and extending from vein CuP to A_1. Hind tarsus with first tarsomere (Figs. 58*m,p*) strongly swollen on basal half, abruptly narrowed at mid length, tapering only very slightly to apex.

Female. Not distinguishable from females of several related species.

Distribution. Eastern Canada (Map 72), south to West Virginia and New Hampshire. Ont., Que., V–VII.

Specimens identified. Canada, 30 ♂♂; United States, 24 ♂♂.

Biology. One specimen was taken in Nova Scotia from dry mixed forest.

Discussion. This form may not be specifically distinct from *P. amplus*, but the following observations suggest that two species are involved: the constant difference in wing microtrichia and shape of the first tarsomere of the hind leg, the almost complete difference in distribution, and the distinctness of the specimen of each form taken at the same locality (Low Bush, Lake Abitibi, Ont.).

Platycheirus jaerensis Nielsen

Figs. 59, 168; Map 76

Platycheirus jaerensis Nielsen, 1971:57.

Length. 9.2–10.1 mm.

Male. Head as in Fig. 10 but with tubercle slightly less prominent. Antenna dark orange, with upper margins of pedicel and of first flagellomere brownish black.

Scutum, scutellum, and pleura subshining with yellow hairs. Lower katepisternal hairs fine, about three-fifths as long as arista. Anepimeron with anterior part haired throughout. Wing with small bare areas near base; most of basal one-third of cell bm bare. Knob of halter yellow to brown. Legs mostly dark yellow-orange; coxae black; trochanters, about basal one-third of fore and mid femora, hind femur except narrow apex, most of hind tibia, and all hind tarsus brown to blackish brown. Fore leg in Fig. 59; trochanter with rather long stiff black setae on entire ventral surface; femur without distinctive hairs or bristles, posteriorly with dense fine black hairs; longest hairs little shorter than femoral diameter; tibia scarcely depressed, slightly and uniformly broadened from base almost to apex, slightly more strongly broadened on apical one-quarter; first tarsomere about twice as long as wide, very slightly wider than tibia, with obscure dorsal keel and with oblique apex; second tarsomere about two-thirds as wide and three-fifths as long as first. Mid femur with shallow concavity at middle of anterior surface (concavity similar to that of *P. amplus* [Fig. 56*e*] but less well developed), preceded and bordered below by short strong straight or slightly curved black setae, followed by short soft black upcurved hairs; mid tibia very slightly arcuate, otherwise scarcely modified, on basal half with an anteroventral tuft of black or partly pale erect or subappressed hairs; longest hair twice as long as tibial diameter. First tarsomere of hind leg scarcely swollen, about five times as long as greatest depth.

Abdomen in Fig. 168. Tergite 2 with rather large semicircular dull yellow spots with straight posterior margins; tergites 3 and 4 each with large subquadrate dull yellow spots lying on anterior margin but well-separated from other margins; tergite 5 dull yellow with black median triangle. Sternite 1 black, other sternites dark yellow.

Female. Probably not distinguishable from females of several similar species.

Distribution. Eastern Canada (Map 76); northern Europe. Que., V.

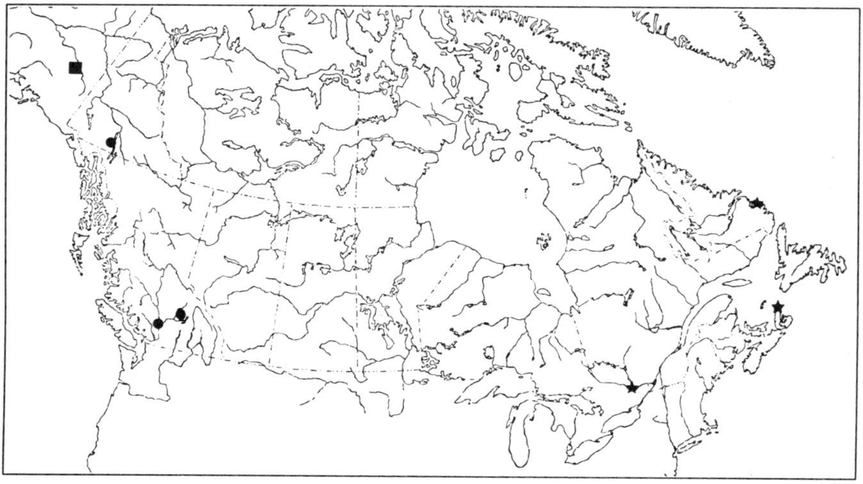

Map 76. Collection localities for *Platycheirus jaerensis* Nielsen (★), for *P. kelloggi* (Snow) (●), and for *P. manicatus* (Meigen) (■).

Specimens identified. Canada, 3 ♂♂; Europe, 1 ♂.

Platycheirus kelloggi (Snow)

Fig. 92; Map 76

Melanostoma kelloggi Snow, 1895:230.
Melanostoma johnsoni Jones, 1917:220.

Length. 7.7–10.1 mm.

Male. Face strongly produced throughout, more than half as wide as head, with small subshining tubercle, otherwise lightly dark gray pruinose, with very faint median groove near upper end in some specimens. Frons large, with dense rather long hairs. Eyes meeting at an angle of about 110°. Antenna black with first flagellomere orange below.

Scutum and scutellum subshining, with long white or yellowish hairs; longest scutellar hairs slightly longer than arista. Pleura very slightly pruinose, with long white or yellowish hairs. Wing extensively bare; cell c bare on basal one-third to three-quarters, cell bm bare except at apex and along about apical half of posterior margin. Knob of halter yellow. Legs with coxae and trochanters black; fore femur orange with blackish posterior stripe and with up to basal half of anterior surface black in some specimens; mid femur orange with posterior surface narrowly black at base or with up to basal half black; hind femur usually black with about apical one-quarter orange but with orange-brown stripe on entire posteroventral surface in some specimens; fore and mid tibia orange with apical one-quarter slightly brownish in some specimens; hind tibia orange at base becoming black toward apex; fore and mid tarsus dark brown to black above with fifth tarsomere usually orange; hind tarsus black. Fore leg (Fig. 92); femur posteriorly with long fine dense mostly pale hairs becoming slightly shorter toward apex and followed by two or three longer and stronger black hairs with strongly curved apices, ventrally near base with row of three or four moderately strong straight yellow or black hairs slightly longer than femoral diameter; tibia posteriorly with many fine pale hairs up to twice as long as tibial diameter; tarsus simple, with very short hairs. Mid femur anteroventrally with two irregular rows of strong stiff white or black hairs; hairs near base almost as long as femoral diameter, near apex appearing as short as fine setulae. Mid femur posteroventrally near base with row of about four long strong black hairs, posteriorly with long dense fine hairs; mid tibia and tarsus with short hairs. Hind femur anteriorly with long fine hairs; hind tibia with fine pale anterodorsal hairs up to twice as long as tibial diameter; first tarsomere of hind leg moderately swollen, about 3.5 times as long as greatest depth.

Abdomen robust, with unusually long hairs. Tergites 2–4 with large subshining silvery gray pruinose spots widely removed from lateral margins; spots on tergite 2 near middle of sclerite; spots on tergites 3 and 4 reaching anterior margins; spots of all tergites usually with at least inner end and commonly with entire surface with dull orange background; tergite 5 black, subshining.

Female. Face with tubercle strongly shining, with pruinosity less uniform than in male. Frons shining with narrow pruinose margin on lower half; pruinosity extending triangularly inward at mid length so that from one-third to half of middle of frons shining. Thoracic hairs shorter than in male but much longer than in related species. Cell c bare or trichose on at most apical one-third; cell bm with at most some microtrichia near apex. Femora and fore and mid tibiae entirely orange; hind tibia orange with up to apical half brown to black; tarsal color as in male. Abdomen broadly oval; markings similar to those of male but those of tergites 3 and 4 and commonly those of tergite 2 entirely bright orange and clearly defined; markings of tergite 2 subtriangular; markings of 3 and 4 subquadrate.

Distribution. Western Canada (Map 76), south to Colorado (2900–3900 m). B.C., VII; Colo., VI–VIII.

Specimens identified. Canada, 2 ♂♂, 3 ♀♀; United States, 8 ♂♂, 77 ♀♀.

Platycheirus latitarsis Vockeroth

Fig. 60; Map 77

Platycheirus latitarsis Vockeroth, 1990:705.

Length. 7.7–8.8 mm.

Male. Very similar to male of *P. amplus*, differing as follows: scutal and scutellar hairs mixed dull yellow and black. Pleural hairs whitish to dull yellow or brown; hairs on posterior margin of anepisternum blackish in some specimens. Anepimeron with hairs not forming dense tuft, extending almost to lower margin. Hairs of lower part of anepisternum at least two-fifths as long as arista. Fore leg in Fig. 60*b*; trochanter with ventral black setae about half as long as basal femoral diameter; first tarsomere little less than twice as long as wide, without dorsal keel; second tarsomere about 1.6 times as wide as long, about nine-tenths as wide as first tarsomere, and as wide as widest part of fore tibia. Mid tibia with anteroventral hair tuft of basal half sparse, partly pale and subappressed in some specimens; longest hairs subequal to tibial diameter; mid tibia also with very short subappressed hairs on apical one-third of ventral surface. Anterior surface of hind tibia (Fig. 60*n*) with more abundant and longer black setae, longest setae about three-quarters as long as tibial diameter; first tarsomere of hind leg less strongly swollen, tapering only slightly to apex.

Female. Probably indistinguishable from females of several similar species.

Distribution. Western Canada (Map 77), south to New Mexico. B.C., VI–VII.

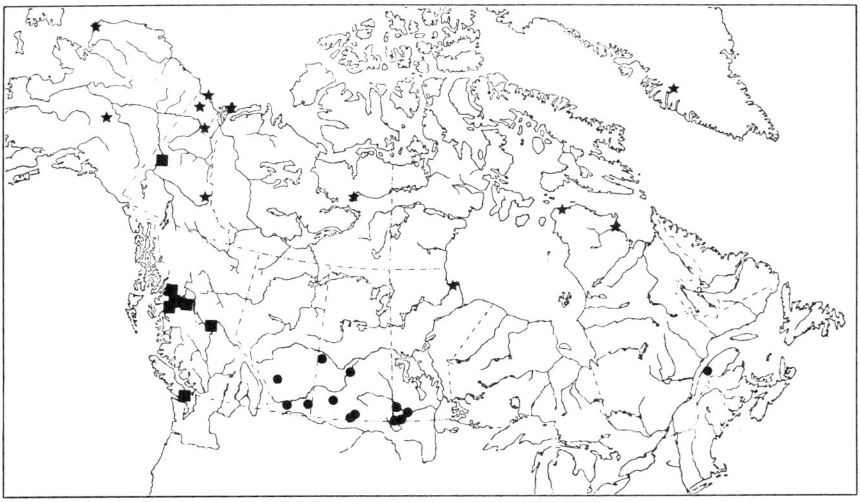

Map 77. Collection localities for *Platycheirus latitarsis* Vockeroth (■), for *P. lundbecki* (Collin) (★), and for *P. luteipennis* (Curran) (●).

Specimens identified. Canada, 9 ♂♂; United States, 4 ♂♂.

Platycheirus latus (Curran)

Fig. 244; Map 78

Melanostoma lata Curran, 1922*b*:276.

Length. 8.7–10.0 mm.

Male. Very similar to male of *P. concinnus*, differing as follows: antenna black.

Thorax without bluish tint. Scutum and scutellum with hairs mostly brown to black, some yellow-brown. Pleura with hairs near upper margin brown to black; other hairs mostly yellowish. Wing with cell c bare at extreme base and with cell bm narrowly bare posteriorly over most of its length. Legs black with narrow apices of femora and bases of tibiae orange-brown.

Markings of tergites with grayish green rather than bluish tint. Surstylus (Fig. 244*b*) with extremely small basal lobe; main arm of surstylus distinctly narrower at mid length than at base and apex. Paramere (Fig. 244*c*) with stout base and closely appressed spine. Aedeagus (Fig. 244*d*) short, stout.

Female. Very similar to female of *P. concinnus*, differing as follows: pruinose spots of frons less distinct, extending only slightly inward. Antenna with only first flagellomere obscurely orange below. Scutum with many black hairs sublaterally; scutellum with mixed black and white hairs. Wing with cell c bare on about basal one-sixth, and with cell bm narrowly bare at base and along most of anterior and posterior margins. Legs black with narrow apices of femora and up to basal one-third of tibiae orange.

Distribution. Yukon Territory, Alberta (Map 78), Washington. Y.T., VI, VII.

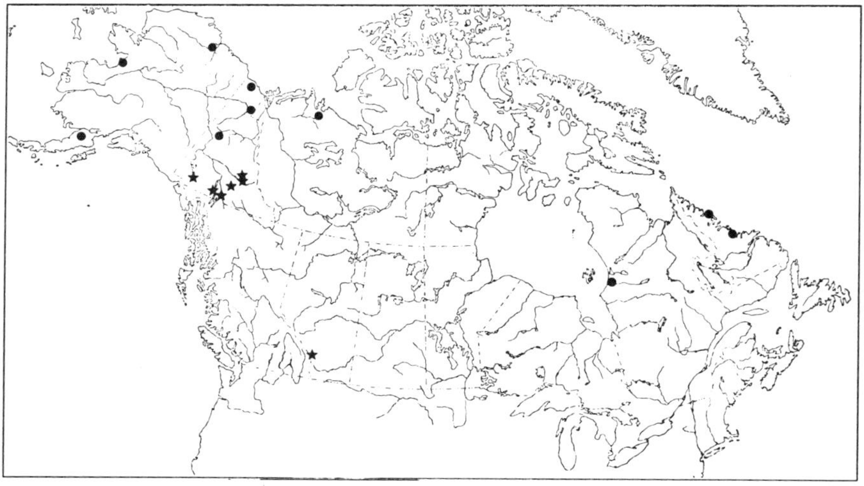

Map 78. Collection localities for *Platycheirus latus* (Curran) (★) and for *P. nigrofemoratus* Kanervo (●).

Specimens identified. Canada, 6 ♂♂, 11 ♀♀; United States, 2 ♂♂.

Discussion. The female holotype cannot be associated with certainty with the males or the other females referred here but, because of distribution, the partly dark scutal hairs, and the occurrence together of such females and males at two Yukon localities, it seems very likely that all these specimens are conspecific.

Platycheirus lundbecki (Collin)

Map 77

Melanostoma lundbecki Collin, 1931:68.
Platycheirus fjellbergi Nielsen, 1974:167.

Length. 5.3–6.5 mm.

Male. Very similar to male of *P. coerulescens*, differing as follows: thorax with few or many brown to black hairs in some specimens. Wing membrane almost entirely trichose, with at most rather obscure bare areas at base of cell c and on anterobasal part of cell bm. Legs much darker, with only base of fore femur, apices of all femora, and bases of all tibiae obscurely yellow-orange. Fore femur with posterior bristles slightly weaker, especially on basal half, and with ventral bristles almost all black; tibia with posterior hairs black, shorter, often subappressed and inconspicuous. Mid femur with anteroventral subbasal bristles black, commonly in irregular double row; mid tibia with posterior hairs black, less conspicuous. First tarsomere of hind leg slightly more swollen, about 3.5 times as long as its greatest depth.

Abdomen with pale spots with subshining metallic grayish or coppery background and dense silvery gray pruinosity.

Female. Not distinguishable from females of several similar species.

Distribution. Alaska, northern Canada, Greenland (arctic and alpine only) (Map 17); northern Europe; Siberia.* Que., VI–VIII.

Specimens identified. Alaska, 4 ♂♂; Canada, 26 ♂♂; Greenland, 17 ♂♂; Europe, 2 ♂♂.

Discussion. A series of males from Peary Land, northern Greenland, are more or less intermediate between specimens of *P. lundbecki* and *P. coerulescens*. The legs are paler than in *P. lundbecki*, with the fore femur dark only posteriorly and the fore and mid tibiae broadly pale basally; the posterior bristles of the fore femur are as in

P. lundbecki but the ventral bristles of the fore and mid femora and the long posterior hairs of the fore and mid tibiae are pale as in *P. coerulescens*. The first tarsomere of the hind leg is strongly swollen, about three times as long as its greatest depth. The wing microtrichia are rather variable, about intermediate between those of the two species. The pale abdominal spots are similar to those of *P. lundbecki*. These specimens possibly represent a distinct species or, more probably, a variant population of *P. lundbecki*.

Platycheirus luteipennis (Curran)

Fig. 246; Map 77

Melanostoma luteipennis Curran, 1925:114.
Melanostoma agens Curran, 1931*b*:253.

Length. 8.7–10.1 mm.

Male. Face produced strongly forward below (as in Fig. 10), sparsely grayish pruinose with poorly defined shining median stripe from upper end of tubercle to lower facial margin, with tubercle rather large but not abrupt above, on upper part with weak median keel flanked on each side by upwardly convergent groove. Antenna black with first flagellomere orange below.

Scutum and scutellum strongly shining, slightly brassy, with short brassy yellow hairs; longest scutellar hairs not more than half as long as arista. Pleura mostly shining, with short yellow to yellow-brown hairs. Wing with cell c bare on about basal one-sixth, with cell bm bare posteriorly on about basal half, and with cell *cup* broadly bare on basal one-third. Knob of halter orange. Legs mostly orange-brown, with fore and hind coxae, basal one-third to half of mid femora, hind femur except apex, usually preapical ring on each tibia, and upper surface of hind tarsus blackish brown to black. Legs simple, without outstanding hairs or bristles; hairs of femora much shorter than femoral diameter. First tarsomere of hind leg slightly swollen, about four times as long as greatest depth.

Abdomen moderately slender. Tergites opaque black with shining lateral margins; tergites 2–4 each with pair of large sublateral coppery or slightly pruinose spots; spots on tergite 2 near middle of segment and broadly separated medially; spots on tergite 3 near anterior margin and narrowly separated; spots of tergite 4 on anterior margin and poorly defined, appearing to cover most of tergite in some specimens; tergite 5 shining, coppery. Surstylus (Fig. 246*b*) with slender subtriangular basal thumblike lobe. Paramere (Fig. 246*c*) slender, with spine widely separated from base. Aedeagus (Fig. 246*d*) short, stout.

Female. Face with keel and grooves of upper part weaker but distinct. Frons shining with narrow elongate pruinose spot on each side of lower half. Thoracic hairs much shorter than in male. Wing with cell c bare on about basal half; cell bm mostly bare with only anterior margin and about apical one-quarter trichose. Abdomen nearly uniformly coppery. Spots of tergites usually less distinct than in male.

Distribution. Central Canada, Quebec (Map 77), south to Montana, Colorado, and North Dakota. Alta., Sask., VI–VIII.

Specimens identified. Canada, 17 ♂♂, 6 ♀♀; United States, 3 ♂♂, 1 ♀.

Platycheirus manicatus (Meigen)

Fig. 52; Map 76

Syrphus manicatus Meigen, 1822:336.

Length. 7.9–10.1 mm.

Male. Similar to male of *P. discimanus*, differing as follows: face produced very strongly forward below, with lower facial margin extending beyond tubercle, with keel and grooves above tubercle obscure in some specimens, moderately gray pruinose with only tubercle shining.

Scutum and scutellum subshining, with distinct olivaceous pruinosity and mixed black and white hairs. Pleura rather densely gray pruinose, with white to pale brown hairs. Basal one-third of fore and mid tibiae yellow; mid tarsus dark brown. Fore leg in Fig. 52*a*; tibia with longer posterior hairs only on apical half; first tarsomere slightly longer than wide with curved lateral margins and very oblique apex, almost 2.5 times as wide as apex of fore tibia; second tarsomere almost as wide as first, with posterior margin about 1.5 times as long as anterior margin. Mid tibia (Fig. 52*h*) with long rather sparse posterodorsal and posterior hairs on apical half; longest hairs about 3.5 times as long as tibial diameter. First tarsomere of hind leg strongly swollen, about three times as long as deep.

Tergites 2–4 each with pair of large elongate subrectangular yellow spots; tergite 5 with small obscure anterolateral yellow spots in some specimens. (Description based on one male from Alaska and several from Europe.)

Female. Face similar to that of male, with lower facial margin produced well beyond tubercle. Frons gray pruinose, shining only immediately above antennae, subshining on ocellar triangle. Scutum weakly but distinctly pruinose, more strongly pruinose laterally, especially on notopleuron. Anterior part of anepimeron haired throughout. Fore tarsus black; legs otherwise colored as in male. Yellow spots of tergites 3 and 4 slightly wider than long. (Description based on several females from Europe.)

Distribution. Alaska (Map 76); Europe; Siberia. Alaska, VI.

Specimens identified. Alaska, 1 ♂; Europe, 43 ♂♂, 27 ♀♀; Siberia, 3 ♂♂.

Biology. Láska and Starý (1980) listed two species of aphids as larval hosts in Czechoslovakia.

Discussion. The female can probably be distinguished from those of other Nearctic species of this genus by the strongly produced face, which has the lower facial margin extending beyond the tubercle, combined with black antennae and a distinctly pruinose scutum. Species of the *P. peltatus* group have the face almost as strongly produced but the lower facial margin extends at most as far forward as the tubercle. The first flagellomere is yellow-orange below, the scutum is distinctly pruinose only laterally, and commonly the lower half of the anterior part of the anepimeron is bare.

Platycheirus modestus Ide

Figs. 9, 84, 169; Map 79

Platycheirus modestus Ide, 1926:155.

Length. 6.2–7.3 mm.

Male. Very similar to male of *P. immarginatus*, differing as follows: face receding below (Fig. 9).

Thoracic hairs entirely yellow; posterior part of anepisternum with hairs dense, uniform, very long, and with crinkly apices. Legs variable in color, similar in color to those of *P. immarginatus* but never with hind leg mostly black and usually with last two tarsomeres of hind leg much paler than first tarsomere. Fore leg setae on ventral surface of trochanter very short, about one-third as long as in *P. immarginatus*; femur posteriorly with only fine yellow hairs, without subbasal white tuft or black bristles; tibia with posteroapical angle slightly longer and more acute and with slightly more conspicuous hairs on basal two-thirds of posterior margin. Mid leg in Fig. 84; femur on apical three-eighths of anteroventral surface with slender black curved hairs twice as long as femoral diameter, otherwise with only very short hairs; tibia with dense subappressed crinkly black hairs about three times as long as tibial diameter on basal three-quarters of anteroventral and ventral surfaces.

Abdomen in Fig. 169. Tergites 2–5 almost entirely yellow-orange, with anterior margin of 2 and posterior margins of 2–5 narrowly black; black anterior margin of tergite 2 with posteromedian triangular projection reaching middle of tergite in some specimens; black posterior margins of tergites 4 and 5 with short broad median triangular projection.

Female. Not distinguishable from females of several similar species.

Distribution. Alaska, Canada (Map 79), south to Colorado, Wisconsin,* and New Hampshire. B.C., VI; Ont., Que., VI–VIII.

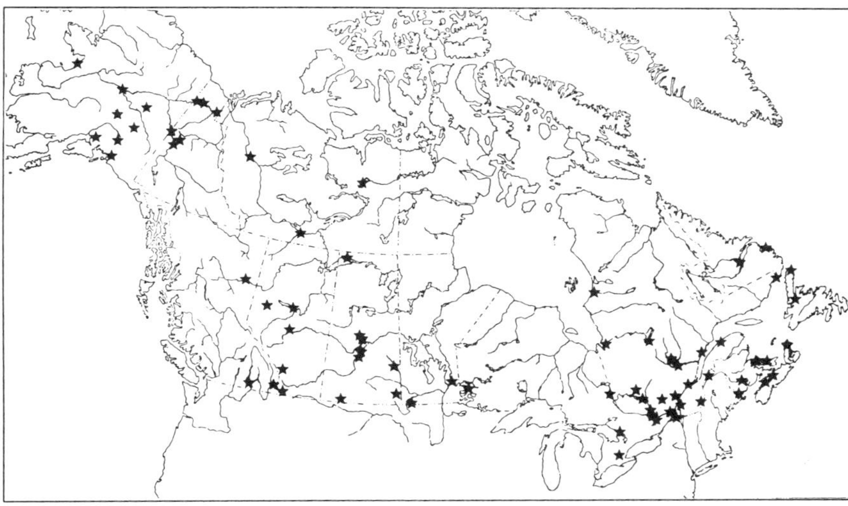

Map 79. Collection localities for *Platycheirus modestus* Ide.

Specimens identified. Alaska, 26 ♂♂; Canada, 265 ♂♂; United States, 37 ♂♂.

Biology. Specimens were collected in Nova Scotia from *Carex* marsh.

Platycheirus nearcticus Vockeroth

Fig. 61; Map 80

Platycheirus nearcticus Vockeroth, 1990:771.
Platycheirus peltatus of American authors (not *Platycheirus peltatus* (Meigen), 1822).

Length. 8.4–10.5 mm.

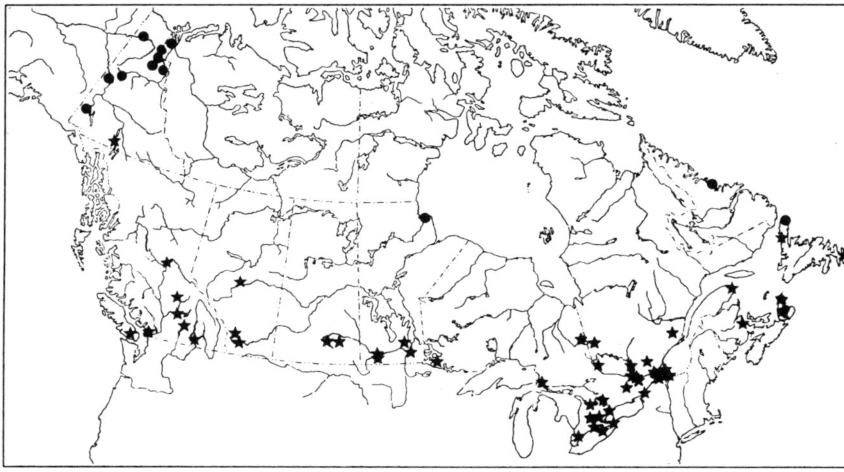

Map 80. Collection localities for *Platycheirus nearcticus* Vockeroth (★) and for *P. nielseni* Vockeroth (●).

Male. Very similar to male of *P. amplus*, differing as follows: antenna usually mostly orange with first flagellomere brown on upper half or less; scape and pedicel brownish above or throughout in some specimens and first flagellomere brown on about upper two-thirds.

Scutum with only pale hairs in some specimens. Scutellar hairs usually entirely pale, rarely some black hairs present near margin. Upper half of anepimeron with moderately dense tuft of upwardly directed pale brown hairs; lower half of anepimeron bare. Longest hairs of lower part of katepisternum from one-sixth to half length of arista. Wing membrane rather extensively bare; cell bm broadly bare basally and bare medially on from half to five-sixths its length. Knob of halter yellow to brown. Legs usually paler; fore and mid femora and first tarsomere of mid leg usually entirely yellow orange. Fore trochanter with ventral black setae about half as long as basal diameter of femur. Mid tibia (Fig. 61*k*) with anteroventral hair tuft on basal half subappressed, usually entirely pale, short, with longest hairs about 1.5 times as long as tibial diameter; hairs of posterior surface yellow to brown. Hind tibia (Fig. 61*n*) with setae of anterior surface moderately dense, rather fine; longest setae about half as long as tibial diameter; first tarsomere of hind leg moderately swollen on about basal one-third, tapering evenly to rather slender apex.

Female. Not distinguishable from females of several related species.

Distribution. Western and southern Canada (Map 80), south to Oregon, New Mexico, and North Carolina. B.C., V–VII; Ont., Que., V–IX.

Specimens identified. Canada, 147 ♂♂; United States, 137 ♂♂.

Biology. Specimens were collected in Nova Scotia and Ontario from dry to moist mixed woodland.

Platycheirus nielseni Vockeroth

Fig. 62; Map 80

Platycheirus nielseni Vockeroth, 1990:712.

Length. 8.4–9.6 mm.

Male. Very similar to male of *P. amplus*, differing as follows: face darker, with dark gray pruinosity.

Scutum with hairs mostly black, paler hairs dark yellow. Scutellar hairs varying from dark yellow to mostly black. Hairs of posterior margin of anepisternum black in some specimens. Upper half of anepimeron with moderately dense tuft of crinkly hairs; lower half of anepimeron bare. Longest hairs of lower part of katepisternum at least half as long as arista. Wing membrane with microtrichia rather sparse at extreme base but without distinct bare areas. Knob of halter yellow to brown. Legs commonly darker, with fore and mid femora almost black on basal one-third of anterior surface and basal two-thirds of posterior surface. Fore trochanter with ventral black setae about three-quarters as long as basal diameter of femur. Mid tibia (Fig. 62*L*) strongly swollen below at apex; anteroventral hair tuft of basal half less dense, mostly pale in some specimens, and with longest hairs about 1.5 times as long as tibial diameter; apical one-third of ventral surface with many black crinkly hairs; hairs at base of apical swelling about as long as tibial diameter. Anterior surface of hind tibia (Fig. 62*n*) with more abundant and slightly longer setae; first tarsomere of hind leg less strongly swollen, tapering only slightly to apex.

Yellow spots of tergite 5 smaller or absent.

Female. Not distinguishable from females of several related species.

Distribution. Northern Canada (Map 80), south to Colorado (2990 m); Europe; Siberia. Y.T., VI; Labr., VI–VII.

Specimens identified. Canada, 32 ♂♂; United States, 1 ♂; Europe, 60 ♂♂; Siberia, 1 ♂.

Platycheirus nigrofemoratus Kanervo

Fig. 67; Map 78

Platycheirus albimanus var. *nigrofemoratus* Kanervo, 1934:122.

Length. 6.2–7.3 mm.

Male. Very similar to male of *P. albimanus*, differing as follows: cell bm with at most very small bare area near base. Fore leg in Fig. 67; tibia nearly uniformly broadened from base to apex; first tarsomere gradually broadened posteriorly from base to apex, with margins slightly divergent throughout, and with apex distinctly arcuate; second tarsomere with both basal and apical margins distinctly arcuate.

Female. Unknown, probably not distinguishable from females of related species.

Distribution. Alaska, northern Canada (Map 78); northern Europe. Que., VI.

Specimens identified. Alaska, 5 ♂♂; Canada, 9 ♂♂; Europe, 1 ♂.

Platycheirus nodosus Curran

Fig. 70; Map 81

Platycheirus nodosus Curran, 1923*b*:272.

Length. 5.3–8.7 mm.

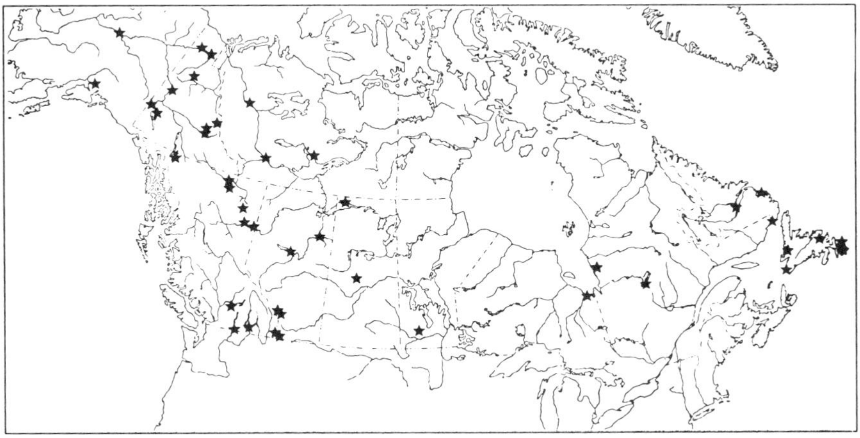

Map 81. Collection localities for *Platycheirus nodosus* Curran.

Male. Face slightly receding below, with very low broad tubercle, in profile as *P. aeratus* (Fig. 11), with uniform dense dark gray pruinosity.

Scutum and scutellum subshining, with slight grayish pruinosity and with pale yellow hairs. Pleura moderately gray pruinose, pleural hairs mostly dark yellow; hairs of posterior margin of anepisternum blackish in some specimens; hairs of lower part of katepisternum straight and rather stout, yellow or black, about one-third as long as arista. Wing membrane entirely trichose. Knob of halter bright yellow. Legs with coxae and trochanters black; fore and mid legs mostly yellow with brown posterior stripe on femora and with fifth tarsomere of fore leg in some specimens, and that of mid leg in all, brown to dark brown; hind leg black with narrow apex of femur and narrow base and apex of tibia black. Fore leg in Fig. 70*a*; trochanter with short stout yellow ventral setae; femur on posterior surface just beyond base with two tufts; each of three closely appressed long dark yellow hairs with black flattened lanceolate tips slightly narrower on basal tuft than on other, otherwise with only very short sparse hairs at most half as long as femoral diameter; tibia almost uniformly broadened from base to three-quarters its length, then slightly narrowed to apex, with posteroapical angle strongly produced, and without outstanding hairs; first tarsomere about two-thirds as wide as widest part of tibia, slightly longer than wide, rather strongly narrowed posteriorly on basal half; last four tarsomeres normal, slightly narrower than first. Mid leg in Fig. 70*e*; femur with short sparse hairs at most two-thirds long as femoral diameter; tibia slender, with rather long fine strongly appressed yellow or black anteroventral hairs on basal two-thirds. First tarsomere of hind leg only slightly swollen, about four times as long as its greatest depth.

Abdomen black with pair of bright yellow-orange spots on each of tergites 2–5. Spots of tergite 2 usually not reaching anterior margin and about six-sevenths as long as tergite; spots of 3 and 4 similar but reaching anterior margins; spots of 5 large leaving only median triangle black.

Female. Not distinguishable from females of several similar species.

Distribution. Alaska, Canada (Map 81), south to Wyoming, New York (555 m), and New Hampshire. B.C., V, VI; Que., VI, VII.

Specimens identified. Alaska, 5 ♂♂; Canada, 138 ♂♂; United States, 4 ♂♂.

Platycheirus normae Fluke

Figs. 85, 159; Map 82

Platycheirus normae Fluke, 1939:366.

Length. 7.1–8.4 mm.

Male. Similar to male of *P. immarginatus*, differing as follows: facial tubercle less prominent.

Thoracic hairs entirely yellow; posterior part of anepisternum with hairs moderately long and dense, uniform, nearly straight. Legs mostly yellow-orange with coxae black and tibia and first tarsomere of fore leg whitish. Fore leg in Fig. 85; trochanter with short broadly triangular ventral process, in anterior view subquadrate rather than subtriangular, with very short pale ventral setulae; fore femur posteriorly with subbasal tuft of three shorter, less densely appressed, yellow rather than white hairs; tuft followed by about six rather long strong yellow hairs on basal half of femur; rest of femur nearly bare; tibia much more strongly broadened, with posterior margin strongly curved and with strong oblique dorsal keel on apical three-eighths; first tarsomere broadened just beyond base, then slightly to strongly constricted and compressed and slightly broadened at apex; tarsomeres 2–4 wide, very short. Mid femur and tibia without conspicuous hairs, setae, or bristles; first tarsomere of mid leg rather strongly compressed, slightly wider at base and apex than at mid length.

Abdomen in Fig. 159. Tergites 2–5 almost entirely yellow-orange; tergite 2 with very narrow black anterior margin; tergites 2–4 with narrow black posterior margin with short median triangular projection; tergite 5 with trace of similar black posterior margin in some specimens. Sternites yellow.

Female. Not distinguishable from females of several similar species.

Distribution. Alaska, Canada (Map 82), south to Wisconsin and Pennsylvania. Ont., Que., VI–VIII.

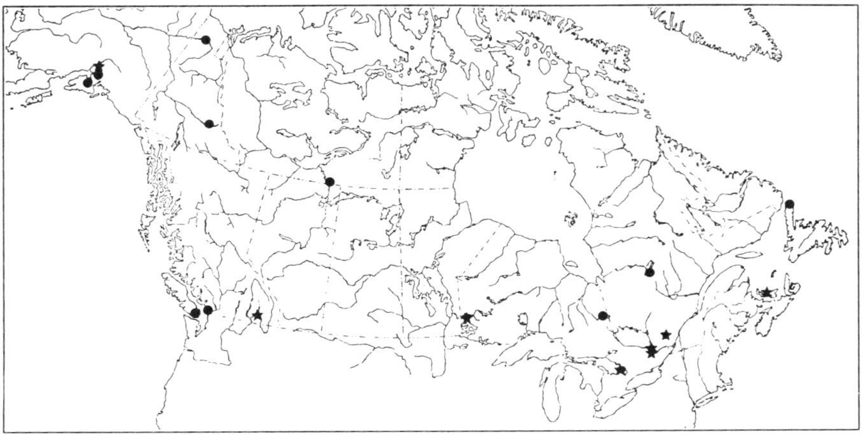

Map 82. Collection localities for *Platycheirus normae* Fluke (★) and for *P. parmatus* Rondani (●).

Specimens identified. Alaska, 1 ♂; Canada, 7 ♂♂; United States, 6 ♂♂.

Biology. Adults were reared during August from larvae feeding on the aphid *Sipha glyceriae* on *Glyceria* sp. in Centre County, Pa., (F.D. Fee, written communication). One male was taken at Beechgrove, Que., during late June with specimens of *P. immarginatus*, *P. scambus*, and *P. thompsoni*, in a wet *Carex–Salix* marsh.

Platycheirus obscurus (Say)

Figs. 268, 270; Map 83

Syrphus obscurus Say, 1824:11.
Melanostoma ? *rostrata* Bigot, 1884:80.
Melanostoma ontario Davidson, 1922:37.
Melanostoma nitidiventris Curran, 1931*b*:252.

Length. 6.8–9.4 mm.

Male. Very similar to male of *P. confusus*, differing as follows: face slightly more strongly produced forward below, with shallower convexity between tubercle and lower facial margin, and with anterior oral margin produced as far, and usually slightly farther, forward than facial tubercle (Fig. 268).

Scutum and scutellum in some specimens with many pale hairs. Pleura with at most some hairs near upper margin black. Wing more extensively bare in some specimens; in eastern specimens cell c bare at extreme base and cell bm bare on about anterobasal one-third; in western specimens bare areas variable in extent, ranging from condition found in eastern specimens to cell c bare or nearly so on basal three-quarters and cell bm with only some microtrichia near apex; (wings most extensively bare in specimens from southern California and Arizona). Legs usually paler, with up to apical half of fore and mid femora, all of fore and mid tibiae and rarely hind tibia, and first and in some second tarsomere of fore and mid legs orange; posterior bristles of fore and mid tibiae slightly weaker.

Abdomen with spots usually more strongly silvery pruinose. Spots on tergite 2 broader and separated by only half width of tergite in some specimens; spots on tergites 3 and 4 distinctly orange in some specimens. Surstylus (Fig. 270) with shorter lobe larger, otherwise as in *P. confusus*.

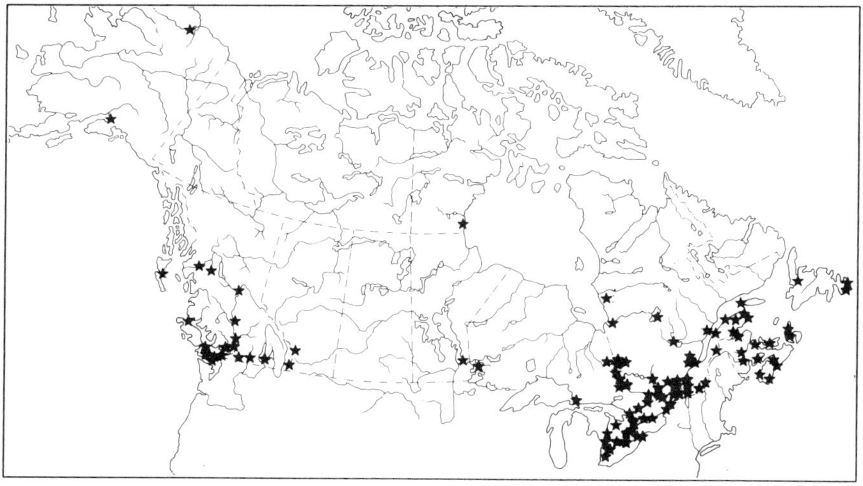

Map 83. Collection localities for *Platycheirus obscurus* (Say).

Female. Very similar to female of *P. confusus*, differing as follows: facial profile as in male of *P. obscurus*. Wing with microtrichia less abundant than in male of *P. obscurus*, almost as variable in eastern as in western specimens; some western specimens with cells c and bm almost entirely bare. Abdominal spots similar to those of male of *P. obscurus* but usually mostly or entirely orange rather than metallic.

Distribution. Alaska, Canada (Map 83), south to California, southern Mexico, Mississippi, and Georgia. B.C., IV–X; Ont., Que., IV–IX.

Specimens identified. Alaska, 1 ♂; Canada, 443 ♂♂, 389 ♀♀; United States, 848 ♂♂, 708 ♀♀; Mexico, 3 ♂♂, 5♀♀.

Biology. Larvae were reared in the laboratory on four species of aphids and developed without diapause (Davidson 1922). Davidson quotes Curran as saying that larvae will feed on rotting chickweed. Males frequently hover at heights of 1–2 m in mixed woodland. The flight season is much longer than that of *P. confusus*; around Ottawa, Ont., many specimens have been taken in late September, which suggests at least two generations; most specimens have been taken in rather dry woods, rather than in the bogs and marshes that are the preferred habitats of *P. confusus*.

Discussion. This species, like *P. confusus*, has a markedly disjunct distribution (Map 83). However, in this case some overlap occurs in the extent of wing microtrichia in eastern and western specimens so that it is less likely that two species are involved. This variation in wing microtrichia makes it difficult to distinguish western specimens of this species from those of *P. confusus*; only the slight differences in facial profile, and in the color of thoracic hairs, the strength of the tibial bristles, and the shape of the surstylus in the male seem to be of value in distinguishing the species in the West.

Platycheirus octavus Vockeroth

Fig. 63; Map 84

Platycheirus octavus Vockeroth, 1990:719.

Length. 7.7–9.6 mm.

Male. Very similar to male of *P. amplus*, differing as follows: anepimeron with moderately dense tuft of hairs on upper half only. Scutellum with some black hairs in some specimens. Wing membrane usually more extensively bare—about basal one-fifth of cell c and up to anterobasal one-third of cell bm mostly bare. First tarsomere of fore leg with distinct and moderately strong dorsal keel on entire length. Mid tibia with anteroventral hairs on basal half pale, subappressed; longest hairs about 1.5 times as long as tibial diameter. Hind tibia with anterior setae about as long as tibial diameter; setae on basal half directed slightly toward apex; setae on apical half directed dorsally (Fig. 63*n*); first tarsomere of hind leg slightly less strongly swollen, tapering evenly to apex.

Female. Probably not distinguishable from females of several related species.

Distribution. British Columbia (Map 84), Washington, Oregon, and Montana. B.C., IV–VI.

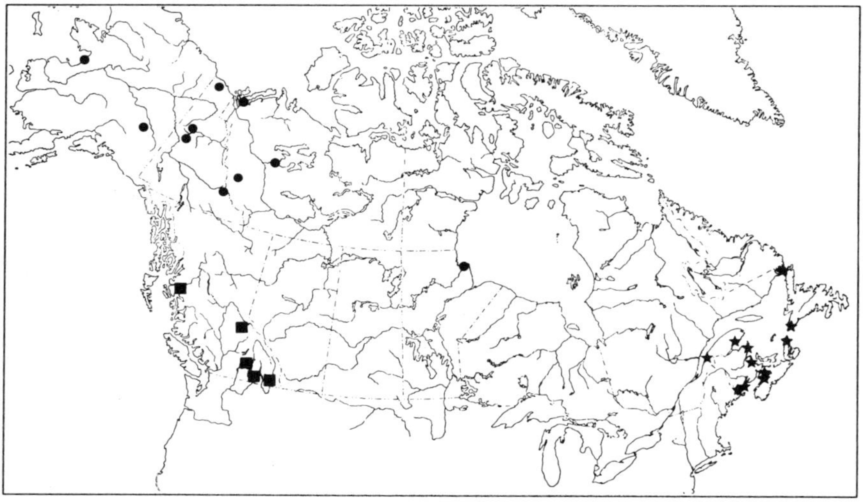

Map 84. Collection localities for *Platycheirus octavus* Vockeroth (■), for *P. orarius* Vockeroth (★), and for *P. pilatus* Vockeroth (●).

Specimens identified. Canada, 8 ♂♂; United States, 11 ♂♂.

Platycheirus orarius Vockeroth

Figs. 86, 170, 238; Map 84

Platycheirus orarius Vockeroth, 1990:720.

Length. 7.9–9.6 mm.

Male. Very similar to male of *P. immarginatus,* differing as follows: thoracic hairs entirely pale except some stiff black hairs present on lower part of katepisternum in some specimens; posterior part of anepisternum with hairs moderately dense, uniform, with scarcely crinkly apices; upper part of katepisternum with at least 12 hairs. Legs yellow-orange with coxae, fore and mid trochanters, first and last two tarsomeres of hind leg, and obscure marks on hind tibia in some specimens dark brown to black; second and third tarsomeres of hind leg brown above in some specimens but distinctly paler than last two tarsomeres. Fore leg in Fig. 86*a*; trochanter with yellow ventral setulae very short; femur without posterior subbasal tuft of pale hairs, with many rather long straight posterior and ventral yellow hairs; row of three or four posterior hairs longer and stronger than in *P. immarginatus* and with one or two black in some specimens; cluster of several longer subbasal ventral hairs black; tibia with posteroapical angle slightly more acute; first tarsomere a little more abruptly widened beyond base, reaching its maximum width at about two-fifths its length rather than beyond mid length; second tarsomere slightly longer than wide. Mid femur (Fig. 86*i*) anteroventrally usually with irregular group of weak black setae at mid length followed by sparse fine wavy black hairs about 1.5 times as long as femoral diameter, ventrally with long hairs of which some black and slightly stronger. Mid tibia with dense, very crinkly, mostly yellow hairs, about as long as tibial diameter, on almost all anteroventral, ventral, and posteroventral surfaces.

Abdomen (Fig. 170) as in paler specimens of *P. immarginatus* but with slightly wider median black line. Surstylus (Fig. 238) with very large dorsobasal lobe.

Female. Not distinguishable from females of several similar species.

Distribution. Eastern Canada (Map 84), south along Atlantic coast to New Jersey. Que., VIII; N.B., VI, VII.

Specimens identified. Canada, 26 ♂♂, 16 ♀♀; United States, 41 ♂♂, 17 ♀♀.

Biology. Thirty-three specimens from four localities are labeled as having been taken in salt marsh; at the type locality, a salt marsh on the coast of New Hampshire, 8 males and 13 females of *P. orarius*, but no other *Platycheirus*, were taken. As all specimens were taken at coastal localities, it is probable that the species is restricted to salt marshes. No other Nearctic species of Syrphinae is known to be so restricted.

Platycheirus oreadis Vockeroth

Figs. 53, 157

Platycheirus oreadis Vockeroth, 1990:720.

Length. 6.2–7.9 mm.

Male. Similar to male of *P. discimanus*, differing as follows: face produced slightly more forward and downward, moderately pale gray pruinose with only tubercle shining.

Thoracic hairs almost all white; scutum with very few black hairs. Base of cell bm with indistinct bare area. Fore and mid tibiae dull yellow; fore and mid tarsi mostly yellow, brownish above towards apex. First tarsomere of fore leg (Fig. 16*a*) strongly broadened beyond base, very oblique apically, 1.2 to 1.3 times as long as wide and about 2.5 times as wide as apex of fore tibia; second tarsomere oblique in position, about 2.5 times as wide as long. Mid tibia (Fig. 53*k*) with most of posterior and posteroventral surface with dense fine rather long hairs; mid tarsus not at all compressed. Hind tibia with most of basal half of anterodorsal surface with fine black progressively longer hairs; longest hairs at least 2.5 times as long as tibial diameter; first tarsomere of hind leg strongly swollen, about three times as long as greatest depth.

Abdomen (Fig. 157) with spots of tergites larger, subshining dark blue, with very faint grayish pruinosity.

Female. Unknown.

Distribution. Colorado, 3900–4020 m. Colo., VII, VIII.

Specimens identified. United States, 2 ♂♂.

Platycheirus parmatus Rondani

Figs. 10, 87; Map 82

Platycheirus parmatus Rondani, 1857:121.
Platycheirus ovalis Becker, 1921:27.
Platycheirus bigelowi Curran, 1927*a*:5.

Length. 9.1–10.1 mm.

Male. Similar to male of *P. jaerensis*, differing as follows: face (Fig. 10) with slightly more prominent tubercle. Antenna black with about lower one-third of first flagellomere orange.

Thoracic hairs longer; hairs of scutum, scutellum, and upper half of pleura black and white mixed. Lower katepisternal hairs as long as arista. Wing with indistinct bare area along middle of basal one-quarter of cell bm. Legs darker; basal four-fifths of fore and mid femora, posterior streak on fore tibia, and most of apical half of mid tibia brown to blackish brown. Fore leg in Fig. 87*a*; trochanter with ventral setae longer and weaker; femur with posterior hairs up to twice as long as femoral diameter; tibia slightly more strongly broadened on apical one-sixth; first tarsomere much wider, about 1.5 times as long as wide, almost twice as wide as apex of fore tibia; second tarsomere very short, subtriangular, about seven-eighths as wide as first. Mid leg in Fig. 87*f*; femur without anterior concavity or stiff setae, basal two-thirds with long fine black anteroventral hairs, apical one-third with similar but shorter reflexed or upwardly curved hairs; tibia with dense soft crinkly black hairs on ventral and posterior surfaces; longest hairs about three times as long as tibial diameter. First tarsomere of hind leg more strongly swollen, about 3.5 times as long as its greatest depth.

Abdomen with yellow spots of tergites 3 and 4 slightly smaller, well-separated from anterior margin; yellow spots of tergite 5 restricted to anterolateral corners.

Female. Not distinguishable from females of several similar species.

Distribution. Alaska, northern Canada (Map 82), south to New Mexico (2740 m); Europe; Siberia; Japan.* B.C., IV; Ont., Que., VI.

Specimens identified. Alaska, 5 ♂♂; Canada, 7 ♂♂; United States, 4 ♂♂; Europe, 9 ♂♂; Siberia, 4 ♂♂.

Biology. Goeldlin (1974) found that this species (as *P. ovalis*) had an obligatory larval diapause of 8–10 months in Switzerland.

Platycheirus peltatoides Curran

Fig. 64; Map 85

Platycheirus peltatoides Curran, 1923*b*:274.

Length. 8.2–10.5 mm.

Male. Very similar to male of *P. amplus*, differing as follows: face produced slightly more strongly forward below.

Scutum and scutellum with hairs mostly black, paler hairs dull yellow. Pleural hairs pale brown to black. Upper half of anepisternum with rather loose tuft of upwardly directed crinkly brown hairs; lower half of anepisternum bare. Longest hairs of lower part of katepisternum at least three-quarters as long as arista. Wing membrane entirely trichose or with very small and indistinct bare areas at base of cells c and bm. Knob of halter dark brown. Fore trochanter with ventral black setae as long as basal diameter of fore femur; first tarsomere of fore leg with moderately strong dorsal keel on most of length (as in Fig. 57*b*); second tarsomere with obscure broad dorsal keel. Mid tibia with anteroventral tuft on basal half scarcely discernible, diffuse, subappressed; hairs pale or dark and shorter than tibial diameter; apical half of posterior surface with short inconspicuous hairs. Anterior surface of hind tibia (Fig. 64) with short appressed black setae on basal one-third to half and with long rather slender upcurved irregularly spaced black setae on apical half to two-thirds; longest setae at least 1.5 times as long as tibial diameter.

Female. Not distinguishable from females of several related species.

Distribution. Alaska, western Canada (Map 85), south to Oregon and Colorado. B.C., V–VIII.

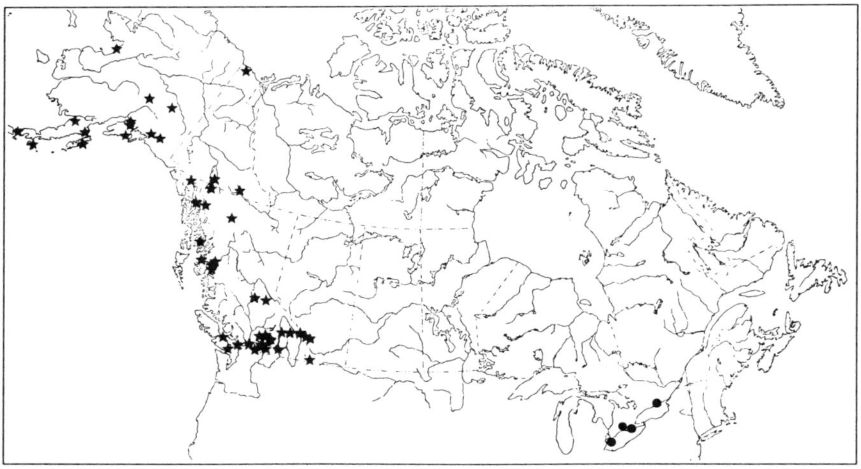

Map 85. Collection localities for *Platycheirus peltatoides* Curran (★) and for *P. scamboides* Curran (●).

Specimens identified. Alaska, 78 ♂♂; Canada, 130 ♂♂; United States, 20 ♂♂.

Platycheirus perpallidus Verrall

Figs. 79, 164; Map 86

Platycheirus perpallidus Verrall, 1901:290.

Length. 5.7–9.1 mm.

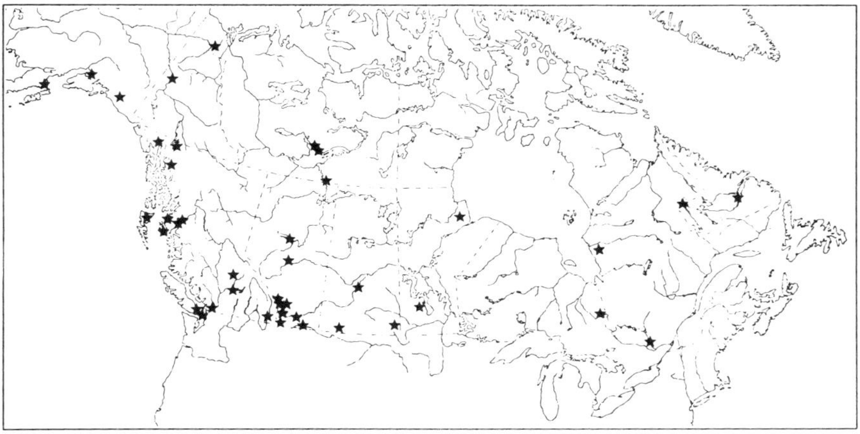

Map 86. Collection localities for *Platycheirus perpallidus* Verrall.

Male. Very similar to male of *P. immarginatus*, differing as follows: thoracic hairs usually entirely pale; posterior part of anepisternum with hairs in some specimens partly brown or black, very dense, long, uniform, with very crinkly apices; some or all of lower katepisternal hairs stiff and black in some specimens. Legs almost as variable in color as in *P. immarginatus* but not as dark as in darkest specimens of that species. Fore leg in Fig. 79*a*; trochanter with moderately long pale setae on ventral surface; femur with posterior subbasal tuft of three long crinkly closely appressed white hairs, otherwise with fine mostly short yellow hairs; tibia with posterior margin slightly straighter, with posteroapical angle slightly longer. Mid leg in Fig. 79*g*; femur anteroventrally on basal two-thirds with irregular row of short stiff rather weak yellow (or partly black in some specimens) setae, on apical one-third with irregular row of fine crinkly black hairs up to 2.5 times as long as femoral diameter, ventrally on basal half with row of long stiff yellow hairs slightly weaker than those of *P. immarginatus*, posteroventrally on apical one-quarter with few long slender black hairs; tibia on basal two-thirds of anteroventral surface with dense fine crinkly black or yellow hairs; hairs near base four times as long as tibial diameter; tibia ventrally on basal half to two-thirds with similar but shorter hairs.

Abdomen (Fig. 164) similar to abdomen of paler specimens of *P. immarginatus*; median and apical black marks of tergites 3 and 4 nearly obsolete in some specimens, rarely with black median line considerably broadened and tergite 2 narrowly black laterally.

Female. Not distinguishable from females of several similar species.

Distribution. Alaska, Canada (Map 86), south to California, Colorado, Michigan, and Massachusetts;* Europe; Siberia; Japan.* B.C., VII, VIII; Ont., Que., VII.

Specimens identified. Alaska, 11 ♂♂; Canada, 93 ♂♂; United States, 24 ♂♂; Europe, 6 ♂♂; Siberia, 1 ♂.

Biology. Metcalf (1917) reared larvae in the laboratory on two species of aphids.

Platycheirus pilatus Vockeroth

Figs. 71, 161; Map 84

Platycheirus pilatus Vockeroth, 1990:724.

Length. 5.3–7.9 mm.

Male. Very similar to male of *P. nodosus*, differing as follows: face with tubercle slightly shining in some specimens.

Thoracic hairs longer; many hairs of anepisternum and anepimeron brown to black; lower katepisternal hairs almost all stiff and black. Last four tarsomeres of fore and mid legs brown. Fore leg in Fig. 71*a*; femur with hairs of basal tuft pale throughout and with apices only very slightly broadened; other hairs more abundant and longer, some slightly longer than tibial diameter; tibia less strongly broadened and with posteroapical angle scarcely produced; first tarsomere almost twice as long as wide, with sides almost parallel. Mid leg in Fig. 71*e*; femur with much longer hairs; many hairs on posteroventral surface 1.5 times as long as femoral diameter; tibia with anteroventral hairs black; hairs on basal one-third erect or suberect.

Abdomen (Fig. 161) with yellow-orange spots slightly or much smaller, variable in size. Spots on tergite 2 usually circular and at most half as long as tergite; spots on tergite 3 at most five-sevenths as long as tergite; spots on tergite 4 at most five-eighths as long as tergite; spots on tergite 5 obscure or absent.

Female. Not distinguishable from females of several similar species.

Distribution. Alaska, northwestern Canada (Map 84), south to Colorado (3230 m). Man., VI, VII.

Specimens identified. Alaska, 3 ♂♂; Canada, 23 ♂♂; United States, 11 ♂♂.

Platycheirus podagratus (Zetterstedt)

Fig. 80; Map 87

Scaeva podagrata Zetterstedt, 1838:606.

Length. 5.7–7.7 mm.

Male. Similar to dark specimens of *P. immarginatus*, differing as follows: face with tubercle slightly less prominent, with sparse gray pruinosity so face almost black.

Thoracic hairs usually yellow to yellow brown; hairs on lower half of posterior margin of anepisternum dark brown to blackish in some specimens; some hairs on lower part of katepisternum black and stiff; posterior part of anepisternum with hairs uniform, rather sparse, wavy; upper part of katepisternum bare or with some hairs. Leg color as in *P. tenebrosus* or with mid leg slightly more extensively yellow. Fore leg in Fig. 80; trochanter with yellow or black setulae below; femur posteriorly with subbasal tuft of two or three long closely appressed wavy white hairs closely followed by several similar black hairs forming loose tuft in some specimens, otherwise with shorter rather sparse black or yellow hairs; tibia gradually broadened on about basal three-fifths, then strongly broadened especially posteriorly and slightly narrowed apically, with posteroapical angle short and broad; first tarsomere rather strongly broadened posteriorly on basal two-fifths to half, then very slightly narrowed to apex, as wide as long; second and third tarsomeres subrectangular, each about 1.5 times as wide as long. Mid femur usually with few to many short weak black setae on basal half of anterior and anteroventral surfaces, otherwise with fine weak mostly black hairs; mid tibia nearly cylindrical, on basal half anteroventrally with fine black suberect wavy hairs about 3.5 times as long as tibial diameter. First tarsomere of hind leg less strongly swollen, about four times as long as its greatest depth.

Abdomen very similar to that of *P. tenebrosus* (Fig. 163), similarly variable but with orange spots of tergites 3 and 4 reaching or nearly reaching anterior margins.

Female. Not distinguishable from females of several similar species.

Distribution. Alaska, western and northern Canada (Map 87), south to Colorado (3230 m) and Maine; Europe; Siberia. B.C., VII; Ont., VI.

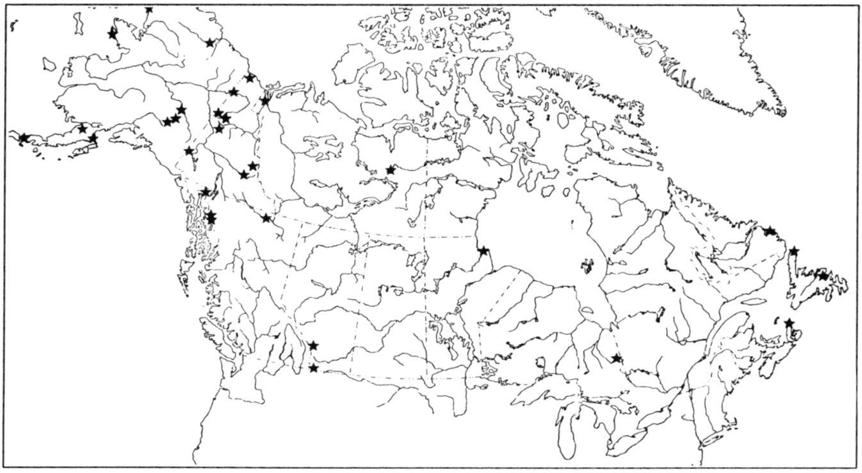

Map 87. Collection localities for *Platycheirus podagratus* (Zetterstedt).

Specimens identified. Alaska, 50 ♂♂; Canada, 47 ♂♂; United States, 10 ♂♂; Europe, 13 ♂♂; Siberia, 2 ♂♂.

Biology. Specimens were taken in Nova Scotia from a damp, open, sphagnum fen at 300 m.

Platycheirus protrusus Vockeroth

Platycheirus protrusus Vockeroth, 1990:726.

Length. 6.1–6.6 mm.

Male. Similar to male of *P. carinatus*, differing as follows: face with distinctly larger and more pointed tubercle near lower margin and without keel above tubercle.

Thoracic hairs mostly black, pale only on katepisternum and on lower part of anepimeron. Knob of halter yellow. Fore femur with only very weak short setae below; fore tibia with posterior bristles slightly weaker; first tarsomere of fore leg with only very short hairs. Mid femur with only very weak short setae below; mid tibia with posterior bristles slightly weaker and less appressed.

Abdomen with spots of tergites 2–4 very faint. Paramere similar to that of *P. setitarsis* (Fig. 239) but slightly more slender and with spine less widely divergent from base.

Female. Similar to male, with large facial tubercle and without keel below antennae. Thoracic hairs shorter, extensively white in some specimens. Tibiae without bristles.

Distribution. Colorado, 4270 m. Colo., VII.

Specimens identified. United States, 1 ♂, 2 ♀♀.

Platycheirus pullatus Vockeroth

Fig. 242; Map 88

Platycheirus pullatus Vockeroth, 1990:726.

Length. 6.2–7.9 mm.

Male. Similar to male of *P. carinatus*, differing as follows: face slightly narrower, upper part with very weak median keel or groove.

Thoracic hairs almost entirely black in northern specimens, mostly white in Alberta specimen. Knob of halter dark brown to blackish. Fore leg with only fine hairs, without outstanding hairs or bristles; tibia and tarsus with hairs much shorter than tibial or tarsal diameter. Mid femur anteroventrally with irregular row of short weak black setae on most of length; mid tibia anteroventrally in some specimens with row of appressed or suberect weak black or pale bristles; bristles near apex about twice as long as tibial diameter; mid tibia posteriorly with short appressed hairs; first tarsomere of mid leg anteroventrally with three or more irregular white or black setae subequal in length to tibial diameter. First tarsomere of hind leg slightly swollen, about 4.5–5.0 times as long as greatest depth.

Abdominal markings similar to *P. carinatus* but spots without orange background, usually with distinctly bluish gray tint, and extending to anterior margins of tergites 2–4. Surstylus (Fig. 242*b*) with broad blunt subbasal process. Paramere (Fig. 242*c*) with heavy base and heavy curved spine. Aedeagus (Fig. 242*d*) with thick dorsobasal process and long strong preapical ventral spine.

Female. Not distinguishable from females of several similar species.

Distribution. Victoria Island, Yukon Territory, Alberta (2190 m) (Map 88). Y.T., VI, VII; Alta., VII.

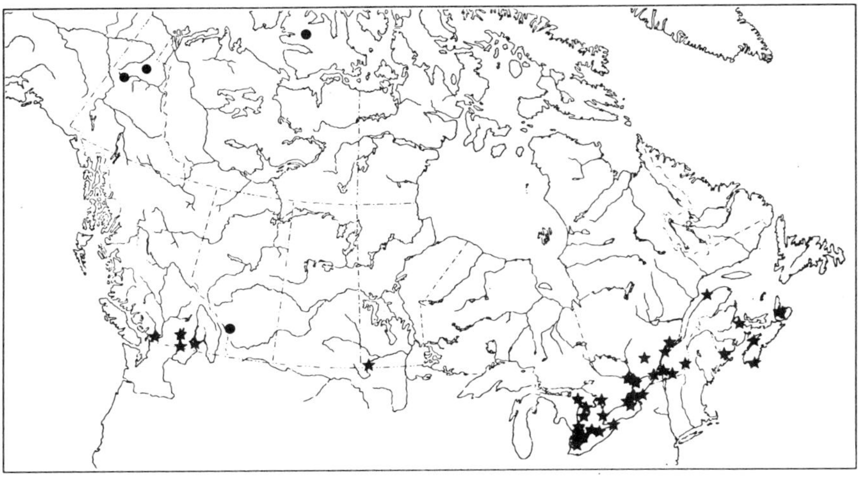

Map 88. Collection localities for *Platycheirus pullatus* Vockeroth (●) and for *P. quadratus* (Say) (★).

Specimens identified. Canada, 16 ♂♂.

Platycheirus quadratus (Say)

Fig. 81; Map 88

Scaeva quadrata Say, 1823:90.
Syrphus fuscanipennis Macquart, 1855:115.

Length. 7.1–9.1 mm.

Male. Similar to male of *P. immarginatus*, differing as follows: first flagellomere very narrowly orange below at base.

Thoracic hairs almost entirely pale, anterior hairs on lower part of katepisternum black, stiff; hairs of posterior part of anepisternum long, dense, uniform, with crinkly apices; hairs of anepimeron and of upper part of katepisternum similar but shorter. Legs colored as in pale specimens of *P. immarginatus*. Fore leg in Fig. 81*a*; femur with posterior subbasal tuft of long wavy closely appressed white hairs, otherwise with short yellow hairs; tibia with posterior margin straight, anterior margin irregular; first tarsomere subtriangular, only slightly longer than wide;

second to fourth tarsomeres only slightly narrower than first, about twice as wide as long. Mid leg in Figs. 81*f*,*i*; femur posteroventrally on basal half with dense tuft of many strong setose hairs; hairs near base black; following hairs orange and slightly longer than femoral diameter; third quarter of femur with shorter and more erect orange setose hairs; fourth quarter with long fine wavy black hairs; femur anteroventrally on basal three-quarters with short fine black hairs, on apical one-quarter with long fine wavy black hairs; tibia rather strongly broadened and depressed, with anterior margin curved and posterior margin nearly straight, with dense erect crinkly black and yellow hairs on basal two-thirds of ventral surface; hairs anteroventrally near base at least three times as long as tibial diameter; first tarsomere broadened and rather strongly depressed, at one-quarter its length as wide as apex of tibia.

Abdomen as in paler specimens of *P. immarginatus*.

Female. Not distinguishable from females of several similar species.

Distribution. Southern Canada (Map 88), south to California, New Mexico, and Florida. B.C., IV, V, VII; Ont., Que., V–VIII.

Specimens identified. Canada, 146 ♂♂; United States, 729 ♂♂.

Biology. Heiss (1938) reared larvae in the laboratory on *Myzus persicae*.

Platycheirus rosarum (Fabricius)

Figs. 172, 241; Map 89

Syrphus rosarum Fabricius, 1787:341.
Pyrophaena rosarum var. *duplicata* Fluke, 1922:228.

Length. 7.4–9.1 mm.

Male. Similar to male of *P. granditarsis*, differing as follows: face slightly receding, with smaller tubercle. Antenna usually with first flagellomere orange to orange-brown below.

Wing in some specimens with faint postmedian cloud on anterior half. Fore and mid femora yellow on about apical half; fore and mid tibiae and tarsi yellow; hind tibia with basal one-third and in some specimens narrow apex yellow; hind tarsus yellow with dorsal surface of first tarsomere and of last one or two tarsomeres in some specimens dark brown. Legs slender, without distinctive hairs or bristles. First tarsomere of hind leg about six times as long as greatest depth.

Abdomen (Fig. 172) mostly black. Tergites 3 and 4 each with pair of yellow posteriorly rounded lateral spots on anterior margin. Sternites black; sternites 3 and 4 each with broad anterior yellow band. Surstylus (Fig. 241*b*) without distinct dorsobasal lobe. Paramere (Fig. 241*c*) with slender base and well-separated long slender spine. Aedeagus (Fig. 241*d*) with strong subbasal dorsal lobe.

Female. Frons shining except for two very small silvery pruinose lateral spots at two-thirds length. Wing cloud often more distinct and slightly larger. Tarsi distinctly wider than in male. Markings of tergites less clearly defined than in male and yellow-orange rather than yellow; tergite 2 black or with pair of rounded yellow spots confluent medially in some specimens; tergite 3 commonly with spots confluent and in some specimens covering most of tergite.

Distribution. Alaska, Canada (Map 89), south to northern California and West Virginia; Europe; Siberia.* B.C., VII; Ont., Que., V–VIII.

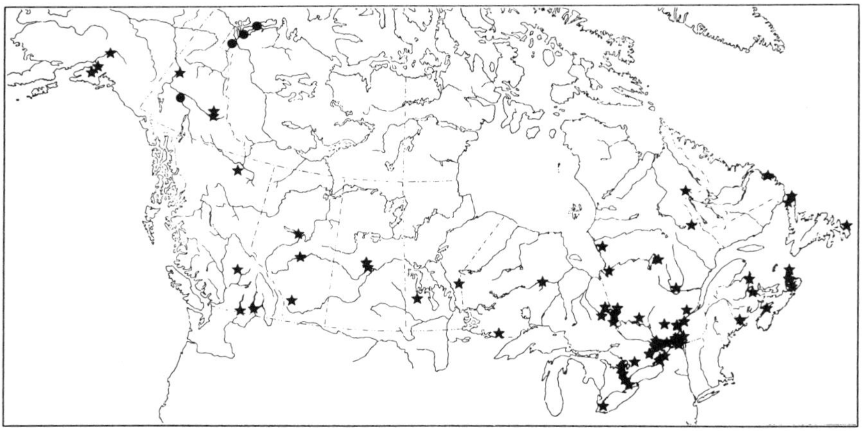

Map 89. Collection localities for *Platycheirus rosarum* (Fabricius) (★) and for *P. rufigaster* Vockeroth (●).

Specimens identified. Alaska, 4 ♂♂; Canada, 210 ♂♂, 295 ♀♀; United States, 21 ♂♂, 37 ♀♀; Europe, 5 ♂♂, 2 ♀♀.

Biology. The species is abundant in sphagnum bogs.

Platycheirus rufigaster Vockeroth

Figs. 171, 240; Map 89

Platycheirus rufigaster Vockeroth, 1990:729.

Length. 7.4–9.1 mm.

Male. Similar to male of *P. granditarsis*, differing as follows: face slightly more pruinose. Antenna with first flagellomere red-orange below.

Thorax, especially pleura, more strongly shining; hairs of disc of scutum and of scutellum mostly black. Legs black, slender, without distinctive hairs or bristles. First tarsomere of hind leg about five times as long as greatest depth.

Abdomen in Fig. 171. Tergite 2 black; tergites 3 and 4 yellow-orange with median line and narrow to broad posterior margin black; black areas poorly defined in some specimens; tergite 5 yellow-orange. Sternites 3–5 yellow-orange; other sternites black. Surstylus (Fig. 240*b*) suboval, without basal lobe. Paramere (Fig. 240*c*) with two long strong spines and one weak slender spine. Aedeagus (Fig. 240*d*) with heavy blunt dorsal subbasal lobe and with strong ventral spicules on apical half.

Female. Frons shining except for two small silvery pruinose lateral spots at two-thirds length. Base of fore and mid tibiae obscurely yellow; fore tarsus distinctly wider than in male. Tergite 1 shining black; lateral margin of tergite 2 obscurely blackish; tergites otherwise yellow-orange. Sternite 1 black; other sternites yellow-orange.

Distribution. Northwestern Canada (Map 30). N.W.T., VII.

Specimens identified. Canada, 9 ♂♂, 3 ♀♀.

Platycheirus rufimaculatus Vockeroth

Map 90

Platycheirus rufimaculatus Vockeroth, 1990:730.

Length. 6.8–8.7 mm.

Male. Very similar to male of *P. concinnus*, differing as follows: thoracic hairs very short; longest scutellar hairs about one-quarter as long as arista. Wing with cell c bare on basal one-eighth to half and cell bm bare or nearly so on basal one-quarter to two-thirds. Fore and mid femora usually brown to black on about basal one-third, mostly orange in some specimens; hind femur brown to black on about basal half to two-thirds; tibiae orange; tarsi orange, partly or entirely brown above in some specimens. Fore and mid femora with posterior hairs much shorter than femoral diameter.

Markings of tergites 3 and 4, and usually of tergite 2, with background partly dull orange in areas very variable in size; tergite 5 in some specimens with pair of obscure orange basal submedian spots. Terminalia as in *P. concinnus* (Fig. 245).

Female. Very similar to female of *P. concinnus*, differing as follows: wing with microtrichia as in male of *P. rufimaculatus*. Femora usually entirely orange or orange-brown, darkened in some specimens basally as in *P. concinnus*. Tergites with orange markings as in male of *P. rufimaculatus* but usually more extensive; orange spots of tergite 5 distinct in some specimens.

Distribution. Southwestern Canada (Map 90), south to California and Utah (2740 m). B.C., VI, VII.

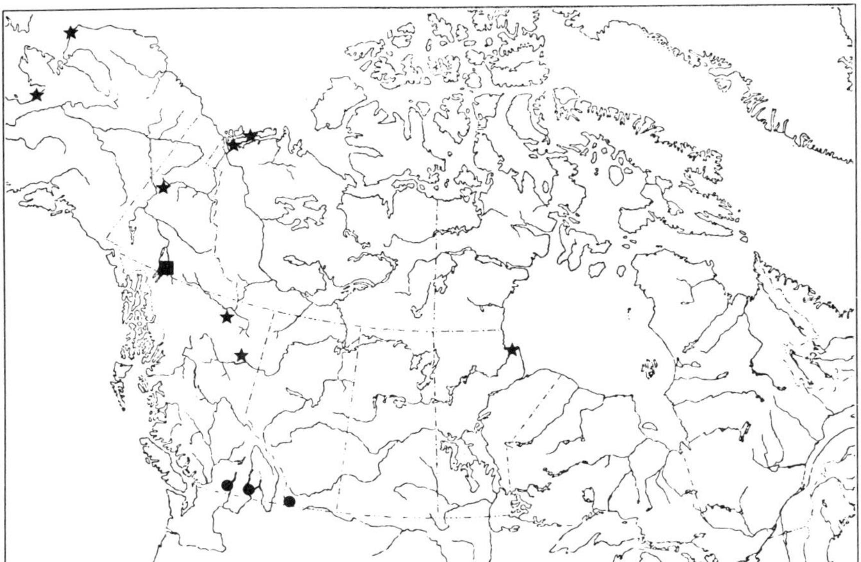

Map 90. Collection localities for *Platycheirus rufimaculatus* Vockeroth (●), for *P. setitarsis* Vockeroth (★), and for *P. sabulicola* Vockeroth (■).

Specimens identified. Canada, 4 ♂♂, 2 ♀♀; United States, 14 ♂♂, 15 ♀♀.

Discussion. As the orange abdominal markings vary in size, they may also be absent; their absence would probably cause a female to be indistinguishable from the female of *P. concinnus*. Males of the two species should always be distinguishable because of the difference in length of thoracic hairs.

Platycheirus russatus Vockeroth

Platycheirus russatus Vockeroth, 1990:731.

Length. 7.8–8.8 mm.

Male. Very similar to male of *P. rufimaculatus* Vockeroth, differing as follows: face with pruinosity slightly more dense and more silvery. Pleura slightly more pruinose; katepisternum between upper and lower patches of hair weakly and almost uniformly pruinose. Fore and mid femora with anterior surface mostly orange-brown, with most of posterior surface darkened. Tergite 1 black; tergite 2 red-orange with anterior margin and posteriorly narrowed lateral margins black; tergites 3–5 entirely red-orange. Terminalia as in *P. concinnus* (Fig. 245).

Female. Unknown.

Distribution. Southern California. Calif., II, VI, X.

Specimens identified. United States, 4 ♂♂.

Platycheirus sabulicola Vockeroth

Fig. 271; Map 90

Platycheirus sabulicola Vockeroth, 1980:731.

Length. 5.2–6.8 mm.

Male. Very similar to male of *P. confusus*, differing as follows: median facial stripe narrower, not extending to lower margin, narrowed above and extending scarcely above tubercle. Angle of junction of eyes 90° or slightly less.

Notopleural hairs usually mostly or entirely white; prescutellar and basal scutellar hairs usually white; pleural hairs usually entirely white, in one specimen almost all black. Wing membrane clear, extensively bare; cell c trichose on at most apical one-fifth; cell bm with only slender patch of microtrichia near apex; cell cua_1 bare on entire width at base. Fore and mid tibia with posterior bristles slightly weaker, present only on apical half or slightly more of tibia.

Spots of tergites without orange background. Surstylus (Fig. 271) with shorter lobe very small, and with longer lobe curved and nearly as broad as base of surstylus.

Female. Very similar to female of *P. confusus* and perhaps not definitely separable from it. Wing microtrichia as in male of *P. sabulicola* or slightly less extensive.

Distribution. Yukon Territory (Map 90). Y.T., VI.

Specimens identified. Canada, 18 ♂♂, 10 ♀♀.

Biology. The type series was taken in a sand dune area without visible water, a habitat very different from those of the similar species *P. confusus* and *P. obscurus*.

Discussion. The much barer wing and markedly different surstylus, as well as the habitat difference, leave little doubt that *P. sabulicola* is specifically distinct from *P. confusus*. The females examined, taken with males, are undoubtedly of *P. sabulicola*; the reduction in extent of wing microtrichia distinguishes them from most females of *P. confusus* (in which cell cua_1 is usually entire trichose). However, a few specimens from southwestern Alberta have this cell with a basal bare area, so their identity is uncertain.

Platycheirus scamboides Curran

Fig. 88; Map 85

Platycheirus scamboides Curran, 1927*a*:6.

Length. 8.2–9.1 mm.

Male. Similar to male of *P. immarginatus*, differing as follows: facial tubercle slightly less prominent, entirely pruinose or very slightly shining. First flagellomere usually distinctly orange below on basal two-thirds.

Posterior part of anepisternum with hairs rather sparse, uniform, nearly straight; upper part of katepisternum with many hairs. Legs yellow with coxae, broad ring on each of hind femur and tibia, and first and last two tarsomeres of hind leg dark brown to black. Fore leg in Fig. 88*a*; ventral surface of fore trochanter with slightly shorter and weaker setae; femur posteriorly without subbasal tuft of white hairs, with rather short yellow hairs and three or four slightly longer and stronger uniformly spaced black hairs, ventrally with short yellow hairs and on basal half row of three or four slightly longer and stronger hairs; tibia with slight constriction on posterior margin at two-thirds its length and with posterior margin of posteroapical angle straight rather than curved slightly inward; first tarsomere slightly wider and nearly parallel sided on apical half rather than on apical one-third. Mid femur (Fig. 88*i*) anteroventrally with row of short stiff almost entirely pale setae on about apical half, posteroventrally with row of three to seven long weak yellow or black bristles on basal half to three-fifths; mid tibia with very short appressed yellow anteroventral and posteroventral hairs. First tarsomere of hind leg only slightly swollen, about 4.5 times as long as its greatest depth.

Abdomen with yellow spots much smaller than shown in Fig. 80. Spots on tergite 2 not reaching anterior or lateral margin; spots on tergites 3 and 4 at most three-quarters as long as tergite and separated by black median line about one-fifth as wide as tergite; spots of tergite 5 separated by median black line.

Female. Not distinguishable from females of several similar species.

Distribution. Southern Ontario (Map 85); Wisconsin to Maine, south to North Carolina. Ont., V, VI, VIII.

Specimens identified. Canada, 4 ♂♂; United States, 63 ♂♂.

Discussion. This form may not differ specifically from *P. scambus*, but the characters given in the key suggest that two species are present. I have seen no intermediate or doubtful specimens.

Platycheirus scambus (Staeger)

Fig. 89; Map 91

Syrphus scambus Staeger, 1843:325.
Platycheirus chaetopodus Williston, 1887:59.

Length. 6.5–9.6 mm.

Male. Similar to *P. immarginatus* and extremely similar to *P. scamboides*, differing from latter as follows: first flagellomere black or obscurely orange below at extreme base.

Lower part of katepisternum with some hairs stiff and black in some specimens. Fore leg with black hairs of posterior surface slightly longer and stronger, almost bristlelike, and with black hairs of ventral surface slightly but distinctly stronger. Mid femur (Fig. 88) anteroventrally with irregular row of 3–15 short strong black (or rarely all yellow) setae followed by one or two long very slender recurved black hairs; mid femur posteroventrally with two to four long strong black bristles; mid tibia with short appressed anteroventral and posteroventral hairs mostly or entirely black. Yellow spots of tergites a little larger; spots of tergite 2 reaching or nearly reaching anterior and lateral margins; spots of tergites 3 and 4 a little longer and a little less broadly separated medially.

Female. Not distinguishable from females of several similar species.

Distribution. Alaska, Canada (Map 91), south to California, New Mexico, and South Carolina;* Europe; Siberia. B.C., V–VII, IX; Ont., Que., V–VIII.

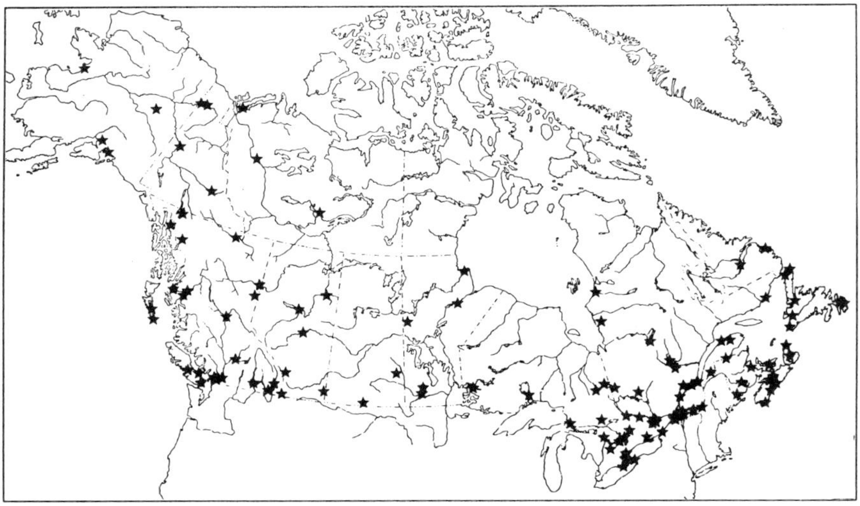

Map 91. Collection localities for *Platycheirus scambus* (Staeger).

Specimens identified. Alaska, 4 ♂♂; Canada, 482 ♂♂; United States, 278 ♂♂; Europe, 15 ♂♂; Siberia, 1 ♂.

Biology. Specimens were taken in Nova Scotia and Quebec from *Carex* marsh.

Platycheirus scutatus (Meigen)

Figs. 68, 236; Map 92

Syrphus scutatus Meigen, 1822:333.

Length. 6.8–8.7 mm.

Male. Similar to male of *P. albimanus*, differing as follows: first flagellomere extensively orange below at base, orange on lower half in some specimens.

Scutum and scutellum with at most very few black hairs. Lower katepisternal hairs soft, white, very short. Wing with small bare areas at base of cell c and along anterior margin of cell bm. Fore leg in Fig. 68*a*; trochanter near apex of ventral surface with dense cluster of about 12 rather long stiff pale setae; femur with basal pale tuft of about five hairs, with two succeeding black tufts more compact, of at least 10 hairs each, and sharply bent preapically, and with many long strong uniform black hairs on rest of posterior surface decreasing in length towards apex of femur; tibia gradually widened on basal five-sevenths, then strongly widened posteriorly with posteroapical angle broadly rounded; posterior margin of tibia with dense soft mostly black hairs about twice as long as tibial width at three-quarters tibial length; first tarsomere with anterior margin straight, with posterior margin oblique at apex, slightly wider than fore tibia; second tarsomere as wide as first and one-sixth its length; third and fourth tarsomeres each slightly longer than second and successively slightly narrower. Mid leg in Fig. 68*h*; coxa with slender anteroapical spur, projecting ventrally, about three-quarters as long as base of coxa and yellow to black; femur with black anteroventral setae finer and less regular, followed by three or four very slender curved black hairs, with row of long rather strong black posteroventral hairs; tibia slightly arcuate, slightly broadened on middle half, with very short dense erect fine pale hairs on basal two-thirds of ventral surface and rather long fine dense suberect black hairs on most of posterior to posteroventral surface. First tarsomere of hind leg moderately swollen, about 3.5 times as long as greatest depth.

Spots of tergites submetallic to yellowish; spots of tergite 2 small and irregular in outline or obsolete in some specimens; spots of tergites 3 and 4 subquadrate but slightly longer laterally, about half as long as tergite; tergite 3 with dense silvery pruinosity anteromedially; tergite 4 with dense silvery pruinosity on most of surface. Terminalia as in Fig. 236.

Female. Not distinguishable from females of several similar species.

Distribution. Alaska, Canada (Map 92), south to Colorado, Pennsylvania, and Massachusetts; Europe; Siberia; Japan.* B.C., IV–VIII; Ont., VI.

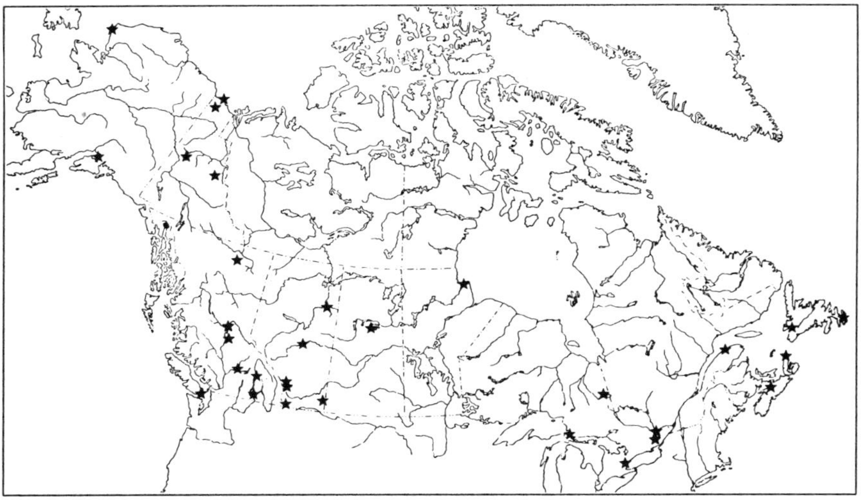

Map 92. Collection localities for *Platycheirus scutatus* (Meigen).

Specimens identified. Alaska, 3 ♂♂; Canada, 50 ♂♂; United States, 57 ♂♂; Europe, 30 ♂♂; Siberia, 3 ♂♂.

Biology. Láska and Starý (1980) reported three species of aphids as larval hosts in Czechoslovakia.

Platycheirus setipes Vockeroth

Figs. 82, 162; Map 93

Platycheirus setipes Vockeroth, 1990:735.

Length. 8.7 mm.

Male. Very similar to male of *P. hispidipes*, differing as follows: facial tubercle slightly more compressed; facial pruinosity less dense so face appears darker.

Hairs of scutum and scutellum mostly black; hairs of pleura all pale. Wing with tiny bare areas; cell bm bare on centre of basal one-fifth; knob of halter dark yellow. Legs mostly dark brown to black; apices of femora, fore tibia except anterior and posterior streaks, base and apex of mid tibia, base of hind tibia, and fore and mid tarsi yellow or whitish. Fore leg in Fig. 82*a*; femur posteriorly with subbasal tuft of four white hairs followed by scattered shorter similar hairs and, on middle one-third, three or four long black bristlelike hairs; tibia strongly and abruptly broadened posteriorly at three-quarters its length, parallel-sided on apical one-eighth; first tarsomere strongly widened on basal one-third, then parallel-sided to apex; second tarsomere little wider than long; third tarsomere subquadrate. Mid femur (Fig. 82*i*) with anteroventral row of about seven short stout black setae on apical two-thirds, followed in one specimen by long fine curved black hair; mid tibia at mid length with one or two strong black posterior hairs about three times as long as tibial diameter. Hind tibia with short hairs.

Abdomen in Fig. 162; tergites 2–4 each with pair of large shining blue-gray spots extending laterally along entire margin in some specimens.

Female. Unknown.

Distribution. Southern British Columbia (Map 93). B.C., VI, VIII.

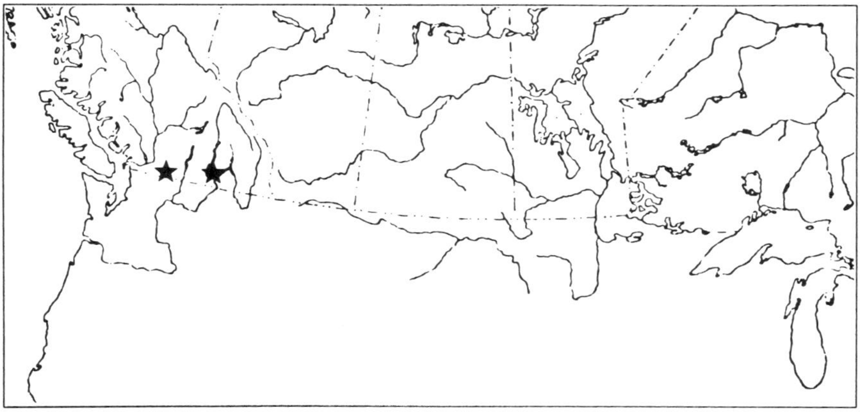

Map 93. Collection localities for *Platycheirus setipes* Vockeroth.

Specimens identified. Canada, 2 ♂♂.

Platycheirus setitarsis Vockeroth

Figs. 98, 239; Map 90

Platycheirus setitarsis Vockeroth, 1990:736.

Length. 7.1–8.2 mm.

Male. Very similar to male of *P. carinatus*, differing as follows: thoracic hairs almost entirely black, pale only on katepimeron and on lower part of katepisternum. Knob of halter brown to black. Fore femur with four to eight black ventral setae on basal three-eighths, all of nearly equal length, all shorter than femoral diameter; fore tibia with posterior bristles not more closely spaced near apex; first tarsomere with only very short hairs. Mid femur with ventral setae not longer than femoral diameter; mid tibia with posterior bristles less strongly appressed; first tarsomere of mid leg with three more or less evenly spaced stiff black anteroventral setae, with last setae at apex, and with all setae at least as long as diameter of tarsomere (Fig. 98).

Spots of tergites subrectangular; spots of tergites 3 and 4 almost half as long as tergites, without orange-brown background. Paramere (Fig. 239) sickle-shaped.

Female. Not distinguishable from females of several similar species.

Distribution. Alaska, northwestern Canada (Map 90). B.C., VI, VII.

Specimens identified. Alaska, 8 ♂♂; Canada, 19 ♂♂.

Platycheirus spinipes Vockeroth

Platycheirus spinipes Vockeroth, 1990:737.

Length. 5.7–7.1 mm.

Male. Very similar to male of *P. stegnus*, differing as follows: face slightly narrower and with slightly finer punctures, shining median black stripe more extensive, usually reaching lower margin of face and extending above upper limit of tubercle.

Scutellum with black hairs at least posteriorly.

Female. Very similar to female of *P. stegnus*, differing in head characters as do males of both species.

Distribution. California to Wyoming, south to southern Mexico (in Durango at 2437 and 2742 m). Colo., VIII, IX.

Specimens identified. United States, 53 ♂♂, 59 ♀♀; Mexico 6 ♂♂.

Platycheirus squamulae (Curran)

Fig. 95; Map 94

Melanostoma squamulae Curran, 1922*b*:275.

Length. 7.3–7.9 mm.

Male. Very similar to male of *P. confusus*, differing as follows: thoracic hairs almost entirely black, only those of katepisternum pale. Wing with membrane distinctly brownish, entirely trichose. Fore tibia with posterior bristles longer and stronger, especially on basal part; longest bristles almost half as long as tibia and beginning very near base of tibia (Fig. 95). Mid tibia with bristles as long as those of fore tibia but slightly weaker. First tarsomere of hind leg apparently slightly more swollen, about four times as long as greatest depth.

Lateral hairs of tergites almost all black, long, and strong; longest hairs on tergite 3 at least half as long as apical width of tergite. Surstylus as in *P. confusus* (Fig. 269).

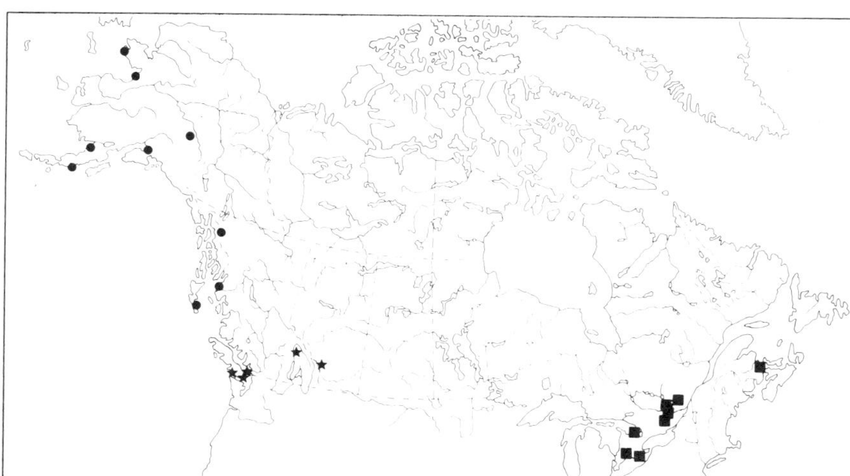

Map 94. Collection localities for *Platycheirus squamulae* (Curran) (★), for *P. tenebrosus* Coquillett (●), and for *P. thompsoni* Vockeroth (■).

Female. Very similar to female of *P. confusus*, differing as follows: scutum sublaterally and scutellum on disc with some black hairs. Wing membrane more distinctly brownish; cell c with very small bare area at base; cell bm entirely trichose.

Distribution. British Columbia (Map 94), Oregon. B.C., IV–VI.

Specimens identified. Canada, 5 ♂♂, 4 ♀♀; United States, 2 ♂♂.

Discussion. This form may be conspecific with *P. confusus* (see discussion under that species).

Platycheirus stegnoides Vockeroth

Fig. 96; Map 95

Platycheirus stegnoides Vockeroth, 1990:738.

Length. 7.7–8.2 mm.

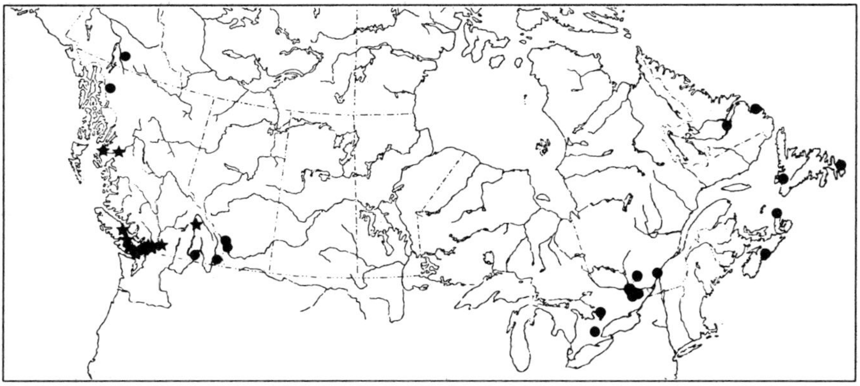

Map 95. Collection localities for *Platycheirus stegnoides* Vockeroth (★) and for *P. thylax* Hull (●).

Male. Very similar to male of *P. hesperius*, differing as follows: face with more distinct punctures, with only tubercle shining (as in Fig. 6).

Upper half of pleura with many black hairs. Wing membrane entirely trichose. Fore tibia with about 11 slightly longer and weaker posterior bristles; bristles near tibial base very short (Fig. 96). Mid tibia with about five long posterior bristles only slightly weaker than those of fore tibia. First tarsomere of hind leg less strongly swollen, about 4.5 times as long as greatest depth.

Spots of tergites submetallic with rather dense shining silvery gray pruinosity.

Female. Face and wing as male. Frons not distinctly shining above antennae, otherwise as in female of *P. hesperius*.

Distribution. British Columbia (Map 95), south to northern California. B.C., VI–VIII.

Specimens identified. Canada, 30 ♂♂, 12 ♀♀; United States, 16 ♂♂, 6 ♀♀.

Platycheirus stegnus (Say)

Fig. 6; Map 96

Syrphus stegnus Say, 1829:163.
Melanostoma tigrinum Osten Sacken, 1877:323.

Length. 6.8–9.1 mm.

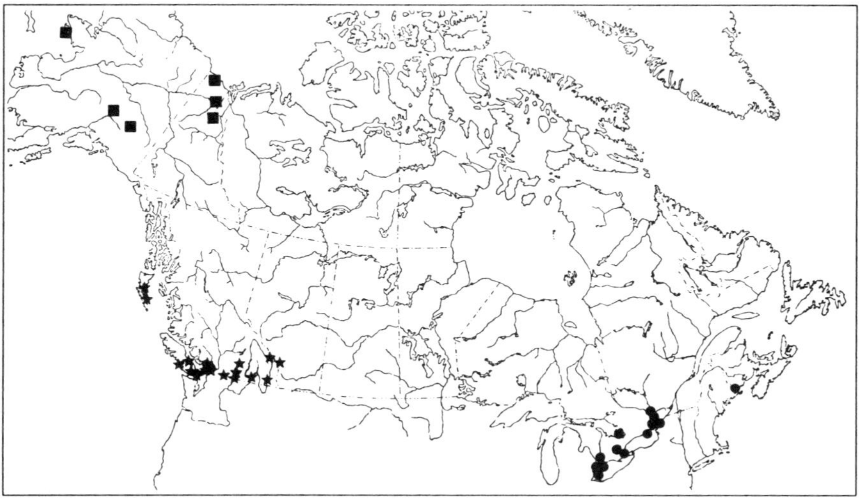

Map 96. Collection localities for *Platycheirus stegnus* (Say) (★), for *P. subordinatus* Becker (■), and for *Xanthogramma flavipes* (Loew) (●).

Male. Very similar to male of *P. stegnoides*, differing as follows: face with slightly larger punctures (Fig. 6). Antenna with first flagellomere more extensively orange below.

Scutal hairs almost entirely pale; scutellar hairs entirely pale or black only on margin; pleural hairs entirely pale. Wing membrane extensively bare on basal half; cell c bare; cell bm bare or with very few microtrichia near apex.

Female. Very similar to female of *P. stegnoides* but with slightly larger facial punctures, with entirely white thoracic hairs, and with wing extensively bare as in male of *P. stegnus*.

Distribution. Southwestern Canada (Map 96), east to Nebraska, south to California, New Mexico, and Mexico. B.C., IV–X.

Specimens identified. Canada, 24 ♂♂, 52 ♀♀; United States, 497 ♂♂, 420 ♀♀; Mexico, 1 ♂, 2 ♀♀.

Biology. Davidson (1922) reared larvae in the laboratory on four species of aphids that developed without diapause but adults were smaller than usual. He suggested either that the aphids were not the normal hosts or that the larvae are normally partly phytophagous.

Platycheirus striatus Vockeroth

Map 97

Platycheirus striatus Vockeroth, 1990:741.

Length. 7.3–11.0 mm.

Male. Very similar to male of *P. luteipennis*, differing as follows: face more densely pruinose, with only tubercle shining, without shining median stripe extending to lower facial margin or well above tubercle, with several striae, grooves, or ridges converging above on upper median part of face but usually without distinct median keel. Antenna with pedicel extensively orange-brown below.

Scutum and scutellum usually slightly less brassy; hairs white to pale yellow and longer, with longest scutellar hairs subequal in length to arista. Pleural hairs slightly longer, white to pale brown. Wing with cell c entirely trichose or bare only on about basal one-tenth, with cell bm entirely trichose or with indistinct bare median line near base, and with cell cu*p* entirely trichose. Legs with femora more extensively blackened and with tarsi darker above. Longer femoral hairs at least as long as femoral diameter.

Abdomen with markings of tergites metallic bluish, brassy, or coppery. Terminalia very similar to those of *P. luteipennis* (Fig. 246) but surstylus with basal lobe broader and bluntly rounded basally; paramere with spine more widely separated from base.

Female. Very similar to female of *P. luteipennis*, differing as follows: face with tubercle shining but without shining median stripe. Thoracic hairs usually white to yellowish rather than brassy. Wing with cell c bare only at extreme base; cell bm trichose with narrow median bare stripe on about basal two-thirds. Legs slightly darker, as in male of *P. striatus*. Abdomen distinctly more robust.

Distribution. Canada (Map 97), south to California, Colorado, and New Hampshire. B.C., VI, VII; Ont., Que., VI, VII.

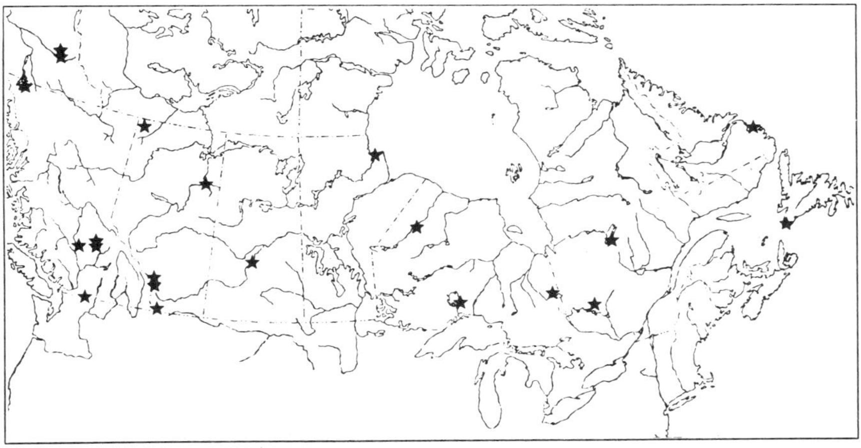

Map 97. Collection localities for *Platycheirus striatus* Vockeroth.

Specimens identified. Canada, 38 ♂♂; United States, 27 ♂♂.

Discussion. This description of the female is based on females collected with males; separation from females of *P. concinnus* and *P. luteipennis* may not always be possible.

Platycheirus subordinatus Becker

Fig. 54; Map 96

Platycheirus subordinatus Becker, 1915:60.

Length. 6.2–7.3 mm.

Male. Very similar to male of *P. discimanus*, differing as follows: face slightly more strongly produced forward below, at least weakly pruinose except on tubercle.

Thoracic hairs white to very pale brown. Knob of halter pale brown to dark brown. Mid tarsus black. Fore leg in Fig. 54*a*; trochanter with fine pale hairs ventrally, without black setulae; tibia with posterior hairs sparse, strong, and nearly straight; first tarsomere rather variable, from 1.5–2.5 times as long as wide; second tarsomere subquadrate to 1.5 times as wide as long. Mid tibia (Fig. 54*k*) with anteroventral hairs of basal half at most three times as long as tibial diameter, with dense fine hairs on basal half to two-thirds of posterior surface; longest hairs three to four times as long as tibial diameter; mid tarsus cylindrical. First tarsomere of hind leg slightly to moderately swollen, 3.5–4.0 times as long as its greatest depth.

Tergite 2 with pale spots obscure or absent; tergites 3 and 4 with spots distinct, varying from densely gray pruinose with only very faint yellowish background to distinctly yellow to dull orange with weak pruinosity.

Female. Not distinguishable from females of several similar species.

Distribution. Alaska, Yukon Territory (Map 96); northern Europe; Siberia. Alaska, Y.T., VI, VII.

Specimens identified. Alaska, 5 ♂♂; Canada, 6 ♂♂; Europe, 2 ♂♂; Siberia, 2 ♂♂.

Discussion. The 15 available males show more variation than is usual in one species, particularly in the shape of the fore tarsus and the markings of the abdomen. These specimens may represent more than one species.

Platycheirus tenebrosus Coquillett

Figs. 83, 163; Map 94

Platycheirus tenebrosus Coquillett, 1900:428.

Length. 5.7–7.8 mm.

Male. Similar to dark specimens of *P. immarginatus*, differing as follows: face with very sparse gray pruinosity, nearly black.

Thoracic hairs extensively black; hairs of anepisternum and anepimeron usually all black; upper part of katepisternum without hairs; lower part of katepisternum with many stiff black hairs and some fine pale hairs. Legs mostly black; following areas yellow to yellow-orange: fore femur except posterior stripe, fore tibia except posterior spot at mid length, fore tarsus, narrow apex of mid and hind femur, base and apex of mid tibia, mid tarsus (upper surface often brown especially on last tarsomere), and narrow base of hind tibia. Fore leg in Fig. 83; trochanter with short pale or black setae on ventral surface; femur with posterior subbasal tuft of long wavy white hairs, otherwise with fine black or rarely yellow hairs; tibia moderately and uniformly broadened on basal three-fifths, then strongly broadened especially posteriorly and slightly narrowed apically, with posteroapical angle short and broad; first tarsomere rather strongly broadened on basal one-third to half, then very slightly narrowed to apex, about 1.3 times as long as wide; second tarsomere subquadrate. Mid femur in some specimens with some weak black or pale anteroventral setae near mid length, otherwise with mostly black hairs; some hairs on posteroventral surface stronger than others; mid tibia almost cylindrical, without conspicuous hairs. First tarsomere of hind leg moderately to rather strongly swollen, from about 3.5–5.0 times as long as greatest depth.

Abdomen in Fig. 163. Tergite 2 with bright to dull yellow spots of variable size not reaching margins and almost obsolete in some specimens; tergite 3 with rounded bright yellow or yellow-orange spots not reaching margins and from half to three-quarters as long as tergite; tergite 4 usually with similar but smaller spots, rarely with spots obsolete; tergite 5 without distinct pale spots, with anterolateral angles submetallic.

Female. Not distinguishable from females of several similar species.

Distribution. Alaska, British Columbia (Map 94), Oregon. B.C., V, VI, VIII.

Specimens identified. Alaska, 48 ♂♂; Canada, 6 ♂♂; United States, 2 ♂♂.

Platycheirus thompsoni Vockeroth

Figs. 72, 160; Map 94

Platycheirus thompsoni Vockeroth, 1990:743.

Length. 7.8–9.0 mm.

Male. Very similar to male of *P. nodosus*, differing as follows: fore and mid tarsi entirely yellow. Fore leg in Fig. 72; femur with single subbasal tuft preceded by single long pale hair in some specimens; hairs of tuft dark yellow with slender dark brown to black lanceolate apices; tibia slightly narrower; first tarsomere about 1.5 times as long as wide. Mid femur with posterior hairs about as long as femoral diameter; mid tibia without long anteroventral hairs.

Yellow-orange spots of tergite 2 reaching anterior margin (Fig. 160).

Female. Not distinguishable from females of several similar species.

Distribution. Southeastern Canada (Map 94), Minnesota to Maine. Ont., Que., V–VII.

Specimens identified. Canada, 24 ♂♂; United States, 7 ♂♂.

Biology. Specimens were collected in Quebec from wet *Carex–Salix* marsh.

Platycheirus thylax Hull

Fig. 55; Map 95

Platycheirus thylax Hull, 1944:78.

Length. 5.6–6.8 mm.

Male. Very similar to male of *P. discimanus*, differing as follows: anterior oral margin slightly less produced, not extending as far forward as tubercle; keel and grooves on upper part of face obscure or absent in some specimens.

Thoracic hairs mostly or entirely black. Knob of halter brown. Fore tibia (Fig. 55*a*) yellowish above on entire length; mid tibia usually dull yellow on about basal one-quarter; first two tarsomeres of mid leg yellow-brown. Mid tibia (Fig. 55*j*) with anteroventral hairs on basal one-third very sparse, not longer than tibial diameter; first tarsomere of mid leg (Fig. 55*j*) slightly compressed, scarcely deeper than wide; second tarsomere very slightly compressed; ventral setulae yellow-brown to dark brown.

Pruinose spots of tergites variable in size, with dull orange-brown background in some specimens.

Female. Not distinguishable from females of several similar species.

Distribution. Canada (Map 95) south to Idaho, Pennsylvania, and Massachusetts. B.C., VII; Ont., Que., IV, V.

Specimens identified. Canada, 50 ♂♂; United States, 4 ♂♂.

Biology. Specimens were collected in Nova Scotia from damp open sphagnum fen at 300 m.

Platycheirus urakawensis (Matsumura)

Fig. 69; Map 98

Melanostoma urakawense Matsumura, 1919:132.

Length. 7.8–9.7 mm.

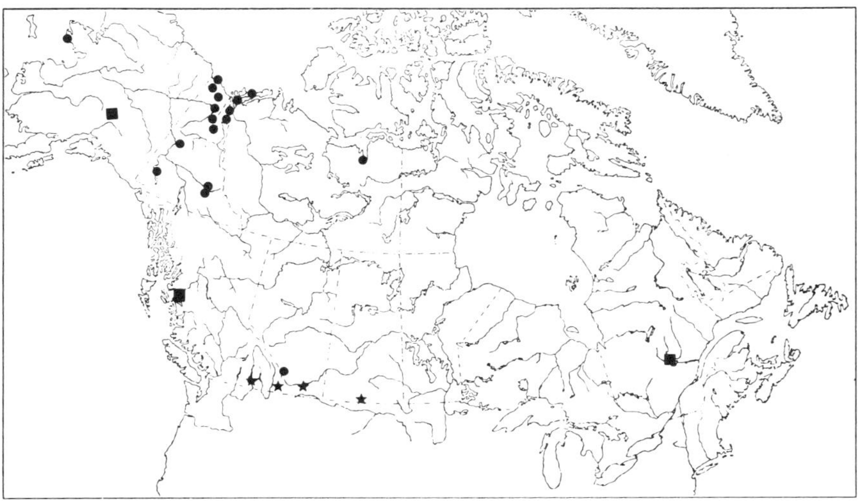

Map 98. Collection localities for *Platycheirus urakawensis* (Matsumura) (■), for *P. willistoni* (Goot) (★), and for *P. yukonensis* Vockeroth (●).

Male. Very similar to male of *P. albimanus*, differing as follows: lower katepisternal hairs all black, stiff. Wing with cell bm entirely trichose. Fore tibia more strongly broadened on apical one-third and with posteroapical angle broadly rounded; first tarsomere strongly broadened posteriorly on about basal two-thirds, sharply angulate at this point, then with straight margin to apex (Fig. 69). Mid tibia at mid length with rather strong black posterior bristle about 3.5 times as long as tibial diameter; first four tarsomeres of mid leg yellow; fifth tarsomere brown above.

Spots of tergite 2 indistinct, silvery pruinose, metallic anteriorly; spots of tergites 3 and 4 distinct, densely silvery pruinose.

Female. Unknown in North America, probably not distinguishable from females of related species.

Distribution. Alaska, British Columbia, Quebec (Map 98); eastern Siberia; Nepal; Japan. B.C., VI.

Specimens identified. Alaska, 1 ♂; Canada, 4 ♂♂; Siberia, 1 ♂; Nepal, 1 ♂; Japan, 4 ♂♂.

Platycheirus varipes Curran

Figs. 90, 167; Map 99

Platycheirus varipes Curran, 1923*a*:65.

Length. 7.1–8.7 mm.

Male. Similar to male of *P. hispidipes*, differing as follows: face slightly less protruding below, with less dense pruinosity. Antenna black. Pleura with mixed black and white hairs; hairs of lower part of katepisternum fine. Wing with very small bare areas near base. Knob of halter blackish brown. Legs mostly dark brown to black, with following yellow or whitish yellow: fore femur except posterodorsal and also anteroventral streak in some specimens, fore tibia except posterior streak, fore tarsus except last tarsomere in some specimens, apex of mid femur, narrow base and apex of mid tibia, first tarsomere of mid leg. Fore leg in Fig. 90; femur posteriorly without subbasal tuft of white hairs, with long strong nearly uniform black hairs; tibia slightly and uniformly broadened on basal two-thirds, then more strongly broadened posteriorly, with short broad posteroapical angle, with moderately long posterior hairs on most of its length; first tarsomere strongly broadened posteriorly on basal two-fifths, then slightly narrowed or parallel-sided to slightly arcuate apex, as long as wide; second tarsomere 1.6 times as wide as long, slightly narrowed toward apex; third tarsomere subrectangular, slightly wider than long. Mid femur with anteroventral row of short stiff pale or black setae on apical three-fifths and usually with slender curved black preapical hair; posteriorly and posteroventrally with long fine mostly black hairs; mid tibia nearly cylindrical, ventrally with dense fine erect wavy black or pale hairs, anteroventrally near base with hairs about 2.5 times as long as tibial diameter. Hind femur with shorter hairs.

Abdomen (Fig. 167) with large gray or bluish gray submetallic spots densely silvery gray pruinose and extending at most obscurely to lateral margins.

Female. Not distinguishable from females of several similar species.

Distribution. Alaska, western and northern Canada, Greenland (Map 99), south to Colorado and Maine. B.C., VI, VII.

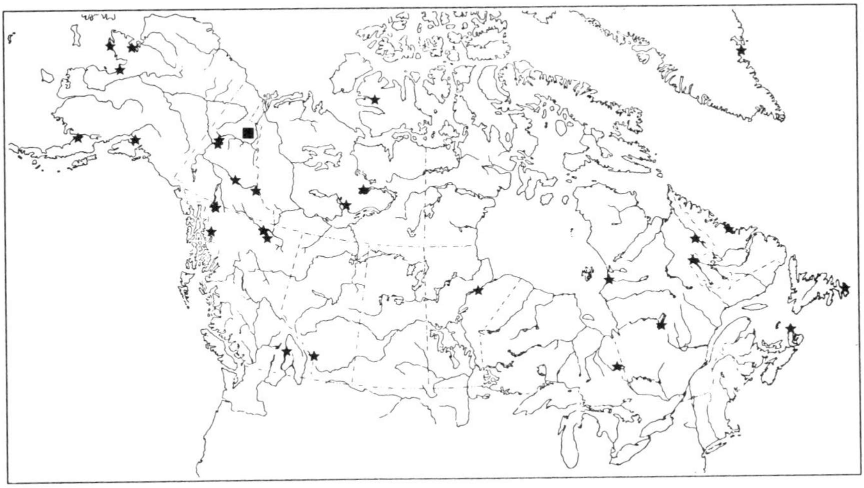

Map 99. Collection localities for *Platycheirus varipes* Curran (★) and for *P. woodi* Vockeroth (■).

Specimens identified. Alaska, 9 ♂♂; Canada, 68 ♂♂; Greenland, 6 ♂♂; United States, 6 ♂♂.

Biology. Specimens were taken in Nova Scotia from *Betula–Abies* woodland.

Platycheirus willistoni Goot

Map 98

Cheilosia rufipes Williston, 1882:306 (preoccupied Macquart, 1828).
Melanostoma willistoni Goot, 1964:219 (new name for *C. rufipes* Williston, 1882).
Melanostoma willistoni Sedman, 1965:575 (new name for *C. rufipes* Williston, 1882).

Length. 7.7 – 10.5 mm.

Male. Very similar to male of *P. concinnus*, differing as follows: thorax usually with slight coppery tinge, especially laterally, with hairs yellowish in some specimens, rather long; longest scutellar hairs subequal in length to arista. Wing with microtrichia very variable; cell c bare on up to basal one-fifth; cell bm bare on from basal one-third to basal seven-eighths. Legs with femora and tibia orange; tarsi orange with third and fourth tarsomeres and, on hind leg, also fifth tarsomere usually brown above.

Abdomen with markings bronzy rather than bluish.

Female. Very similar to female of *P. concinnus* but legs colored as in male of *P. willistoni*.

Distribution. Southwestern Canada (Map 98), south to California and Colorado. B.C., VI, VIII.

Specimens identified. Canada, 10 ♂♂, 4 ♀♀; United States, 5 ♂♂, 6 ♀♀.

Discussion. This species may be based on specimens of *P. concinnus* with entirely orange femora and tibiae.

Platycheirus woodi Vockeroth

Fig. 99; Map 99

Platycheirus woodi Vockeroth, 1990:747.

Length. 7.9 mm.

Male. Similar to male of *P. pullatus*, differing as follows: facial tubercle more abrupt and protruding, subacute; upper part of face with very weak median keel.

Thorax subshining, distinctly less pruinose; thoracic hairs mostly white (as in Alberta specimen of *P. pullatus*). Knob of halter pale yellow-brown. Fore tibia posteriorly and posteroventrally on entire length with dense fine mostly pale yellow hairs about twice as long as tibial diameter; first tarsomere of fore leg with similar but slightly shorter hairs (Fig. 99). Mid femur anteroventrally without black setae, with long fine white hairs on basal half; mid tibia anteroventrally with only very short hairs, posteriorly on middle one-third with suberect fine pale hairs slightly longer than tibial diameter; first tarsomere of mid leg without distinct long anteroventral setae.

Abdomen strongly shining, with pruinose lateral spots of tergites 2–4 very obscure.

Female. Unknown.

Distribution. Yukon Territory, 792 m (Map 99). Y.T., VII.

Specimens identified. Canada, 1 ♂.

Biology. The specimen was taken along the top of a very dry, nearly flat, barren, dolomite ridge.

Platycheirus yukonensis Vockeroth

Map 98

Platycheirus yukonensis Vockeroth, 1990:747.

Length. 7.9–8.9 mm.

Male. Very similar to male of *P. carinatus*, differing as indicated in couplet 49 of "Key to Nearctic species of *Platycheirus*."

Female. Not distinguishable from females of several similar species.

Distribution. Alaska, northwestern and western Canada (high boreal and alpine) (Map 98). Y.T., VI, VII.

Specimens identified. Alaska, 1 ♂; Canada, 82 ♂♂.

Genus *Scaeva* Fabricius

Diagnosis. Species robust with very clear wings, lunulate pale yellow abdominal spots and strongly swollen male frons. Length 11.1–15.7 mm.

Description. Eye densely haired; upper two-thirds of eye of male with extensive area of almost uniformly enlarged facets. Frons of male very broad, strongly swollen and with many erect black hairs; eye angle about 140°; frons of female only slightly swollen but very broad, at vertex about three-tenths head width. Face in both sexes moderately swollen, very pale yellow, with narrow dark median stripe on lower half.

Scutum shining black, obscurely yellowish laterally. Scutellum pale translucent yellow-brown. Ventral scutellar fringe complete, long. Pleura subshining black. Anterior anepisternum, meron and metapleuron bare. Upper and lower katepisternal hair patches broadly joined posteriorly, otherwise narrowly separated, with lower patch extending anteriorly to upper anterior corner of katepisternum. Metasternum bare. Vein R_{4+5} broadly and shallowly dipped into cell r_{4+5}. Microtrichia greatly reduced, present but very short and sparse only on about apical and posterior halves of wing (Fig. 21). Hind coxa without posteromedial apical hair tuft.

Abdomen (Fig. 154) oval, flattened, strongly margined from just beyond base of tergite 2 to apex of tergite 5. Tergites 2–5 each with pair of oblique lunulate pale yellow spots not reaching margins of tergites. Sternites black with narrow to broad pale yellow incisures.

Distribution. One Holarctic species; four species in Andes from Ecuador southward; about 10 Palaearctic and northern Oriental species.

Biology. In Europe *Scaeva pyrastri* is multivoltine without larval diapause and overwinters as an adult (Schneider 1947, 1948). Jones (1922) gave 24 days as the average period from egg to adult in the laboratory. Many authors have reported mass flights of the species, especially across the English Channel and through passes in the Alps and Pyrenees. Heiss (1938) listed 17 species of aphids on which larvae have been reared in North America. An additional larval host record in the CNC is *Cinara ponderosae*. Láska and Starý (1980) listed 24 species of aphid hosts from Czechoslovakia.

Scaeva pyrastri (Linnaeus)

Figs. 21, 154; Map 100

Musca pyrastri Linnaeus, 1758:594.
Scaeva affinis Say, 1823:93.
Scaeva unicolor Curtis, 1834:509.

Length. 11.1–15.7 m.

Male and female. Characters as given for genus. Wing as in Fig. 21. Abdomen as in Fig. 154.

Distribution. Alaska, Canada (Map 100), south to Arizona and Arkansas;* Europe; Asia; North Africa.* B.C., V–X.

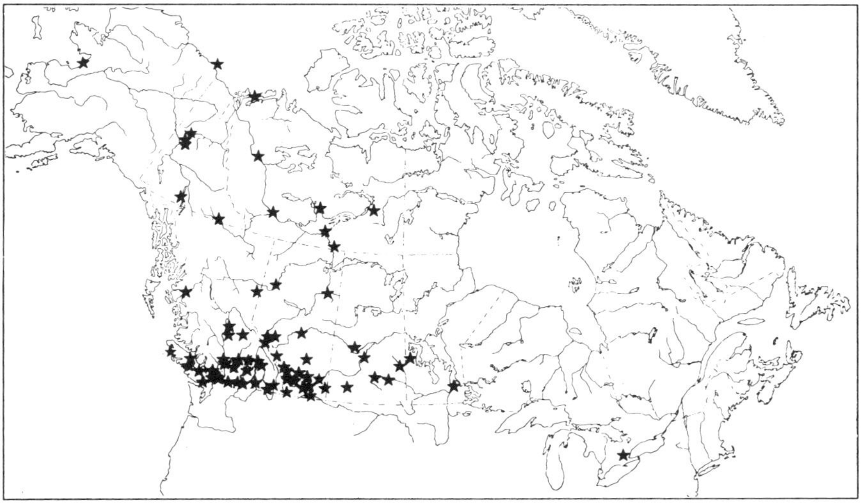

Map 100. Collection localities for *Scaeva pyrastri* (Linnaeus).

Specimens identified. Alaska, 1 ♂; Canada, 114 ♂♂, 290 ♀♀; United States, 14 ♂♂, 76 ♀♀; Europe, 13 ♂♂, 44 ♀♀; Israel, 1 ♀; Siberia, 1 ♂.

Genus *Sphaerophoria* Lepeletier & Serville

Diagnosis. Species slender to very slender, small- to medium-sized, with extremely large hemispherical male terminalia, with bright yellow markings on head and thorax and usually on abdomen but with abdomen almost entirely black in some species. Length 5.6–12.0 mm.

Description. Eye bare. Frons of male yellow, rarely with pair of black spots above antennae; frons of female with median black stripe, in *S. contigua* Macquart with weak nearly longitudinal rugosity. Face with tubercle rather prominent, not concave below, receding only slightly to lower facial margin (Fig. 24); face usually yellow, with median area brown to blackish in some species, with distinct entire brown to black median stripe only in *S. novaeangliae* Johnson.

Scutum black, shining, with broad yellow lateral stripe ending abruptly at transverse suture or extending broadly or narrowly to scutellum. Pleura black with bright yellow markings. Ventral scutellar fringe short and sparse, absent on median one-quarter or more of scutellum. Anterior anepisternum bare. Upper and lower katepisternal hair patches distinctly separated but with former usually triangularly extended ventrally about three-quarters of distance from upper to lower margin of sclerite. Meron and metapleuron bare; metasternum with some hairs. Wing membrane with small bare areas on basal one-third. Hind coxa without posteromedial apical hair tuft. Hind femur in male of *S. scripta* (Linnaeus) with posteroventral hairs short, stiff, black, and setulose (Fig. 31), in female similar but weaker; femoral hairs in other species weak (Fig. 30). Legs mostly yellow; femora blackish basally in some specimens or mostly blackish; tibiae brownish in some specimens; fore and mid tarsi usually yellow to brown and with apical tarsomere not darker than basal tarsomere, fore and mid tarsi distinctly blackish apically only in *S. philanthus* Meigen; hind tarsus brown to black.

Abdomen slender, unmargined, in male parallel sided or slightly constricted near mid length, in female parallel-sided or slightly broadened near mid length; tergite 5 of male with bluntly rounded posterior extension on right side. Tergites 2–4 usually each with entire or divided yellow band usually reaching lateral margins, only in *S. pyrrhina* Bigot extending to anterior or posterior tergite margins in some specimens; tergite 5 usually with very varied yellow markings, rarely black; in some specimens (especially southern specimens) tergites beyond 2 or 3 mostly or entirely yellow-orange or reddish; rarely tergites entirely or almost entirely black. Sternites yellow, without distinct markings, with short to long hairs. Male terminalia in Figs. 254–266. Epandrium extremely large, almost as wide as tergite 5, usually slightly longer than wide and wider posteriorly but in *S. cranbrookensis* Curran almost circular; cerci usually completely surrounded by sclerotization of epandrium (Fig. 266*f*), in membranous emargination of epandrium only in *S. novaeangliae* (Fig. 261*f*). Surstylus (Figs. 254–266) large and elaborate, usually clearly divided into three lobes; dorsal lobe slightly swollen apically, rounded, with dense mass of long coarse yellow hairs; ventral lobe flattened, elongate in some specimens, commonly apically divided, sparsely haired but usually with densely setulose longitudinal ridge on inner surface; two ventral lobes slightly to strongly asymmetrical in some specimens; inner lobe slender, rather weak, sparsely haired, compressed or depressed, arising just below or beside median margin of dorsal lobe (inner lobe lacking in some species); in *S. novaeangliae* dorsal and ventral lobes moderately distinct but not shaped as described here (Fig. 261).

Distribution. Of 13 species in Nearctic region, 1 widespread Palaearctic species, *S. scripta* (Linnaeus), known only from southwestern Greenland; other 12 species occurring in Canada, 2 extending as far south as southern Mexico. About 25 species occurring in Palaearctic region; several of these, plus about eight others, in Oriental region; one of these species extends southward from Korea through Indonesia to Australia. Two endemic species in South Africa.

Note: Records, marked with an asterisk, of southern limits of distribution (e.g., those for *S. cranbrookensis* Curran) are from the excellent distribution maps of the Nearctic species given by Knutson (1973) and are certainly reliable.

Discussion. Knutson (1972, 1973) showed that females of at least three species have sexually aberrant forms with strong similarities. The specimens lack spermathecae; abdominal segments 6 or 7, or both, are enlarged and strongly sclerotized; the sclerites of segment 7 may be fused into a ring. More obviously, the yellow bands of tergites 2–5 are divided sublaterally by extensions from the posterior black band so each tergite has a central yellow band and, except in very dark specimens, lateral yellow margins (Figs. 198, 199). Knutson referred to this marking as the "cleoae" pattern, from *S. cleoae* Metcalf, an unidentifiable species based on a series of such females. These specimens can be identified only in those species in which the head and thorax show distinctive characters.

The keys and descriptions given here are modified from Knutson (1973).

Biology. Knutson (1973) summarized most available information; I mention only a few points here. Several and probably all species are multivoltine, at least in southern Canada. Dušek and Láska (1974*a*) found that two species in Czechoslovakia overwinter as diapausing larvea. Bańkowska (1964) reported that, at 20°–22°C, larval development averaged 18, 18, and 20 days for three species but 25 days for *S. menthastri* (Linnaeus). Goeldlin (1974) found larvae of several European species only on herbaceous plants; he found a single larva on a tree. The larvae were active mostly at night. Knutson reported many aphid hosts for larvae of six Nearctic species; Láska and Starý reported additional aphid hosts for two of these and for two other European species. *Sphaerophoria quadrituberculata* Bezzi was reared from larvae feeding on nymphs of a thysanopteran, *Cercothrips afer*, in South Africa (Stuckenberg 1954). Although adults of *S. scripta* (Linnaeus) have often been reported as strongly migratory in Europe, this behavior has not been reported for other species.

Key to New World species of *Sphaerophoria*

1. Scutum with lateral yellow stripe beginning at humerus and ending abruptly at transverse suture; postalar callus at most slightly yellowish (Figs. 28, 29) 2

   Scutum with lateral yellow stripe beginning at humerus and extending onto postalar callus, obscure on posterior half in some species but always visible immediately behind suture 8

2. Face with broad shining brown to black median stripe particularly well developed above tubercle (Figs. 24, 25). Anteroventral portion of anepimeron brown to black (as in Fig. 29). Male with epandrium membranous behind cerci (Fig. 261*f*); dorsal lobe of surstylus with long slender posterolateral process (Figs. 261*a*,*f*). Widespread .... ***novaeangliae* Johnson**

   Face entirely yellow or with at most faint dark median stripe not extending to antennae. Anteroventral portion of anepimeron black or yellow. Male with epandrium uniformly sclerotized around cercal opening (Fig. 266); dorsal lobe of surstylus very varied but without long slender posterolateral process (Figs. 259, 264, 266) 3

3. Male (eyes meeting on frons) 4

   Female (eyes widely separated on frons) 6

4. Anteroventral portion of anepimeron yellow (Fig. 28). Tergites with yellow bands undivided (Figs. 176–178). Dorsal lobe of surstylus with compact anteromedially directed apical tuft of straight hairs (Figs. 259*d*,*e*). Widespread .... ***contigua* Macquart**

   Anteroventral portion of anepimeron black (Fig. 29). Tergites with yellow bands undivided or divided (Figs. 185–187, 191–193). Dorsal lobe of surstylus with fringe of dorsally curved hairs along entire margin, without compact apical tuft (Figs. 264, 266) 5

5. Dorsal lobe of surstylus longer than broad, subacute apically, its outer surface shining and with very short hairs (Fig. 266*f*). Western Canada .... ***sulphuripes* (Thomson)**

Dorsal lobe of surstylus not longer than broad, obtuse apically, its outer surface not shining and with abundant long hairs (Fig. 264*e*). Widespread .... ***pyrrhina* Bigot**

6. Dark frontal stripe slightly narrowed anteriorly, with bronze or purple lustre, and with margins often brownish and indistinct anteriorly (Fig. 26). Tergites 3 and 4 each with broad, usually narrowly divided yellow band, with three black spots or narrow irregular black band on anterior margin and with three black spots or broader black band on posterior margin (Figs. 188–190). Widespread .... ***pyrrhina* Bigot**

Frontal stripe black, without metallic lustre, broadened or at least parallel-sided anteriorly (Figs. 23, 27). Tergites 3 and 4 each with narrow and usually undivided yellow band (Figs. 179, 194–196) or entirely black 7

7. Anteroventral portion of anepimeron yellow (Fig. 28). Frontal stripe at most slightly broadened anteriorly; area between ocelli and antennae with fine longitudinal wrinkles on either side of median groove (Fig. 23). Widespread .... ***contigua* Macquart**

Anteroventral portion of anepimeron black (Fig. 29). Frontal stripe strongly broadened and Y-shaped anteriorly; frons without wrinkles (Fig. 27). Western Canada .... ***sulphuripes* (Thomson)**

8. Male (eyes meeting on frons) 9

Female (eyes widely separated on frons) 17

9. Hind femur with dense strong black setulae posteroventrally on apical three-quarters (Fig. 31). Dorsal lobe of surstylus elongate, with broad triangular bare flattened area at base (Fig. 263). Greenland .... ***scripta* (Linnaeus)**

Hind femur without distinct setulae on posteroventral surface (Fig. 30). Dorsal lobe of surstylus short or protruding, triangular in dorsal view, with inner margin haired to base (Figs. 254–258, 260, 262, 265). Not in Greenland 10

10. Apical tarsomere of fore and mid legs light brown to black and distinctly darker than basal tarsomere, or all tarsomeres black. Ventral lobe of surstylus with flattened preapical process on dorsal margin (Fig. 262); inner lobe of surstylus present, acute apically. Widespread .... ***philanthus* (Meigen)**

Apical tarsomere of fore and mid legs paler than basal tarsomere, or all tarsomeres unicolorous yellow to pale brown. Ventral lobe of surstylus usually more slender (Figs. 254–258, 260), if broad then with apex bluntly rounded (Fig. 265); inner lobe of surstylus present and acute or blunt apically, or absent 11

11. Hairs on sternite 2 shorter than width of hind femur. Dorsal lobe of surstylus long (Figs. 257, 260); inner lobe absent 12

Hairs on sternite 2 as long as or longer than width of hind femur. Dorsal lobe of surstylus shorter (Figs. 254–256, 258, 265); inner lobe present 13

12. Lateral yellow scutal stripe broad, bright yellow behind suture; anteroventral portion of anepimeron yellow (as in Fig. 28). Dorsal lobe of surstylus with fringe of short sparse hairs; ventral lobe with apex obliquely truncate (Fig. 257). Widespread . . . . ***brevipilosa*** **Knutson**

Lateral yellow scutal stripe narrowed and darkened behind suture; anteroventral portion of anepimeron black (as in Fig. 29). Dorsal lobe of surstylus with fringe of long dense hairs; ventral lobe with stout fingerlike apical process (Fig. 260). Widespread . . . . ***longipilosa*** **Knutson**

13. Abdomen broad, with yellow bands commonly notched but not divided (Fig. 181); black anterior and posterior margins of tergites with distinct bluish lustre at least laterally. Epandrium nearly circular in outline. Surstylus as in Fig. 258. Western Canada . . . . ***cranbrookensis*** **Curran**

Abdomen narrower (Fig. 173), with yellow bands divided or undivided. Epandrium longer than broad. Surstylus as in Figs. 254–256, 265 14

14. Anteroventral lobes of surstyli slightly to strongly asymmetrical (Fig. 256). Tergites with yellow bands undivided. Widespread . . . . ***asymmetrica*** **Knutson**

Anteroventral lobes of surstyli symmetrical (Figs. 254, 255, 265). Tergites with yellow bands divided or undivided . . . 15

15. Ventral lobe of surstylus broad with lower margin not strongly constricted preapically and with apex shallowly emarginate (Fig. 265). Tergites with yellow bands undivided. Western Canada . . . . ***weemsi*** **Knutson**

Ventral lobe of surstylus narrower, with lower margin strongly constricted preapically and with apex ending in slender fingerlike process (Figs. 254, 255). Tergites with yellow bands divided or undivided 16

16. Ventral lobe of surstylus with ventral margin nearly straight and with two apical processes differing only slightly in length (Fig. 254). Widespread .... ***bifurcata*** **Knutson**

Ventral lobe of surstylus with ventral margin curved and with apical processes markedly different in length (Fig. 255). Widespread .... ***abbreviata*** **Zetterstedt**

17. Hind femur with hairs on posteroventral surface distinctly longer and stronger than other hairs. Greenland .... ***scripta*** **(Linnaeus)**

Hind femur without longer and stronger hairs on posteroventral surface. Not in Greenland 18

18. Abdomen broad, oval. Tergites 2 and 3 with very narrow arcuate divided yellow bands; tergite 4 almost entirely black (Fig. 180). Group of black hairs usually present lateral to base of antenna but not on yellow areas of front above level of antennae. Western Canada ...... ***cranbrookensis*** **Curran**

Abdomen narrower, almost parallel sided, with color pattern not as above. Black hairs, if present on yellow areas of frons, not restricted to area lateral to base of antenna 19

19. Black frontal stripe forked or incised toward anterior (Fig. 22). Tergites with yellow bands short, arcuate, narrowly divided or incised posteriorly. Widespread .... ***bifurcata*** **Knutson**

Black frontal stripe not forked or incised anteriorly. Tergites with yellow bands long, almost straight, and undivided or short, straight, and broadly divided 20

20. Hairs on sternite 2 shorter than width of hind femur. Tergites with yellow bands entire. Widespread ....
.... ***brevipilosa*** **Knutson**
.... ***longipilosa*** **Knutson**

Some hairs on sternite 2 as long as or longer than width of hind femur. Tergites with yellow bands entire or divided 21

21. Face almost straight between antennae and tubercle. Yellow part of frons with some black hairs. Tergites with yellow bands entire. Western Canada .... ***weemsi*** **Knutson**

Face distinctly concave between antennae and tubercle. Yellow part of frons with or without black hairs. Tergites with yellow bands entire or divided. Widespread ....
.... ***abbreviata*** **Zetterstedt**
.... ***asymmetrica*** **Knutson**
.... ***philanthus*** **(Meigen)**

Clé des espèces du Nouveau Monde de *Sphaerophoria*

1. Scutum orné d'une rayure latérale jaune, prenant naissance à l'humérus et se terminant brusquement à la suture transversale; calus postalaire tout au plus légèrement jaunâtre (fig. 28, 29) 2

   Scutum orné d'une rayure latérale jaune qui s'étend de l'humérus jusqu'au calus postalaire, peu visible chez certaines espèces dans la portion postérieure mais toujours visible immédiatement derrière la suture 8

2. Face garnie d'une large rayure médiane lustrée brune à noire, particulièrement bien formée au-dessus du tubercule (fig. 24, 25). Portion antéro-ventrale de l'anépimère brune à noire (comme dans la fig. 29). Chez le mâle, épandrium membraneux derrière le cerque (fig. 216*f*); lobe dorsal du surstylus pourvu d'un appendice postéro-latéral long et mince (fig. 261*a*,*f*). Espèce répandue . . . . ***novaeangliae* Johnson**

   Face entièrement jaune ou avec tout au plus une rayure médiane foncée peu visible qui n'atteint pas les antennes. Portion antéro-ventrale de l'anépimère noire ou jaune. Chez le mâle, épandrium uniformément sclérotisé autour de l'ouverture cercale (fig. 266); lobe dorsal du surstylus très variable, mais dépourvu d'un appendice postéro-latéral long et mince (fig. 259, 264, 266) 3

3. Mâle (yeux contigus sur le front) 4

   Femelle (yeux largement séparés sur le front) 6

4. Portion antéro-ventrale de l'anépimère jaune (fig. 28). Tergites ornés de bandes jaunes continues (fig. 176–178). Lobe dorsal du surstylus garni d'une touffe apicale et compacte faite de poils droits, dirigée en position antéro-médiane (fig. 259*d*,*e*). Espèce répandue . . . . ***contigua* Macquart**

   Portion antéro-ventrale de l'anépimère noire (fig. 29). Tergites ornés de bandes jaunes continues ou non (fig. 185–187, 191–193). Lobe dorsal du surstylus orné d'une frange de poils recourbés vers l'arrière tout le long du bord; aucune touffe apicale compacte (fig. 264, 266) 5

5. Lobe dorsal du surstylus plus long que large, subaigu dans sa portion apicale, à face externe lustrée et garnie de poils très courts (fig. 266*f*). Ouest du Canada . . . . ***sulphuripes* (Thompson)**

   Lobe dorsal du surstylus pas plus long que large, obtus dans sa portion apicale; surface externe non lustrée, recouverte

d'abondants poils longs (fig. 264*e*). Espèce répandue ....
.... ***pyrrhina*** **Bigot**

6. Rayure frontale foncée légèrement rétrécie vers l'avant, avec reflet bronze ou pourpre, avec des bords souvent brunâtres et peu apparents sur la face antérieure (fig. 26). Tergites 3 et 4 ornés tous deux d'une large bande jaune en général étroitement divisée, avec trois taches noires ou une étroite bande noire irrégulière sur le bord antérieur et trois taches noires ou une bande noire plus large sur le bord postérieur (fig. 188–190). Espèce répandue .... ***pyrrhina*** **Bigot**

Rayure frontale noire, sans reflet métallique, élargie ou tout au moins ayant des côtés parallèles antérieurement (fig. 23, 27). Tergites 3 et 4 tous deux ornés d'une étroite bande jaune, habituellement continue (fig. 179, 194–196), ou entièrement noirs 7

7. Portion antéro-ventrale de l'anépimère jaune (fig. 28). Rayure frontale tout au plus légèrement élargie vers l'avant; zone entre les ocelles et les antennes soulignée de fines rides longitudinales de chaque côté du sillon médian (fig. 23). Espèce répandue .... ***contigua*** **Macquart**

Portion antéro-ventrale de l'anépimère noire (fig. 29). Rayure frontale en forme de Y et fortement élargie sur la face antérieure; front non ridé (fig. 27). Ouest du Canada ....
.... ***sulphuripes*** **(Thomson)**

8. Mâle (yeux contigus sur le front) 9

Femelle (yeux largement séparés sur le front) 17

9. Fémur postérieur orné, sur les trois quarts apicaux, de soies postéro-ventrales noires, denses et fortes (fig. 31). Lobe dorsal du surstylus allongé, ayant une large zone triangulaire glabre et aplatie à la base (fig. 263). Groënland ....
.... ***scripta*** **(Linnaeus)**

Fémur postérieur dépourvu de soies distinctes sur la face postéro-ventrale (fig. 30). Lobe dorsal du surstylus court ou en saillie, triangulaire en vue dorsale, poilu sur le bord interne jusqu'à la base (fig. 254–258, 260, 262, 265). Pas au Groënland
.... 10

10. Article apical du tarse des pattes antérieure et médiane brun pâle à noir et nettement plus foncé que l'article basilaire, ou tous les articles de couleur noire. Lobe ventral du surstylus garni d'un appendice pré-apical aplati sur le bord dorsal (fig. 262); lobe interne du surstylus, pointu dans sa portion apicale. Espèce répandue .... ***philanthus*** **(Meigen)**

Article apical du tarse des pattes antérieure et médiane plus pâle que l'article basilaire, ou tous les articles de même couleur, jaune à brun pâle. Lobe ventral du surstylus habituellement plus mince (fig. 254–258, 260); lorsque le lobe est large, son apex est nettement arrondi (fig. 265); lobe interne du surstylus aigu ou émoussé dans sa portion apicale, ou absent 11

11. Sternite 2 orné de poils plus courts que la largeur du fémur postérieur. Lobe dorsal du surstylus long (fig. 257, 260); aucun lobe interne 12

Sternite 2 orné de poils aussi longs ou plus longs que la largeur du fémur postérieur. Lobe dorsal du surstylus plus court (fig. 254–256, 258, 265); lobe interne présent 13

12. Scutum portant une large rayure latérale jaune, de couleur plus vive derrière la suture; portion antéro-ventrale de l'anépimère jaune (comme dans la fig. 28). Lobe dorsal du surstylus frangé de poils courts et clairsemés; lobe ventral ayant un apex tronqué (fig. 257). Espèce répandue .... ***brevipilosa* Knutson**

Scutum orné d'une rayure latérale jaune, rétrécie et foncée derrière la suture; portion antéro-ventrale de l'anépimère noire (comme dans la fig. 29). Lobe dorsal du surstylus frangé de poils longs et denses; lobe ventral pourvu d'un robuste appendice apical digitiforme (fig. 260). Espèce répandue .... ***longipilosa* Knutson**

13. Abdomen large orné de bandes jaunes souvent échancrées mais non divisées (fig. 181); lisérés antérieur et postérieur des tergites noirs ayant un reflet bleuâtre distinct, du moins sur les côtés. Épandrium de profil presque circulaire. Surstylus comme dans la fig. 258. Ouest du Canada .... ***cranbrookensis* Curran**

Abdomen plus étroit (fig. 173) ayant des bandes jaunes continues ou non. Épandrium plus long que large. Surstylus comme dans les fig. 254–256, 265 14

14. Lobes antéro-ventraux des surstyli de légèrement à fortement asymétriques (fig. 256). Tergites ornés de bandes jaunes continues. Espèce répandue .... ***asymmetrica* Knutson**

Lobes antéro-ventraux des surstyli symétriques (fig. 254, 255, 265). Tergites ornés de bandes jaunes continues ou non 15

15. Lobe ventral du surstylus large avec un bord inférieur peu rétréci dans la portion pré-apicale et un apex légèrement échancré (fig. 265). Tergites ornés de bandes jaunes continues. Ouest du Canada .... ***weemsi* Knutson**

Lobe ventral du surstylus plus étroit ayant un bord inférieur fortement rétréci dans la portion pré-apicale et un apex se terminant par un mince appendice digitiforme (fig. 254, 255). Tergites ornés de bandes jaunes continues ou non 16

16. Lobe ventral du surstylus ayant un bord ventral presque droit et deux appendices apicaux diffèrent seulement légèrement en longueur (fig. 254). Espèce répandue .... ***bifurcata*** **Knutson**

Lobe ventral du surstylus ayant un bord ventral incurvé et deux appendices apicaux de longueur nettement différentes (fig. 255). Espèce répandue ...... ***abbreviata*** **Zetterstedt**

17. Fémur postérieur garni sur la face postéro-ventrale de poils nettement plus longs et forts que les autres poils. Groënland .... ***scripta*** **(Linnaeus)**

Fémur postérieur dépourvu de poils plus longs et plus forts sur la face postéro-ventrale. Pas au Groënland 18

18. Abdomen large et ovale. Tergites 2 et 3 ornés de bandes jaunes très étroites, arquées et non continues; tergite 4 presque entièrement noir (fig. 180). Groupe de poils noirs en général sur les côtés, à la base de l'antenne, mais non sur les zones jaunes du front au-dessus du niveau des antennes. Ouest du Canada .... ***cranbrookensis*** **Curran**

Abdomen plus étroit ayant des côtés presque parallèles, de couleurs différentes du précédent. Lorsque des poils noirs ornent les zones jaunes du front, ceux-ci ne se limitent pas aux parties situées à côté de la base des antennes 19

19. Rayure frontale noire fourchue ou incisée antérieurement (fig. 22). Tergites ornés de bandes jaunes courtes, arquées, étroitement divisées ou incisées sur la face postérieure. Espèce répandue .... ***bifurcata*** **Knutson**

Rayure frontale noire ni fourchue, ni incisée antérieurement. Tergites soulignés de longues bandes jaunes presque droites et continues, ou de bandes courtes, droites et largement divisées 20

20. Poils du sternite 2 plus courts que la largeur du fémur postérieur. Tergites ornés de bandes jaunes complètes. Espèces répandues .... ***brevipilosa*** **Knutson**
.... ***longipilosa*** **Knutson**

Quelques poils du sternite 2 aussi longs ou plus longs que la largeur du fémur postérieur. Tergites ornés de bandes jaunes complètes ou divisées 21

21. Face presque droite entre les antennes et le tubercule. Partie jaune du front garnie de quelques poils noirs. Tergites portant des bandes jaunes complètes. Ouest du Canada .... ***weemsi* Knutson**

Face nettement concave entre les antennes et le tubercule. Partie jaune du front ayant ou non des poils noirs. Tergites portant des bandes jaunes, complètes ou divisées. Espèces répandues .... ***abbreviata* Zetterstedt**
.... ***asymmetrica* Knutson**
.... ***philanthus* (Meigen)**

Sphaerophoria abbreviata Zetterstedt

Fig. 254; Map 101

Sphaerophoria abbreviata Zetterstedt, 1859:6007.

Length. 6.3–10.2 mm.

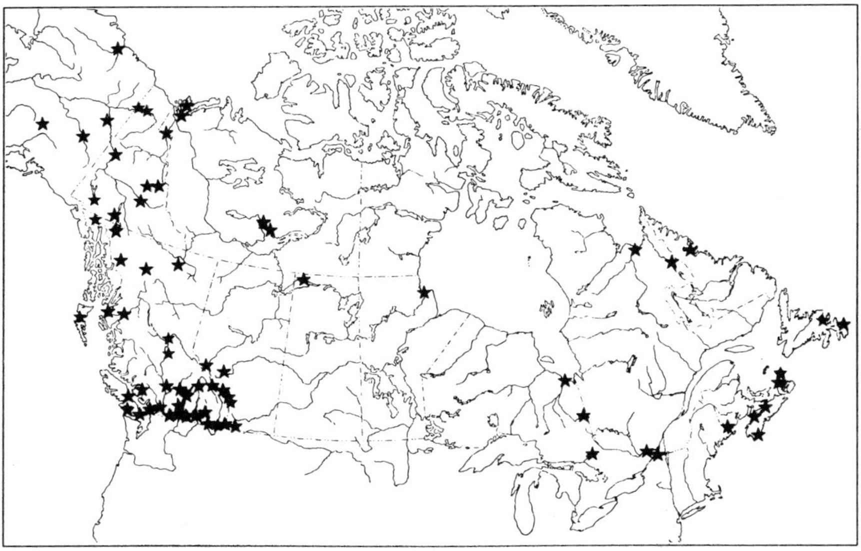

Map 101. Collection localities for *Sphaerophoria abbreviata* Zetterstedt.

Male. Differs from male of *S. philanthus* as follows: fore and mid tarsi nearly concolorous yellow to orange, not darker than tibiae; apical tarsomere as pale as or paler than basal tarsomere.

Hairs of sternite 2 sparse, longer than width of hind femur, straight or barely crinkled apically. Surstylus (Fig. 254) with dorsal lobe slightly shorter; inner lobe tapering uniformly to broad and bluntly rounded apex; ventral lobe slightly more slender with blunt toothlike preapical posterior process projecting scarcely above margin of lobe and with slender fingerlike apical process extending far beyond posterior process.

Female. Not distinguishable from females of *S. philanthus* and *S. asymmetrica*.

Distribution. Alaska, Canada (Map 101), south to Oregon, Colorado, and North Carolina; Europe; Asia.* B.C., V–VIII; Ont., Que., V–IX.

Specimens identified. Alaska, 2 ♂♂; Canada, 225 ♂♂; United States, 15 ♂♂; Europe, 10 ♂♂.

Biology. Bańkowska (1964) and Láska and Starý (1980) recorded two species of aphids as larval hosts in central Europe.

Sphaerophoria asymmetrica Knutson

Figs. 156, 173; Map 102

Sphaerophoria asymmetrica Knutson, 1973:28.

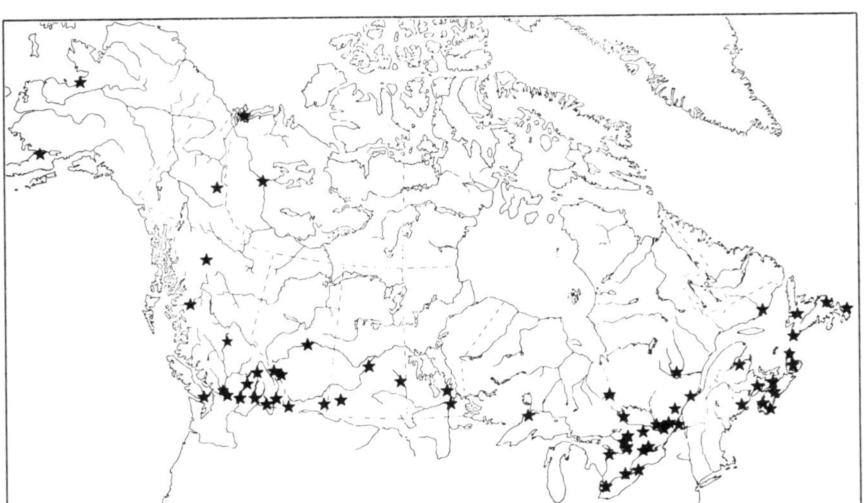

Map 102. Collection localities for *Sphaerophoria asymmetrica* Knutson.

Length. 6.8–9.0 mm.

Male. Differs from male of *S. abbreviata* as follows: tergites 2–4 with yellow bands broad, entire (Fig. 173). Surstyli (Fig. 256) with ventral lobes slightly to strongly asymmetrical, variable in form, usually with stout dorsal preapical process and with longer stout to moderately slender thumblike apical process, rarely without distinct preapical process and with bluntly tapering or thumblike apical process.

Female. Not distinguishable from females of *S. abbreviata* and *S. philanthus*.

Distribution. Alaska, Canada (Map 102), south to Colorado and Georgia. B.C., VII–IX; Ont., Que., V–VIII.

Specimens identified. Alaska, 3 ♂♂; Canada, 211 ♂♂; United States, 34 ♂♂.

Sphaerophoria bifurcata Knutson

Figs. 22, 255; Map 103

Sphaerophoria bifurcata Knutson, 1973:37.

Length. 7.1–8.0 mm.

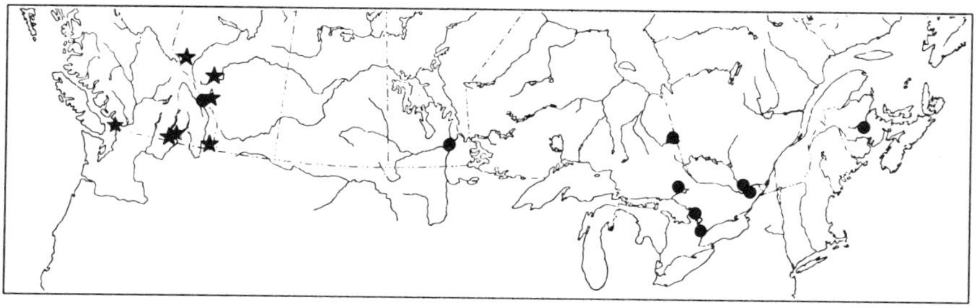

Map 103. Collection localities for *Sphaerophoria bifurcata* Knutson (●) and for *S. cranbrookensis* Curran (★).

Male. Frons with many black hairs at least on upper part.
Scutum with lateral yellow stripe extending to scutellum, narrowed behind suture in some specimens. Legs mostly yellow; fore and mid tarsi brownish in some specimens; apical tarsomere as pale as or paler than basal tarsomere.

Tergites 2–4 with yellow bands narrow, usually entire, narrowly divided in some specimens. Sternite 2 with hairs long, sparse, straight. Surstylus (Fig. 255) with dorsal lobe depressed, with long hairs along ventral margin; inner lobe present, variable in shape; ventral lobe broad, abruptly constricted very near apex, with short preapical toothlike posterior process and short fingerlike apical process.

Female. Frons with median stripe very broad, extending to antennal bases, bifurcate apically (Fig. 22); yellow areas of frons with many black hairs. Tergites 2–4 with yellow bands narrow, usually entire on 2 and 3, usually divided on 4.

Distribution. Southern Canada (Map 103), south to Oregon,* Michigan,* and Maine. B.C., VI; Ont., Que., V, VI.

Specimens identified. Canada, 2 ♂♂, 8 ♀♀; United States, 1 ♂.

Sphaerophoria brevipilosa Knutson

Figs. 174, 175, 257; Map 104

Sphaerophoria brevipilosa Knutson, 1973:38.

Length. 6.3–8.1 mm.

Male. Frons with some black hairs on upper part only.

Scutum with lateral yellow stripe extending to scutellum. Legs mostly yellow; fore and mid legs with apical tarsomere as pale as or paler than basal tarsomere.

Tergites 2–4 with broad undivided yellow bands; tergites beyond 2 entirely reddish in some specimens (Figs. 174, 175). Sternite 2 with hairs sparse and shorter than width of hind femur. Surstylus (Fig. 257) with dorsal lobe very short, with many rather short marginal hairs and with median portions very short and rounded; inner lobe absent; ventral lobes very short and broad, slightly asymmetrical, each tapering slightly to obliquely truncate apex.

Female. (*Note*: description from Knutson [1973] probably based on specimens of both *S. brevipilosa* and *S. longipilosa*.) Black frontal stripe short, usually acute, brownish and indistinct apically. Abdomen narrow, parallel-sided. Yellow bands of tergites 2–4 very broad, entire, slightly arcuate; tergites 5 and 6 extensively yellow, each with pair of anterolateral and pair of posterolateral black spots and incomplete narrow black median stripe. All sternites with hairs shorter than width of hind femur.

Distribution. Southern Canada (Map 104), south to Oregon,* Colorado, and Maryland.* B.C., V, VII; Ont., Que., V–VIII.

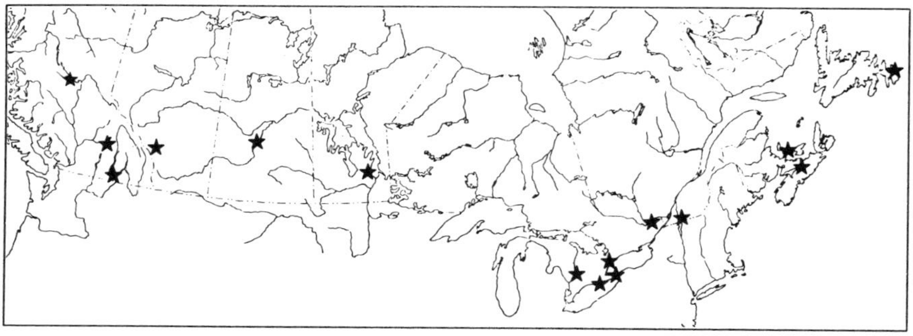

Map 104. Collection localities for *Sphaerophoria brevipilosa* Knutson.

Specimens identified. Canada, 18 ♂♂; United States, 2 ♂♂.

Biology. Knutson (1973) recorded *Myzus cerasi* as a larval host.

Sphaerophoria contigua Macquart

Figs. 23, 28, 176–179, 259; Map 105

Syrphus cylindricus Say, 1824:11 (preoccupied Fabricius, 1781).
Sphaerophoria contigua Macquart, 1847:78.
Sphaerophoria fulvicauda Bigot, 1884:104.

Length. 6.5–8.8 mm.

Male. Scutum with lateral yellow stripe ending abruptly at suture; postalar callus usually yellowish. Anteroventral portion of anepimeron yellow (Fig. 28). Legs usually yellow, rarely brownish, tarsi darker than rest of legs.

Abdomen in Figs. 176–178. Tergites 2–4 with yellow bands entire; anterior and posterior margins of 2–4 usually black; tergite 2 rarely and tergites 3 and 4 commonly with margins reddish to yellowish making pattern very obscure. Surstylus (Fig. 259) rather variable, with dorsal lobe much longer than broad, bluntly rounded apically, with long dense compact brush of hairs; inner lobe slender, tapering to subacute apex; ventral lobe broad, tapering to broad subacute apex.

Female. Face usually yellow, rarely with broad light brown median stripe throughout. Frons (Fig. 23) with many wrinkles fine, irregular, convergent anteriorly, on either side of shallow median groove; black frontal stripe well-defined laterally, parallel-sided to slightly broadened apically, truncate anteriorly or with narrow apicolateral extensions. Legs yellow, tarsi darkened in some specimens. Abdomen in Fig. 179. Tergites 2–5 with undivided arcuate yellow bands, commonly incised posteromedially on 5; tergite 6 with median basal black spot and sublateral basal black spots in some specimens; black areas of tergites reddish to yellow in some specimens making pattern obscure.

Distribution. Canada (Map 105), south to California, Mexico (Chiapas), and Florida.* B.C., V–VII; Ont., Que., V–X.

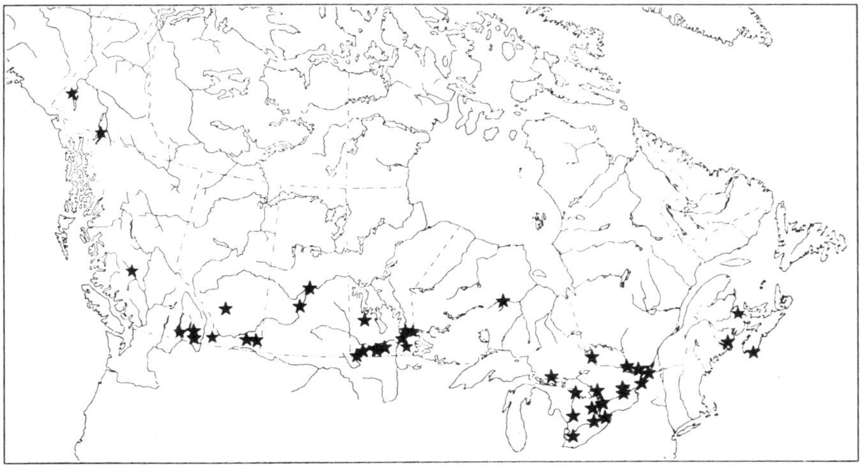

Map 105. Collection localities for *Sphaerophoria contigua* Macquart.

Specimens identified. Canada, 126 ♂♂, 126 ♀♀; United States, 124 ♂♂, 81 ♀♀; Mexico, 11 ♂♂, 18 ♀♀.

Biology. Knutson (1973) listed 26 species of aphids from which *S. contigua* has been reared.

Sphaerophoria cranbrookensis Curran

Figs. 180, 181, 258; Map 103

Sphaerophoria cranbrookensis Curran, 1921*b*:173.

Length. 7.5–9.5 mm.

Male. Scutum with lateral yellow stripe extending to scutellum, indistinct behind suture in some specimens. Dark areas of thorax deep shining blue. Legs mostly yellow, femoral bases darkened in some specimens; fore and mid tarsi at most slightly darkened, with apical tarsomere not darker than basal tarsomere.

Abdomen (Fig. 181) broad, slightly narrowed at end of segment 2, with sides deep shining blue. Tergites 2 and 3 with entire, narrow to moderately broad, reddish yellow bands notched medially and narrowed laterally; tergite 4 with similar very narrow band. Sternite 2 with moderately dense long straight hairs. Epandrium robust, almost circular in outline. Surstylus (Fig. 258) with dorsal lobe very short, with many long hairs; inner lobe broad basally, tapering to acute apex; ventral lobe rather slender, scarcely constricted preapically, with short slender posteroapical process and longer anteroapical process.

Female. Black frontal stripe very broad, truncate apically. Scutellar hairs black. Abdomen (Fig. 180) very broad. Tergites 2–4 with very narrow, usually divided, yellow bands; bands on tergite 4 present in some specimens only laterally or entirely absent; tergite 5 with rather large to tiny oblique submedian yellow spots and lateral yellow margins.

Distribution. Western Canada (Map 103), south to Oregon* and Montana.* B.C., V, VI.

Specimens identified. Canada, 4 ♂♂, 3 ♀♀.

Sphaerophoria longipilosa Knutson

Figs. 182, 260; Map 106

Sphaerophoria longipilosa Knutson, 1973:41.

Length. 7.5–8.2 mm.

Male. Frons with many dark hairs, especially beside antennal bases.

Scutum with lateral yellow stripe extending to scutellum, strongly reduced behind suture in some specimens. Legs extensively darkened;

fore and mid tarsi yellow to dark brown but apical tarsomere not darker than basal tarsomere.

Abdomen (Fig. 182) slender. Yellow bands on tergites 2–4 narrow, arcuate, entire or narrowly divided. Sternite 2 with hairs yellow to brown, shorter than width of hind femur. Surstylus (Fig. 260) with dorsal lobe small, rounded apically, with narrow dense row of very long hairs along ventral margin; inner lobe absent; ventral lobe rather broad, scarcely tapering, with blunt toothlike posterior preapical process and long stout apical fingerlike process.

Female. Not distinguishable from female of *S. brevipilosa*.

Distribution. Southern Canada (Map 106), south to Minnesota* and Rhode Island;* Colorado. B.C., V; Ont., Que., V, VI.

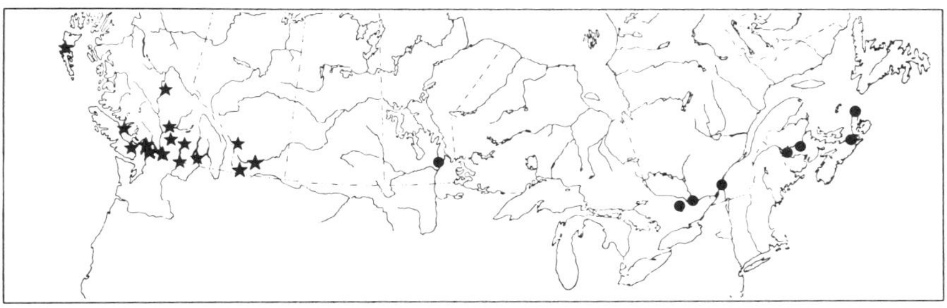

Map 106. Collection localities for *Sphaerophoria longipilosa* Knutson (●) and for *S. sulphuripes* (Thomson) (★).

Specimens identified. Canada, 13 ♂♂; United States, 1 ♂.

Sphaerophoria novaeangliae Johnson

Figs. 24, 25, 183, 184, 261; Map 107

Spherophoria novaeangliae Johnson, 1916:76.

Length. 6.8–8.2 mm.

Male. Frons with pair of small dark spots above antennae. Face (Fig. 24) with distinct black to brown median stripe particularly broad and dark above tubercle.

Scutum with lateral yellow stripe ending abruptly at transverse suture. Anepimeron usually brown anteroventrally and yellow only near upper margin, rarely entirely black.

Abdomen (Fig. 183) slender. Tergites 2–4 each with yellow band entire, moderately broad; bands of tergites 3 and 4 narrowed laterally, with band on latter not reaching margin in some specimens. Epandrium (Fig. 261*a*) with cercal emargination open making area between cerci and posterior margin of tergite membranous. Surstylus (Figs. 261*a,f*) with dorsal lobe broad, rather swollen, with many short hairs on outer surface and some slightly longer marginal hairs, produced ventrolaterally into slender tapering nearly bare and slightly incurved process; inner lobe absent; ventral lobe not clearly separated from dorsal lobe, very short, fingerlike, and tapering.

Female. Frons (Fig. 25) smooth, with small dark median spot above antennae joined to broad median black frontal stripe in some specimens. Face with median stripe distinct but narrowed below in some specimens. Tergites 2–4 (Fig. 184) with yellow bands entire, slightly arcuate, not narrowed laterally; tergite 5 mostly black, usually with triangular anterolateral yellow spot confluent with yellow posterior margin in some specimens.

Distribution. Alaska, Canada (Map 107), south to Colorado, Wisconsin,* and Massachusetts. B.C., VI, VIII; Ont., Que., V–VII.

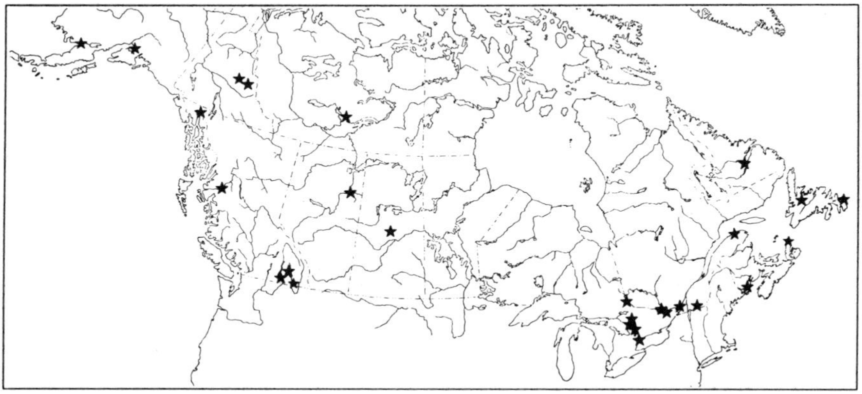

Map 107. Collection localities for *Sphaerophoria novaeangliae* Johnson.

Specimens identified. Alaska, 1 ♂; Canada, 21 ♂♂, 29 ♀♀; United States, 10 ♂♂, 9 ♀♀.

Sphaerophoria philanthus (Meigen)

Figs. 30, 262; Map 108

Syrphus philanthus Meigen, 1822:327.
Sphaerophoria dubia Zetterstedt, 1849:3162.
Sphaerophoria nigritarsi Fluke, 1930:143.
Sphaerophoria robusta Curran, 1930*b*:62.

Length. 7.6–9.6 mm.

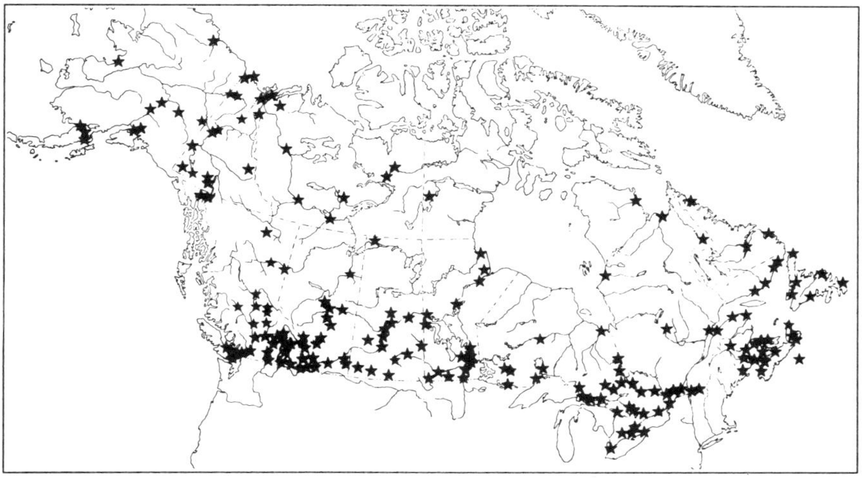

Map 108. Collection localities for *Sphaerophoria philanthus* (Meigen).

Male. Scutum with lateral yellow stripe extending to scutellum but narrowed behind suture in some specimens. Legs mostly yellow to brown; coxae and femoral bases black in some specimens; fore and mid tarsi with apical two or three tarsomeres darker than tibia; fifth tarsomere usually much darker than first tarsomere, rarely all tarsomeres dark; hind tarsus uniformly dark brown to black. Hind femur with only weak setae posteriorly (Fig. 30).

Tergites 2–4 each usually with broad entire yellow band; bands reduced in some specimens or even divided medially; tergite 5 with two broad to narrow longitudinal yellow marks partly joined in some specimens; posterior dark band of tergite 3 and dark areas of tergites 4 and 5 reddish in some specimens. Hairs of sternite 2 usually dense, long, and crinkled at tips, but occasionally shorter, sparser, and almost straight as in *S. abbreviata*. Surstylus (Fig. 262) with dorsal lobe prominent, slightly compressed, with dense long yellowish to brownish hairs over most of surface; inner lobe broad basally, strongly constricted at mid length and tapering to acute apex; ventral lobe very broad, rounded below, with flattened preapical posterior process truncated apically and projecting distinctly above margin of lobe and with slender fingerlike apical process projecting only slightly beyond posterior process.

Female. Black frontal stripe very variable. Legs yellow; fore and mid tarsi usually slightly darkened; hind tarsi dark. Abdomen moderately broad. Tergites 2–4 with yellow bands usually entire, rarely divided. Hairs of sternite 2 long (from Knutson 1973).

Distribution. Alaska, Canada (Map 108), south to California, Texas, and Florida; Europe; Asia. B.C., IV–IX; Ont., Que., V–IX.

Specimens identified. Alaska, 86 ♂♂; Canada, 876 ♂♂; United States, 171 ♂♂; Europe, 15 ♂♂; Japan, 17 ♂♂.

Biology. Knutson (1973) listed five species of aphids from which *S. philanthus* has been reared in North America or Europe.

Sphaerophoria pyrrhina Bigot

Figs. 26, 185–190, 264; Map 109

Sphaerophoria pyrrhina Bigot, 1884:101.
Sphaerophoria guttulata Hull, 1942:20.

Length. 5.6–6.9 mm.

Male. Scutum with lateral yellow stripe ending abruptly at suture; postcallus black. Anteroventral portion of anepimeron black (as in Fig. 29). Legs usually yellow; femora darkened at base in some specimens; tarsi usually darkened.

Abdomen in Figs. 185–187. Yellow band of tergite 2 usually divided; yellow band of tergite 3 usually entire but medially emarginate on anterior and posterior margins; tergite 4 mostly yellow, and with one or three black marks on each of anterior and posterior margins or with black marks extensive and covering all of mid line; tergite 5 mostly yellow except for median black line and in some specimens paired anterolateral and posterolateral black spots. Surstylus (Fig. 264) with dorsal lobe scarcely longer than broad, bluntly rounded apically, with moderately dense fringe of fine hairs on apical and lateral margins and moderately long hairs on dorsal surface; inner lobe absent; ventral lobe slender and directed more posteriorly than in *S. contigua*.

Female. Black frontal stripe parallel-sided or narrowed anteriorly, often not reaching antenna, its margins often brownish and poorly defined (Fig. 26). Frons smooth; black area around ocellar tubercle with strong bronze or purple lustre. Postalar callus more or less yellowish. Yellow bands of tergites 2–4 broad, entire or divided, usually reaching anterior margins sublaterally; tergites 5 and 6 with anteromedian black stripe and usually with anterolateral and posterolateral black spots (Figs. 188–190).

Distribution. Canada (Map 109), south to California, Mexico (D.F.),* Colorado, and North Dakota.* B.C., V, VII, VIII, X; Ont., Que., VII.

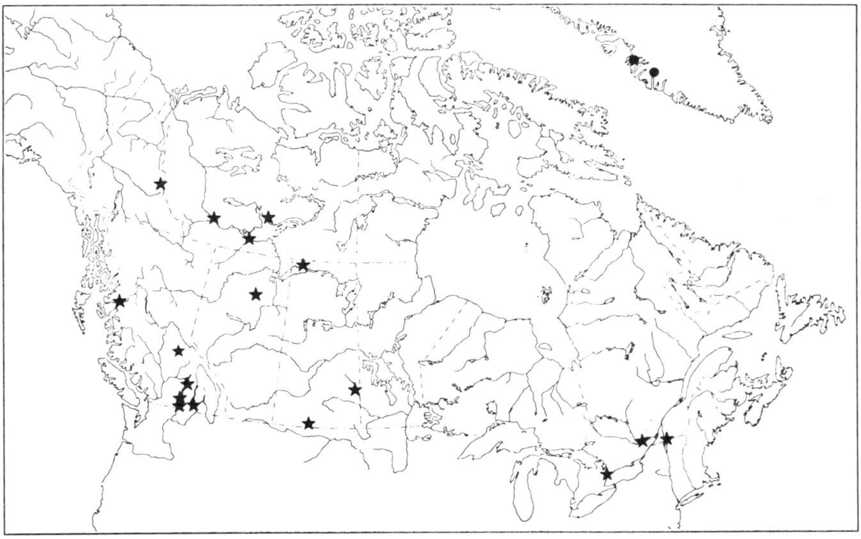

Map 109. Collection localities for *Sphaerophoria pyrrhina* Bigot (★) and for *S. scripta* (Linnaeus) (●).

Specimens identified. Canada, 20 ♂♂, 12 ♀♀; United States, 24 ♂♂, 21 ♀♀.

Sphaerophoria scripta (Linnaeus)

Figs. 31, 197, 263; Map 109

Musca scripta Linnaeus, 1758:594.
Sphaerophoria strigata Staeger, 1845:362.

Length. 9.0–12.0 mm.

Male. Face with diffuse brownish median stripe and narrowly black ventral margin.

Scutum with lateral yellow stripe clearly extending to scutellum but narrower and darker behind transverse suture. Base of fore and mid femora, most of hind femur, and all tarsi darkened; rest of legs yellow. Hind femur with strong black setulae posteroventrally on apical three-quarters (Fig. 31).

Yellow bands of tergites 2–4 narrow, constricted or divided medially, almost absent in some specimens (Fig. 197). Surstylus (Fig. 263) with dorsal lobe long, broadly rounded apically, with basal half flat and bare above and apical half with many moderately long hairs; inner lobe broad basally, curved downward apically and tapering strongly to acute apex; ventral lobe very broad, not narrowed apically.

Female. Face as in male. Black frontal stripe broad, parallel-sided, apically trilobate, or truncate. Legs yellow with only coxae and tarsi dark. Hind femur with anteroventral hairs distinctly longer and stronger than other hairs. Abdomen slender. Tergites 2–5 with narrow divided yellow bands.

Distribution. Southwestern Greenland (Map 109); Iceland;* Europe; Asia. Greenland, VI–VIII.

Specimens identified. Greenland, 24 ♂♂, 24 ♀♀; Europe, 87 ♂♂, 100 ♀♀; Israel, 2 ♂♂; Lebanon, 14 ♂♂, 1 ♀.

Biology. Knutson (1973) listed 10 species of aphids recorded as larval hosts in Europe and Asia. Láska and Starý (1980) recorded 10 additional aphid hosts from Czechoslovakia.

Discussion. Nearctic records of *S. scripta* and *S. strigata* (e.g., Wirth et al. 1965) were probably based on misidentifications of *S. philanthus* or other species.

Sphaerophoria sulphuripes (Thomson)

Figs. 27, 29, 191–196, 266; Map 106

Syrphus sulphuripes Thomson, 1869:500.
Syrphus infumata Thomson, 1869:501.
Sphaerophoria dubia Bigot, 1884:101.
Sphaerophoria pachypyga Bigot, 1884:104.
Sphaerophoria melanosa Williston, 1887:106.

Length. 6.2–8.3 mm.

Male. Face usually yellow, tubercle and lower margin darkened in some specimens.

Scutum with lateral yellow stripe ending abruptly at suture; postalar callus black. Anepimeron black on lower half or more (Fig. 29). Legs yellow; femora blackened basally in some specimens; tibiae brownish apically in some specimens.

Abdomen in Figs. 191–193. Tergites 2–4 with yellow bands usually entire but reduced in some specimens, in very dark specimens each band forming pair of small yellow spots. Surstylus (Figs. 266*a,f*) with dorsal lobe much longer than broad, tapering to subacute apex, with downward-directed fringe on lateral margin and with dorsal surface shining and with very short hairs; inner lobe absent; ventral lobe very broad, obliquely truncate apically.

Female. Face yellow. Frons (Fig. 27) smooth, without brassy or purple lustre; black frontal stripe greatly broadened apically, usually well-defined. Abdomen in Figs. 194–196. Tergites 2–4 usually with narrow but entire yellow bands; bands reduced in some specimens and medially divided or entirely absent; tergite 5 usually with divided yellow band and yellow posterior margin, rarely entirely black; tergite 6 yellow with three separate or confluent black spots anteriorly.

Distribution. Southwestern Canada (Map 106), south to Mexico (Baja California)* and Arizona. B.C., VI–IX.

Specimens identified. Canada, 15 ♂♂, 14 ♀♀; United States, 73 ♂♂, 52 ♀♀.

Biology. Knutson (1973) listed two species of aphids from which *S. sulphuripes* has been reared.

Sphaerophoria weemsi Knutson

Fig. 265; Map 110

Sphaerophoria weemsi Knutson, 1973:38.

Length. 6.4–7.8 mm.

Male. Frons usually with many black hairs.

Scutum with lateral yellow stripe extending to scutellum, narrowed behind suture in some specimens. Legs mostly yellow; fore and mid tarsi unicolorous yellow to pale brown.

Tergites 2–5 with yellow bands broad, entire. Sternite 2 with yellow hairs long, sparse, and straight. Surstylus (Fig. 265) with dorsal lobe depressed, with long hairs along ventral margin; inner lobe present, variable in shape; ventral lobes broad, only slightly asymmetrical, not constricted near apex, with very broad U-shaped apical emargination.

Female. Frons with median black stripe very broad, extending almost to antennae, truncate but not bifurcate anteriorly; yellow areas of frons with many black hairs. Tergites 2–4 with broad, slightly curved, entire yellow bands.

Distribution. Western Canada (Map 110), south to California* and Colorado.* B.C., VI, VII.

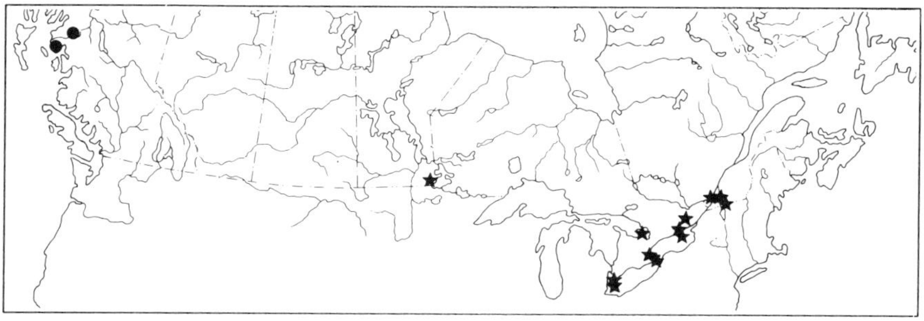

Map 110. Collection localities for *Sphaerophoria weemsi* Knutson (●) and for *Syrphus knabi* Shannon (★).

Specimens identified. Canada, 2 ♂♂; United States, 1 ♀.

Genus *Syrphus* Fabricius

Diagnosis. Species robust. Scutum shining or subshining, yellowish or grayish pruinose laterally. Tergites with well-defined yellow bands or pairs of spots. Length 7.2–13.3 mm.

Description. Eyes bare or with dense and distinct hairs. Frons of male yellow to dark gray pruinose on about upper half to two-thirds, yellow or mostly dark on about lower one-third to half; frons of female subshining to shining black on upper one-quarter to one-third, yellow-gray pruinose on about middle one-third to half, yellow or mostly dark on about lower one-fifth to one-third. Face yellow or with narrow brown median stripe.

Scutum shining to subshining, blackish green; sides of scutum and all pleura distinctly yellowish or grayish pruinose. Ventral scutellar fringe complete. Anterior anepisternum bare. Upper and lower katepisternal hair patches narrowly joined posteriorly. Meron, metapleuron, and metasternum bare. Wing entirely trichose or partly bare on basal half. Lower calypter with long erect pale hairs on posteromedian part of upper surface. Hind coxa with tuft of hairs at posteromedial apical angle (as in Fig. 46).

Abdomen oval, moderately broad, slightly convex above, distinctly margined from about middle of tergite 2 to apex of tergite 5 or 6. Tergite 2 with pair of yellow spots reaching lateral margins; tergites 3 and 4 with well-defined entire or divided bright yellow bands usually reaching lateral margins. Sternites yellow or with very variable obscure or distinct dark markings.

Distribution. Of 11 species occurring in Nearctic region, 5 Holarctic; 1 other species in Europe, 1 in eastern Siberia, 1 or perhaps several in Oriental region, and about 8 more in Neotropical region, especially in Chile.

Biology. Several of the species, probably all, are multivoltine, at least in southern Canada. *S. torvus* required 19 days from egg to adult at 21°C in the laboratory. *S. ribesii* and *S. vitripennis* overwinter as diapausing larvae in Czechoslovakia (Dušek and Láska, 1974*a*).

Larval hosts include many species of aphids on herbs, shrubs, and deciduous and coniferous trees. Both free-living and gall-forming aphids are attacked. It is likely that all species are polyphagous.

Key to Nearctic species of *Syrphus*

1. Wing with cell bm bare on anterior one-quarter or more 2

   Wing membrane entirely trichose 13

2. Setulae of ventral surface of first tarsomere of mid leg entirely yellow to orange. About lower half of frons of male and lower one-third of frons of female yellow with pair of dark spots above antennal bases in some species 3

   At least some preapical setulae of ventral surface of first tarsomere of mid leg brown to black. Lower part of frons yellow or blackish 11

3. Yellow bands of tergites 3 and 4 broadly divided medially; margins of tergites narrowly yellow throughout (Fig. 203). Boreal regions .... ***sexmaculatus* (Zetterstedt)**

Yellow bands of tergites 3 and 4 entire; margins of tergites yellow only at ends of yellow spots or bands or entirely black (Fig. 201) 4

4. First tarsomere of hind leg mostly or entirely orange above, much paler than following three tarsomeres. Scutum strongly yellow pruinose laterally. Hind femur yellow or with broad brown preapical ring. Cell c bare on about basal one-quarter. Eastern Canada .... ***knabi*** **Shannon**

First tarsomere of hind leg dark brown to black above, not paler than following three tarsomeres. Scutum not distinctly yellow pruinose laterally. Hind femur black except at apex, entirely yellow, or yellow with faint darker preapical ring. Cell c varying from almost entirely trichose to almost entirely bare. Western Canada 5

5. Male (eyes meeting on frons) 6

Female (eyes widely separated on frons) 9

6 Hind femur yellow. Cell c trichose on at most apical one-sixth. Western United States .... ***currani*** **Fluke**

Hind femur black with apex narrowly yellow. Cell c mostly trichose or trichose only on apical one-fifth 7

7. Cell c bare on at least basal four-fifths; cell bm bare at apex over almost entire width although with some microtrichia preapically; cell *cup* bare subbasally over its entire width. Frons often with pair of brown to blackish spots above antennae. Yellow bands of tergites 3 and 4 extending narrowly or broadly to lateral margins in some specimens. Southwestern United States, Mexico .... ***sonorensis*** **Vockeroth**

Cell c bare on at most basal one-fifth; cell bm trichose apically over most of its width; cell *cup* broadly trichose posteriorly over its entire length. Frons yellow above antennae. Yellow bands of tergites 3 and 4 not extending to lateral margins 8

8. Cell cua_1 narrowly bare basally and posterobasally; cell bm with at most some scattered microtrichia along posterior margin of basal half, narrowly trichose posteriorly on apical half. Western Canada .... ***intricatus*** **Vockeroth**

Cell cua_1 entirely trichose; cell bm narrowly trichose along posterior margin except on basal one-fifth in some specimens. Western Canada .... ***opinator*** **Osten Sacken** (in part)

9. Hind femur black with apex narrowly yellow. Cell c trichose on at most apical one-eighth. Southwestern United States, Mexico .... ***sonorensis*** **Vockeroth**

Hind femur yellow, at most obscurely darkened preapically. Cell c with microtrichia variable 10

10. Cell c trichose on about apical one-eighth; cell bm with at most small compact group of microtrichia near apex; cell cua_1 bare over its entire width at extreme base. Western United States .... ***currani* Fluke**

Cell c densely or moderately densely trichose on at least apical one-quarter and with scattered microtrichia on preceding half or more; cell bm with sparse but more widely distributed microtrichia near apex; cell cua_1 trichose medially at extreme base, bare only anteromedially and posteromedially. Western Canada .... ***intricatus* Vockeroth**

11. Lower half of frons of male and lower one-third of frons of female bright yellow. Yellow bands of tergites 3 and 4 not reaching lateral margins (as in Fig. 201) or reaching them only narrowly. Female with hind femur yellow. Western Canada .... ***opinator* Osten Sacken** (in part)

Frons blackish throughout. Yellow bands of tergites 2 and 3 reaching lateral margins narrowly or broadly. Female with hind femur either mostly yellow or mostly black 12

12. Hind femur black on about basal two-thirds. Widespread ... . ***vitripennis* Meigen**
.... ***rectus* Osten Sacken** (male only)

Hind femur yellow on basal half, usually partly brownish on apical half. Eastern Canada ....
.... ***rectus* Osten Sacken** (female only)

13. Eye with numerous hairs at least as long as distance between their bases (hairs much longer and more abundant in male than in female). Yellow bands of tergites 3 and 4 entire. Both sexes with hind femur black on about basal three-quarters. Widespread .... ***torvus* Osten Sacken**

Eye virtually bare, at most with some short hairs much shorter than distance between their bases. Yellow bands of tergites 3 and 4 entire or divided. Male with hind femur extensively black or entirely yellow; female with hind femur yellow on at least basal half 14

14. Lateral margins of tergites narrowly but continuously yellow (Fig. 200). Male with hind femur entirely yellow or black on up to basal half; female with hind femur yellow. Widespread, mostly boreal regions .... ***attenuatus* Hine**

Lateral margins of tergites yellow only at ends of yellow spots or bands (Fig. 202). Male with hind femur black on about basal three-quarters; female with hind femur entirely yellow or darkened on apical half. Widespread ..... ***ribesii* (Linnaeus)**

Clé des espèces néarctiques de *Syrphus*

1. Aile pourvue d'une cellule bm glabre sur au moins le quart antérieur 2

Aile ayant une membrane entièrement velue 13

2. Face ventrale du premier article du tarse de la patte médiane portant des soies entièrement jaune à orange. Environ la moitié inférieure du front chez le mâle et le tiers inférieur chez la femelle sont de couleur jaune ornée, chez certaines espèces, d'une paire de taches foncées au-dessus de la base des antennes 3

Face ventrale du premier article du tarse de la patte médiane au moins quelques soies pré-apicales brunes à noires. Partie inférieure du front jaune ou noirâtre 11

3. Bandes jaunes sur les tergites 3 et 4 largement séparées sur le plan médian; étroits lisérés jaunes sur l'ensemble des tergites (fig. 203). Régions boréales .... ***sexmaculatus* (Zetterstedt)**

Bandes jaunes complètes sur les tergites 3 et 4; lisérés jaunes, uniquement aux extrémités des taches ou bandes jaunes, ou entièrement noirs (fig. 201) 4

4. Premier article du tarse de la patte postérieure largement ou totalement orange sur le dessus et beaucoup plus pâle que les trois articles qui suivent. Scutum orné d'une pruinosité jaune sur les côtés. Fémur postérieur jaune ou souligné d'un large anneau pré-apical brun. Cellule c glabre sur environ un quart de la base. Est du Canada .... ***knabi* Shannon**

Premier article du tarse de la patte postérieure brun foncé à noir sur le dessus, pas plus pâle que les trois suivants. Scutum dépourvu de bande jaune pruineuse distincte sur les côtés. Fémur postérieur noir sauf à l'apex, entièrement jaune, ou jaune avec un léger anneau pré-apical plus foncé. Cellule c variant de presque entièrement velue à presque entièrement glabre. Ouest du Canada 5

5. Mâle (yeux contigus sur le front) 6

Femelle (yeux largement séparés sur le front) 9

6. Fémur postérieur jaune. Cellule c velue sur au plus un sixième apical. Ouest des États-Unis .... ***currani*** **Fluke**

Fémur postérieur noir, un peu jaune à l'apex. Cellule c largement velue ou velue seulement sur un cinquième apical 7

7. Cellule c glabre sur au moins les quatre cinquièmes de la partie basilaire; cellule bm glabre à l'apex sur presque toute sa largeur, bien qu'il y ait certaines microchètes dans la portion pré-apicale; cellule *cup* glabre sur toute sa largeur dans la région subbasilaire. Front souvent orné d'une paire de taches brun à noirâtre au-dessus des antennes. Bandes jaunes sur les tergites 3 et 4 qui, chez certains spécimens, s'étendent largement ou étroitement jusqu'aux bords latéraux. Sud-ouest des États-Unis, Mexique .... ***sonorensis*** **Vockeroth**

Cellule c glabre sur au plus un cinquième basilaire; cellule bm velue sur presque toute sa largeur, dans la portion apicale; cellule *cup* fortement velue à la partie arrière sur toute sa longueur. Front jaune au-dessus des antennes. Bandes jaunes sur les tergites 3 et 4 n'atteignant pas les bords latéraux 8

8. Cellule cua_1 ayant une étroite zone glabre dans les parties basilaire et postéro-basilaire; cellule bm portant tout au plus quelques microchètes dispersées le long de la marge postérieure de la moitié basilaire, étroitement velue à la partie arrière dans la portion apicale. Ouest du Canada .... ***intricatus*** **Vockeroth**

Cellule cua_1 entièrement velue; cellule bm étroitement velue le long de la marge postérieure sauf, chez certains spécimens, sur un cinquième de la base. Ouest du Canada .... ***opinator*** **Osten Sacken** (partim)

9. Fémur postérieur noir ayant un peu de jaune à l'apex. Cellule c velue sur au plus un huitième apical. Sud-ouest des États-Unis, Mexique .... ***sonorensis*** **Vockeroth**

Fémur postérieur jaune, foncé tout au plus dans la région pré-apicale. Cellule c pourvue de microchètes variables 10

10. Cellule c velue sur environ un huitième apical; cellule bm avec tout au plus un petit groupe compact de microchètes près de l'apex; cellule cua_1 glabre sur toute sa largeur à l'extrémité basilaire. Ouest des États-Unis .... ***currani*** **Fluke**

Cellule c fortement ou modérément velue, tout au moins sur un quart de l'apex, et garnie de microchètes clairsemées sur au moins la moitié précédente; cellule bm garnie de rares microchètes plus largement distribuées près de l'apex; cellule

cua_1 velue à l'extrémité basilaire de sa portion médiane, glabre seulement dans les zones antéro-médiane et postéro-médiane. Ouest du Canada .... ***intricatus*** **Vockeroth**

11. Front dont la moitié inférieure chez le mâle et le tiers inférieur chez la femelle sont de couleur jaune clair. Bandes jaunes sur les tergites 3 et 4 n'atteignant pas les bords latéraux (comme dans la fig. 201) ou ne les atteignant qu'étroitement. Fémur postérieur jaune chez la femelle. Ouest du Canada .... ***opinator*** **Osten Sacken** (partim)

Front entièrement noirâtre. Bandes jaunes sur les tergites 2 et 3 s'étendant largement ou étroitement jusqu'aux bords latéraux. Fémur postérieur de la femelle surtout jaune ou surtout noir 12

12. Fémur postérieur noir sur environ les deux tiers basilaires. Espèces répandues .... ***vitripennis*** **Meigen**
.... ***rectus*** **Osten Sacken** (mâle seulement)

Fémur postérieur jaune sur la moitié basilaire, en général partiellement brunâtre sur la moitié apicale. Est du Canada
.... ***rectus*** **Osten Sacken** (femelle seulement)

13. Oeil garni de nombreux poils, d'une longueur au moins égale à la distance entre leur base (les poils sont beaucoup plus longs et abondants chez le mâle que chez la femelle). Bandes jaunes complètes sur les tergites 3 et 4. Chez les deux sexes, fémur postérieur noir sur environ les trois quarts basilaires. Espèce répandue .... ***torvus*** **Osten Sacken**

Oeil presque nu, orné tout au plus de quelques poils beaucoup plus courts que la distance entre leur base. Bandes jaunes sur les tergites 3 et 4 complètes ou divisées. Chez le mâle, fémur postérieur largement noir ou entièrement jaune; chez la femelle, il est jaune, au moins dans la moitié basilaire ... 14

14. Tergites bordés de liserés jaunes, étroits mais continus (fig. 200). Chez le mâle, le fémur postérieur est tout jaune, ou noir au plus dans la moitié basilaire; chez la femelle, il est jaune. Espèce répandue, surtout dans les régions boréales ...
.... ***attenuatus*** **Hine**

Tergites bordés de liserés jaunes seulement aux extrémités des taches ou des bandes jaunes (fig. 202). Chez le mâle, fémur postérieur noir sur environ les trois quarts basilaires; chez la femelle, il est entièrement jaune, ou foncé dans la moitié apicale. Espèce répandue .... ***ribesii*** **(Linnaeus)**

Syrphus attenuatus Hine

Fig. 200; Map 111

Syrphus attenuatus Hine, 1922:144.
Syrphus hinei Fluke, 1933:73.

Length. 8.1–12.8 mm.

Male. Eye nearly bare. Upper half of frons dark, gray pruinose; lower half of frons dark to bright yellow, slightly darkened medially in some specimens. Face yellow.

Wing membrane entirely trichose. Fore and mid femora with basal one-tenth to two-fifths black, otherwise yellow, with at least some black hairs posteriorly; hind femur entirely yellow or with up to basal half black, with few to many black hairs anteriorly on apical half. First tarsomere of mid leg with many black setulae below.

Tergites 3 and 4 each with yellow band deeply emarginate posteriorly; or one or both tergites with narrowly or broadly separated yellow spots; bands or spots extending narrowly or broadly to tergite margins; tergites with rather broad continuous yellow margins (Fig. 200).

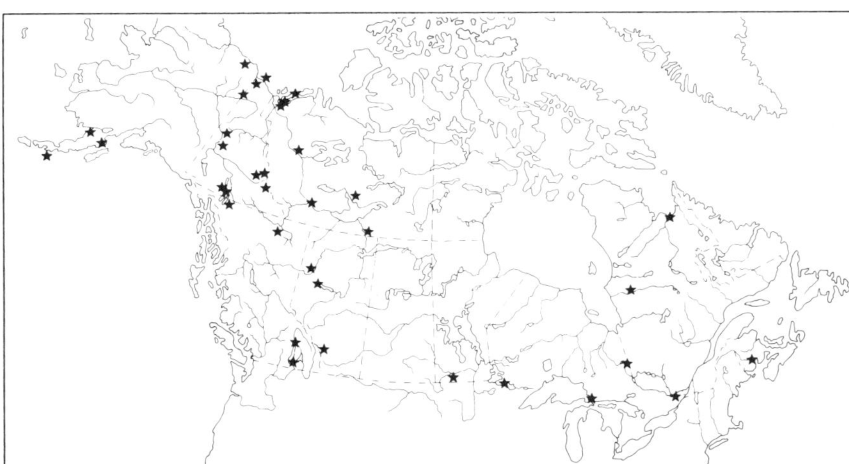

Map 111. Collection localities for *Syrphus attenuatus* Hine.

Female. Lower one-third of frons bright yellow. Femora entirely yellow; mid and hind femora with some black hairs posteriorly in some specimens; hind femur with few to many black hairs anteriorly on apical half.

Distribution. Alaska, Canada (Map 111), south to Colorado, Wisconsin,* and Pennsylvania; northern Europe. B.C., IV–VII; Que., VI, VII.

Specimens identified. Alaska, 8 ♂♂, 46 ♀♀; Canada, 27 ♂♂, 52 ♀♀; United States, 1 ♂, 1 ♀; Europe, 4 ♂♂, 5 ♀♀.

Syrphus currani Fluke

Syrphus currani Fluke, 1939:365.

Length. 11.4–12.5 mm.

Male. Frons yellow, yellow-gray pruinose on upper half, shining on lower half. Face yellow.

Wing membrane extensively bare; cells c, br, and bm almost entirely bare; cell *cup* broadly bare basally and anteriorly; cells adjoining br, bm, and *cup* with small bare areas basally or anteriorly, or both. Femora yellow, mostly yellow-haired; hind femur with some black hairs anteriorly on apical half in some specimens. First tarsomere of mid leg with only yellow setulae below.

Tergites 3 and 4 with yellow bands emarginate posteriorly, not reaching lateral margins or reaching them only very narrowly.

Female. Frons shining black on about upper one-third, black with yellow to gray pruinosity on about middle one-third, shining yellow on about lower one-third. Wing membrane as in male or slightly less trichose. Tergites 3 and 4 with yellow bands reaching lateral margins very narrowly to broadly.

Distribution. Oregon, Idaho, Nevada, Utah, Colorado.

Specimens identified. United States, 3 ♂♂, 3 ♀♀.

Syrphus knabi Shannon

Map 110

Syrphus knabi Shannon, 1916:200.

Length. 7.2–12.9 mm.

Male. Eye bare. Frons yellow and yellow pruinose on about upper two-thirds, yellow and subshining and usually with pair of brown anterior spots on about lower one-third. Face yellow. First flagellomere slightly longer and more tapered apically than that of other species.

Scutum more broadly and distinctly yellow pruinose laterally than in other species. Wing membrane extensively bare; cell c bare at base; cells br and bm almost entirely bare; cell c*u*p bare basally and anteriorly; most cells adjacent to cells br, bm, and c*u*p with small bare areas. Fore and mid femora yellow, yellow-haired; hind femur yellow with apical half mostly brownish and with few to many black hairs anteriorly. First tarsomere of mid leg with only yellow setulae below. First tarsomere of hind leg mostly or entirely orange above.

Tergites 3 and 4 with long scarcely emarginate yellow bands extending over lateral margins in nearly their full length.

Female. Frons black and sparsely pruinose on at most upper one-quarter, otherwise mostly yellow, densely yellow pruinose and commonly slightly darkened medially, anteriorly narrowly shining yellow and with pair of brown anterior spots.

Distribution. Southeastern Canada (Map 110), south to Texas and Florida. Ont., Que., V–VIII.

Specimens identified. Canada, 11 ♂♂, 19 ♀♀; United States, 23 ♂♂, 34 ♀♀.

Biology. Larvae have been found on pulverulent species of aphids in galls or rolled leaves (Heiss 1938).

Syrphus intricatus Vockeroth

Map 112

Syrphus intricatus Vockeroth, 1983:177.

Length. 8.1–10.4 mm.

Male. Eye nearly bare. Frons yellow, gray pruinose on about upper half, shining below, without dark spots above antennae. Face yellow.

Wing extensively bare on about basal half, but with cell c mostly trichose (see key). Fore and mid femora black on about basal one-third; hind femur black with up to apical one-third yellow; tibiae yellow, with hind tibia obscurely darkened toward apex; fore and mid tarsi yellow, with hind tarsus dark brown to black above; first tarsomere of mid tarsus with only yellow setulae below.

Yellow bands of tergites 3 and 4 entire, not extending to lateral margins.

Female. Frons shining yellow on about lower one-third, without dark spots above antennae. Wing slightly more extensively bare than in male; cell c in some specimens with only scattered microtrichia on basal three-quarters. Legs and abdomen as in male.

Distribution. Southwestern Canada (Map 112), south to California and Colorado. B.C., V–IX.

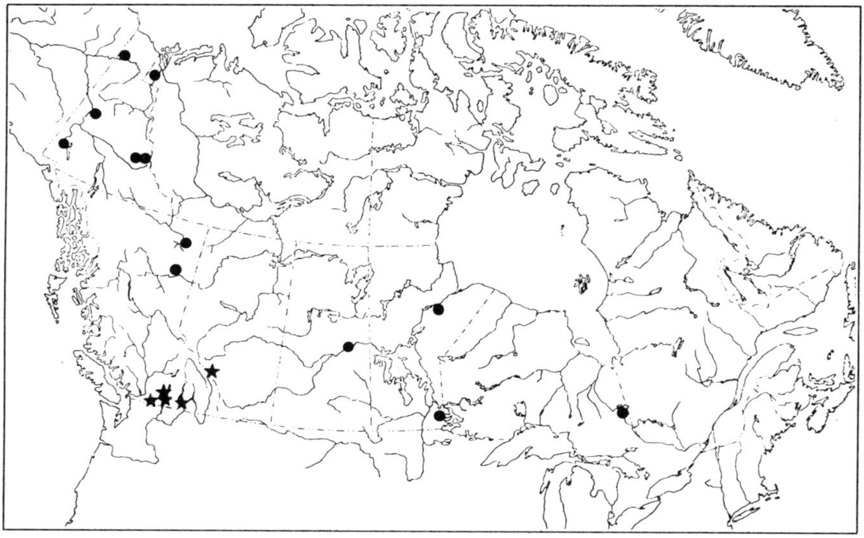

Map 112. Collection localities for *Syrphus intricatus* Vockeroth (★) and for *S. sexmaculatus* (Zetterstedt) (●).

Specimens identified. Canada, 6 ♂♂, 12 ♀♀; United States, 7 ♂♂, 10 ♀♀.

Biology. A specimen from Westbridge, B.C., was reared from a larva or pupa collected on *Pinus ponderosa*.

Discussion. This species can be readily distinguished from the other two western species with yellow setulae below the first tarsomere of the mid leg by the much more extensively trichose cell c (see key). However, occasional males of *S. opinator*, in which these setulae are usually partly or entirely black, apparently have only yellow setulae. *S. intricatus* and *S. opinator* may be a single variable species. However, two species may be present, because specimens from the only locality (Robson, B.C.) from which a series of both forms is available can be readily distinguished: *S. opinator* with black setulae and cell cua_1 entirely trichose, *S. intricatus* with yellow setulae and cell cua_1 partly bare basally. Setulae are variable in color in one species. The distribution of microtrichia in cell cua_1 is constant within each species but differs in the two species.

Syrphus opinator Osten Sacken

Fig. 201; Map 113

Syrphus opinator Osten Sacken, 1877:327.

Length. 8.8–11.7 mm.

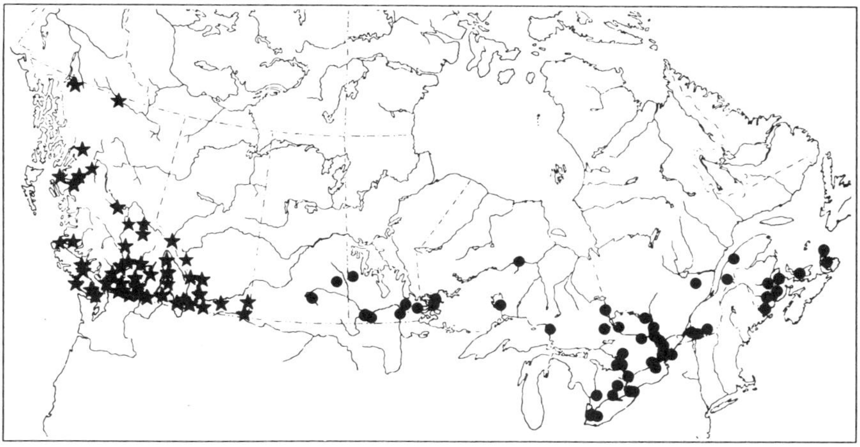

Map 113. Collection localities for *Syrphus opinator* Osten Sacken (★) and for *S. rectus* Osten Sacken (●).

Male. Frons entirely yellowish, gray pruinose on upper half or less, subshining yellow below. Face yellow.

Wing membrane with bare areas as follows: most of cell r in front of spurious vein, about anterior half of cell bm, narrow stripe behind CuP. Fore and mid femora with basal one-third black, apical two-thirds yellow, with yellow hairs posteriorly; hind femur mostly black, apical one-third to one-fifth yellow, with at most some black hairs anteroventrally on apical half. First tarsomere of mid leg usually with many black setulae below, rarely with only yellow setulae.

Tergites 3 and 4 each with entire yellow band slightly and broadly emarginate posteriorly and usually well-separated from lateral margins but reaching them anterolaterally in some specimens (Fig. 201).

Female. Front subshining black on about upper one-third, densely yellow-gray pruinose on middle one-third or slightly more, subshining yellow on lower one-third. Wing with cell r mostly bare behind spurious vein. Hind femur yellow, with few to many black hairs anteriorly on apical half. First tarsomere of mid leg with at least some black setulae below.

Distribution. Western Canada (Map 113), south to California, New Mexico, Texas,* and Minnesota. B.C., V–XI.

Specimens identified. Canada, 227 ♂♂, 327 ♀♀; United States, 192 ♂♂, 164 ♀♀.

Syrphus rectus Osten Sacken

Map 113

Syrphus rectus Osten Sacken, 1875*a*:140.
Syrphus transversalis Curran, 1921*a*:155.

Length. 7.2–10.3 mm.

Male. Not definitely separable from male of *S. vitripennis* but apparently with bare areas of wing slightly more extensive (cell bm commonly only very narrowly trichose posteriorly on basal half) and with yellow bands of tergites 3 and 4 commonly reaching margins over half or more of their length.

Female. Similar to female of *S. vitripennis*, usually differing as male in wing microtrichia and tergite markings. Fore and mid femora entirely yellow or obscurely brownish at base; hind femur entirely yellow or with obscure or distinct brown ring on apical half. First tarsomere of mid leg with many black setulae below.

Distribution. Central and eastern Canada (Map 113), south to Colorado,* Texas, Mississippi, and North Carolina. Ont., Que., V–X.

Specimens identified. Canada, 191 ♀♀; United States, 73 ♀♀.

Biology. Larvae have been found on many species of aphids (Heiss 1938).

Syrphus ribesii (Linnaeus)

Figs. 1, 202; Map 114

Musca ribesii Linnaeus, 1758:593.
Syrphus philadelphicus Macquart, 1842:153.
Syrphus ribesii var. *vittafrons* Shannon, 1916:202.
Syrphus similis Jones, 1917:224 (preoccupied Blanchard, 1852).
Syrphus bigelowi Curran, 1924*b*:288.
Syrphus ribesii var. *jonesi* Fluke, 1949:41 (new name for *S. similis* Jones, 1917).
Syrphus autumnalis Fluke, 1954:3.

Length. 8.1–13.3 mm.

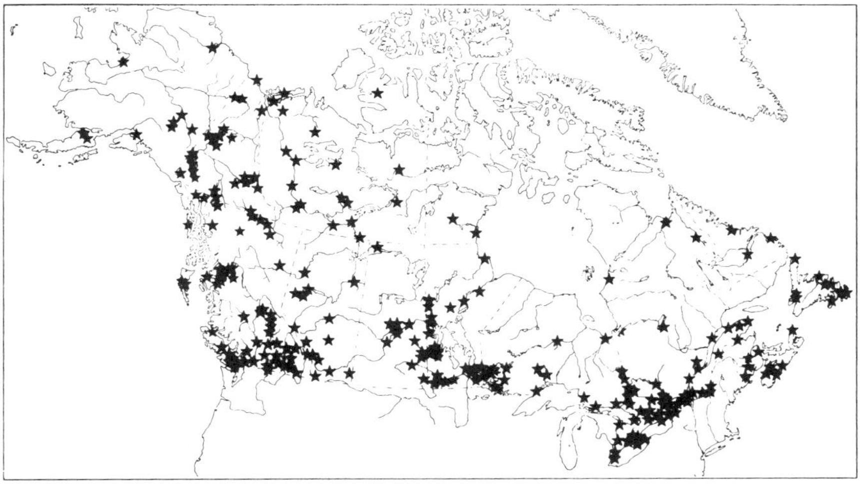

Map 114. Collection localities for *Syrphus ribesii* (Linnaeus).

Male. (Fig. 1). Eye nearly bare. Frons usually dark throughout but yellow on about lower one-fifth in some specimens. Face yellow or with obscure to distinct median brown stripe.

Wing membrane entirely trichose. Fore and mid femora black on about basal one-third, otherwise yellow, usually with many black hairs posteriorly; hind femur black on about basal two-thirds, otherwise yellow, with few to many black hairs anteriorly. First tarsomere of mid leg with many black setulae below.

Tergites 3 and 4 each with yellow band of rather variable length extending narrowly or broadly to lateral margins, bands usually entire but narrowly divided medially in some specimens (Fig. 202).

Female. Mid and hind femora black only at extreme base, usually with some black hairs posteriorly; hind femur usually entirely yellow, darkened preapically in some specimens, with many black hairs on apical one-third or more of anterior surface.

Distribution. Alaska, Canada (Map 114), south to California, Mexico (Chiapas), Central America,* and Mississippi; Europe; Asia. B.C., V–X; Ont., Que., IV–IX.

Specimens identified. Alaska, 13 ♂♂, 113 ♀♀; Canada, 606 ♂♂, 1373 ♀♀; United States, 42 ♂♂, 94 ♀♀; Mexico, 2 ♂♂; Europe, 58 ♂♂, 76 ♀♀; Siberia, 1 ♂.

Biology. Larvae have been found on several species of *Macrosiphum* and *Aphis* (Heiss 1938); also on *Pemphigus populicaulis*, *Cinara hottesi*, *Cinara carolina*, *Cinara lasiocarpae* (records from CNC). Láska and Starý (1980) listed about 16 species of aphid hosts from Czechoslovakia.

Discussion. *Syrphus ribesii* shows more variation in color of frons, antenna, and face and in extent and shape of abdominal markings than do other Nearctic species. Fluke (1954) treated as five species specimens that I consider as belonging to *S. ribesii*.

Syrphus sexmaculatus (Zetterstedt)

Fig. 203; Map 112

Scaeva sexmaculatus Zetterstedt 1838:603.

Length. 8.9–11.0 mm.

Male. Eye bare. Lower half to two-thirds of frons shining yellow. Face yellow.

Wing membrane with very restricted bare areas in front of spurious vein and behind CuP, with about anterior half of cell bm bare on middle half of its length. Fore and mid femora brown on about basal one-quarter, otherwise yellow, with yellow hairs posteriorly; hind femur yellow or with basal two-thirds brownish yellow or brown, with at most very few black hairs anteroventrally near apex. First tarsomere of mid leg with only yellow setulae below.

Tergites 3 and 4 each with pair of well-separated yellow spots narrowed laterally and not reaching margins; margins of tergites continuously but narrowly yellow (Fig. 203).

Female. Frons with upper half shining black, middle one-third yellow gray pruinose, lower one-third shining yellow. Wing with cell bm bare except narrow posterior margin and broad apex. Femora entirely yellow; hind femur anteriorly with only yellow hairs or with few to many black hairs on apical half. Spots of tergites 3 and 4 reaching tergite margins anterolaterally in some specimens.

Distribution. Alaska, Canada (Map 112); northern Europe. Y.T., VII, VIII; Ont., VII.

Specimens identified. Alaska, 1 ♀; Canada, 10 ♀♀; Europe, 8 ♂♂, 6 ♀♀.

Syrphus sonorensis Vockeroth

Syrphus sonorensis Vockeroth, 1983:178.

Length. 9.1–11.3 mm.

Male. Eye nearly bare. Frons yellow, gray pruinose on about upper half, shining below, commonly with pair of small to large brown to blackish spots above antennae. Face yellow.

Scutum not distinctly yellow pruinose laterally. Wing extensively bare on about basal half (see key). Fore and mid femora black on about basal one-third; hind femur black with apex narrowly yellow. Tibiae yellow, with hind tibia slightly darkened on about apical half. Fore and mid tarsi yellow; hind tarsus dark brown to black above. First tarsomere of mid leg with only yellow setulae below.

Yellow bands of tergites 3 and 4 entire, not extending to lateral margins or reaching them very narrowly or over about half their length.

Female. Frons shining yellow on about lower one-third; spots above antennae, if present, usually small and brown, rarely blackish. Wing slightly more extensively bare than in male (see “Key to Nearctic species of Syrphus”). Legs and abdomen as in male.

Distribution. Arizona, New Mexico, Mexico (Durango and Tabasco). Ariz., III, VI–IX, XI.

Specimens identified. United States, 9 ♂♂, 20 ♀♀; Mexico, 11 ♂♂, 35 ♀♀.

Discussion. This species differs from other Nearctic species with yellow setulae below the first tarsomere of the mid leg in having the hind femur of both sexes almost entirely black. The distribution apparently does not overlap that of any other of these species. Other Nearctic species with the hind femur mostly black in both sexes either have the eye haired (*S. torvus* Osten Sacken) or cell c and posterior half of cell bm mostly trichose (*S. vitripennis* Meigen).

Syrphus torvus Osten Sacken

Map 115

Syrphus torvus Osten Sacken, 1875*a*:139.

Length. 8.6–13.3 mm.

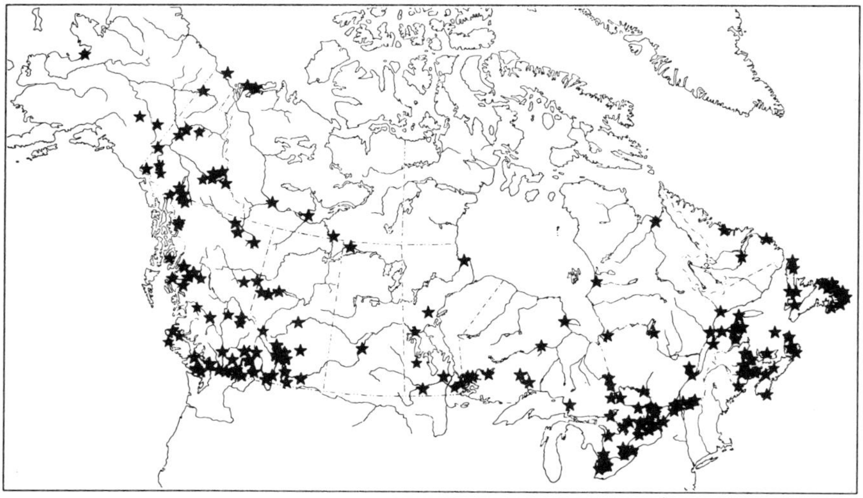

Map 115. Collection localities for *Syrphus torvus* Osten Sacken.

Male. Eye densely haired; hairs much longer than distance between their bases. Frons dark throughout. Face yellow.

Wing membrane entirely trichose. Fore and mid femora black on about basal one-third, otherwise yellow, with scattered black hairs on most of posterior surface; hind femur black on about basal two-thirds, otherwise yellow, with mostly black hairs anteriorly on apical one-thirds. First tarsomere of mid leg with many black setulae below.

Tergites 3 and 4 each with entire yellow band reaching lateral margin on about half its length.

Female. Eye with hairs less dense and shorter than in male but still very distinct.

Distribution. Alaska, Canada (Map 115), south to California, Louisiana, and Mississippi; Europe; Asia. B.C., IV–XI; Ont., Que., IV–IX.

Specimens identified. Alaska, 1 ♂, 23 ♀♀; Canada, 757 ♂♂, 696 ♀♀; United States, 97 ♂♂, 192 ♀♀; Europe, 19 ♂♂, 23 ♀♀; Siberia, 1 ♂; Nepal, 12 ♂♂.

Biology. Larvae have been found on several species of aphids (Heiss 1938), on *Cinara carolina*, and on *Euceraphis* sp. on *Alnus rubra* (records from CNC.). Láska and Starý (1980) listed five species of aphid hosts from Czechoslovakia.

Syrphus vitripennis Meigen

Map 116

Syrphus vitripennis Meigen, 1822:308.

Length. 7.6–11.4 mm.

Male. Frons dark, densely yellow-gray pruinose on upper half, mostly subshining black on lower half, with restricted yellow areas immediately above antenna. Face yellow.

Wing membrane with bare areas similar to those of *S. opinator* but slightly less extensive. Fore and mid femora black on basal one-third, otherwise yellow, with yellow hairs on posterior surface; hind femur brown to black with apical one-quarter yellow, with at most some scattered black hairs on apical one-quarter of anterior surface. First tarsomere of mid leg with black setulae below at least near apex.

Tergites 3 and 4 with yellow bands entire, broadly and shallowly emarginate posteriorly, reaching lateral margins on about half their length.

Female. Upper one-third of frons shining black, most of lower two-thirds densely yellow-gray pruinose with black and yellow areas on lower part of frons as in male.

Distribution. Alaska, Canada (Map 116), south to California,* Colorado, Massachusetts, and North Carolina;* Europe; Asia. B.C., V–X; Ont., Que., V–IX.

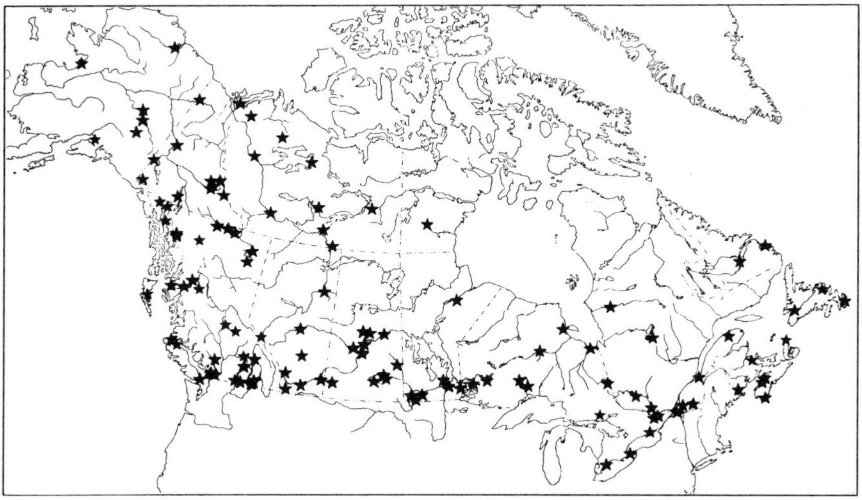

Map 116. Collection localities for *Syrphus vitripennis* Meigen.

Specimens identified. Alaska, 15 ♀♀; Canada, 370 ♀♀; United States, 23 ♀♀; Europe, 60 ♂♂, 68 ♀♀; Siberia, 1 ♂.

Biology. Larvae have been found on *Euceraphis* sp. on *Alnus rubra* (record from CNC.). Láska and Starý (1980) listed about 29 species of aphid hosts from Czechoslovakia.

Discussion. The description of the male is based on European specimens, because males cannot be definitely distinguished from those of *S. rectus*. The distribution of *S. vitripennis* is much more extensive than that of *S. rectus* (Maps 113, 116).

Genus *Toxomerus* Macquart

Diagnosis. Small species with bright yellow thoracic markings and complex yellow and black abdominal patterns. Length 4.9–9.2 mm.

Description. Eye bare, posteriorly with distinct emargination above level of antennal insertion (Fig. 17). Frons yellow in male, with broad black median stripe in female. Face yellow, with large tubercle beginning immediately below insertion of antennae (Fig. 17). Antenna yellow in male, with first flagellomere brownish above in female.

Scutum dark, with broad bright to dull yellow lateral stripe, with disc slightly to strongly pruinose, obscurely or distinctly striped. Scutellum entirely yellow or with disc black. Ventral scutellar fringe sparse but complete. Pleura black, subshining to pruinose, with at least posterior part of anepimeron, large rounded spot on upper part of katepimeron, and all of metapleuron bright yellow. Anterior anepisternum with erect hairs on much of surface in male, with hairs shorter and fewer in number in female. Upper and lower katepisternal hair patches broadly separated. Metanepisternum with many short erect hairs. Meron and metasternum bare. Wing membrane with moderate bare areas on basal one-quarter to one-third. Hind coxa without posteromedial apical hair tuft. Legs simple or with hind femur and tibia of male swollen and arcuate and femur with strong ventrolateral subbasal process (Figs. 100, 101).

Abdomen flattened, with weak margin on tergites 2–4 or 5. Tergites black with extensive and complex yellow markings; at least tergites 3 and 4 with yellow markings extending broadly to anterior margins of tergite. Male terminalia large, with sclerotized haired triangular process arising from fused surstylar apodemes and projecting posteriorly between bases of surstyli (Vockeroth and Thompson, 1987, fig. 102).

Distribution. Genus restricted to New World; 4 species occurring in Canada, 13 more in United States from Illinois and Virginia southward, and about 140 in Neotropical region.

Biology. Confusion exists concerning the larval food of *Toxomerus* spp. Heiss (1938) and earlier authors have reliably reported that larvae of *T. politus* (Say) feed on corn plants (*Zea mays*), apparently both on pollen and leaf tissue. They pupate, head downward, on various parts of the plant. *Toxomerus marginatus* (Say) is reported to feed on both pea aphids and the bloom of pea leaves (Fluke 1929), but these records, as well as the statement by Folsom (1909) that larvae of the two species, as well as those of *T. geminatus* (Say), are predaceous, should be verified. I have reared, in the laboratory, several specimens of *T. occidentalis* Curran from eggs to adults on Chinese cabbage infested with aphids and without pollen, but I observed no larval feeding. Many Neotropical species, if predaceous, almost certainly have hosts other than aphids, as very few species of that group are native to the Neotropical region.

Key to Canadian species of *Toxomerus*

1. Margin of abdomen narrowly but continuously yellow (Fig. 206). Widespread .... ***marginatus* (Say)**

   Margin of abdomen alternately yellow and black 2

2. Scutellum yellow. Male with hind femur and tibia slender, not arcuate; hind femur without ventral process. Female with abdomen broadly rounded apically. Eastern Canada .... ***politus* (Say)**

   Scutellum black with narrow yellow margin. Male with hind femur swollen and arcuate and with ventrolateral subbasal process, with hind tibia arcuate and with strongly broadened apex (Figs. 100, 101). Female with abdomen subacute apically (Fig. 205) 3

3. Occurring from Manitoba east to New Brunswick. Ventrolateral process of hind femur of male with protruding apex and with long slender horizontal spur (Fig. 100) .... ***geminatus* (Say)**

   Occurring in British Columbia. Ventrolateral process of hind femur of male without protruding apex and with at most short stout horizontal spur (Fig. 101) ..... ***occidentalis* Curran**

Clé des espèces canadiennes de *Toxomerus*

1. Bord de l'abdomen jaune, étroit mais continu (fig. 206). Espèce répandue .... ***marginatus* (Say)**

   Bord de l'abdomen, jaune et noir en alternance 2

2. Scutellum jaune. Chez le mâle, fémur et tibia postérieurs minces et non arqués; pas d'appendice ventral sur le fémur postérieur. Portion apicale de l'abdomen nettement arrondie chez la femelle. Est du Canada .... ***politus* (Say)**

   Scutellum noir avec un étroit liséré jaune. Chez le mâle, fémur postérieur renflé, arqué et pourvu d'un appendice subbasilaire ventro-latéral; tibia postérieur arqué et fortement élargi à l'apex (fig. 100, 101). Portion apicale de l'abdomen subaiguë chez la femelle 3

3. De l'est du Manitoba au Nouveau-Brunswick. Chez le mâle, appendice ventro-latéral du fémur postérieur ayant un apex en saillie et pourvu d'un long et mince éperon horizontal (fig. 100) .... ***geminatus* (Say)**

En Colombie-Britannique. Chez le mâle, appendice ventro-latéral du fémur postérieur sans apex en saillie, pourvu tout au plus d'un éperon horizontal court et épais (fig. 101) .... ***occidentalis* Curran**

Toxomerus geminatus (Say)

Figs. 17, 100, 204, 205; Map 117

Scaeva geminata Say, 1823:92.
Eumerus privernus Walker, 1852:225.
Syrphus interrogans Walker, 1852:238.
Toxomerus notatus Macquart, 1855:113.

Length. 6.1–7.6 mm.

Male. Disc of scutum slightly brownish pruinose, with narrow but distinct blue-gray median stripe and faint gray submedian stripe. Propleuron and anepimeron entirely black. Scutellum with disc black; margin broadly yellow. Fore and mid legs with coxae mostly or entirely black; femora rarely brownish above; fore and mid legs otherwise yellow; hind coxa yellow; femur mostly black with basal two-fifths and narrow apex yellow; tibia yellow with subapical and also subbasal brown ring in some specimens; tarsus dark brown above. Hind femur (Fig. 100) strongly swollen, arcuate, with subbasal posteroventral process rather variable in shape but always with protruding apex and slender horizontal spur, and with short black setulae on middle half of posteroventral surface; hind tibia arcuate, stout, apically strongly broadened and excavated below.

Abdomen (Fig. 204) with black markings in some specimens slightly more or less extensive than shown but always reaching lateral margins on tergites 2–4. Sternite 4 slightly longer than wide, narrowed, bluntly rounded and slightly swollen anteriorly, broadly and shallowly emarginate posteriorly; sternite 8 with large shining black area.

Female. Hairs of anterior anepisternum extremely short, scarcely discernible. Hind femur very slightly arcuate, yellow with preapical dark brown ring incomplete below in some specimens; hind tibia straight, with faint preapical pale brown ring; hind tarsus pale brown above. Abdomen (Fig. 205) strongly tapering apically. Tergites 6 and 8 well-exposed. Sternites flat.

Distribution. Eastern Canada (Map 117), south to Texas and Florida. Ont., Que., V–X.

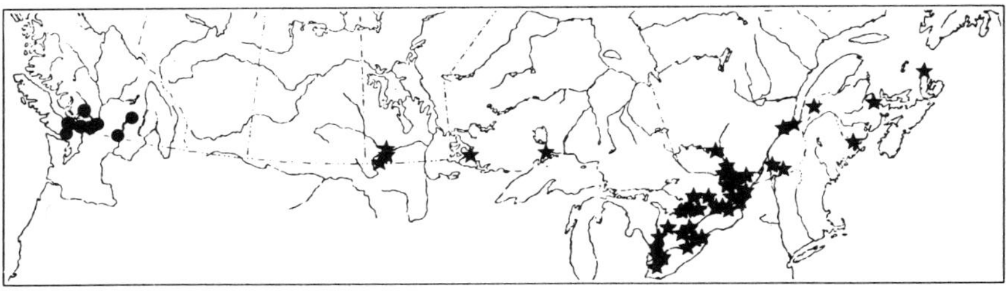

Map 117. Collection localities for *Toxomerus geminatus* (Say) (★) and for *Toxomerus occidentalis* Curran (●).

Specimens identified. Canada, 310 ♂♂, 134 ♀♀; United States, 309 ♂♂, 289 ♀♀.

Biology. Described for genus.

Toxomerus marginatus (Say)

Fig. 206; Map 118

Scaeva marginata Say, 1823:92.
Syrphus limbiventris Thomson, 1869:495.

Length. 4.9–5.7 mm.

Male. Disc of scutum with rather dense olivaceous pruinosity and with very narrow gray median stripe. Scutellum yellow, with disc darkened in some specimens. Propleuron with small but distinct yellow spot above fore coxa; anepimeron black. Legs yellow, simple; hind tarsus brownish above.

Abdomen (Fig. 206) with lateral margin entirely narrowly yellow; dark markings in some specimens more or less extensive than shown. In extreme cases tergites 3–5 with only obscure brownish markings. Sternite 4 wider than long, rectangular, flat; sternite 8 yellow.

Female. Anterior anepisternum with short but distinct hairs. Abdomen scarcely tapering. Tergites 6 and 8 usually retracted or only slightly exposed.

Distribution. Canada (Map 118), south to Peru;* Bermuda; Hawaii. B.C., V–VIII; Ont., Que., V–IX, XI.

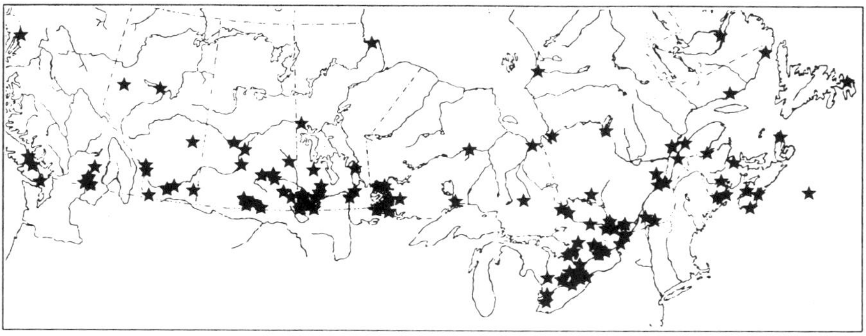

Map 118. Collection localities for *Toxomerus marginatus* (Say).

Specimens identified. Canada, 390 ♂♂, 484 ♀♀, United States, 309 ♂♂, 289 ♀♀; Bermuda, 6 ♂♂; Hawaii, 5 ♂♂, 1 ♀; Mexico, 44 ♂♂, 56 ♀♀.

Biology. Described for genus.

Toxomerus occidentalis Curran

Fig. 101; Map 117

Toxomerus occidentalis Curran, 1922*a*:258.

Length. 6.2–7.8 mm.

Male. Differing from male of *T. geminatus* only in having apex of ventrolateral hind femoral process truncate instead of protruding, with spur short and thick or very rarely absent (Fig. 101). Yellow margins of scutum and scutellum narrower in some specimens than in *T. geminatus*. Fore and mid femora extensively brown in some specimens. Hind femur almost entirely black.

Black areas of abdomen more extensive in some specimens.

Female. Not definitely distinguishable from female of *T. geminatus*. Hind leg commonly more extensively darkened. Margin of abdomen commonly narrowly entirely black from middle of tergite 2 to end of tergite 4.

Distribution. British Columbia (Map 117), south to California, Utah, Colorado,* and Texas.* B.C., V–X.

Specimens identified. Canada, 72 ♂♂, 48 ♀♀; United States, 132 ♂♂, 52 ♀♀.

Biology. Described for genus.

Discussion. This form may be only a variant of *T. geminatus*; but the apparent lack of either form over a wide area from Saskatchewan and Alberta southward at least to Texas, and the fact that males of the two populations can be readily separated, indicate that the two forms may be specifically distinct.

Toxomerus politus (Say)

Fig. 207; Map 119

Scaeva polita Say, 1823:88.
Syrphus cingulatulus Macquart, 1850:459.
Syrphus hecticus Jaennicke, 1867:398.

Length. 7.0–9.0 mm.

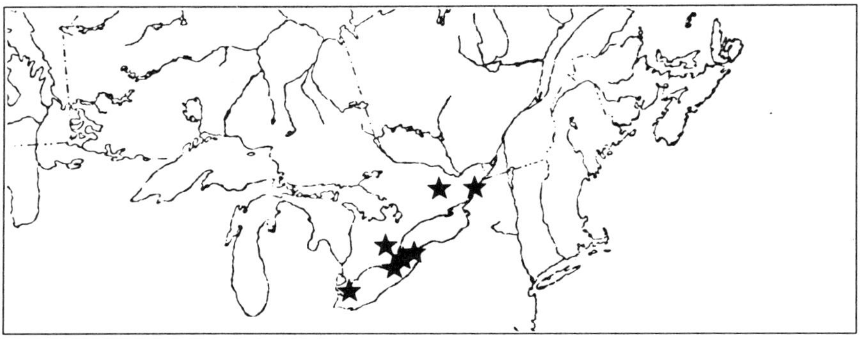

Map 119. Collection localities for *Toxomerus politus* (Say).

Male. Scutum with dense olivaceous pruinosity so lateral stripe dull yellow; disc with distinct blue-gray median stripe and with obscure brownish lateral stripe. Scutellum dull yellow. Pleura pruinose; propleuron with large yellow spot above fore coxa; anepimeron yellow on about upper half. Legs simple, yellow. Hind tarsus brown above towards apex.

Abdomen (Fig. 207) with dark markings black or more commonly brown, much reduced in some specimens. Dark posterior bands of tergites 3 and 4 always reaching lateral margins; submedian stripes of these tergites always extending laterally at their anterior ends. Sternite 4 wider than long, rectangular; flat; sternite 8 usually yellow, extensively darkened in some specimens.

Female. Anterior anepisternum with short but distinct hairs. Abdomen scarcely tapering. Tergites 6 and 8 usually well-exposed.

Distribution. Southeastern Canada (Map 119), south to Argentina (Jujuy) and Brazil (Santa Catarina). Ont., Que., VIII, IX.

Specimens identified. Canada, 20 ♂♂, 8 ♀♀; United States, 14 ♂♂, 18 ♀♀; Neotropical region, 74 ♂♂, 81 ♀♀.

Biology. Described for genus. Larvae are known to feed on pollen and leaf tissues of corn (*Zea mays*); it is unlikely that they take animal food as well.

Genus *Xanthogramma* Schiner

Diagnosis. Species rather slender with bright yellow scutal margins, pleural markings, and abdominal bands. Abdomen flattened above and nearly parallel sided. Length 7.3–12.3 mm.

Description. Eye bare. Face bright yellow. Frons long. Antenna inserted at about middle of head height. Face moderately retreating and slightly narrowed below.

Scutum black with broad well-defined bright yellow lateral margins. Scutellum with basal half black, apical half bright yellow. Ventral scutellar fringe absent. Pleura black, shining, with bright yellow markings. Upper and lower katepisternal hair patches widely separated. Anterior anepisternum, meron, metapleuron, and metasternum bare. Wing with bare areas near base of cells br, r_1, bm, and cu*p*. Hind coxa without posteromedial apical hair tuft.

Abdomen (Fig. 208) rather flat, nearly parallel sided, with strong margin from middle of tergite 2 to end of tergite 5. Tergite 1 black; tergite 2 with pair of yellow spots; tergites 3 and 4 each with entire subbasal bright yellow band and with undivided submetallic or reddish subapical band in some specimens; tergite 5 yellow anterolaterally and apically. Sternites black; sternites 2–4 bright yellow anteriorly.

Distribution. One Nearctic species, several Palaearctic and Oriental species.

Biology. Vimmer (1933) recorded the pupa of an unnamed European species from an ant nest. Dixon (1960) recorded the larva of the European *X. pedissequum* (Harris) from a nest of *Lasius niger* containing *Trama* sp. Heiss (1938) suggested that the larva of *X. flavipes* is probably saprophytic.

Xanthogramma flavipes (Loew)

Fig. 208; Map 96

Doros flavipes Loew, 1863*b*:318.

Length. 7.3–12.3 mm.

Male and female. Characters as given for genus. Abdomen as in Fig. 208.

Distribution. Eastern Canada (Map 96), south to Nebraska,* Illinois, and North Carolina. Ont., Que., V–VIII.

Specimens identified. Canada, 54 ♂♂, 41 ♀♀; United States, 13 ♂♂, 38 ♀♀.

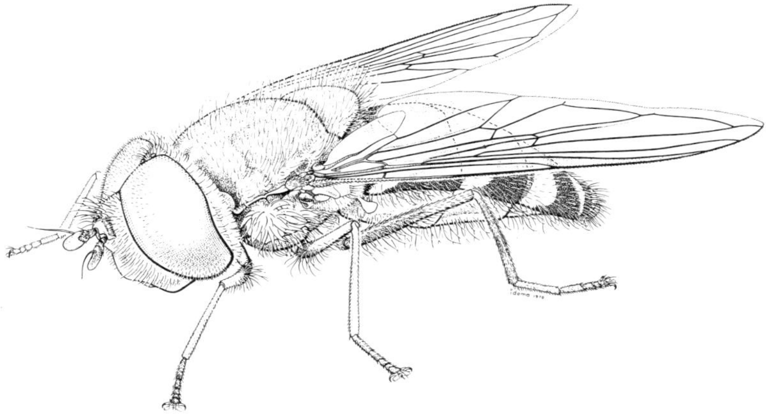

1 Syrphus ribesii ♂

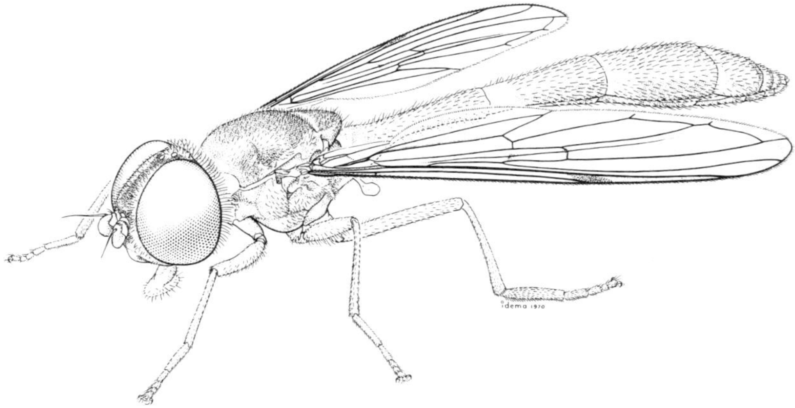

2 Baccha elongata ♀

Figs. 1, 2. Whole specimens.

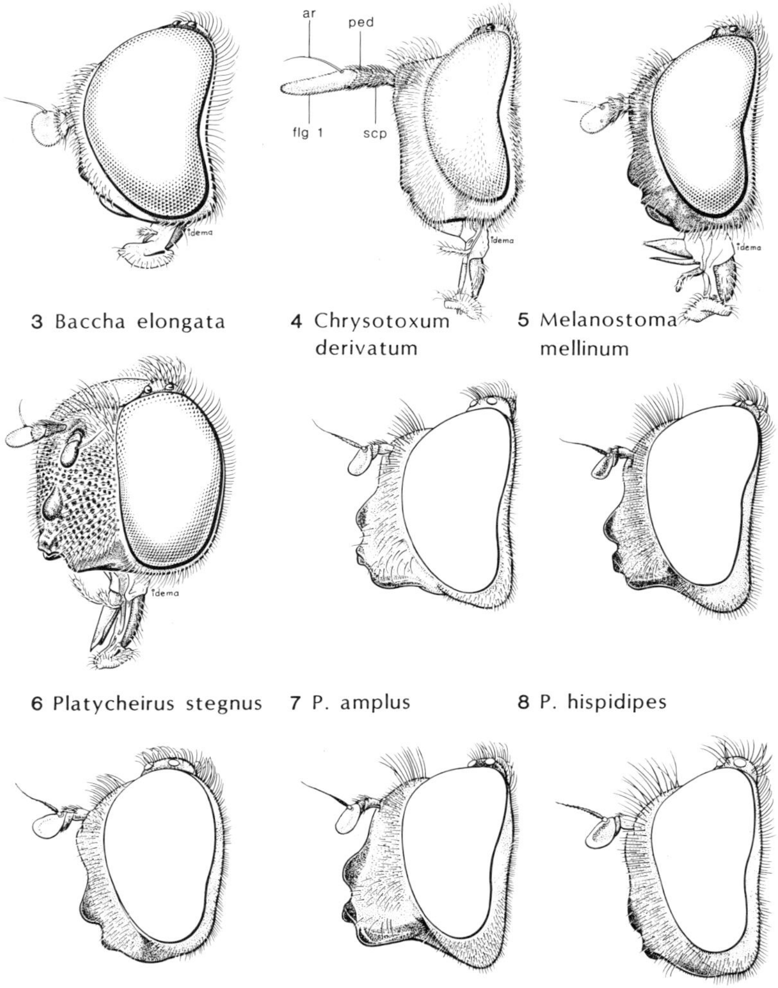

Figs. 3–11. Heads of males. (Abbreviations: ar, arista; flg, flagellomere; ped, pedicel; sc, scape.)

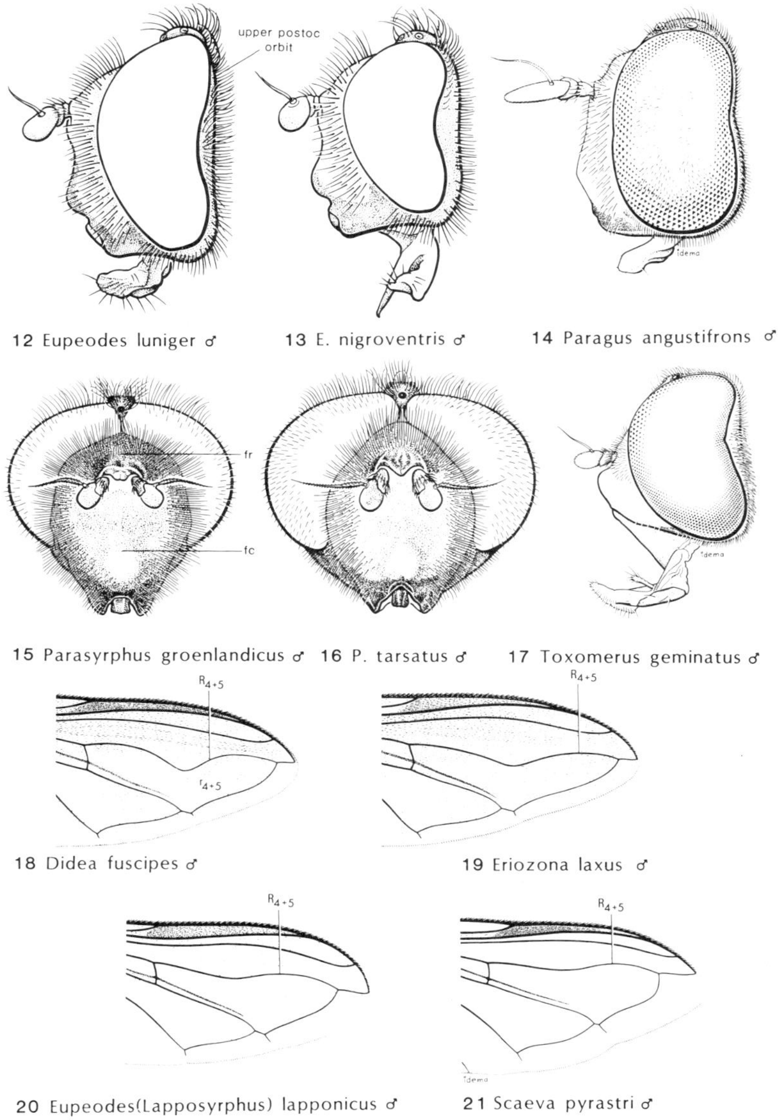

Figs. 12–21. Heads and distal parts of wing, showing distribution of microtrichia. (Abbreviations: fc, face; fr, frons.)

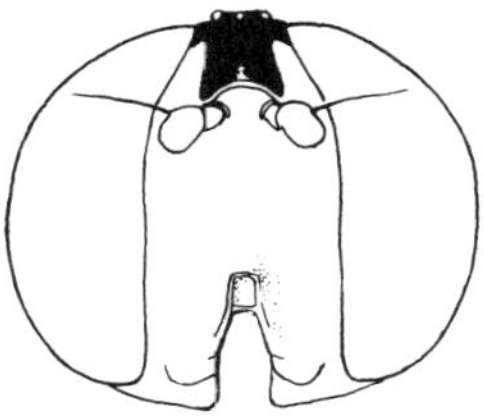

22 Sphaerophoria bifurcata ♀

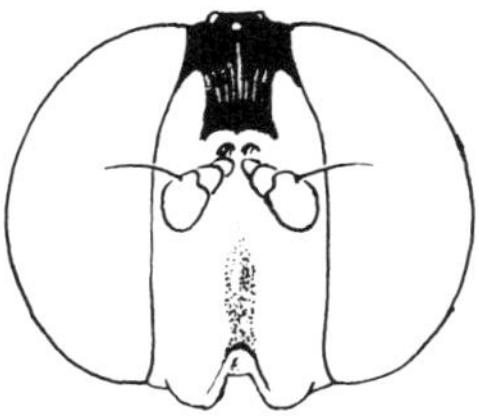

23 S. contigua ♀

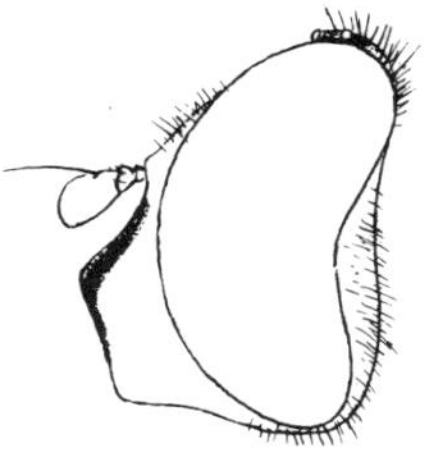

24 S. novaeangliae ♂

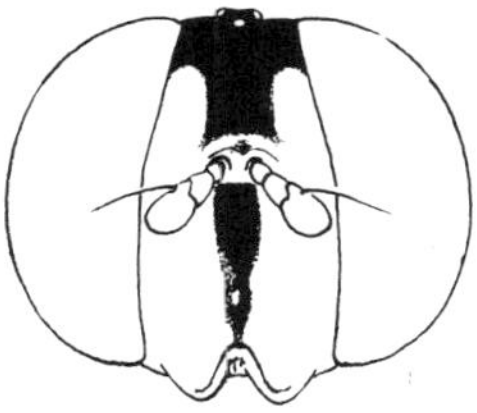

25 S. novaeangliae ♀

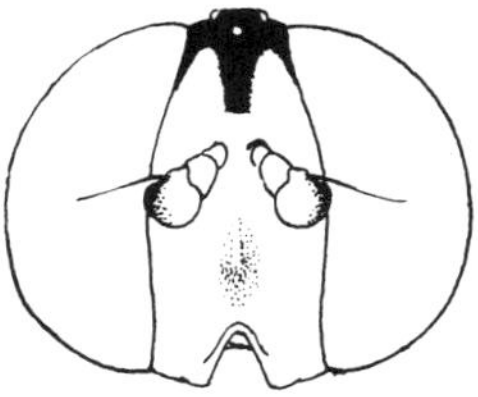

26 S. pyrrhina ♀

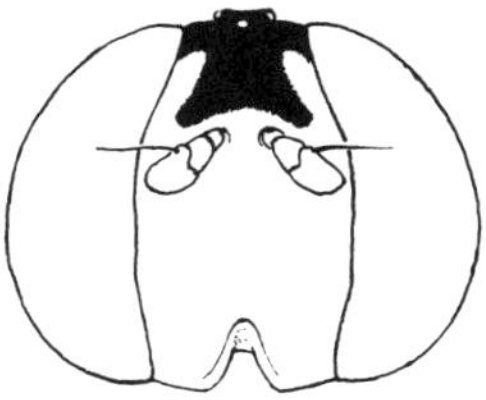

27 S. sulphuripes ♀

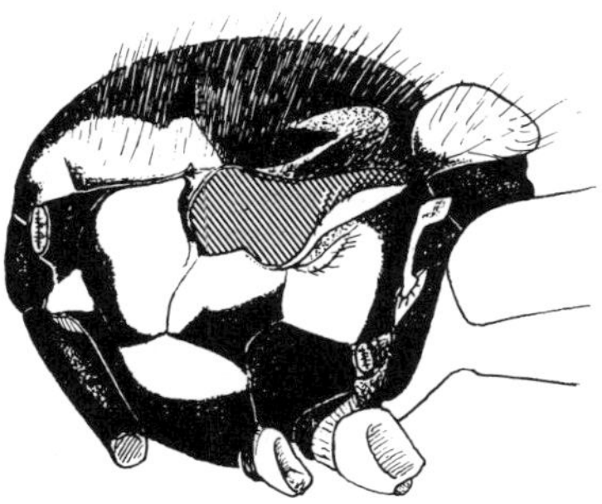

28 S. contigua

29 S. sulphuripes

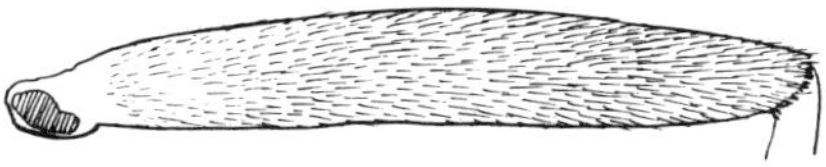

30 S. philanthus ♂

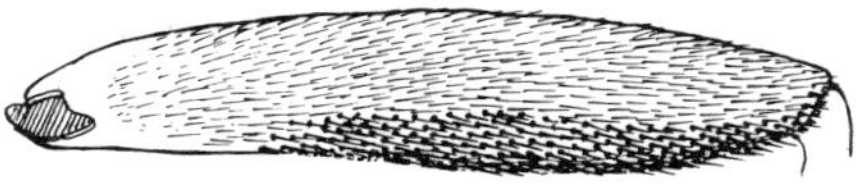

31 S. scripta ♂

Figs. 22–31. Heads, thoraces, and posterior views of hind femora.

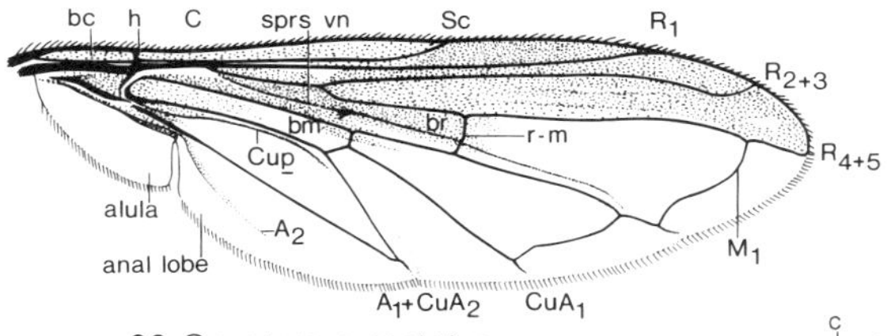

32 Ocyptamus costatus

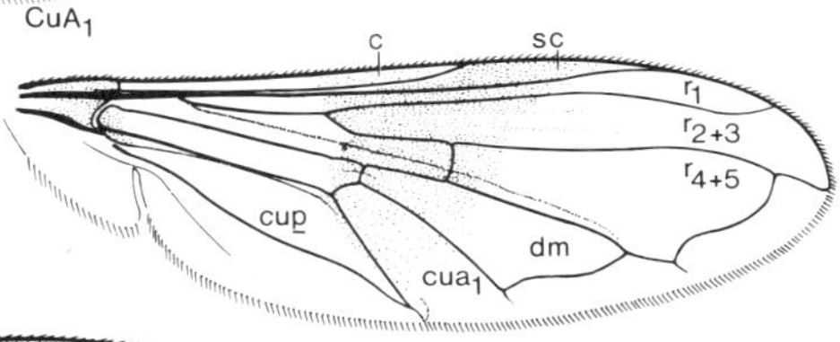

33 O. fascipennis

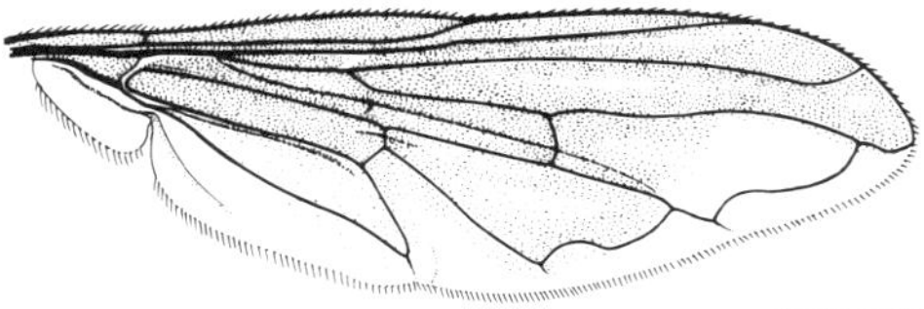

34 O. fuscipennis

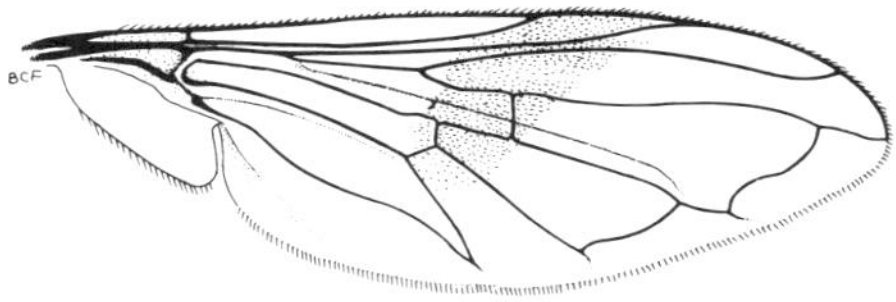

35 O. lemur

Figs. 32–35. Wings.

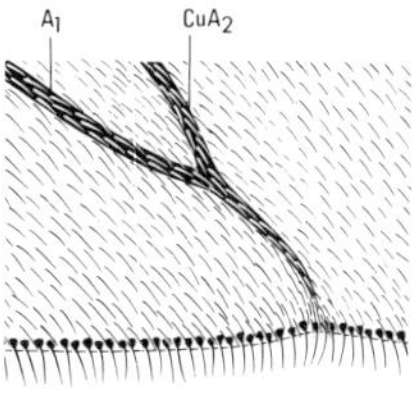

36 Meliscaeva cinctella ♂

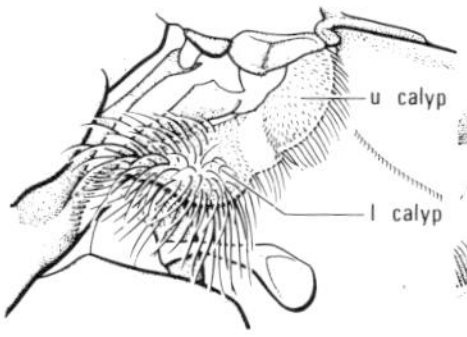

37 Syrphus ribesii ♂

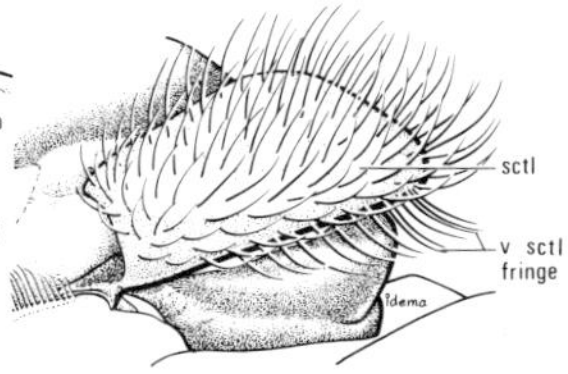

38 Allograpta obliqua ♂

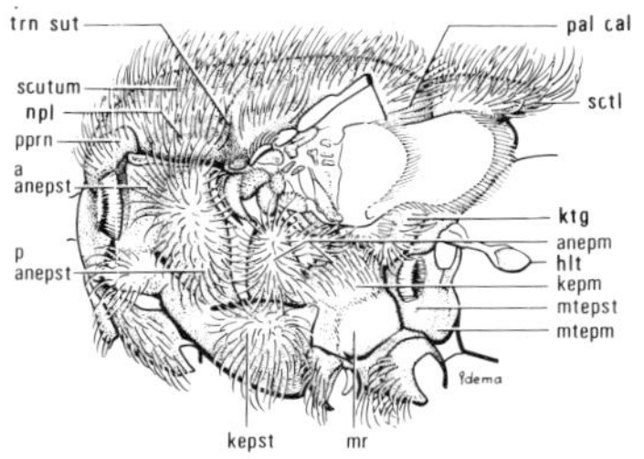

39 Eristalinus aeneus (Scopoli) ♂

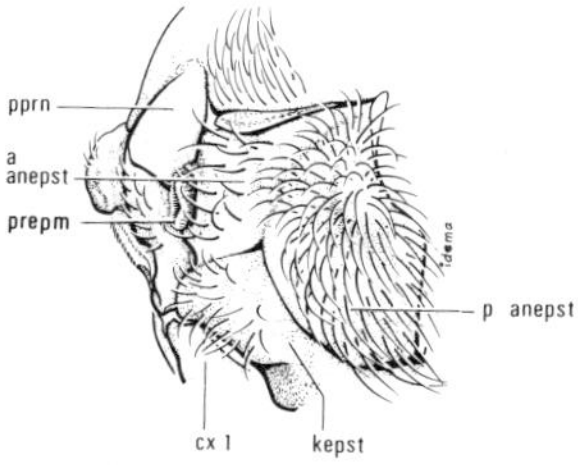

40 Meliscaeva cinctella ♂

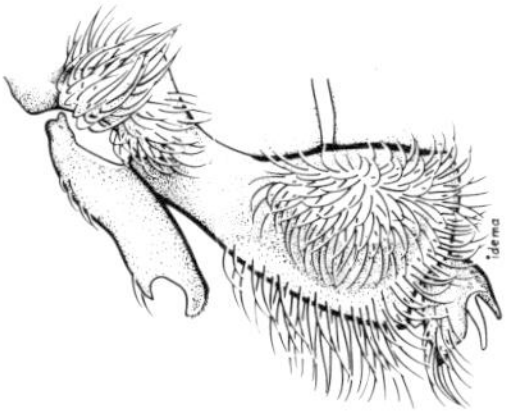

41 Epistrophe (Epistrophe) grossulariae ♂

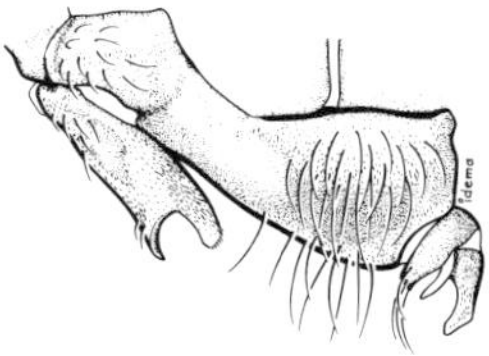

42 E. (Epistrophella) emarginata ♂

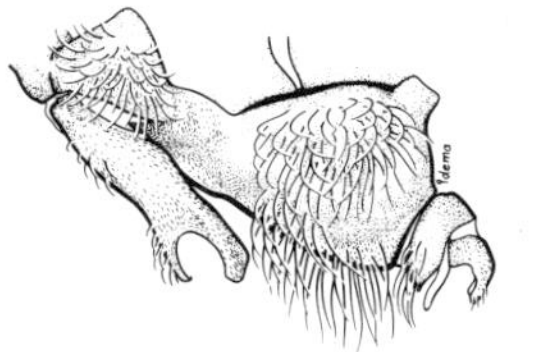

43 Eupeodes americanus ♂

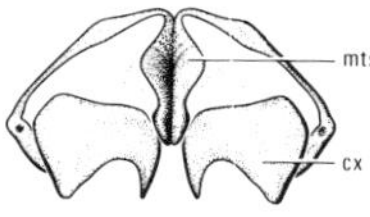

44 Melanostoma mellinum ♂

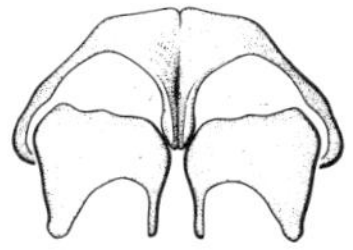

45 Platycheirus quadratus ♂

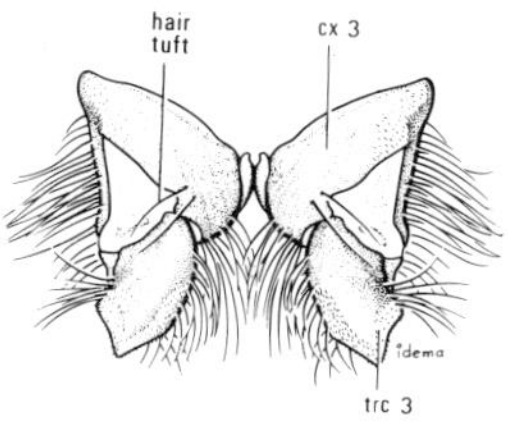

46 Parasyrphus tarsatus ♂

Figs. 36–46. Details of wing and thorax: (36) part of posterior margin of wing; (37) calypters; (38) scutellum, lateral; (39) thorax, lateral; (40) anterior part of thorax; (41–43) anepisternum, fore and mid coxae; (44, 45) metasternum and hind coxa; (46) hind coxae and trochanters. (Abbreviations: a anepst, anterior anepisternum; anepm, anepimeron; cx, coxa; hlt, halter; kepm, katepimeron; kepst, katepisternum; ktg, katatergite; l calyp, lower calypter; mtepm, metepimeron; mr, meron; mtepst, metepisternum; mtst, metasternum; pal cal, postalar callus; p anepst, posterior anepisternum; pprn, postpronotum; prepm, proepimeron; sctl, scutellum; trn sut, transverse suture; u calyp, upper calypter; v sctl fring, ventral scutellar fringe.)

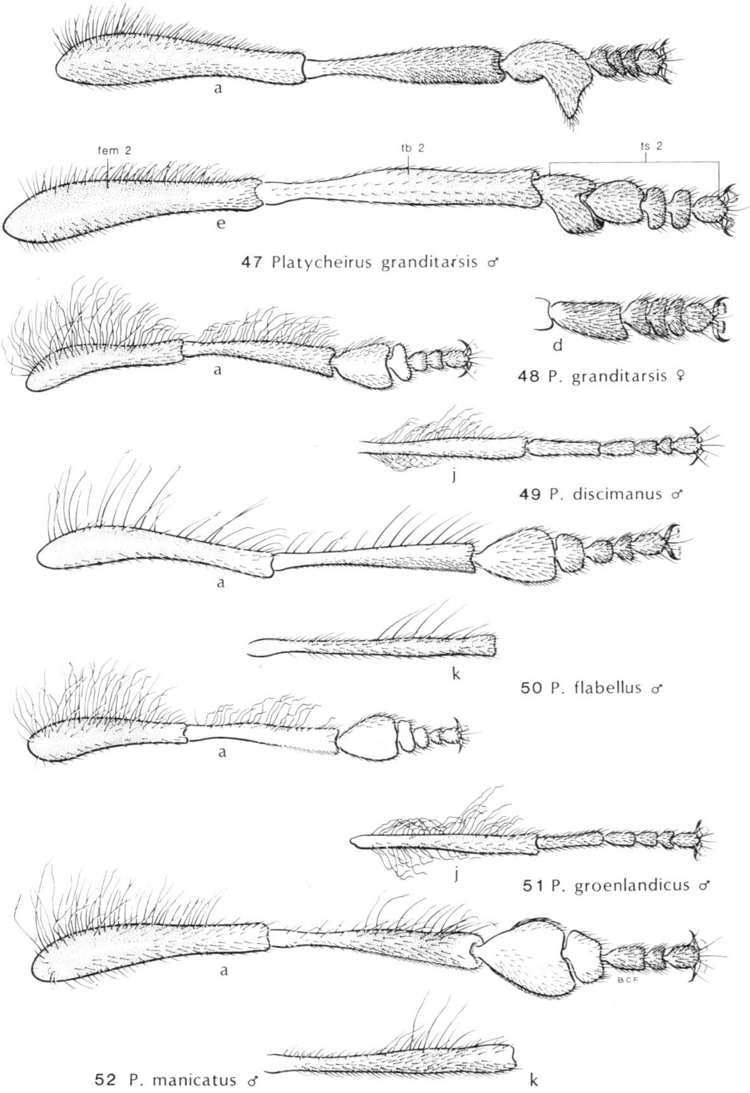

Figs. 47–52. Legs. (Adscripts: *a*, femur, tibia, and tarsus of left fore leg, dorsal; *b*, tibia and tarsus of left fore leg, dorsal; *c*, tibia of left fore leg, dorsal; *d*, tarsus of left fore leg, dorsal; *e*, femur, tibia, and tarsus of left mid leg, dorsal; *f*, femur, tibia, and first tarsomere of left mid leg, dorsal; *g*, femur, tibia, and tarsus of left mid leg, anterior; *h*, coxa, trochanter, and femur of left mid leg, anterior; *i*, femur of left mid leg, anterior; *j*, tibia and tarsus of left mid leg, dorsal; *k*, tibia of left mid leg, dorsal; *l*, tibia of left mid leg, anterior; *m*, tibia and tarsus of left hind leg, dorsal; *n*, tibia of left hind leg, dorsal; *p*, first tarsomere of left hind leg, anterior.) (Abbreviations: fem, femur; tb, tibia; ts, tarsus.)

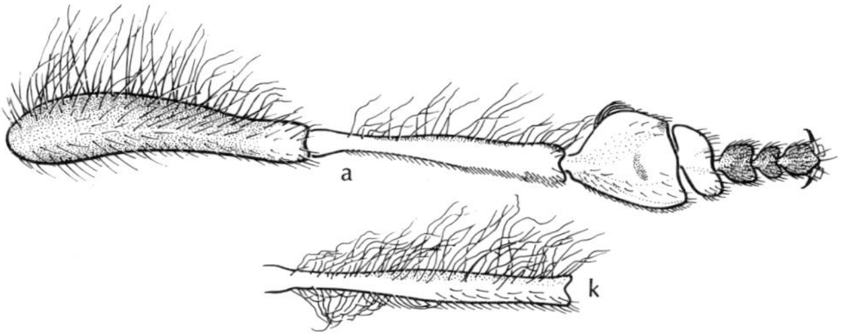

53 Platycheirus oreadis ♂

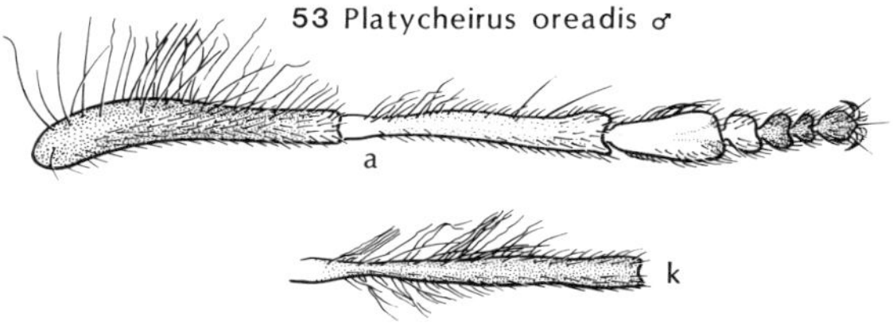

54 P. subordinatus ♂

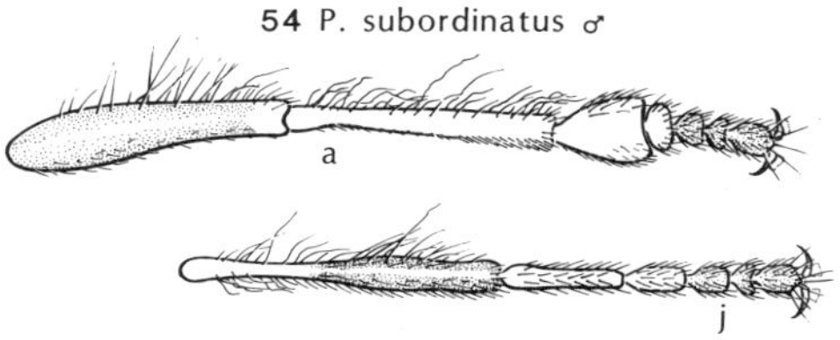

55 P. thylax ♂

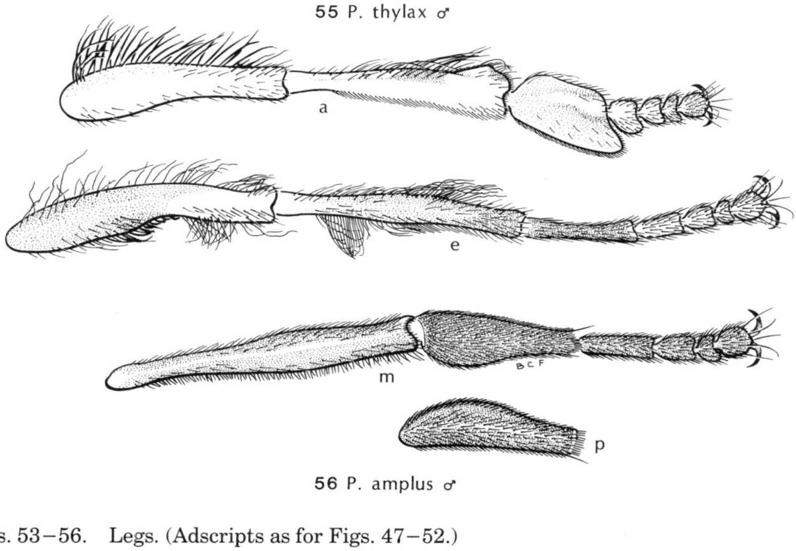

56 P. amplus ♂

Figs. 53–56. Legs. (Adscripts as for Figs. 47–52.)

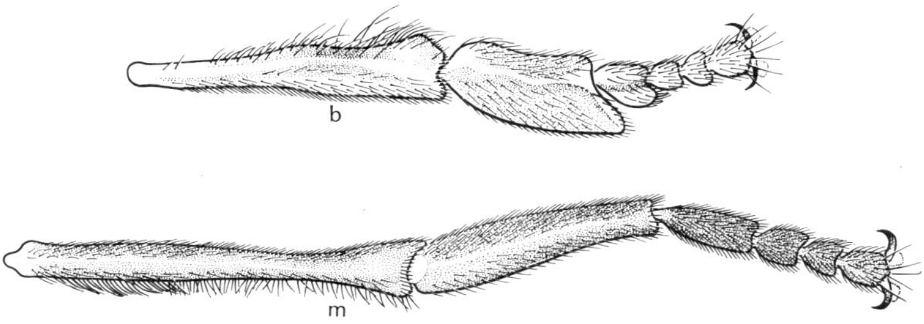

57 Platycheirus holarcticus ♂

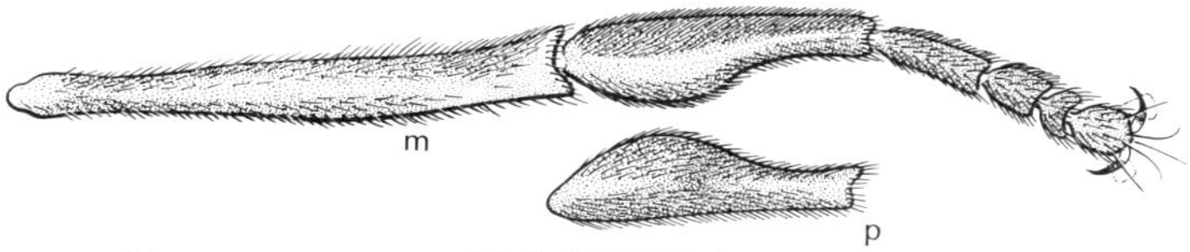

58 P. inversus ♂

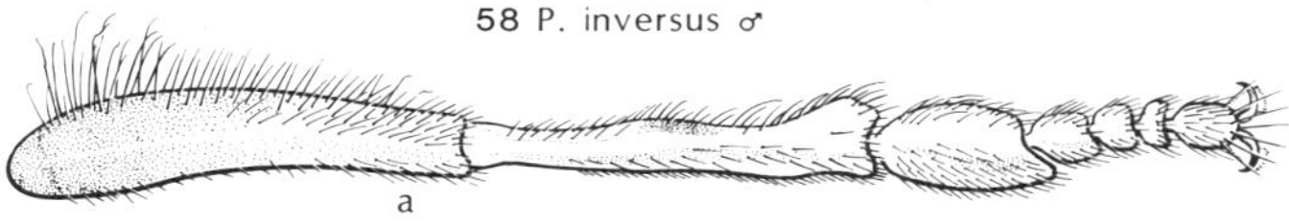

59 P. jaerensis ♂

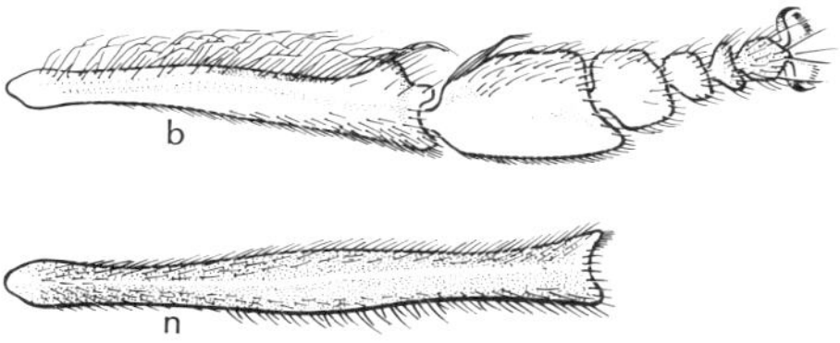

60 P. latitarsis ♂

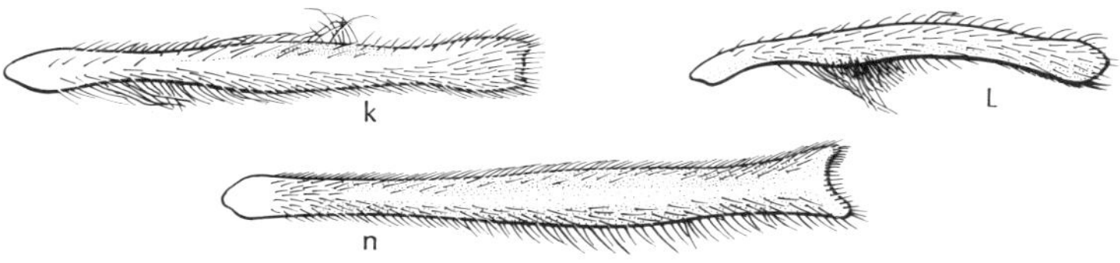

61 P. nearcticus ♂

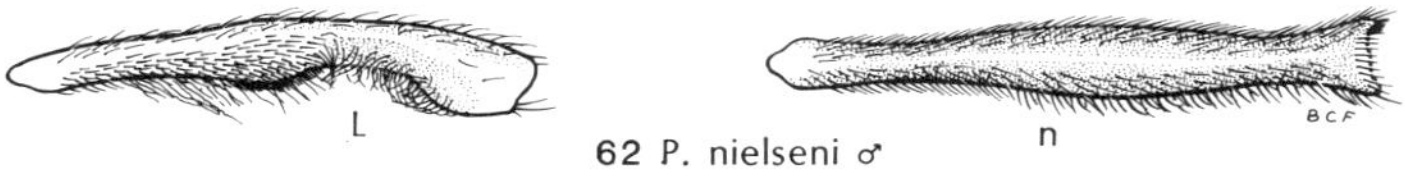

62 P. nielseni ♂

Figs. 57–62. Legs. (Adscripts as for Figs. 47–52.)

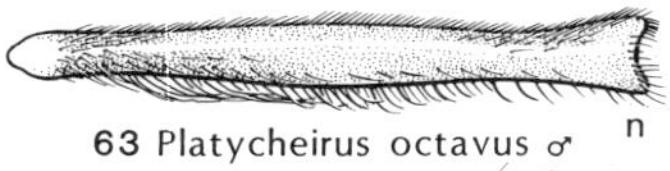

63 Platycheirus octavus ♂

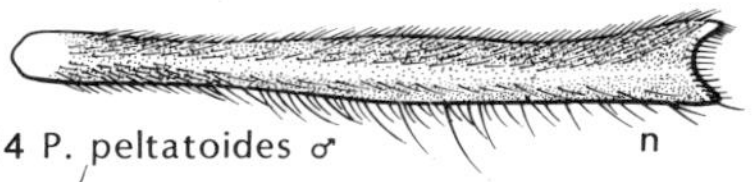

64 P. peltatoides ♂

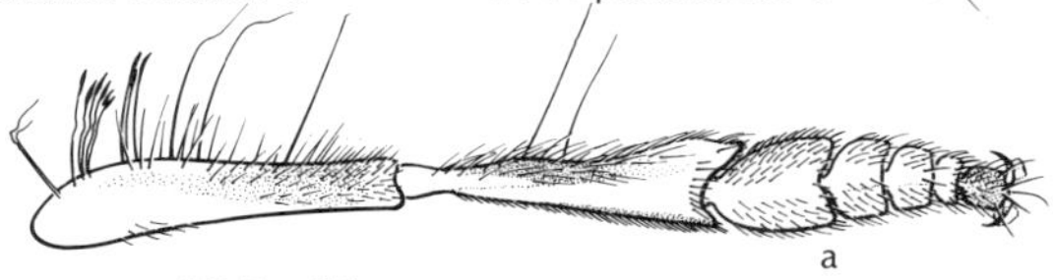

65 P. albimanus ♂

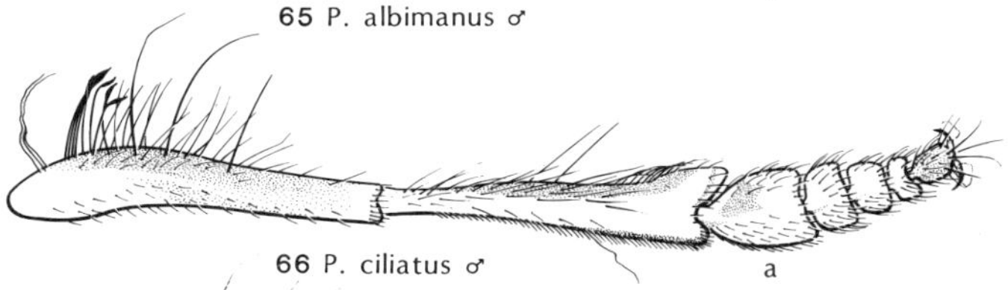

66 P. ciliatus ♂

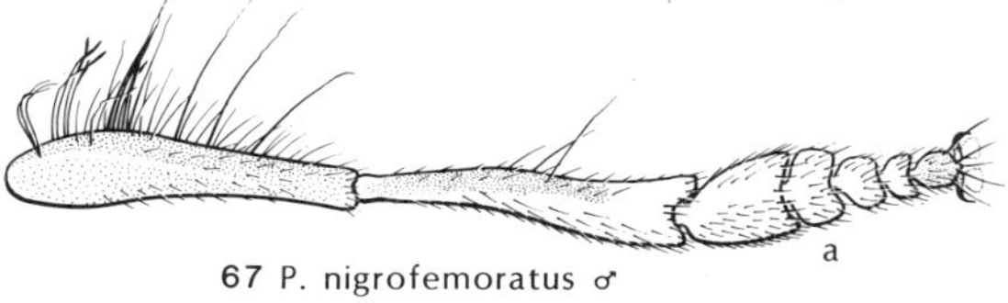

67 P. nigrofemoratus ♂

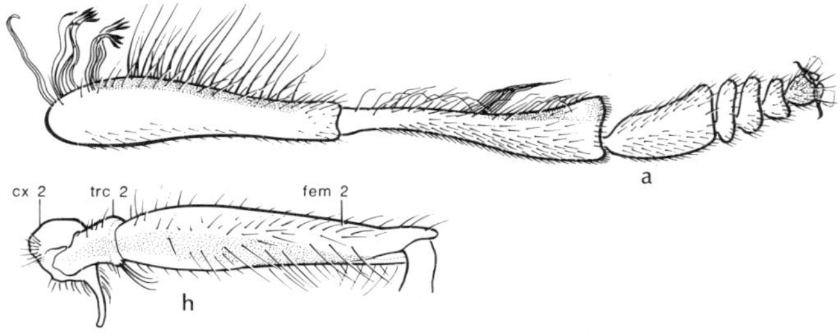

68 P. scutatus ♂

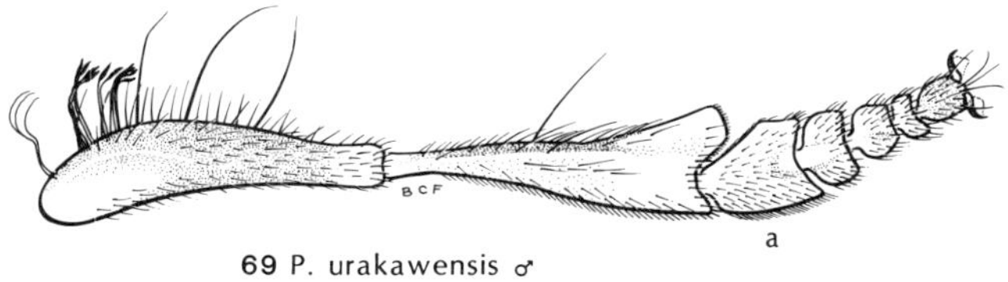

69 P. urakawensis ♂

Figs. 63–69. Legs. (Adscripts as for Figs. 47–52.) (Abbreviations: cx, coxa; fem, femur; tr, trochanter.)

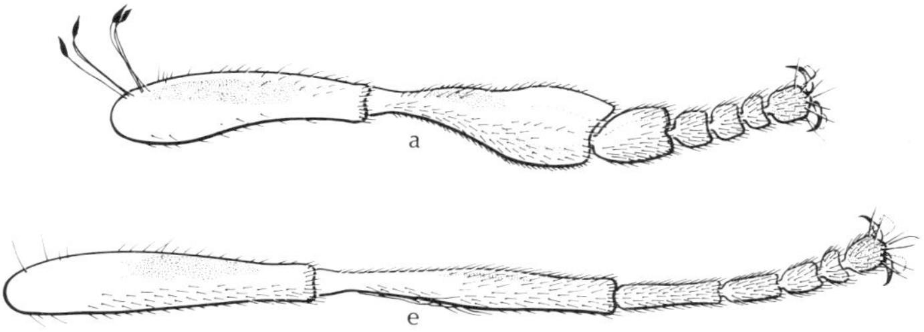

70 Platycheirus nodosus ♂

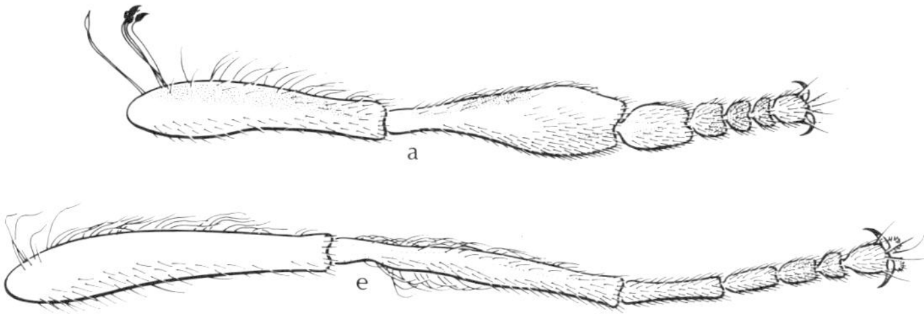

71 P. pilatus ♂

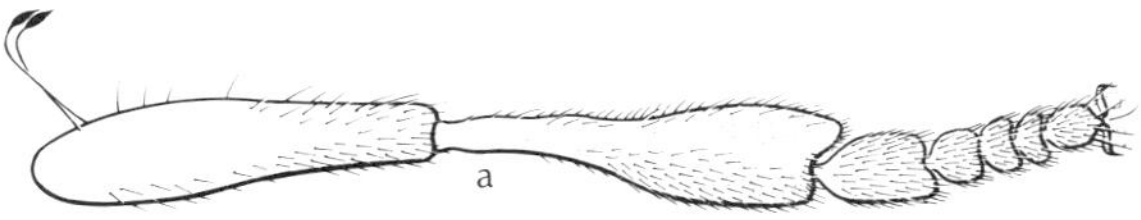

72 P. thompsoni ♂

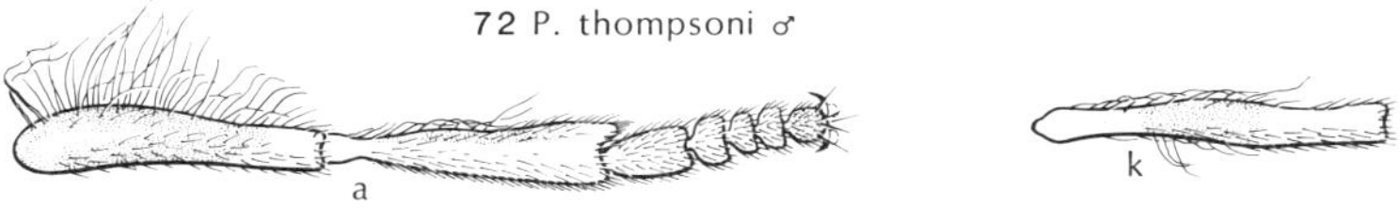

73 P. aeratus ♂

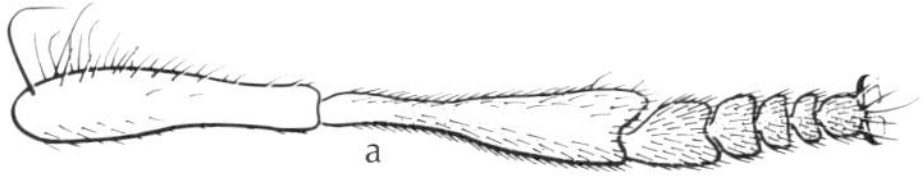

74 P. angustatus ♂

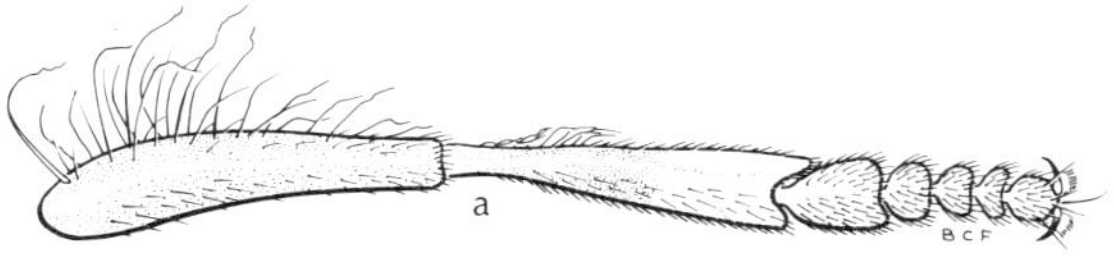

75 P. clypeatus ♂

Figs. 70–75. Legs. (Adscripts as for Figs. 47–52.)

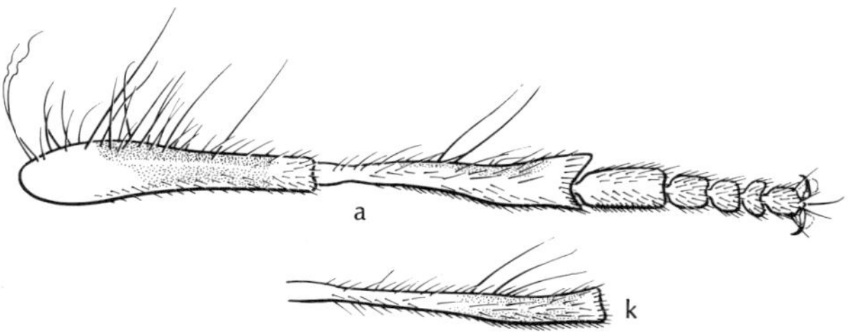

76 Platycheirus hispidipes ♂

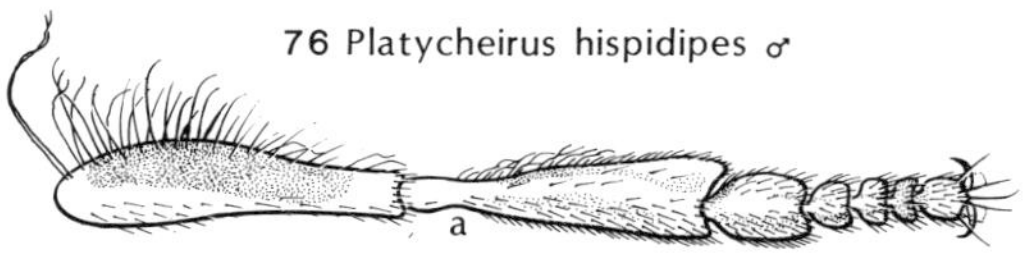

77 P. hyperboreus ♂

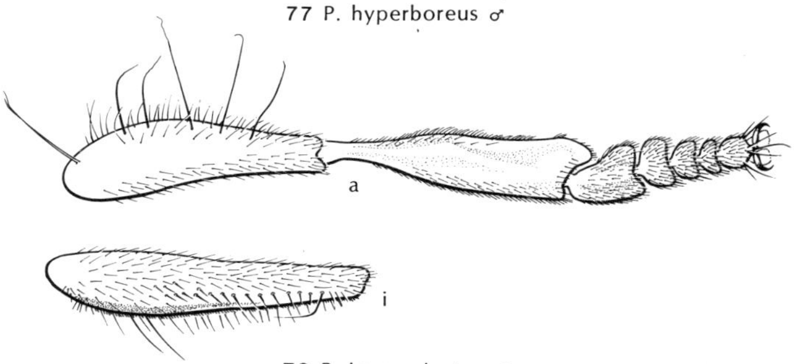

78 P. immarginatus ♂

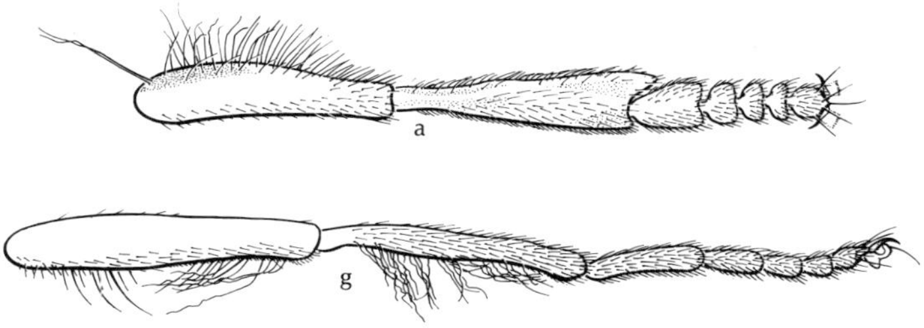

79 P. perpallidus ♂

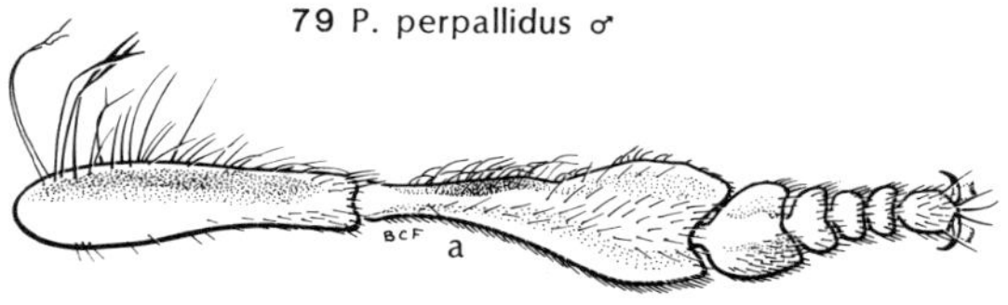

80 P. podagratus ♂

Figs. 76–80. Legs. (Adscripts as for Figs. 47–52.)

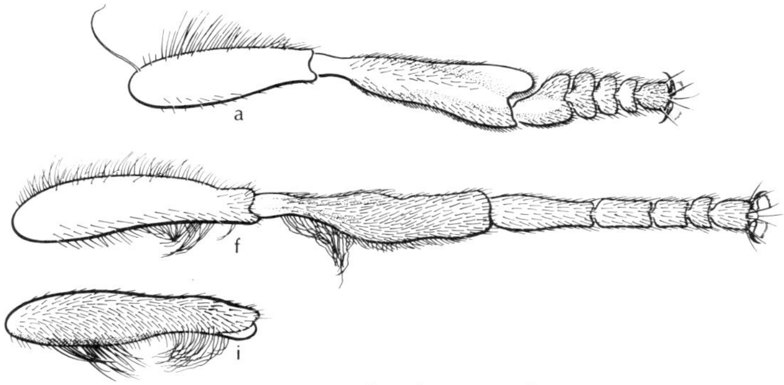

81 Platycheirus quadratus ♂

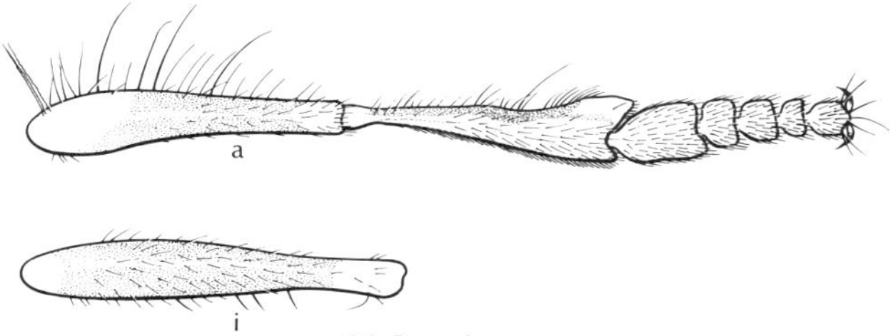

82 P. setipes ♂

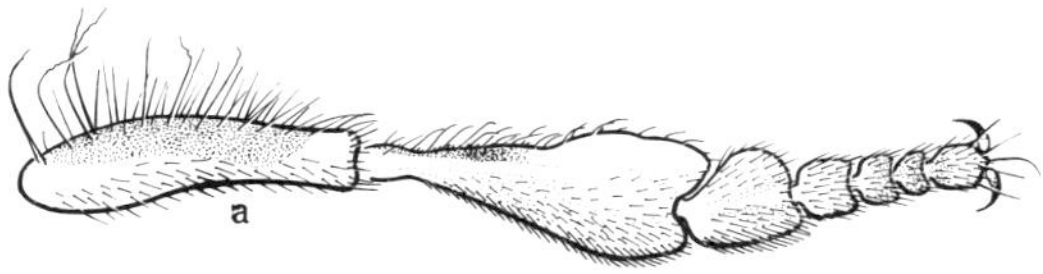

83 P. tenebrosus ♂

84 P. modestus ♂

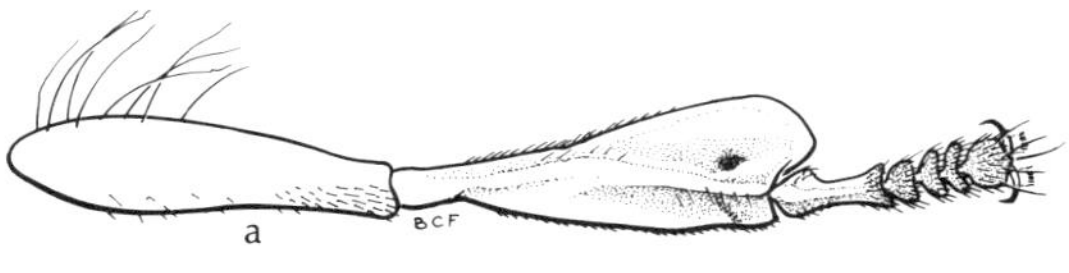

85 P. normae ♂

Figs. 81–85. Legs. (Adscripts as for Figs. 47–52.)

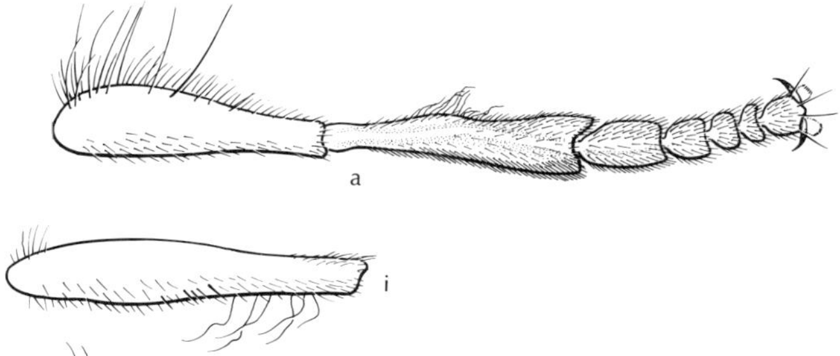

86 Platycheirus orarius ♂

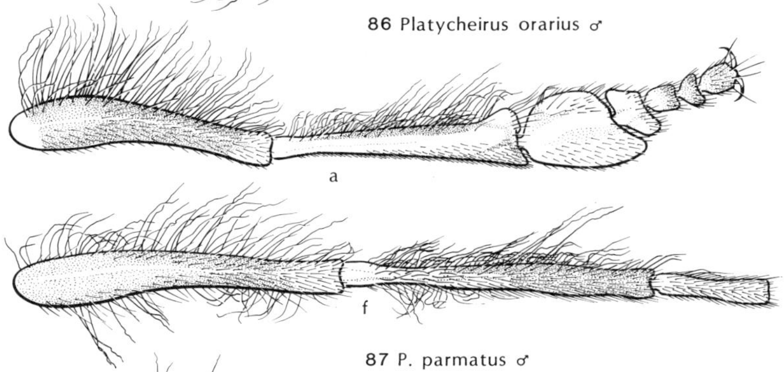

87 P. parmatus ♂

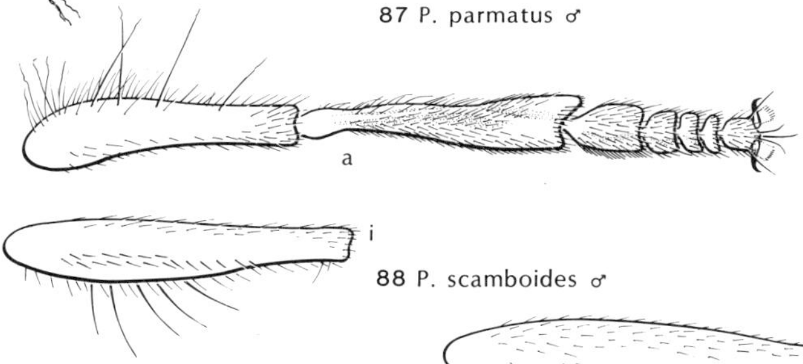

88 P. scamboides ♂

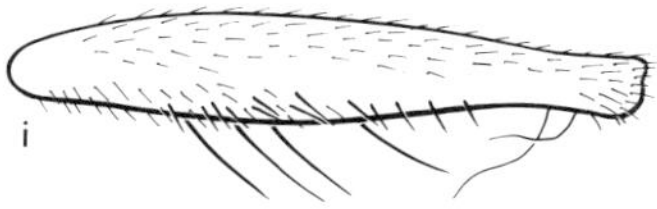

89 P. scambus ♂

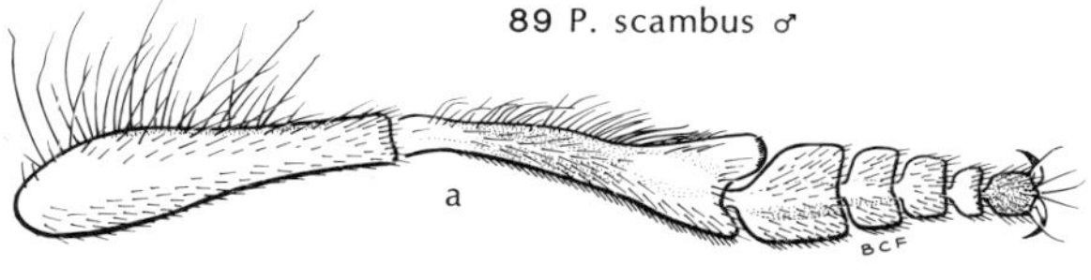

90 P. varipes ♂

Figs. 86–90. Legs. (Adscripts as for Figs. 47–52.)

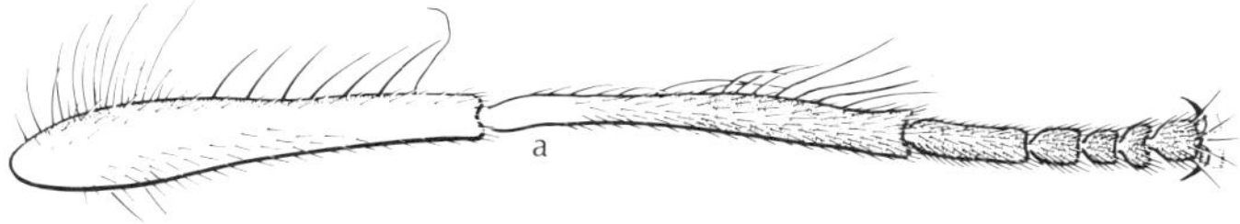

91 Platycheirus coerulescens ♂

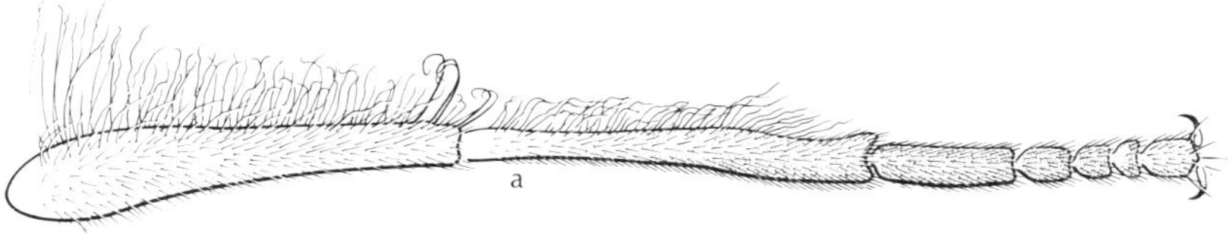

92 P. kelloggi ♂

93 P. confusus ♂

94 P. hesperius ♂

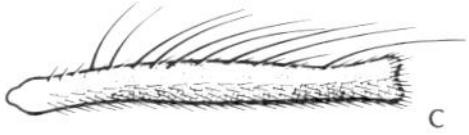

95 P. squamulae ♂

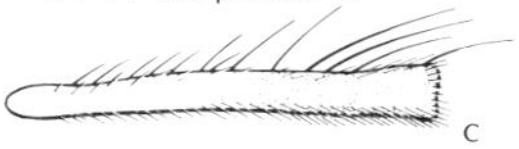

96 P. stegnoides ♂

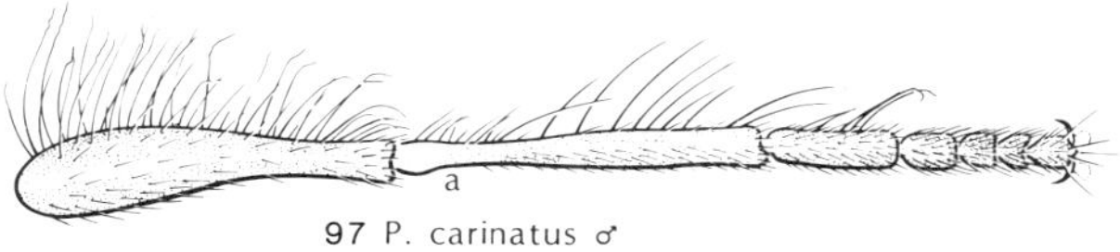

97 P. carinatus ♂

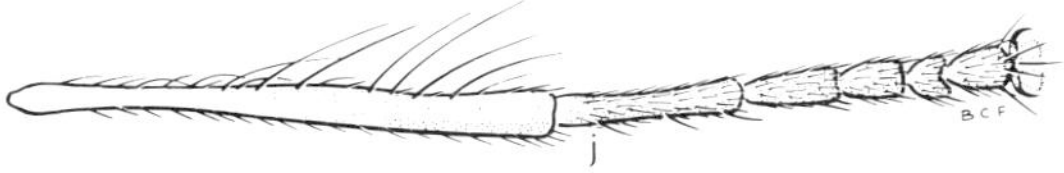

98 P. setitarsis ♂

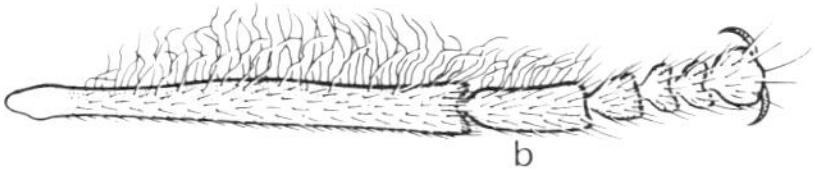

99 P. woodi ♂

100 Toxomerus geminatus ♂

101 T. occidentalis ♂

Figs. 91–101. Legs. (Adscripts as for Figs. 47–52, plus *q*, femur and tibia of left hind leg, posterior.)

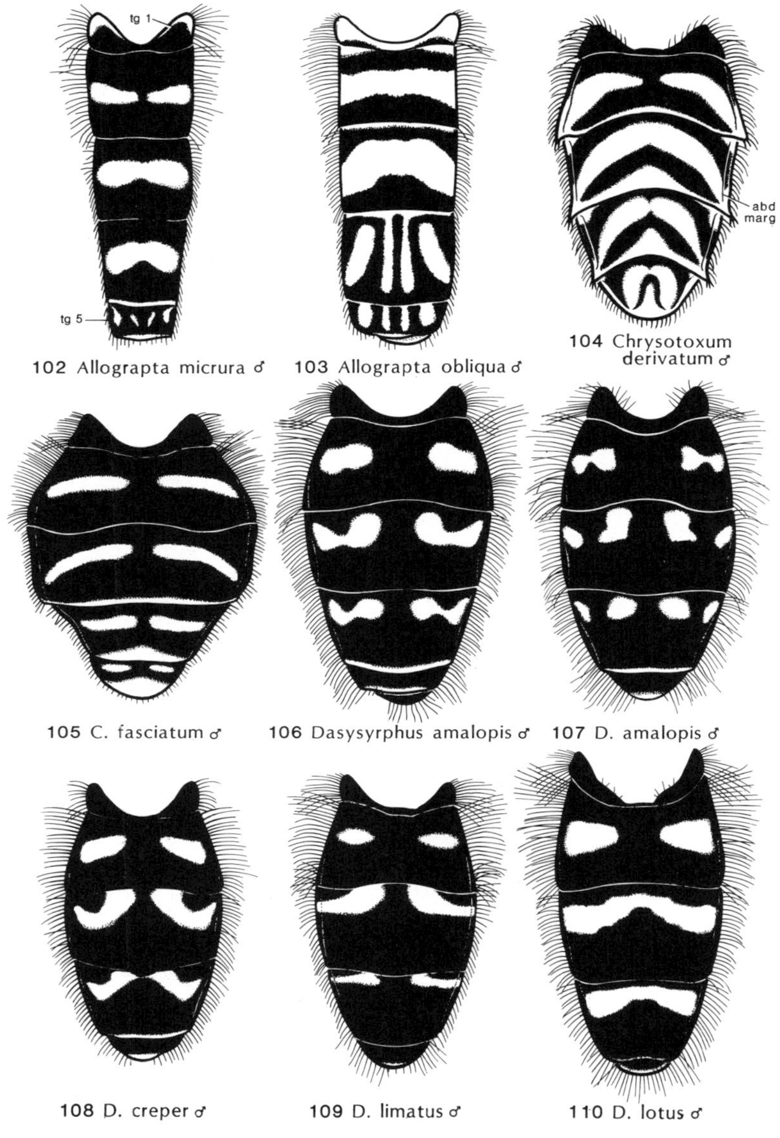

Figs. 102–110. Abdomens. (Abbreviations: abd mg, abdominal margin; tg, tergite.)

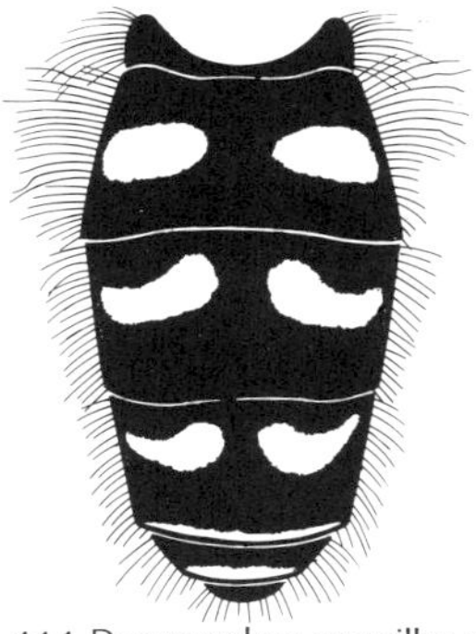

111 Dasysyrphus pauxillus ♂

112 D. pauxillus ♀

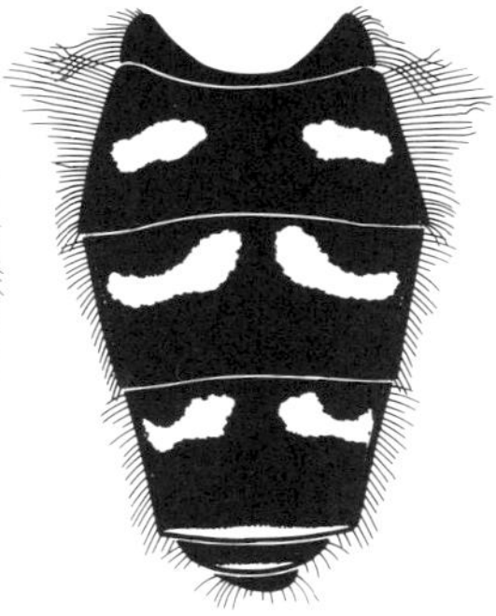

113 D. pinastri ♂

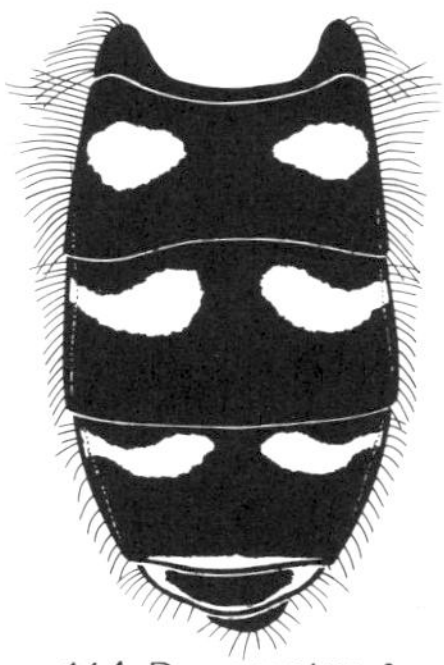

114 D. venustus ♂

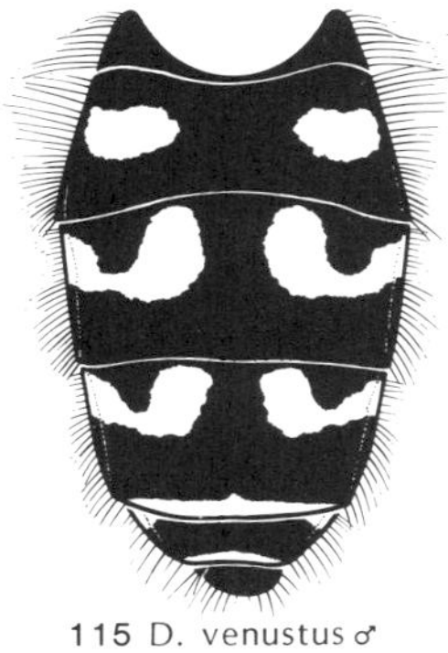

115 D. venustus ♂

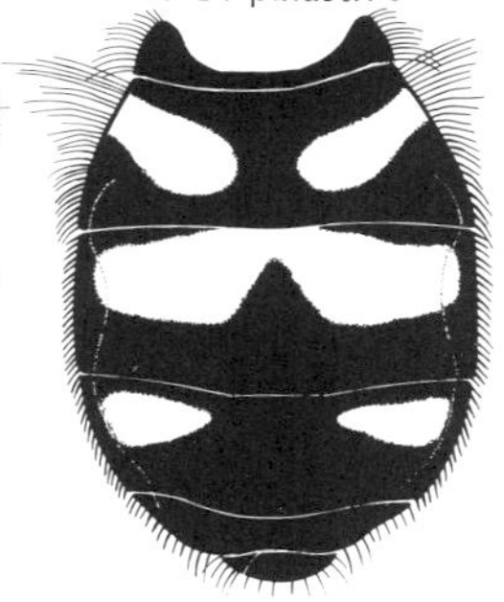

116 Didea alneti ♂

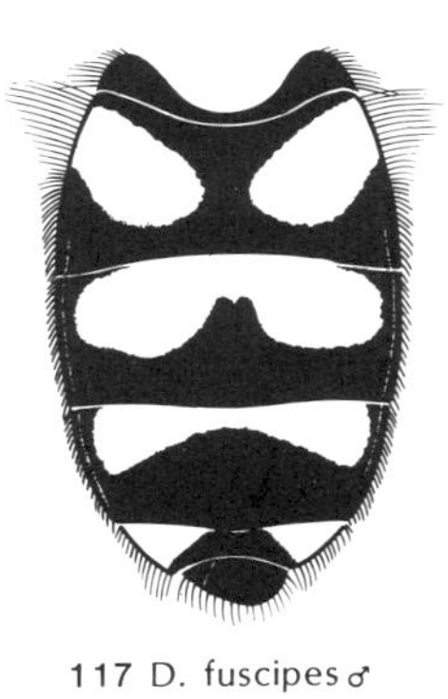

117 D. fuscipes ♂

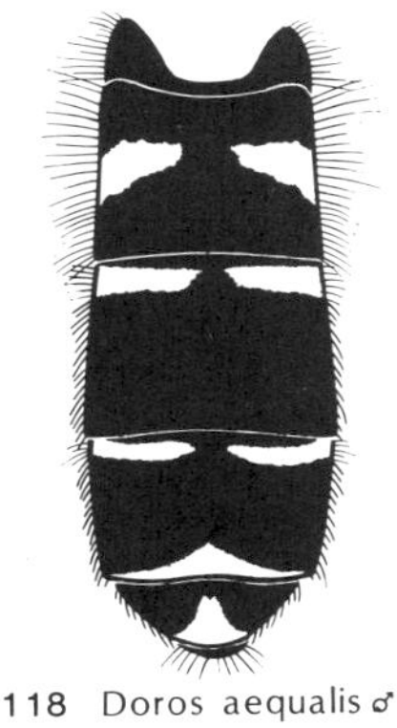

118 Doros aequalis ♂

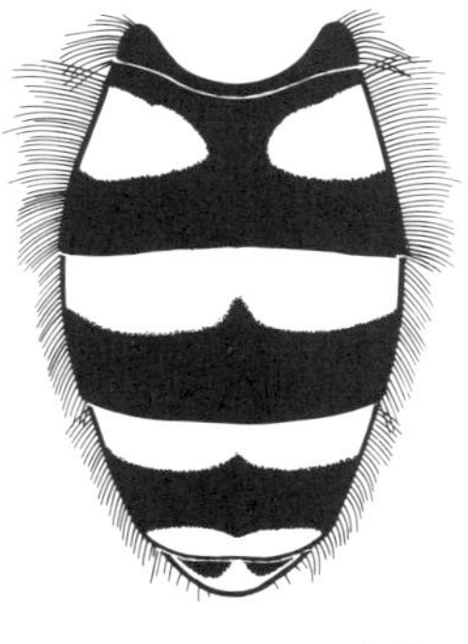

119 Epistrophe grossulariae ♂

Figs. 111–119. Abdomens.

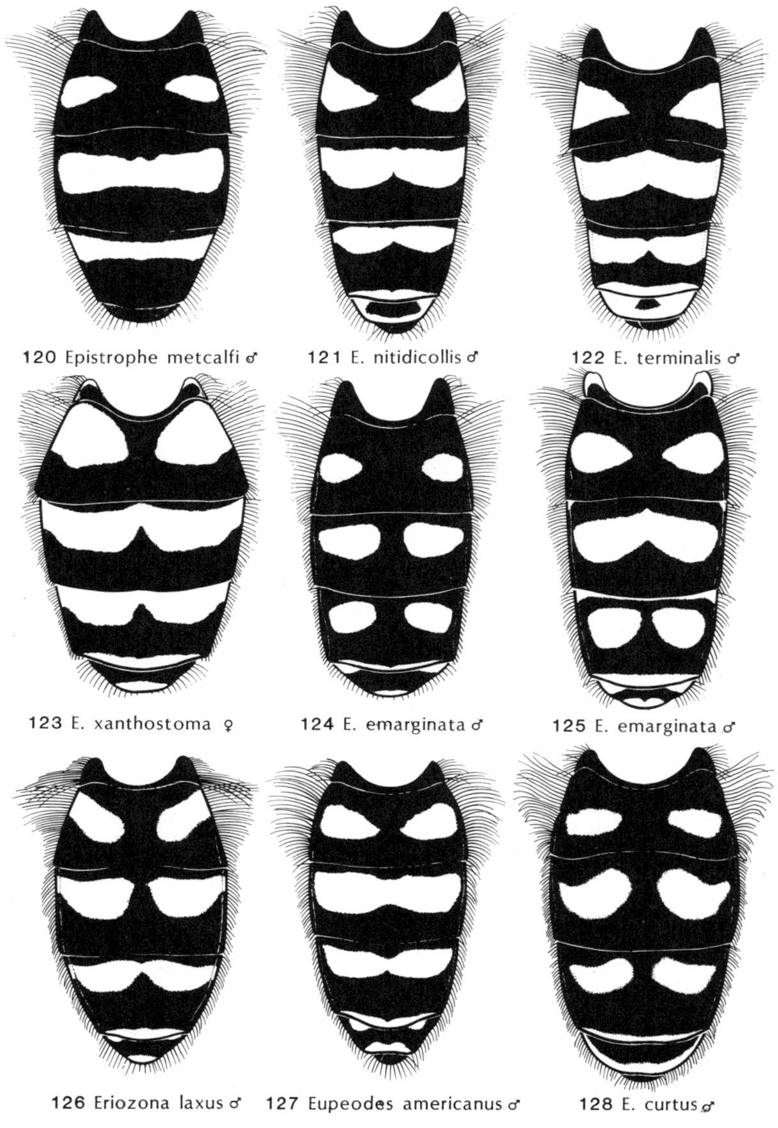

Figs. 120–128. Abdomens.

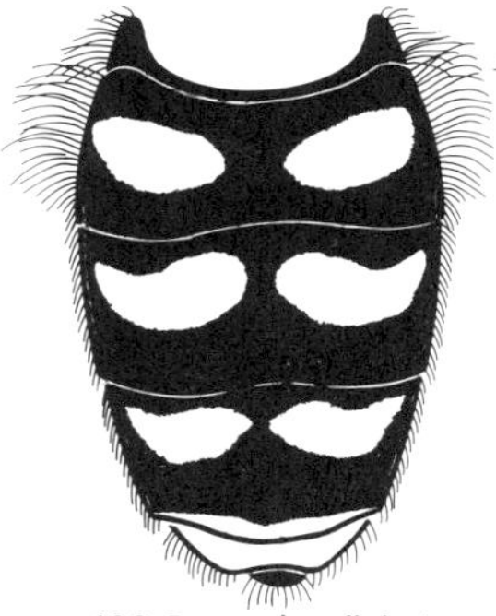

129 Eupeodes flukei ♂

130 E. fumipennis ♂

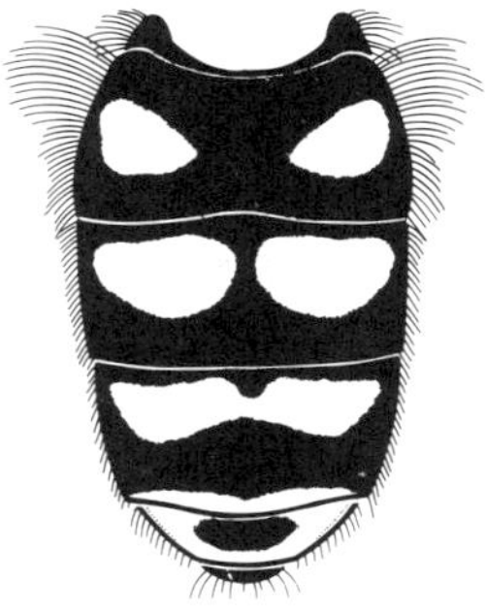

131 E. latifasciatus ♂

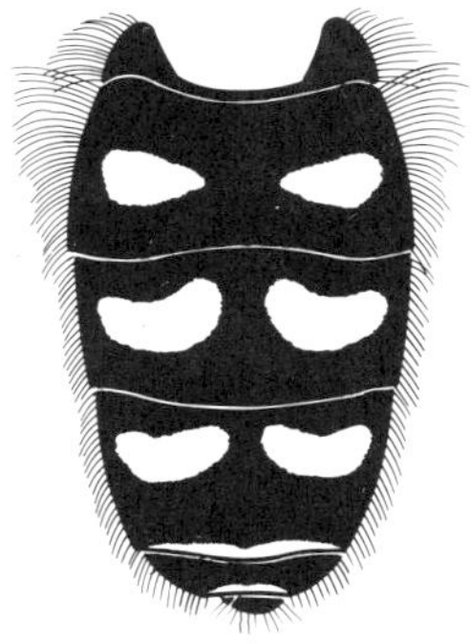

132 E. luniger ♂

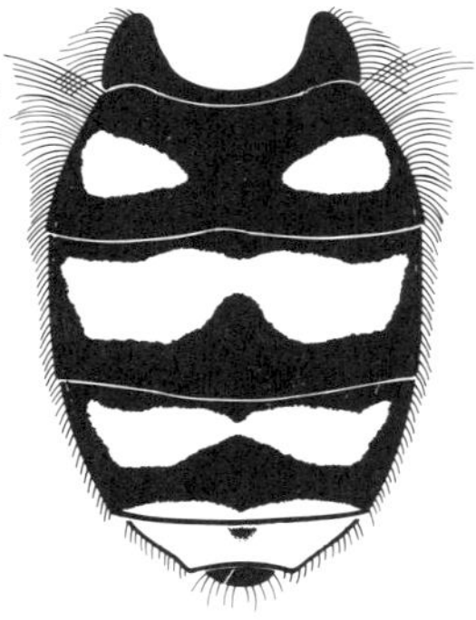

133 E. montivagus ♂

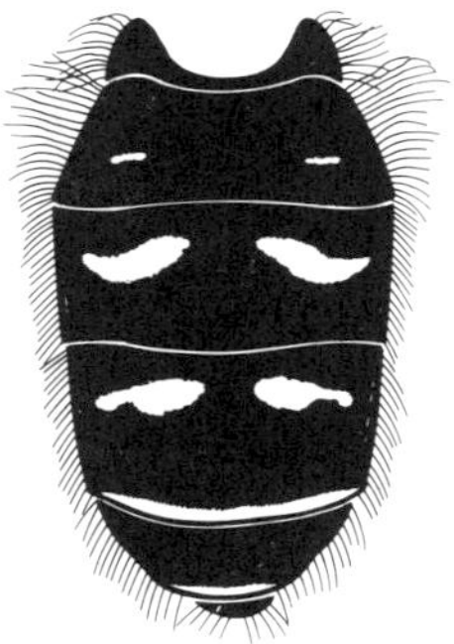

134 E. nigroventris ♂

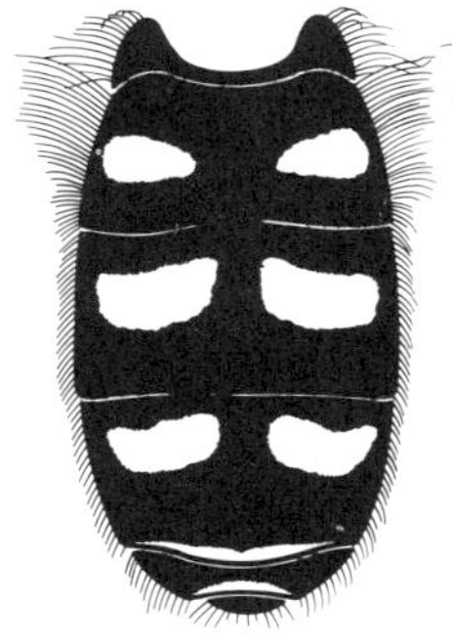

135 E. perplexus ♂

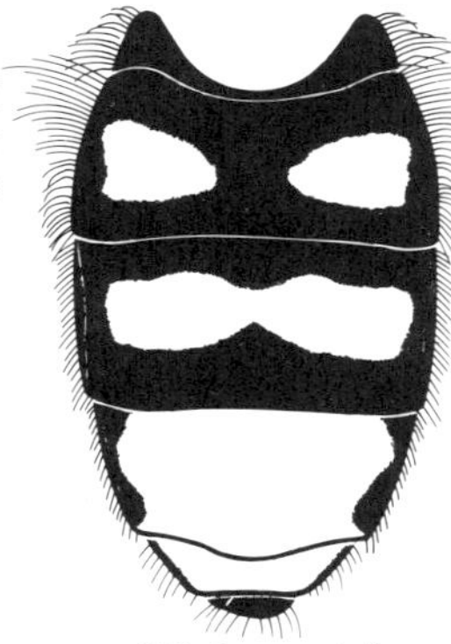

136 E. snowi ♂

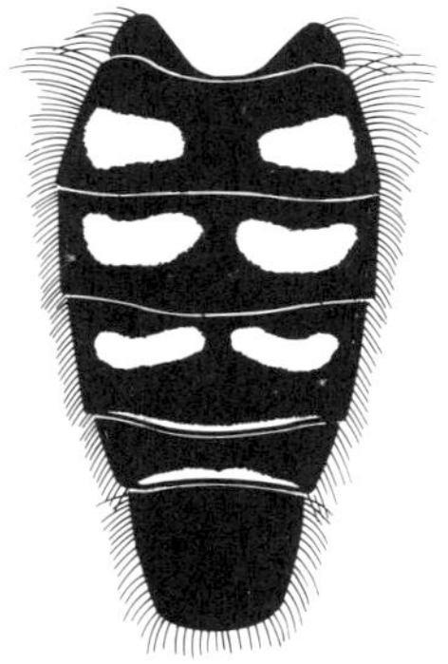

137 E. volucris ♂

Figs. 129–137. Abdomens.

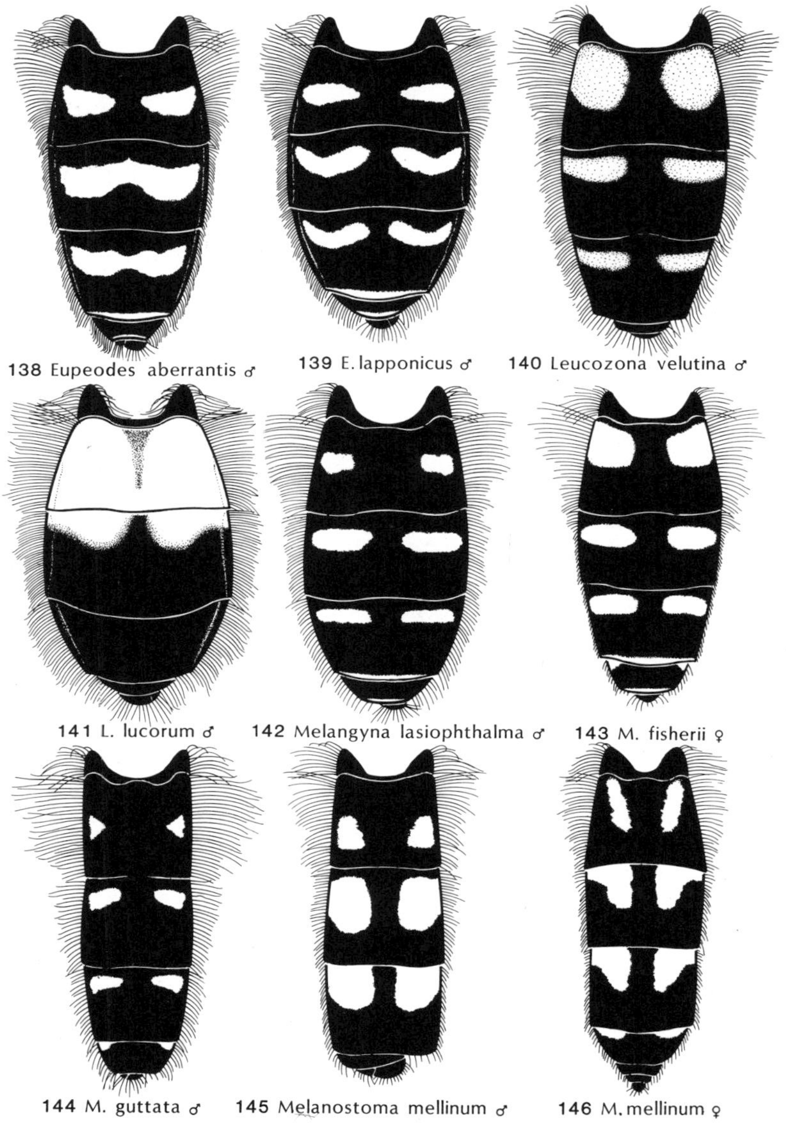

Figs. 138–146. Abdomens.

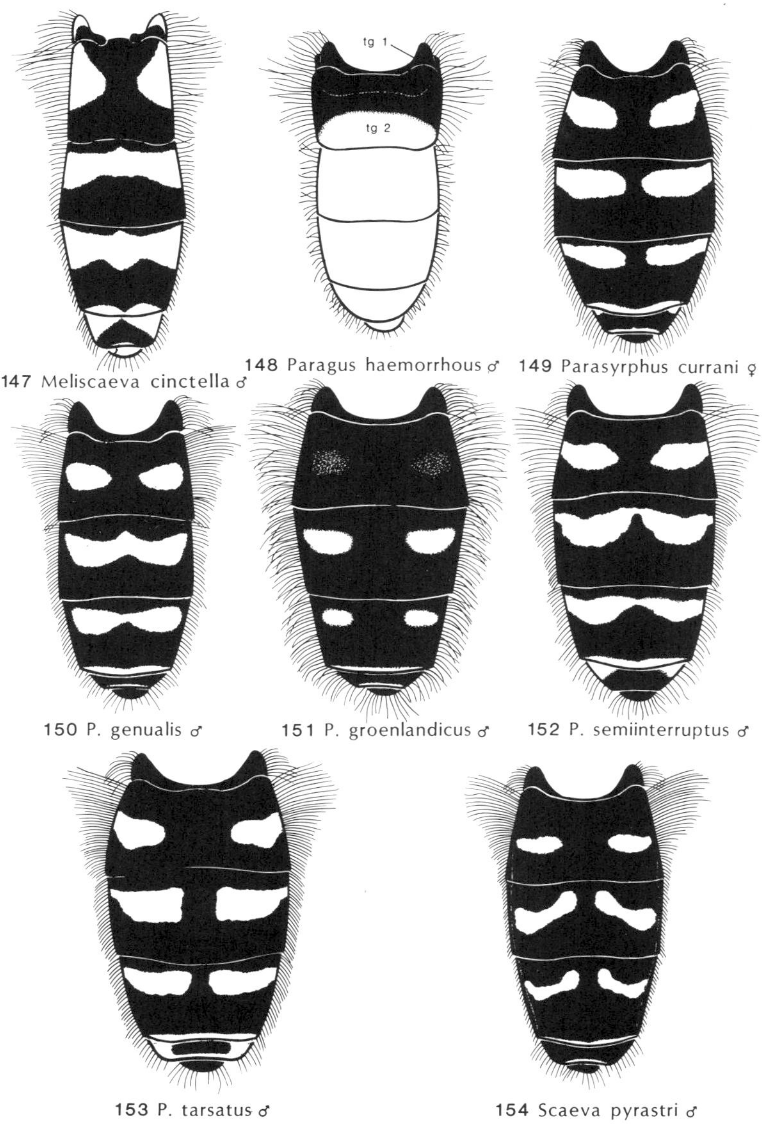

Figs. 147–154. Abdomens. (Abbreviation: tg, tergite.)

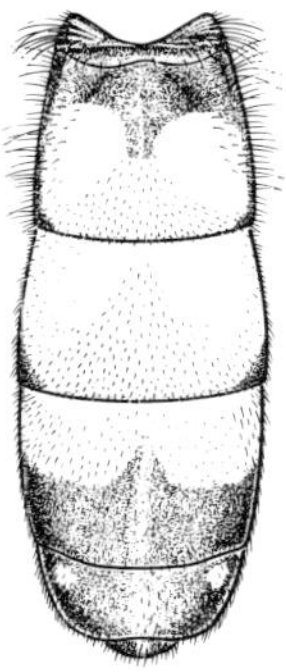

155 Platycheirus granditarsis ♂

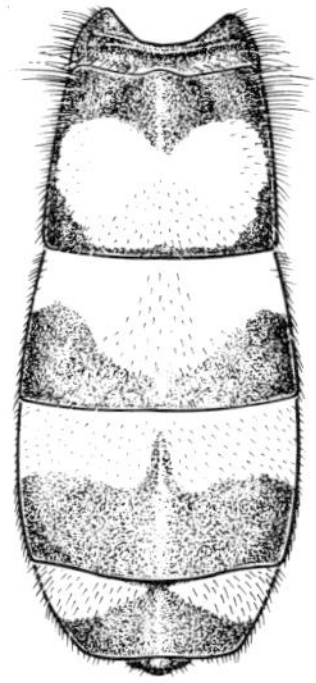

156 P. granditarsis ♀

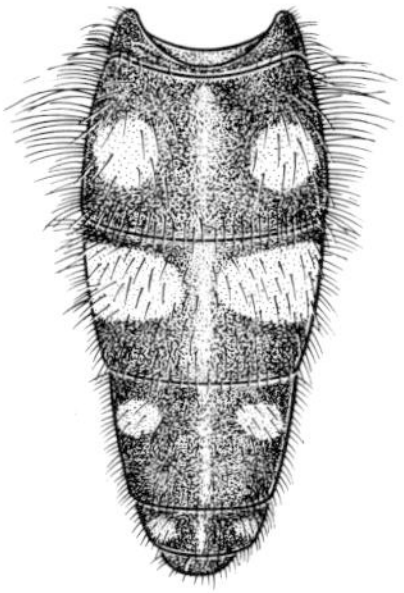

157 P. oreadis ♂

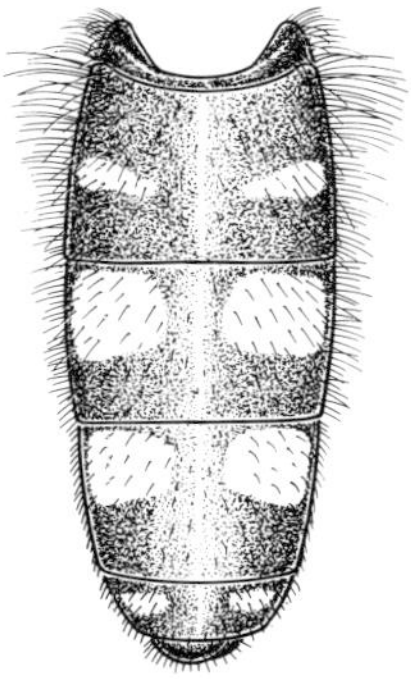

158 P. holarcticus ♂

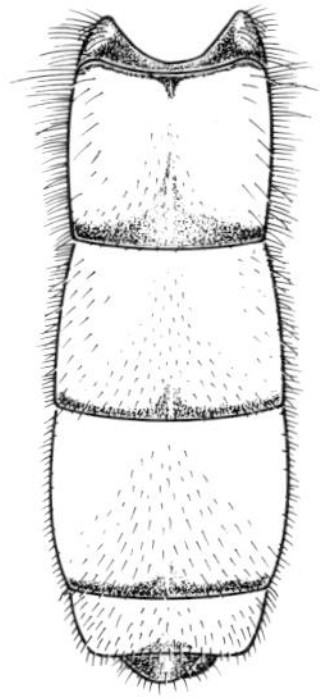

159 P. normae ♂

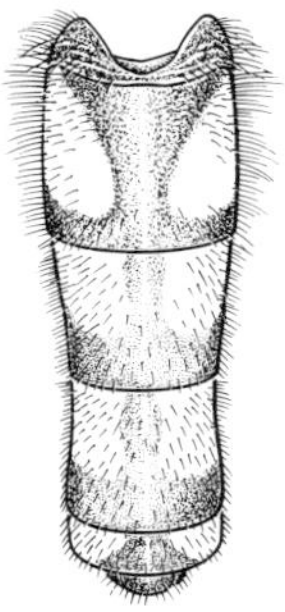

160 P. thompsoni ♂

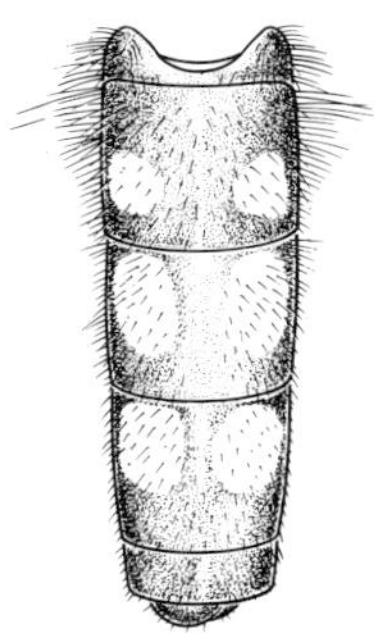

161 P. pilatus ♂

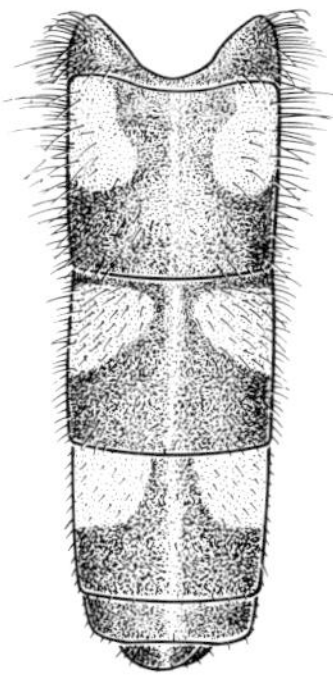

162 P. setipes ♂

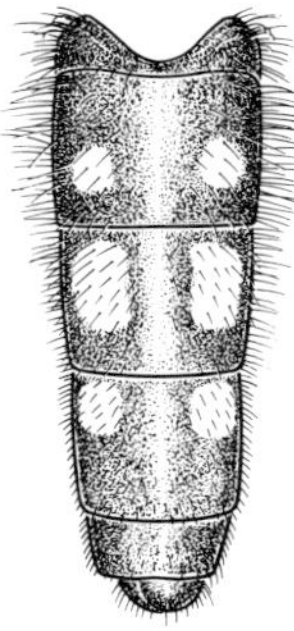

163 P. tenebrosus ♂

Figs. 155–163. Abdomens.

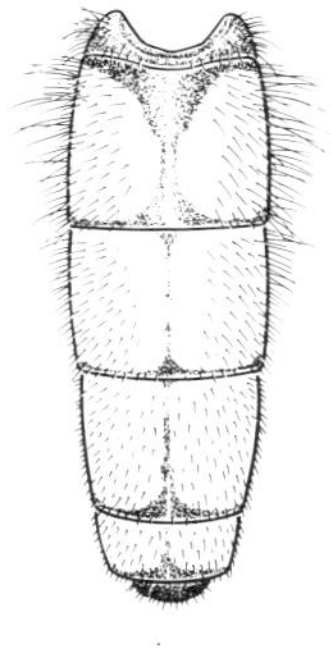

164 Platycheirus perpallidus ♂

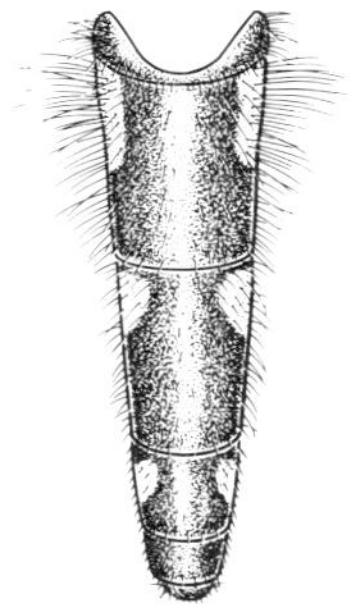

165 P. hispidipes ♂

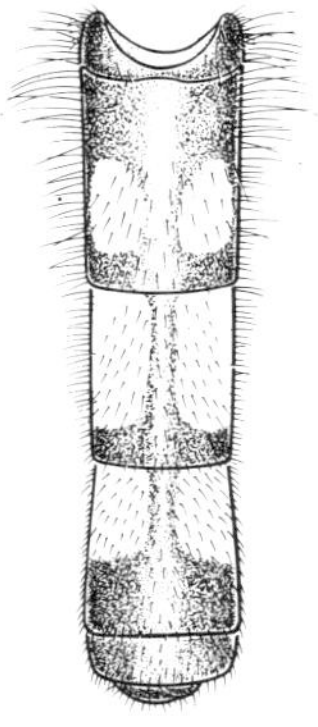

166 P. angustatus ♂

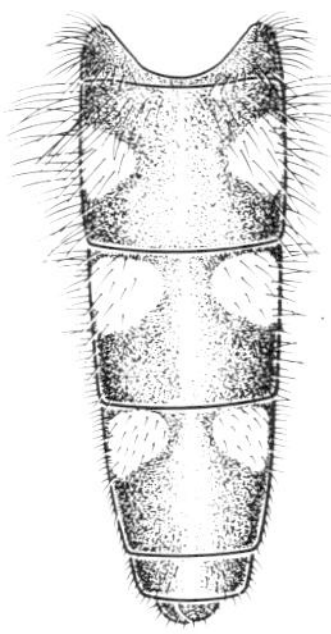

167 P. varipes ♂

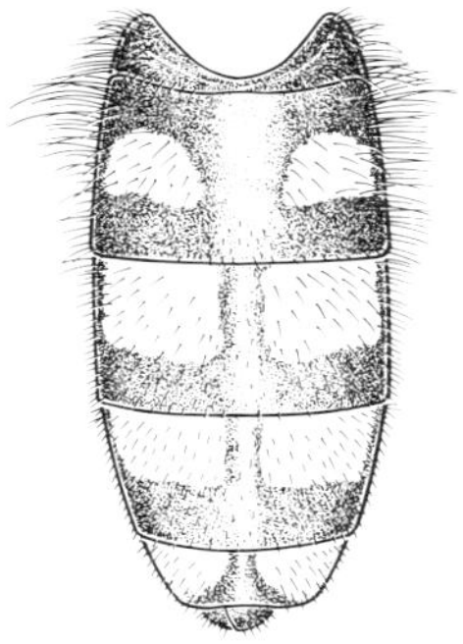

168 P. jaerensis ♂

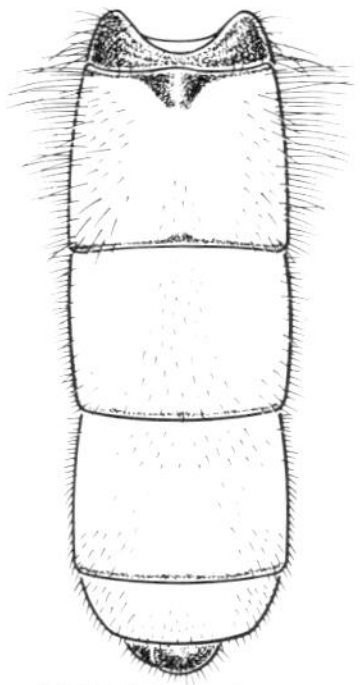

169 P. modestus ♂

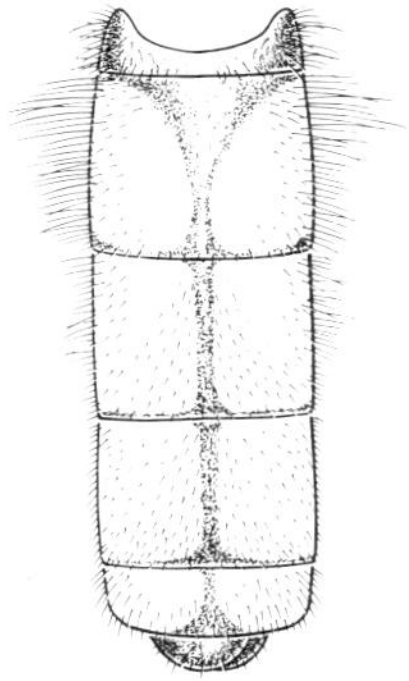

170 P. orarius ♂

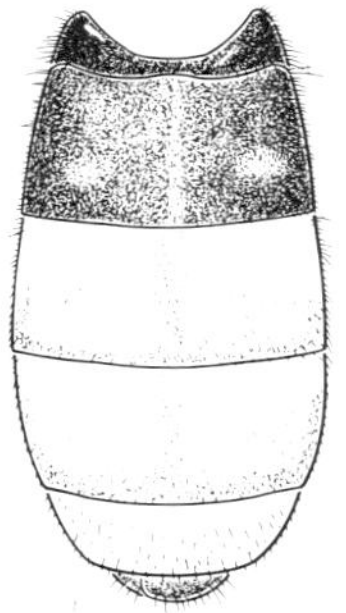

171 P. rufigaster ♂

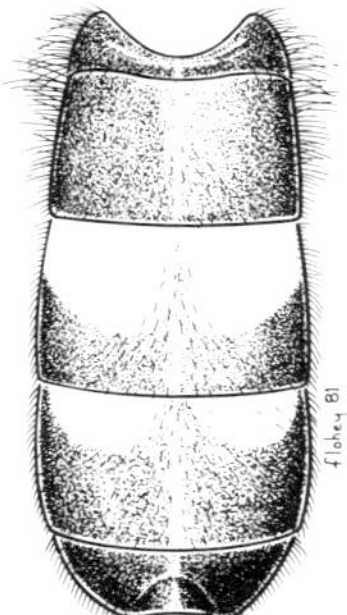

172 P. rosarum ♂

Figs. 164–172. Abdomens.

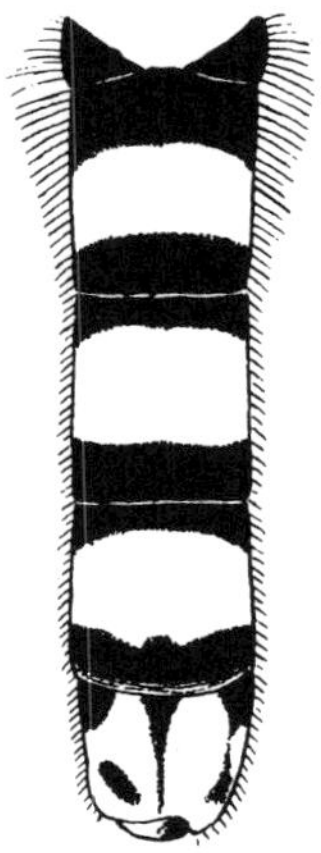

173 Sphaerophoria asymmetrica ♂

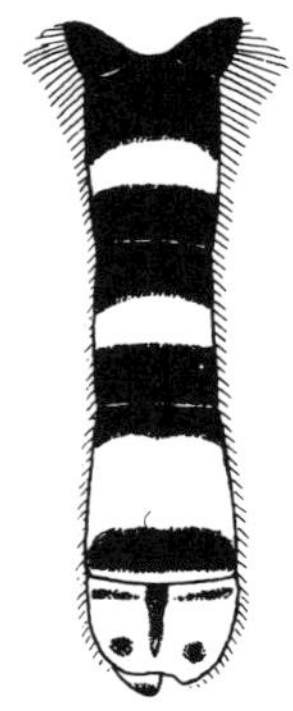

174 S. brevipilosa ♂

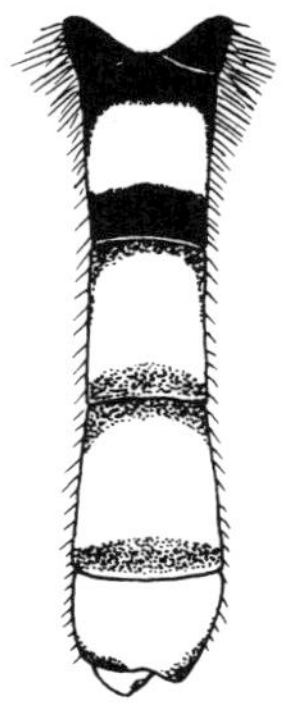

175 S. brevipilosa ♂

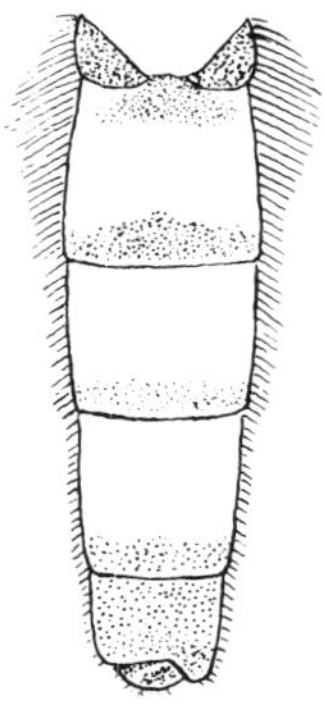

176 S. contigua ♂

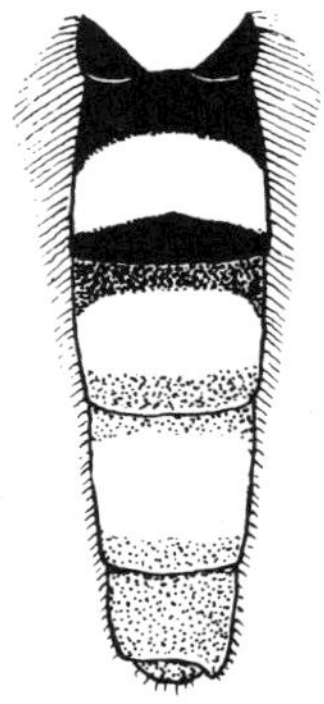

177 S. contigua ♂

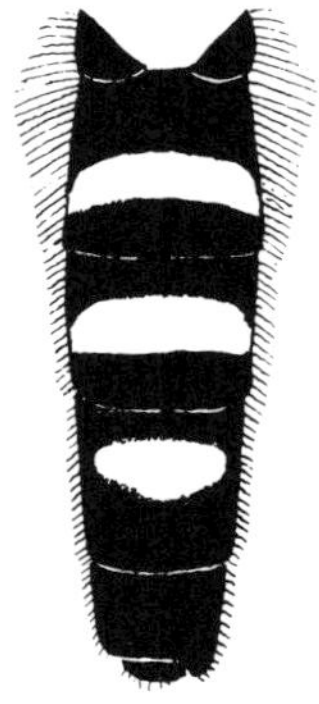

178 S. contigua ♂

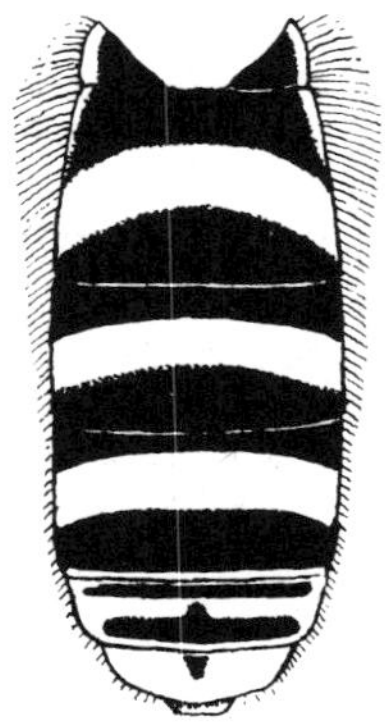

179 S. contigua ♀

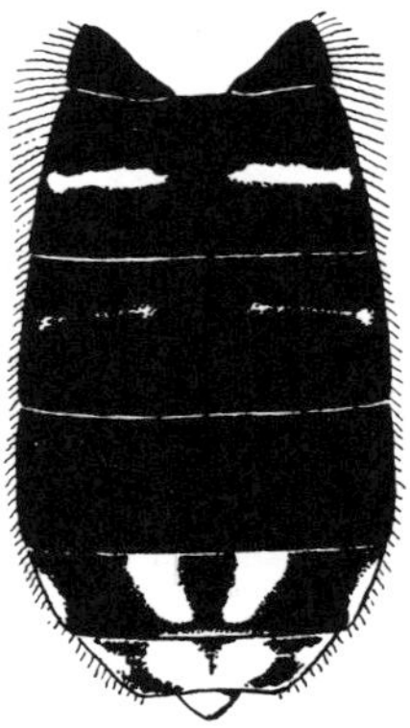

180 S. cranbrookensis ♀

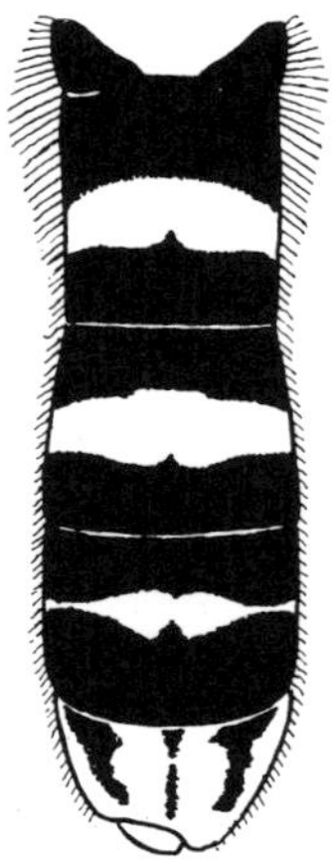

181 S. cranbrookensis ♂

Figs. 173–181. Abdomens.

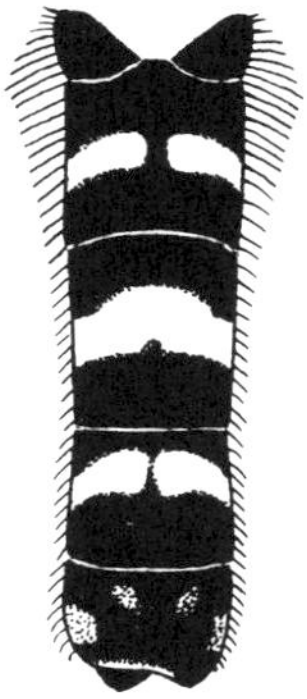

182 Sphaerophoria longipilosa ♂

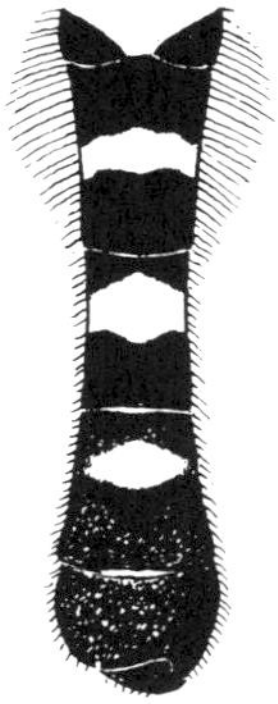

183 S. novaeangliae ♂

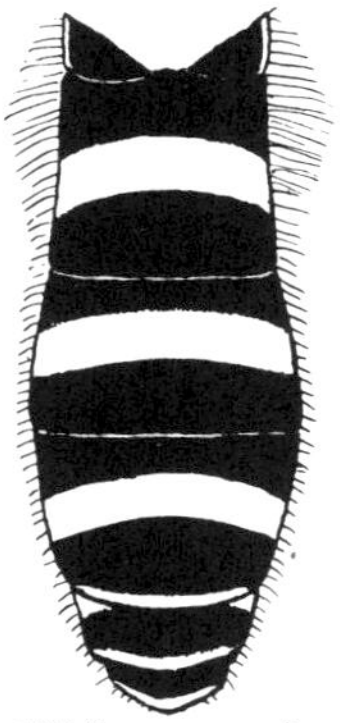

184 S. novaeangliae ♀

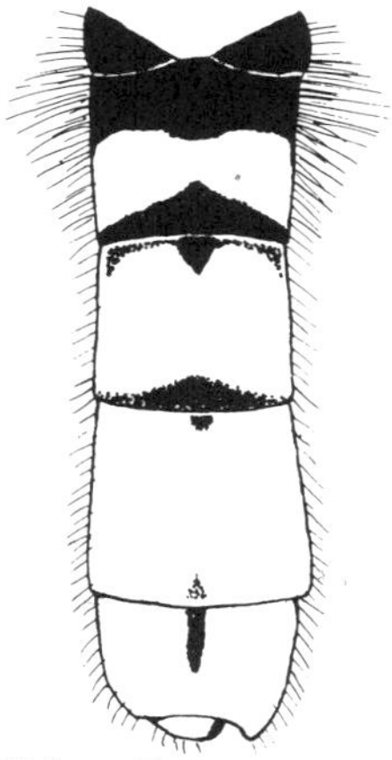

185 S. pyrrhina ♂

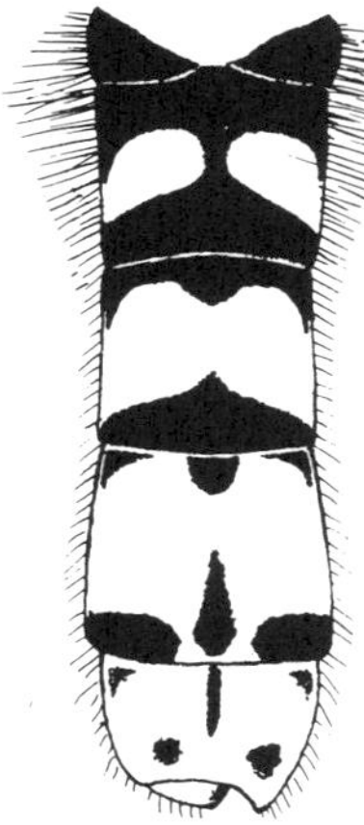

186 S. pyrrhina ♂

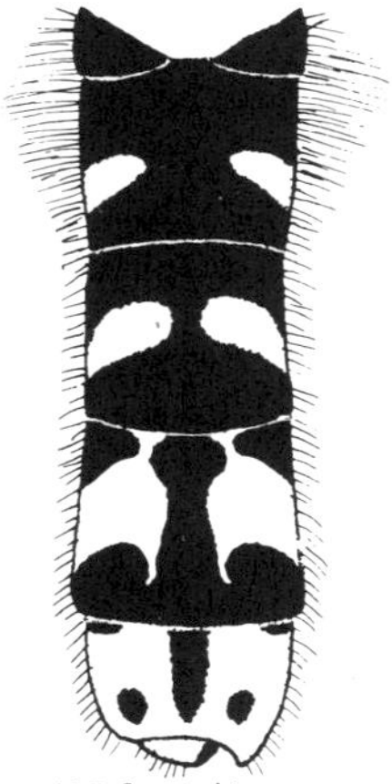

187 S. pyrrhina ♂

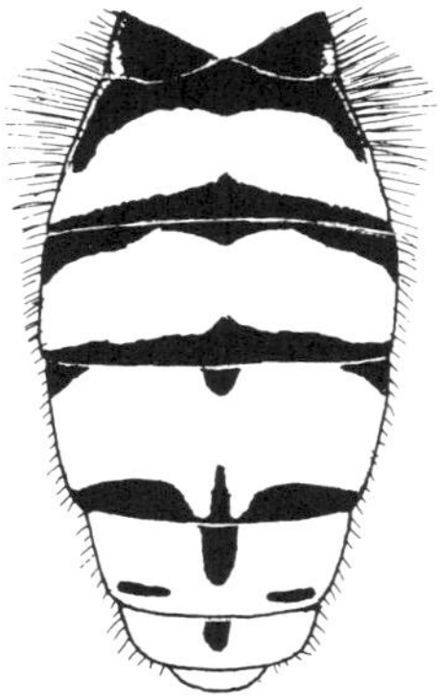

188 S. pyrrhina ♀

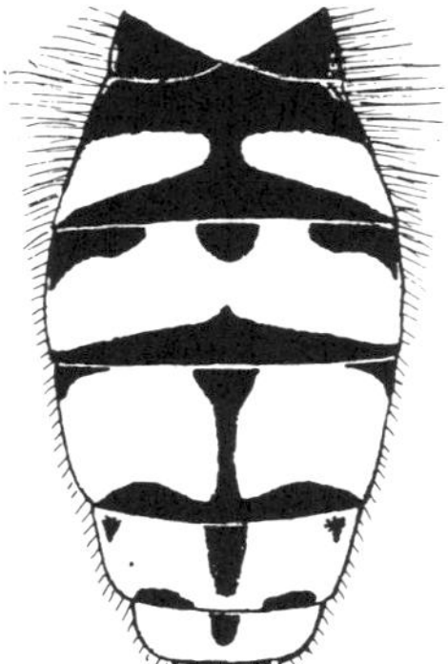

189 S. pyrrhina ♀

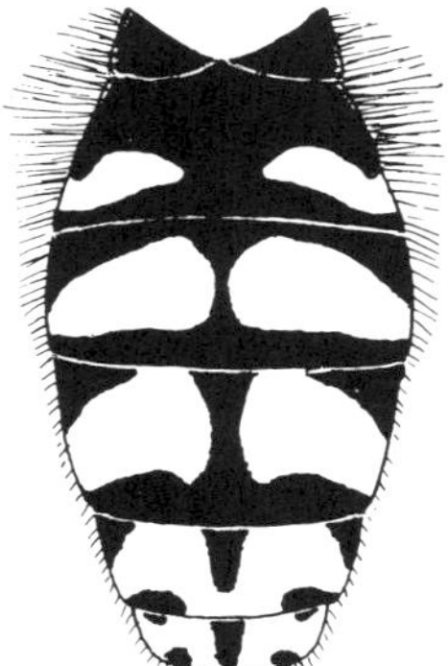

190 S. pyrrhina ♀

Figs. 182–190. Abdomens.

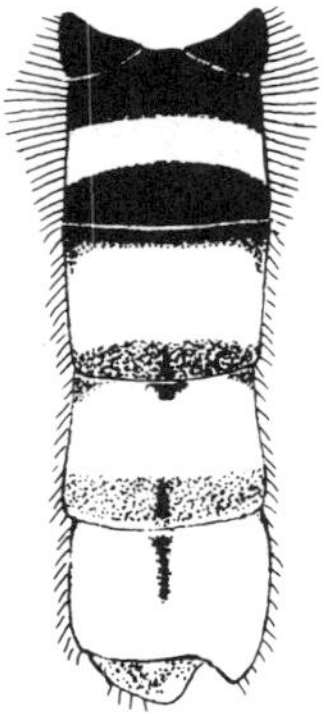

191 Sphaerophoria sulphuripes ♂

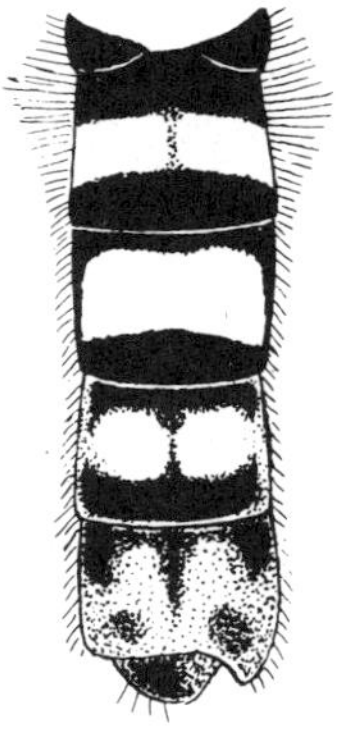

192 S. sulphuripes ♂

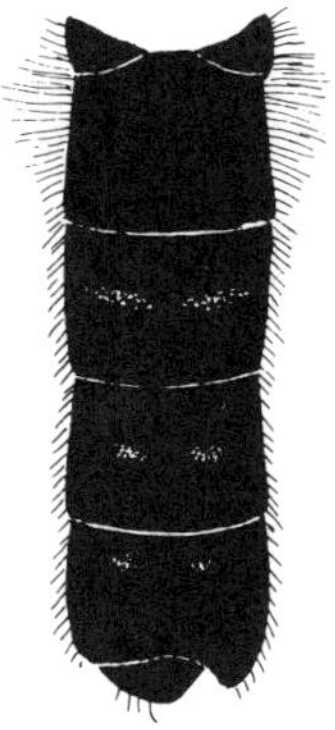

193 S. sulphuripes ♂

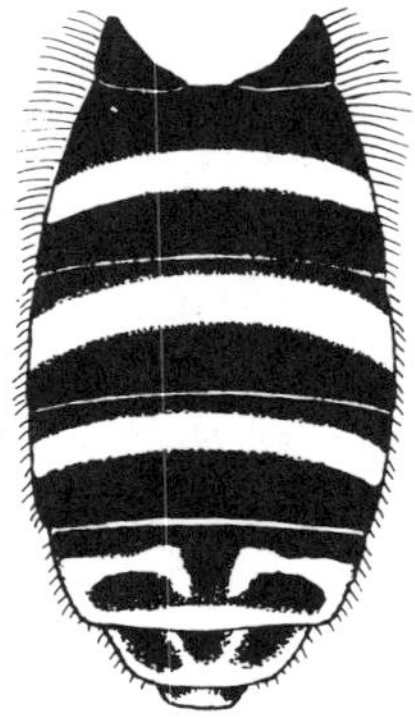

194 S. sulphuripes ♀

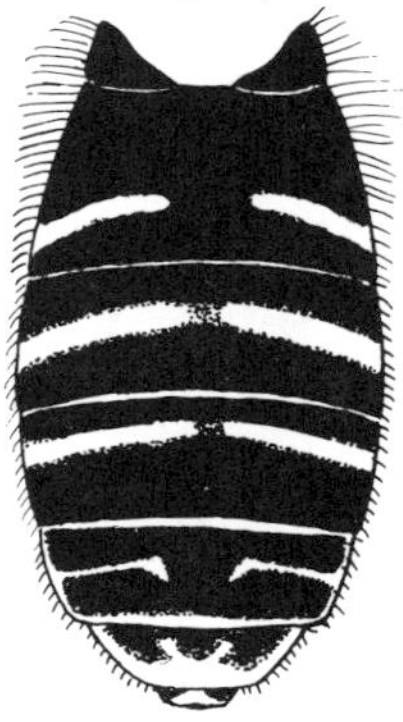

195 S. sulphuripes ♀

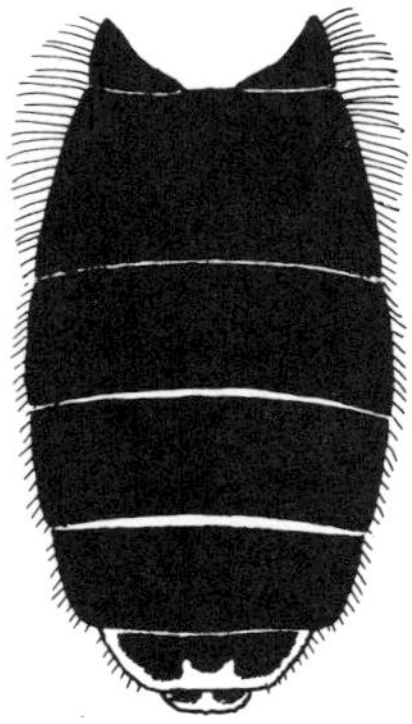

196 S. sulphuripes ♀

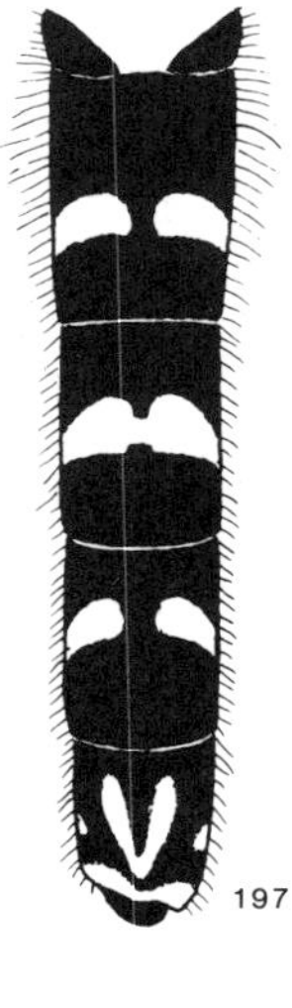

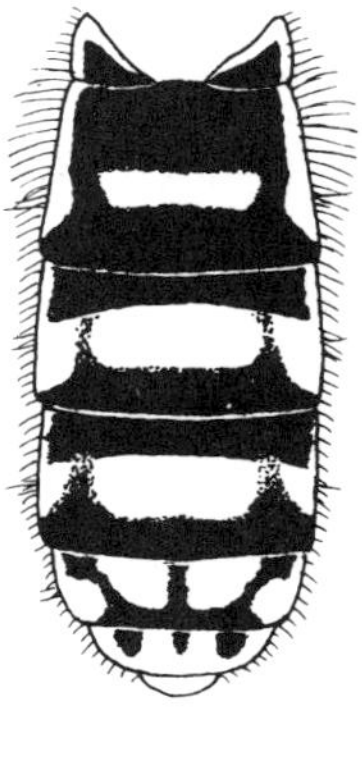

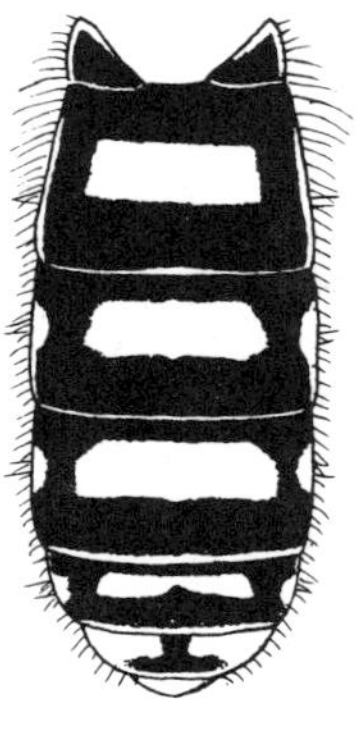

197 S. scripta ♂ **198** S. sp. ♀ "cleoae" pattern **199** S. contigua ♀ "cleoae" pattern

Figs. 191–199. Abdomens.

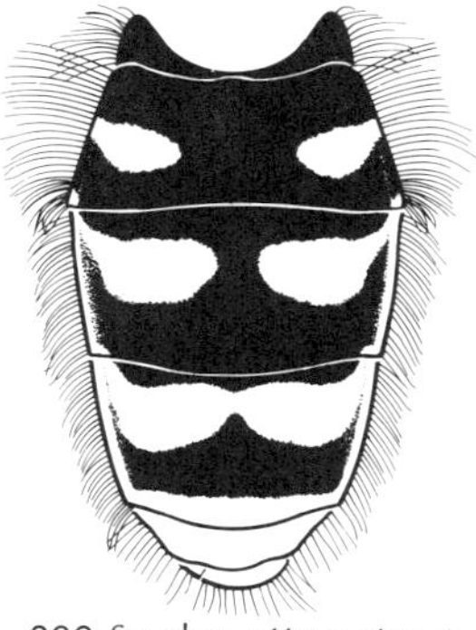

200 Syrphus attenuatus ♂

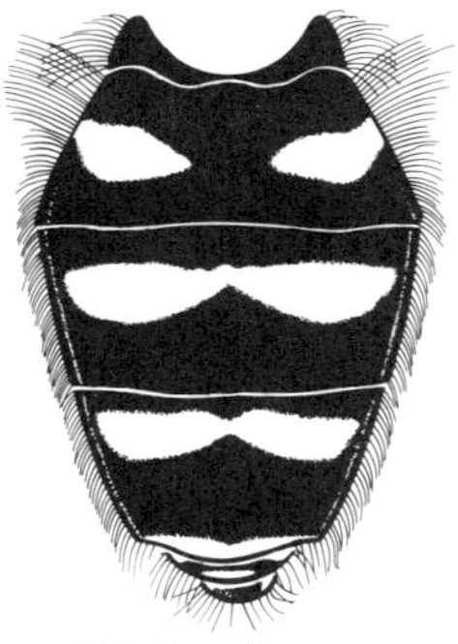

201 S. opinator ♂

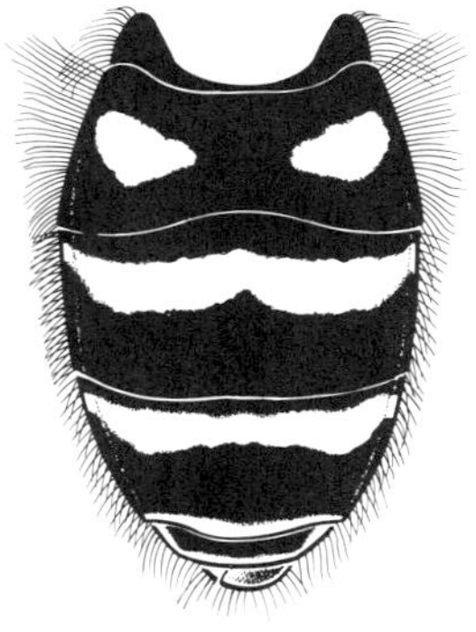

202 S. ribesii ♂

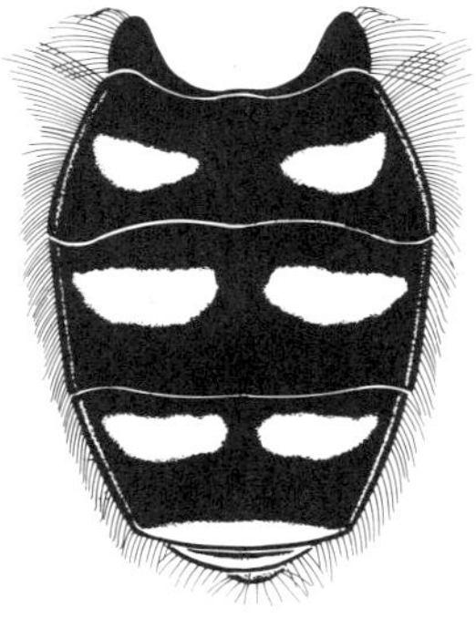

203 S. sexmaculatus ♂

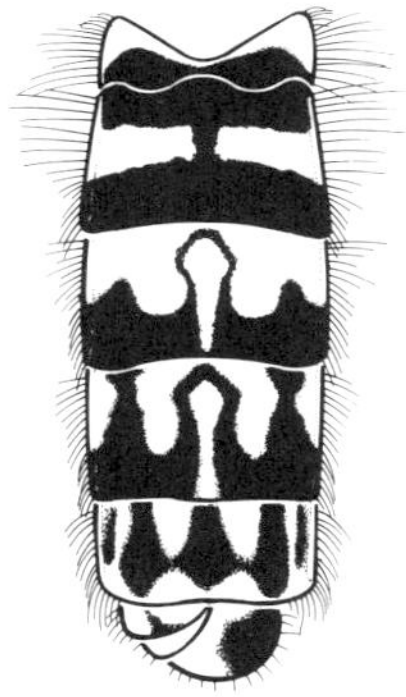

204 Toxomerus geminatus ♂

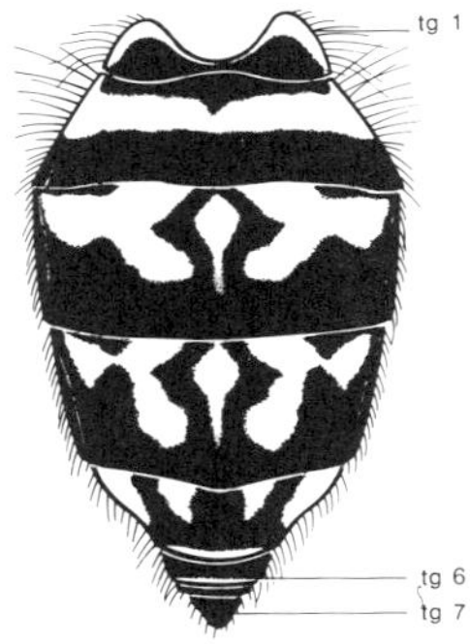

205 T. geminatus ♀

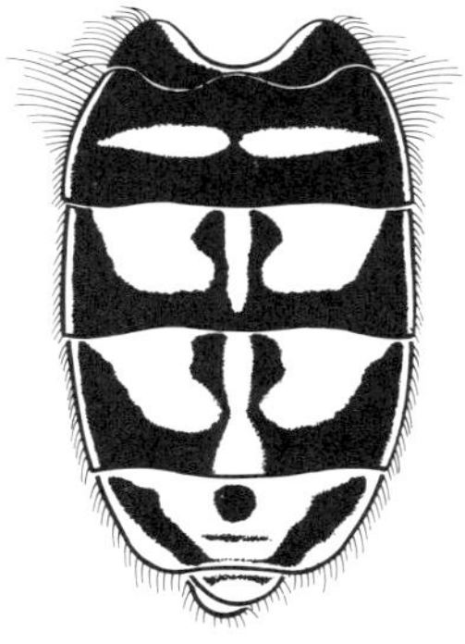

206 T. marginatus ♀

207 T. politus ♂

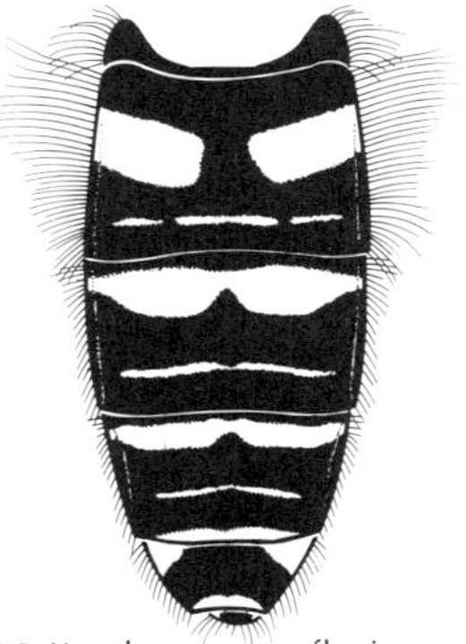

208 Xanthogramma flavipes ♂

Figs. 200–208. Abdomens.

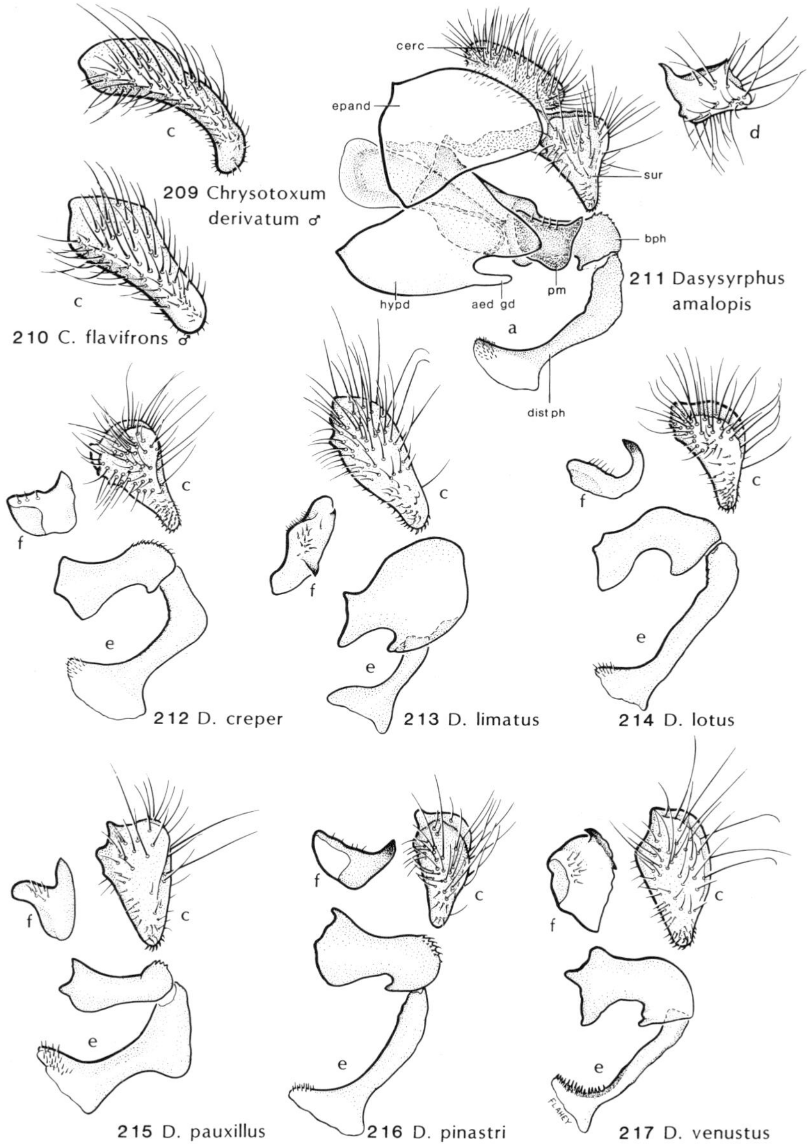

Figs. 209–217. Male terminalia. (Adscripts: *a*, terminalia, left lateral; *c*, left surstylus, lateral; *d*, left surstylus, dorsal; *e*, aedeagus, lateral; *f*, left paramere, lateral.) (Abbreviations: aed b, aedeagal base; aed gd, aedeagal guide; cerc, cercus; dist seg aed, distal segment of aedeagus; epand, epandrium; hypd, hypandrium; pm, paramere; sur, surstylus.)

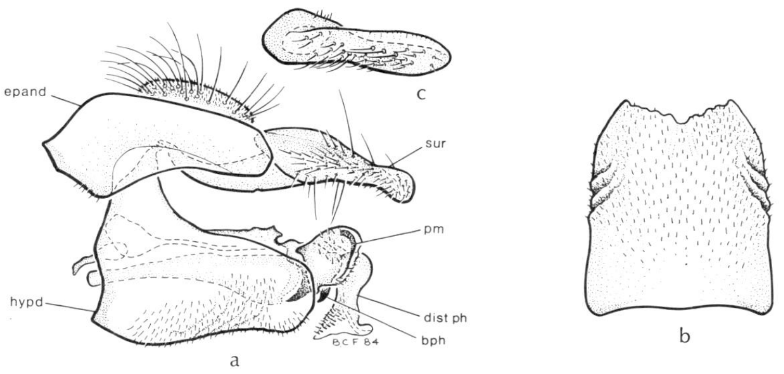

218 Eupeodes americanus

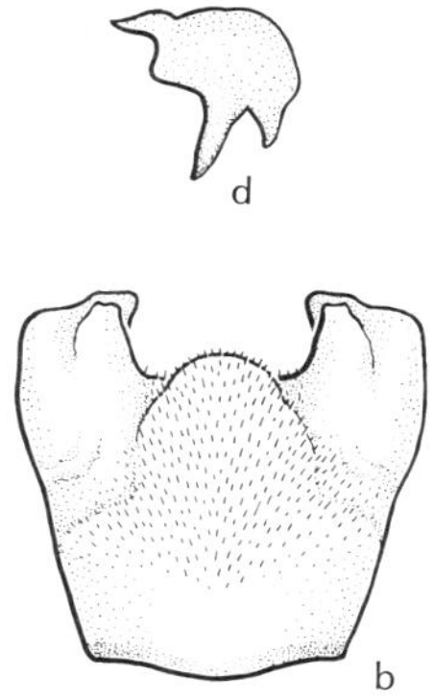

219 E. curtus

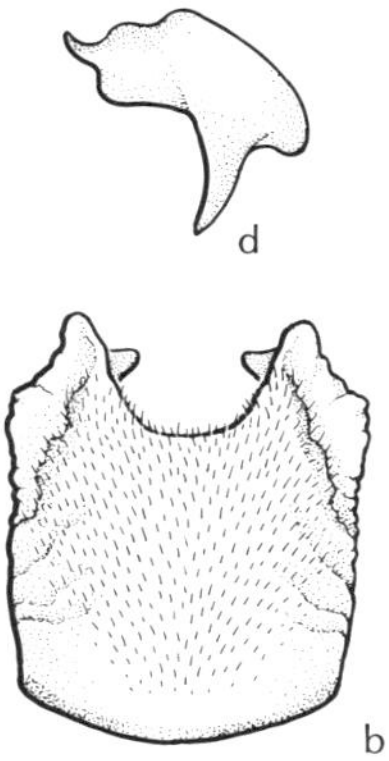

220 E. latifasciatus

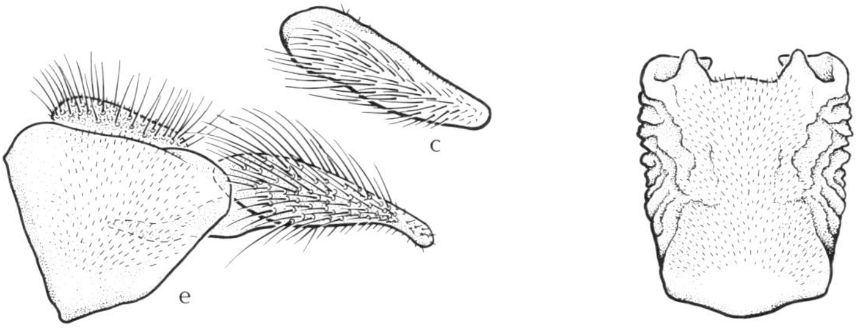

221 E. fumipennis

Figs. 218–221. Male terminalia. (Adscripts: *a*, terminalia, left lateral; *b* hypandrium, ventral; *c*, left surstylus, dorsal; *d*, aedeagal base, left lateral; *e*, epandrium, cercus and surstylus, left lateral. (Abbreviations: aed, aedeagus; dist seg aed, distal segment of aedeagus; epand, epandrium; hypd, hypandrium; pm, paramere; sur, surstylus.)

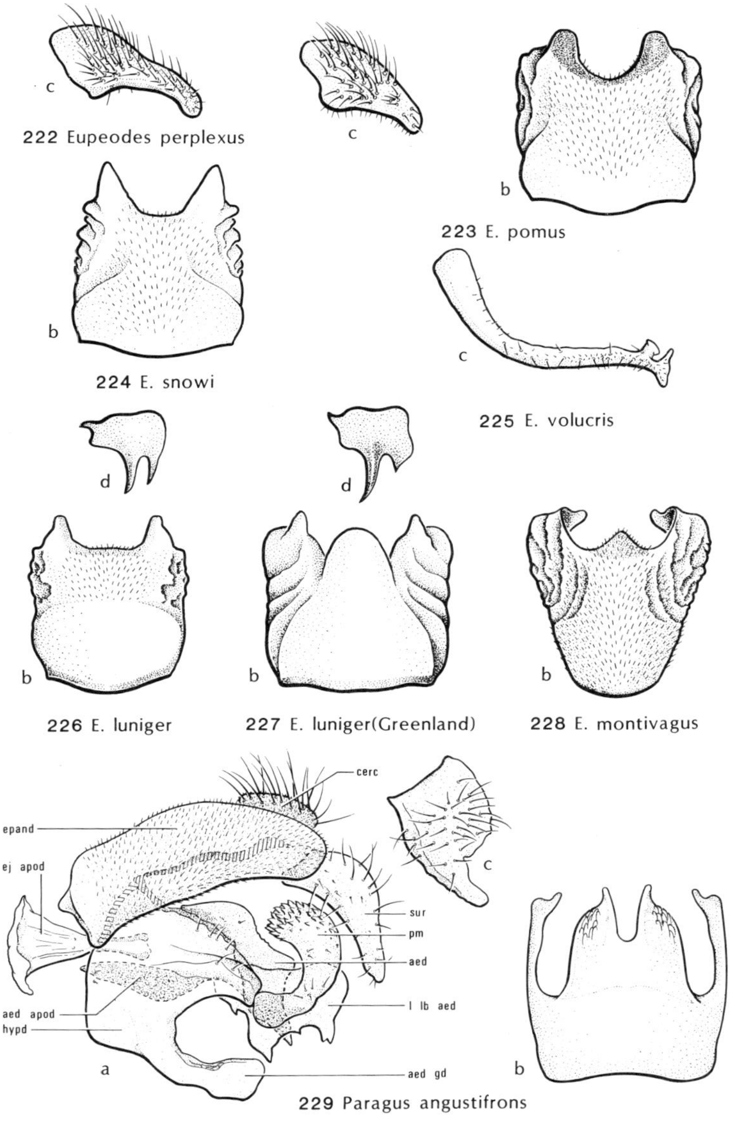

Figs. 222–229. Male terminalia. (Adscripts as for Figs. 218–221.) (Abbreviations: aed, aedeagus; aed apod, aedeagal apodeme; aed gd, aedeagal guide; cerc, cercus; ej apod, ejaculatory apodeme; epand, epandrium; hypd, hypandrium; l lb aed, lateral lobe of aedeagus; pm, paramere; sur, surstylus.)

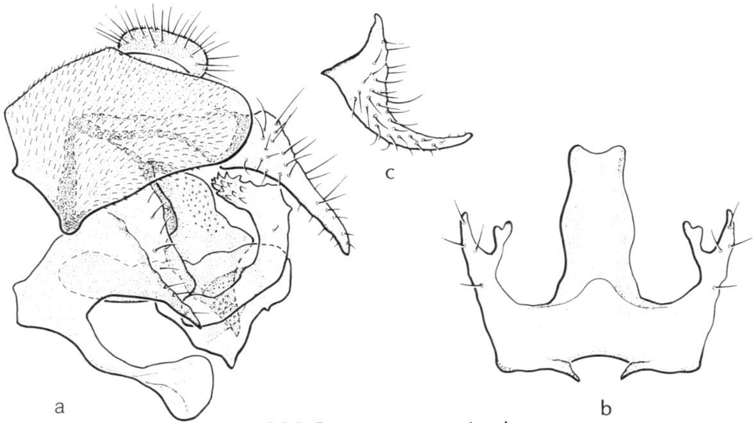

230 Paragus angustistylus

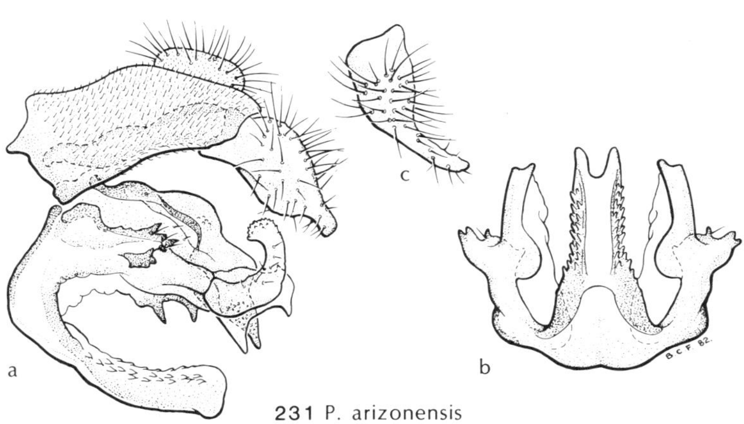

231 P. arizonensis

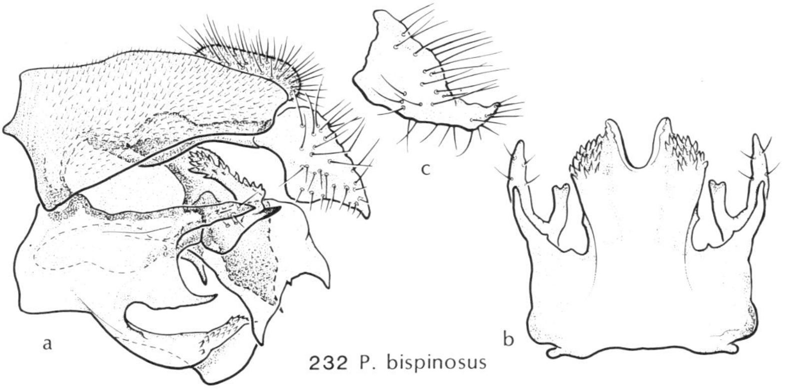

232 P. bispinosus

Figs. 230–232. Male terminalia. (Adscripts as for Figs. 218–221.)

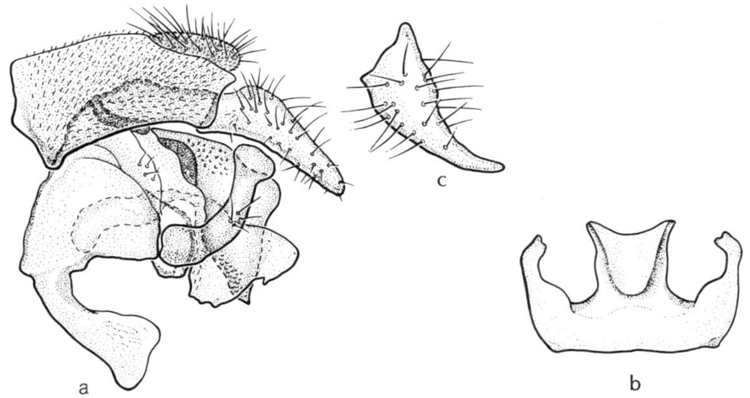

233 P. cooverti

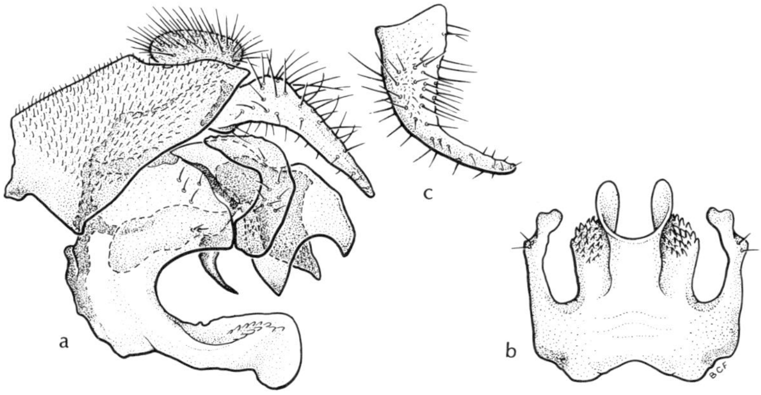

234 P. longistylus

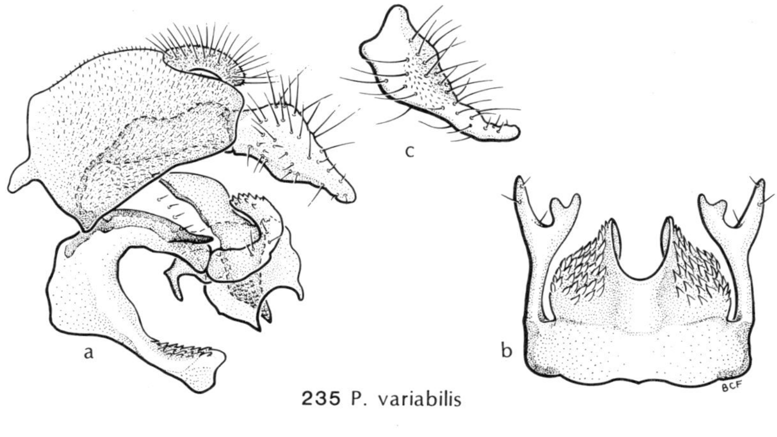

235 P. variabilis

Figs. 233–235. Male terminalia of *Paragus*. (Adscripts as for Figs. 218–221.)

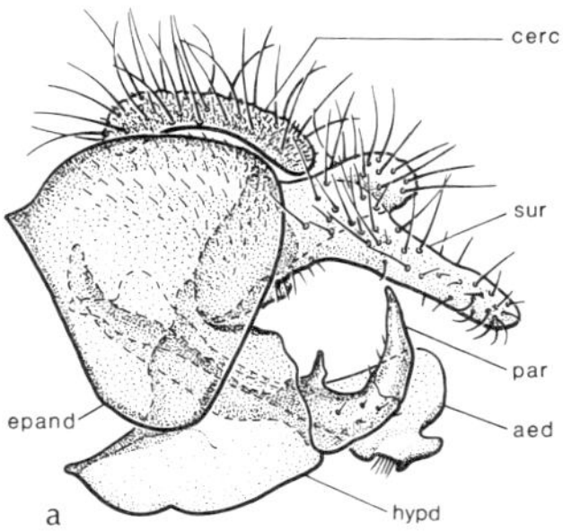

236 Platycheirus scutatus

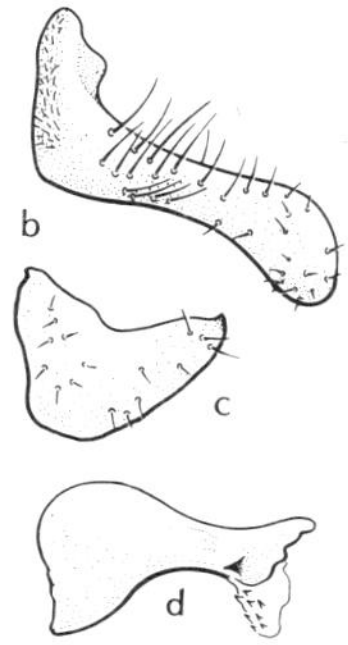

237 P. granditarsis

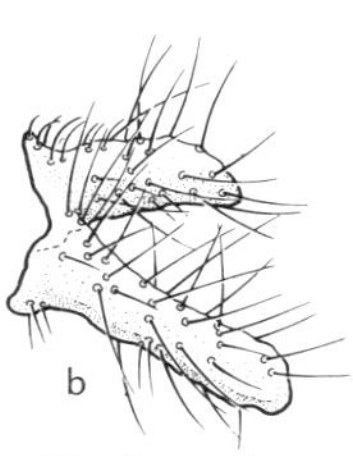

238 P. orarius

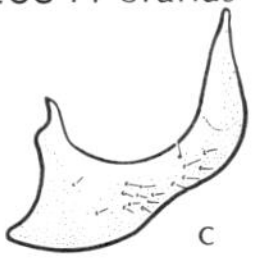

239 P. setitarsis

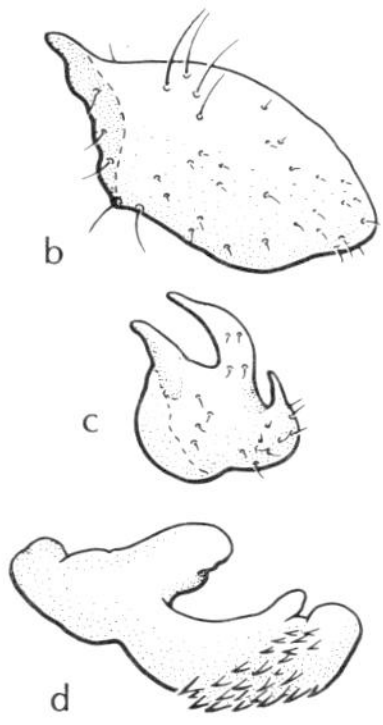

240 P. rufigaster

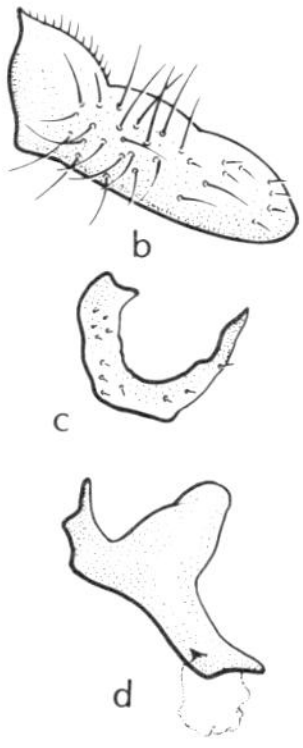

241 P. rosarum

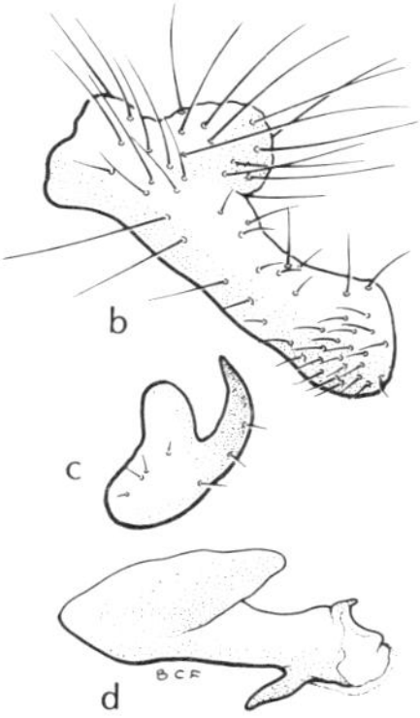

242 P. pullatus

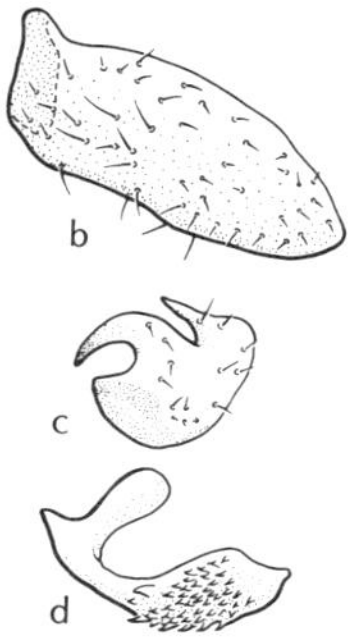

243 P. coracinus

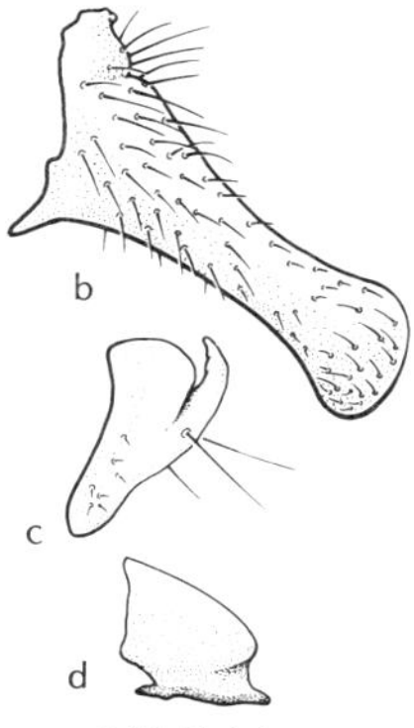

244 P. latus

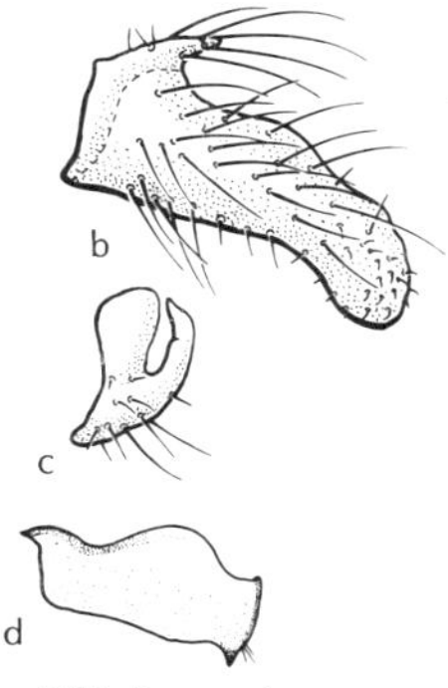

245 P. concinnus

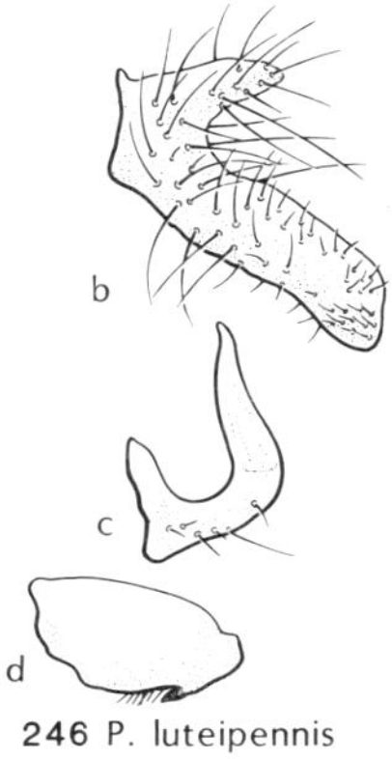

246 P. luteipennis

Figs. 236–246. Male terminalia. (Adscripts: *a*, terminalia, left lateral; *b*, left surstylus, dorsolateral; *c*, left paramere, lateral; *d*, aedeagus, left lateral.) (Abbreviations: aed, aedeagus; cerc, cercus; epand, epandrium; hypd, hypandrium; pm, paramere; sur, surstylus.)

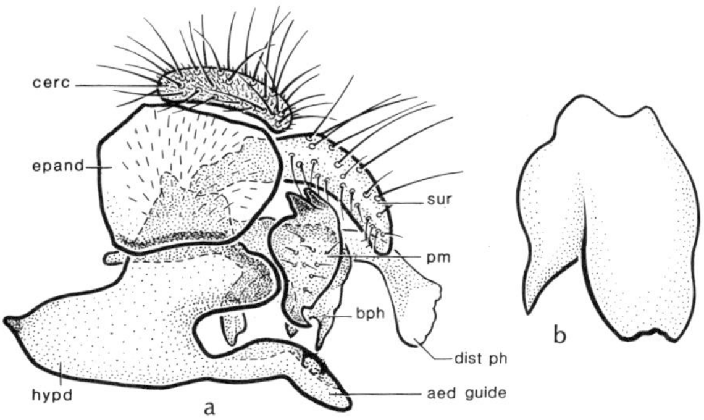

247 Parasyrphus n.sp.

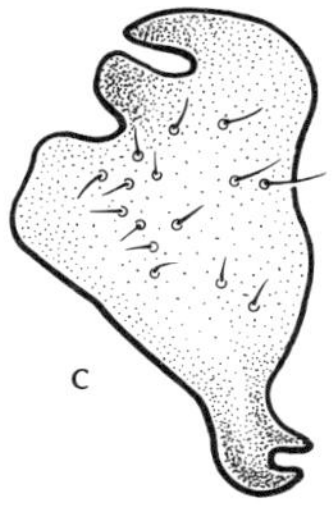

248 P. currani

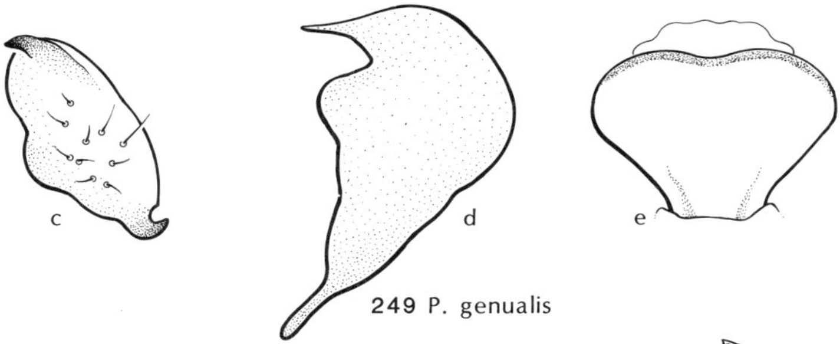

249 P. genualis

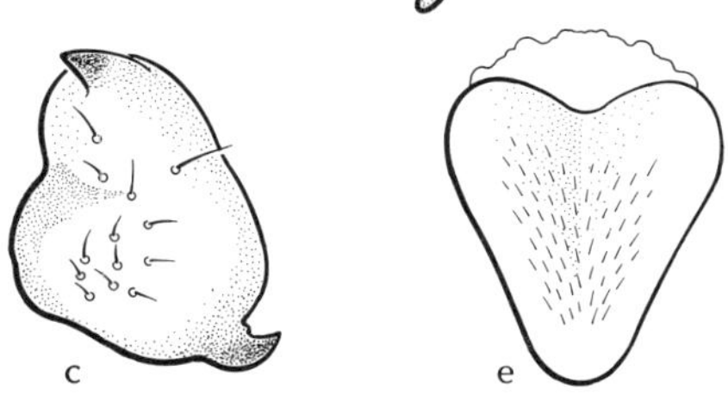

250 P. insolitus

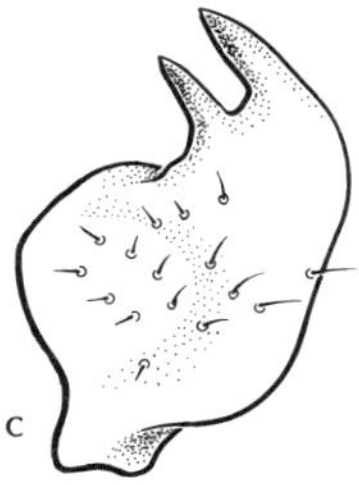

251 P. macularis

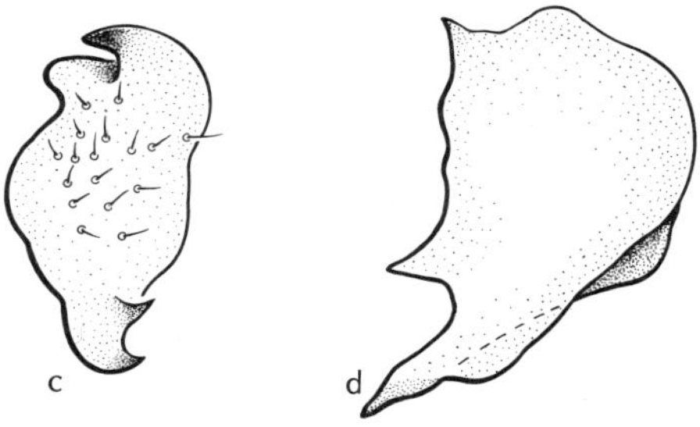

252 P. relictus

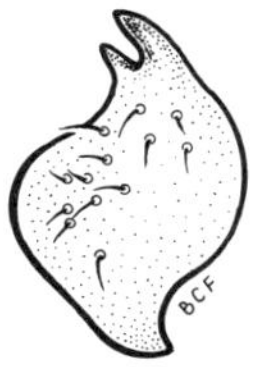

253 P. semiinterruptus

Figs. 247–253. Male terminalia. (Adscripts: *a*, terminalia, left lateral; *b*, aedeagal base, left ventrolateral; *c*, left paramere, lateral; *d*, aedeagal base, left lateral; *e*, distal segment of aedeagus, dorsal.) (Abbreviations: aed b, aedeagal base; cerc, cercus; dist seg aed, distal segment of aedeagus; epand, epandrium; hypd, hypandrium; pm, paramere; sur, surstylus.)

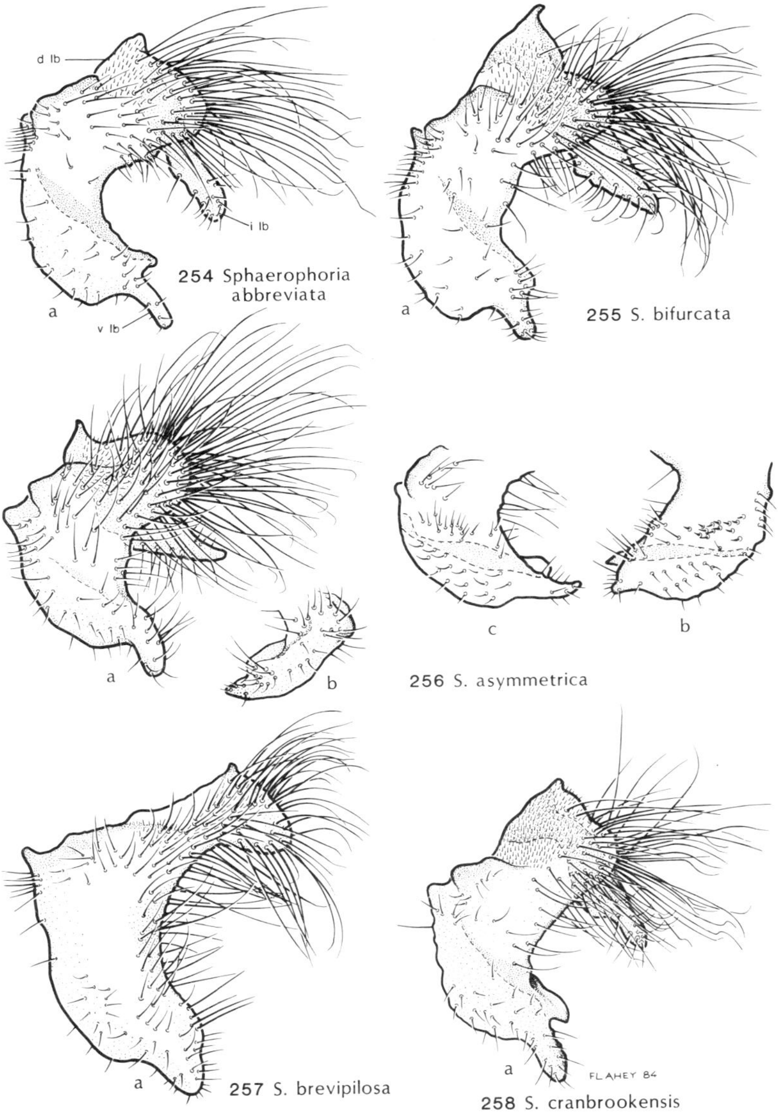

Figs. 254–258. Male terminalia. Note: Two figures of *S. asymmetrica* on left from one specimen, two on right from another specimen. (Adscripts: *a*, left surstylus, lateral; *b*, ventral lobe of right surstylus, lateral; *c*, ventral lobe of left surstylus, lateral.) (Abbreviations: d lb, dorsal lobe; i lb, inner lobe; v lb, ventral lobe.)

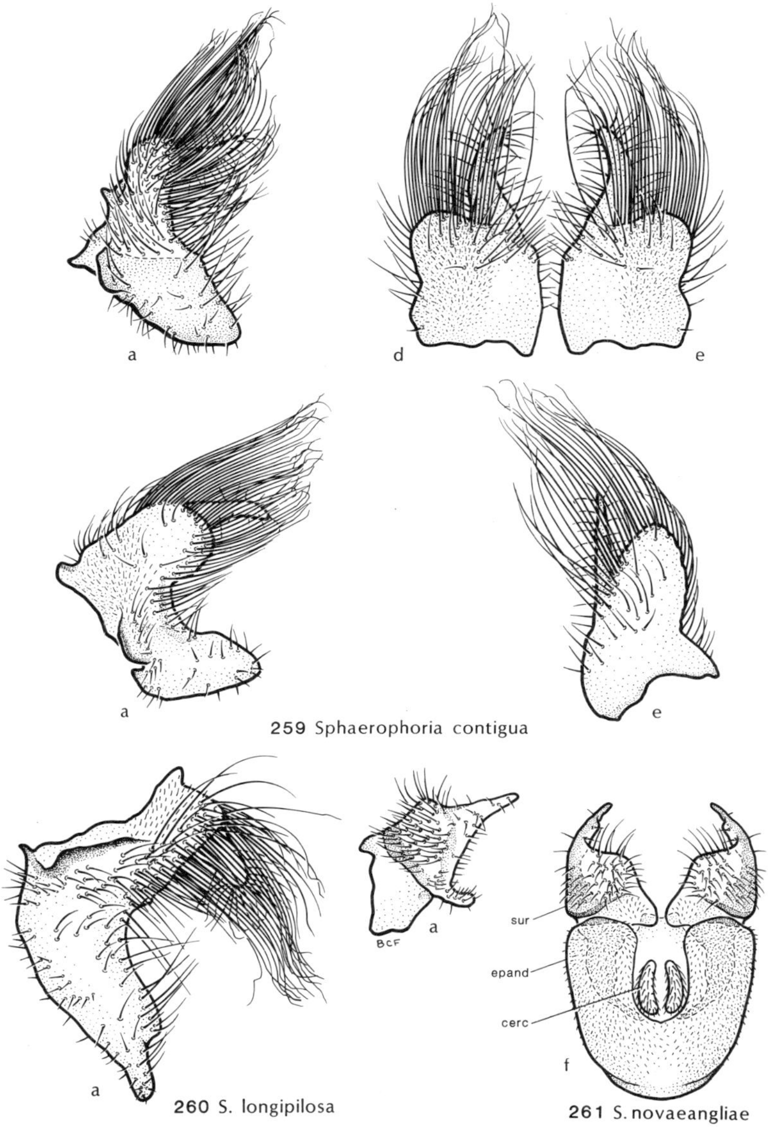

Figs. 259–261. Male terminalia. Note: One specimen of *S. contigua* shown in upper row, another specimen in lower row. (Adscripts: *a*, left surstylus, lateral; *d*, right surstylus, dorsal; *e*, left surstylus, dorsal; *f*, epandrium, cerci and surstylus, dorsal.) (Abbreviations: cerc, cercus; epand, epandrium; sur, surstylus.)

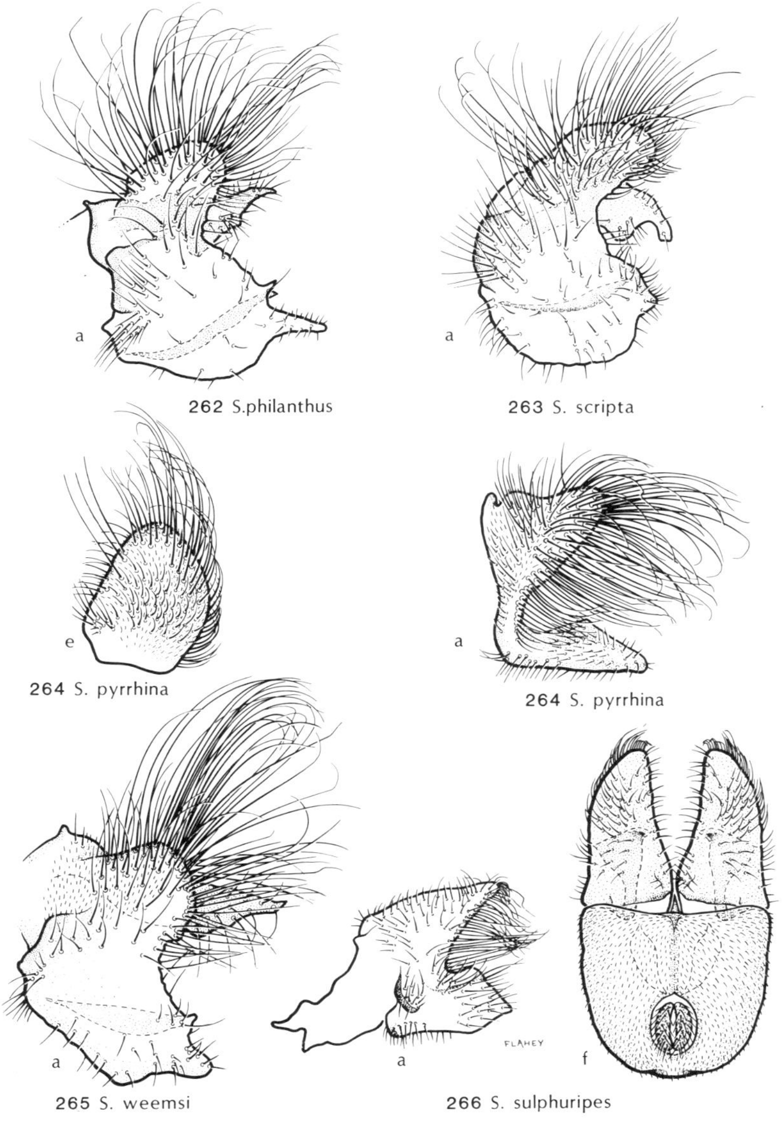

Figs. 262–266. Male terminalia of *Sphaerophoria*. (Adscripts as for Figs. 259–261.)

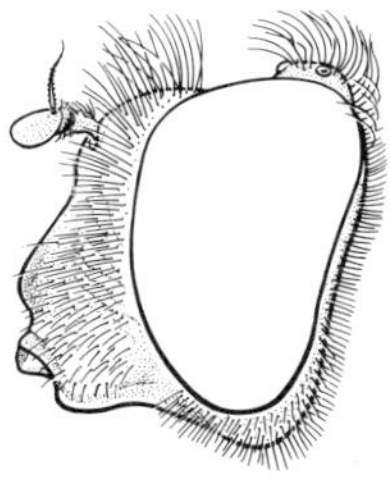

267 Platycheirus confusus

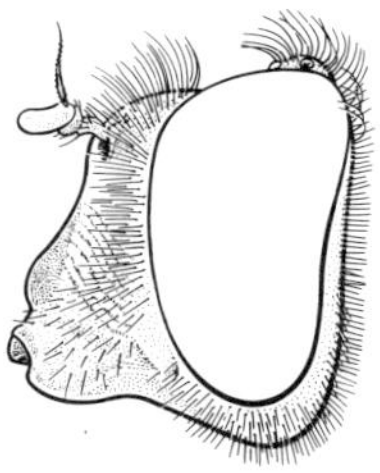

268 P. obscurus

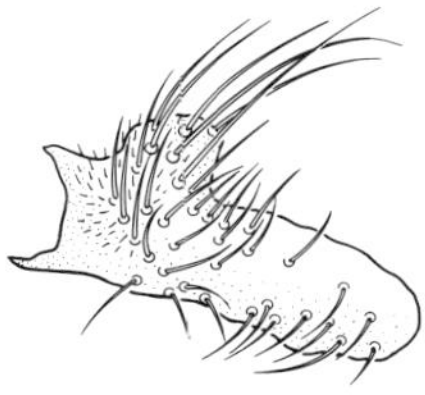

269 P. confusus

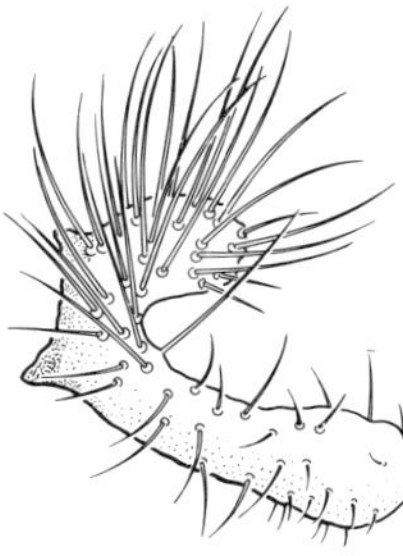

270 P. obscurus

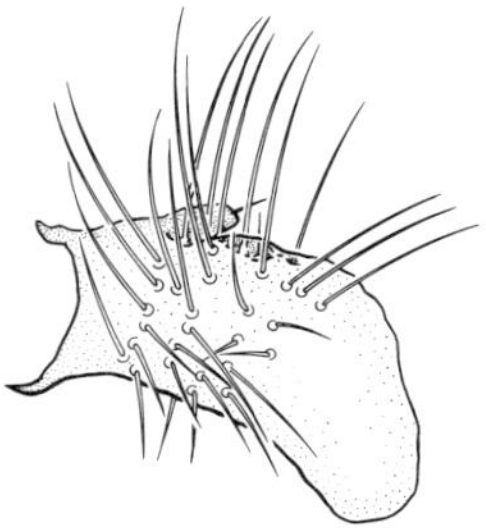

271 P. sabulicola

Figs. 267–271. Heads of males and surstyli.

Glossary[7]

aedeagal guide The slender to broad median process extending posteriorly from a shallow to very deep rounded posterior notch on the ventral surface of the **hypandrium** (lingula of previous authors) (Figs. 211*a*, 229*a*).

aedeagus The male organ used for the transfer of sperm; in most Syrphini clearly divided into an aedeagal base (**basiphallus**) and an articulated and usually mostly membranous distal portion (**distiphallus**) (Figs. 211, 218); in Bacchini, Paragini, and Toxomerini not divided, simple and nearly cylindrical in some, complex and with prominent lateral lobes in others (Figs. 211, 229, 236).

alula A subtriangular apically rounded membranous lobe at base of the posterior margin of the wing (Fig. 32).

anal lobe A posterobasal portion of the wing behind vein A_1 (Fig. 32).

anepimeron A **sclerite** of thoracic **pleura** lying below the wing base; always at least partly haired in Syrphidae (Fig. 39).

anepisternum A **sclerite** of thoracic **pleura** lying between the anterior spiracle and **anepimeron** and divided into a convex, haired, posterior portion and a flat, haired or bare, anterior portion (Fig. 39).

arista A bristlelike structure arising near the base of the dorsal surface of the first **flagellomere** (Fig. 4).

bare (of sclerite) Without obvious erect hairs.
(of wing membrane) Without **microtrichia**.

calypter (pl. **calypteres**) One of the two large, semimembranous lobes between the base of the wing and the thorax; when the wing is folded in over the abdomen, the outer lobe (**upper calypter**) lies above the inner lobe (**lower calypter**) (Fig. 37).

cell An area of the wing bounded by **veins**; named usually from the vein along its anterior margin; abbreviation written in lower case (e.g., cell r_{4+5}); cells shown in Fig. 33.

cercus One of a pair of weak, haired lobes (**cerci**) lying on either side of the anal opening; in the male set in a posterodorsal notch in the **epandrium** or, rarely, surrounded by the epandrium (Figs. 211, 229, 261).

coxa (pl. **coxae**) The first segment of each leg (Figs. 40, 68*h*).

---

7 For a fuller discussion of most terms see McAlpine (1981).

dichoptic With compound eyes distinctly and usually broadly separated on the upper surface of the head (as in Fig. 22).
discal On the central part of a **sclerite** or other structure.

emarginate Notched.
epandrium The large shell-like **sclerite**, composed mostly or entirely of tergite 9, covering the **aedeagus** and associated structures and bearing the articulated **surstyli** posterolaterally; in a resting position its dorsal surface directed ventrally below the apex of the abdomen (Figs. 211, 229, 236, 247).
eye angle The anterior angle of contact of the compound eyes of the male.

face The anterior part of the head bounded above by the antennal insertions and laterally by the compound eyes (Fig. 15).
fascicle A cluster or bundle of closely placed structures (usually hairs or bristles) appearing almost like a single structure.
femur (pl. **femora**) The third segment of each leg (Figs. 31, 47*d*, 68*h*).
flagellomere A subdivision of the **flagellum**; in Syrphidae the flagellum consists of a large first flagellomere (third antennal segment of previous authors) and a bristlelike **arista** made up of two or three additional flagellomeres (Fig. 4).
flagellum The third segment of each antenna; in almost all Diptera subdivided into **flagellomeres**.
frons The anterodorsal surface of the head; triangular in male and bounded by compound eyes and antennal insertions; subrectangular in female and bounded above by upper margin of head, laterally by compound eyes, and below by antennal insertions (Figs. 15, 26).

half-band Half of a medially divided pale band of an abdominal **tergite**.
halter (pl. **halteres**) The stalked knob arising from each side of the dorsolateral part of the thorax behind the wing (Fig. 39).
holoptic With compound eyes touching or nearly so on the upper surface of the head (as in Figs. 6, 15).
hypandrium The troughlike **sclerite** composed mostly or entirely of **sternite** 9, lying below the **epandrium**, supporting the **aedeagus** (which projects medially from its posterior end), and bearing the **parameres** at its posterodorsal lateral angles (Figs. 211, 229, 236, 247).

incisure Slightly depressed or infolded line of junction between successive abdominal **tergites** or **sternites**.

katatergite The strongly convex pleural **sclerite** lying between the **anepimeron** and the base of the **halter**; with minute dense pile but without erect hairs in Nearctic Syrphinae (Fig. 39).
katepimeron The convex, nearly transverse **sclerite** lying below the **anepimeron** and in front of the posterior spiracle; always haired in Syrphinae (Fig. 39).
katepisternum The large subtriangular pleural **sclerite** lying below the **anepisternum** and the **anepimeron** and between the fore and mid **coxae**, and extending to the mid-ventral line of the thorax; with ventral hairs and usually with dorsal hairs in Syrphinae (Fig. 39), the two groups of hairs separated or partly confluent.

lunulate Crescent-shaped.

margined abdomen With a shallow longitudinal depression just inside the lateral margin of each **tergite** (Figs. 104–110, 124–139).
membrane The transparent portion of a wing.
meron The pleural **sclerite** lying below the **katepimeron** and between the mid and hind **coxae** (Fig. 39).
metapleuron The lower posterior portion of the thoracic **pleura** lying behind the **meron**, below the posterior spiracle, and above the hind **coxa**, divided by a faint vertical line into an anterior **metepisternum** and a posterior **metepimeron** (Fig. 39).
metasternum The ventral surface of the thorax between the mid and hind **coxae**; bare or with rather long hairs.
metepimeron The posterior division of the **metapleuron**; usually bare but with fine hairs in some species (Fig. 39).
metepisternum The anterior division of the **metapleuron**; usually bare but with fine hairs in some species (Fig. 39).
microtrichia The minute hairlike structures on the wing **membrane**.

notopleural callus The swollen posterior part of the **notopleuron**.
notopleuron The anterolateral portion of **scutum** lying between the **postpronotum** and **transverse suture**; usually slightly depressed anteriorly and slightly swollen posteriorly (Fig. 39).

ommatidium (pl. **ommatidia**) A single facet of the eye.
opercula The anterior sections of the **puparium** loosened or detached when the adult emerges.

paramere A structure articulated with (or rarely fused with) the lateral apex of the **hypandrium** and lying beside the base of the aedeagus (Figs. 211, 218, 229, 236, 247).
pedicel The second segment of each antenna (Fig. 4).
petiolate Slender near the base and broadened toward the apex.

pleura (pl. **pleurae**) The lateral surface of the thorax.

porrect Projecting forward.

postalar callus The slightly swollen posterolateral part of the **scutum** lying between the wing base and the **scutellum** (Fig. 39).

postmetacoxal bridge The sclerotized posterior surface of the thorax below the base of the abdomen and above the hind **coxae** joining the **metepimera** of the two thoracic **pleurae**.

postocular orbit The lateral surface of head behind each compound eye (Fig. 12).

postpronotum The anterolateral **sclerite** of the dorsal surface of the thorax; bare in Nearctic Syrphinae (Fig. 40).

postsutural Behind the transverse suture of the **scutum**.

presutural In front of the transverse suture of the **scutum**.

proepimeron The **sclerite** at the anteroventral corner of the lateral surface of the thorax, immediately above the fore **coxa**; always haired in Syrphinae (Fig. 40).

pruinose Dull because of minute hairlike outgrowths of the surface looking like fine dust (pollinose or dusted of previous authors).

pruinosity The minute hairlike outgrowths that give a dull appearance to a surface.

puparium The hardened, retracted, and nonmobile skin of the mature larva inside which pupation and adult development occur.

scape The first segment of the antenna (Fig. 4).

sclerite A more or less well defined plate making up part of the body surface, e.g., the **anepisternum** of the thoracic **pleura** or the **tergite** of an abdominal segment.

scutellum The semicircular **sclerite**, with convex upper and posterior surface and flat lower surface, projecting at posterodorsal part of thorax (Figs. 38, 39).

scutum The slightly convex dorsal surface of the thorax in front of the **scutellum** but not including the **postpronotum** at each anterolateral corner (Fig. 39).

setulae Short stiff hairs.

sternite A **sclerite** forming the ventral surface of an abdominal segment.

subcranial cavity The cavity on underside of head into which mouthparts are retracted.

surstylus (pl. **surstyli**) The main clasper of the male **terminalia**, articulated with the dorsolateral apical margin of the **epandrium** (Figs. 211, 218, 229, 236, 247, 261).

tarsomere The secondary division of a **tarsus** (always five in Syrphidae).

tarsus (pl. **tarsi**) The fifth (last apparent) segment of each leg; in Syrphidae secondarily divided into five **tarsomeres** (Fig. 47*d*).

tergite A **sclerite** forming the dorsal surface of an abdominal segment (Figs. 102, 148).
terminalia The segments and associated structures at the apex of the abdomen, modified for copulation in males, and for copulation and oviposition in females.
tibia (pl. **tibiae**) The fourth segment of each leg (Fig. 47*d*).
transverse suture (of scutum) A weakly impressed line across the surface of the **scutum** beginning on each side just behind the **notopleuron** (Fig. 39).
trichose Bearing **microtrichia** (of wing **membrane**).
trochanter The second segment of each leg (Fig. 68).

unmargined abdomen **Tergites** curving smoothly under without a longitudinal depression near the lateral margin of each. (Figs. 102, 103, 155–199).
upper postocular orbit The area between the upper half of the eye and the slightly concave posterior surface of the head (Fig. 13).

vein A riblike stiffening along the anterior margin or on the surface of wing (Fig. 32); abbreviation written in upper case (e.g., vein R_{4+5}).
ventral scutellar fringe A single or multiple row of downward pointing hairs along the posterior margin of the flat ventral surface of the **scutellum**, extending slightly onto ventral surface in some specimens (Fig. 38).

References

Andersson, H. 1970. Taxonomic notes on the genera *Platycheirus* and *Melanostoma* (Dipt. Syrphidae) with lectotype designations. Entomologica scand. 1:236–240.

Aubert, J., P. Goeldlin, and J.-P. Lyon. 1969. Essais de marquage et de reprise d'insectes migrateurs en automne 1968. Mitt. schweiz. ent. Ges. 42:140–166.

Bańkowska, R. 1964. Studien über die paläarktischen Arten der Gattung *Sphaerophoria* St. Farg. et Serv. (Diptera, Syrphidae). Annls zool., Warsz. 22:285–353.

Bansch, R. 1966. On prey-seeking behaviour of aphidophagous insects. Pages 123–128, 1 pl. *in* I. Hodek, ed. Ecology of aphidophagous insects. Academia, Prague.

Barlow, C.A. 1961. On the biology and reproductive capacity of *Syrphus corollae* Fab. in the laboratory. Entomologia exp. appl. 4:91–100.

Becker, T. 1915. Cyclorrhapha Aschiza. Pages 60–62 *in* T. Becker, H. Dziedzicki, J. Schnabl, and J. Villeneuve. Résultats scientifiques de l'expédition des freres Kuznecov (Kouznetzov) a l'Oural arctique en 1909, sous la direction de H. Backlund. Zap. imp. Akad. Nauk Ser. 8, 28(7):1–67, 3 pls.

Becker, T. 1921. Neue Dipteren meiner Sammlung. Mitt. zool. Mus. Berl. 10:3–93.

Bigot, J.M.F. 1883. Diptères nouveaux ou peu connus. 22e partie, 32: Syrphidi (2e partie). Espèces nouvelles, no. I. Annls Soc. ent. Fr. ser. 6, 3:315–356.

Bigot, J.M.F. 1884. Diptères nouveaux ou peu connus. 24e partie, 32: Syrphidi (2e partie). Espèces nouvelles, no. III. Annls Soc. ent. Fr. ser. 6, 4:73–116.

Bombosch, S. 1957. Erfahrungen über Laboratoriumszuchten aphidivorer Schwebfliegen (Dipt. Syrphidae). TagBer. dt. Akad. LandwWiss. Berl. 11:160–162.

Bombosch, S. 1959. Ein Beitrag zur Frage biozönoseschonender Schädlingsbekämpfung in Rübenbau. Pages 975–977 *in* Verh. int. PflSchutz Kongr., Hamburg 1957.

Bombosch, S. 1962. Ueber den einfluss der Nahrungsmenge auf die Entwicklung von *Syrphus corollae* Fabr. (Dipt. Syrphidae). Z. angew. Ent. 50:40–45.

Boyle, R.M.D., and B.J.R. Philogene. 1983. The native pollinators of an apple orchard: variations and significance. J. hort. Res. 58:355–363.

Brower, J.V.Z., and L.P. Brower. 1965. Experimental studies of mimicry. 8. Further investigations of honeybees (*Apis mellifera*) and their dronefly mimics (*Eristalis* spp.). Am. Nat. 99:173–187.

Brower, L.P., J.V.Z. Brower, and P.W. Westcott. 1960. Experimental studies of mimicry. 5. The reaction of toads (*Bufo terrestris*) to bumblebees (*Bombus americanorum*) and their robberfly mimics (*Mallophora bomboides*), with a discussion of aggressive mimicry. Am. Nat. 94:343–356.

Burns, D.P., and D.E. Donley. 1970. Biology of the tuliptree scale, *Toumeyella liriodendri* (Homoptera: Coccidae). Ann. ent. Soc. Am. 63:228–235.

Campbell, R.E., and W.M. Davidson. 1924. Notes on aphidophagous Syrphidae of southern California. Bull. Sth. Calif. Acad. Sci. 23:3–9, 59–71.

Chandler, A.E.F. 1966. Some aspects of host plant selection in aphidophagous Syrphidae. Pages 113–115 *in* I. Hodek, ed. Ecology of aphidophagous insects. Academia, Prague.

Chandler, A.E.F. 1968. A preliminary key to the eggs of some of the commoner aphidophagous Syrphidae (Diptera) occurring in Britain. Trans. R. ent. Soc. Lond. 120:199–217.

Chandler, A.E.F. 1969. Locomotory behaviour of first instar larvae of aphidophagous Syrphidae (Diptera) after contact with aphids. Anim. Behav. 17:673–678.

Clausen, C.P. 1915. Mealy bugs of citrus trees. Bull. Calif. agric. Exp. Stn 258:19–48.

Coe, R.L. 1953. Diptera. Family Syrphidae. Handbk Ident. Br. Insects 10(1):1–98.

Cole, F.R. 1969. The flies of western North America. University of California Press, Berkeley and Los Angeles. xi + 693 pp.

Collin, J.E. 1931. The Oxford University expedition to Greenland, 1928. Diptera (Orthorrhapha Brachycera and Cyclorrhapha) from Greenland. Ann. Mag. nat. Hist. ser. 10, 7:67–91.

Coquillett, D.W. 1900. Papers from the Harriman Alaska expedition. IX. Entomological results (3): Diptera. Proc. Wash. Acad. Sci. 2:389–464.

Curran, C.H. 1921*a*. A revision of *Syrphus* species belonging to the *ribesii* group (Dipt.). Can. Ent. 53:152–160.

Curran, C.H. 1921*b*. New species of Syrphidae (Diptera). Can. Ent. 53:171–176.

Curran, C.H. 1922*a*. A new western syrphid (Diptera). Can. Ent. (1921) 53:258–260.

Curran, C.H. 1922*b*. New species of Canadian Syrphidae (Diptera). Pt. I. Can. Ent. (1921) 53:275–276.

Curran, C.H. 1922*c*. New and little-known Canadian Syrphidae (Diptera). Can. Ent. 54:94–96.

Curran, C.H. 1923*a*. Two undescribed syrphid flies from New England. Occ. Pap. Boston Soc. nat. Hist. 5:65–67.

Curran, C.H. 1923*b*. Our North American *Leucozona*, a variety of *lucorum* (Syrphidae, Diptera). Can. Ent. 55:38.

Curran, C.H. 1923*c*. The *Stenosyrphus sodalis* group (Syrphidae, Diptera). Can. Ent. 55:59–64.

Curran, C.H. 1923*d*. Changes of names. Can. Ent. 55:74.

Curran, C.H. 1923*e*. New cyclorrhaphous Diptera from Canada. Can. Ent. 55:271–279.

Curran, C.H. 1924*a*. Synopsis of the genus *Chrysotoxum* with notes and descriptions of new species (Syrphidae, Diptera). Can. Ent. 56:34–40.

Curran, C.H. 1924*b*. A new Canadian syrphid (Diptera). Can. Ent. 56:288.

Curran, C.H. 1924*c*. New species of Syrphidae. Occ. Pap. Boston Soc. nat. Hist. 5:79–82.

Curran, C.H. 1925. Contribution to a monograph of the American Syrphidae from north of Mexico. Kans. Univ. Sci. Bull. (1924) 15:7–216.

Curran, C.H. 1926. New nearctic Diptera mostly from Canada. Can. Ent. 58:81–89.

Curran, C.H. 1927*a*. Synopsis of males of the genus *Platycheirus* St. Fargeau and Serville with descriptions of new Syrphinae (Diptera). Am. Mus. Novit. 247:1–13.

Curran, C.H. 1927*b*. Notes on Syrphidae (Diptera). Can. Ent. 59:205–207.

Curran, C.H. 1930*a*. New Diptera from North and Central America. Am. Mus. Novit. 415:1–16.

Curran, C.H. 1930*b*. Report on the Diptera collected at the Station for the Study of Insects, Harriman Interstate Park, N.Y. Bull. Am. Mus. nat. Hist. (1931) 61:21–115.

Curran, C.H. 1931*a*. Four new Diptera in the Canadian National Collection. Can. Ent. 63:93–98.

Curran, C.H. 1931*b*. Some new North American Diptera. Can. Ent. 63:249–254.

Curtis, J. 1834. British entomology: being illustrations and descriptions of the genera of insects found in Great Britain and Ireland. Vol. 11. Curtis, London. Pls. 482–529.

Davidson, W.M. 1916. Economic Syrphidae in California. J. econ. Ent. 9:454–457.

Davidson, W.M. 1922. Notes on certain species of *Melanostoma* (Diptera: Syrphidae). Trans. Am. ent. Soc. 48:35–47, 1 pl.

Degeer, C. 1776. Mémoires pour servir à l'histoire des Insectes. Vol. 6. Hosselberg, Stockholm. 523 pp., 30 pls.

Dixon, T.J. 1960. Key to and descriptions of the third instar larvae of some species of Syrphidae (Diptera) occurring in Britain. Trans. R. ent. Soc. Lond. 112:345–379.

Dušek, J., and J. Křístek. 1967. Zur Kenntnis der Schwebfliegenlarven (Diptera, Syrphidae) in den Gallen der Pappelblattläuse (Homoptera, Pemphigidae). Z. angew. Ent. 60:124–136.

Dušek, J., and P. Láska. 1960. Weitere unbekannte Syrphiden-larven. Čas. čsl. Spol. ent. 57:371–380, 5 pls.

Dušek, J., and P. Láska. 1961. Beitrag zur Kenntnis der Schwebfliegen-Larven (Syrphidae, Diptera) III. Přírodov. Čas. slezký 22:513–541, 1 pl.

Dušek, J., and P. Láska. 1962. Beitrag zur Kenntnis einiger Syrphiden-Larven (Diptera, Syrphidae). Čas. čsl. Spol. ent. 59:348–356, 7 pls.

Dušek, J., and P. Láska. 1964. A contribution to distinguishing European species of the subgenus *Syrphus* Fabricius (Diptera, Syrphidae) according to male genitalia and larvae. Čas. čsl. Spol. ent. 61:58–70.

Dušek, J., and P. Láska. 1966. Occurrence of syrphid larvae on some aphids. Pages 37–38 *in* I. Hodek, ed. Ecology of aphidophagous insects. Academia, Prague.

Dušek, J., and P. Láska. 1967. Versuch zum Aufbau eines natürlichen Systems mitteleuropäischer Arten der Unterfamilie Syrphinae (Diptera). Acta sci. nat. Brno 1:349–390.

Dušek, J., and P. Láska. 1974*a*. Overwintering and spring emergence of some common species of aphidophagous syrphids (Syrphidae, Diptera). Folia Fac. Sci. nat. Univ. Purkynianae brunensis, 15, Biologia 43:71–75.

Dušek, J., and P. Láska. 1974*b*. Influence of temperature during pupal development on the colour of syrphid adults (Syrphidae, Diptera). Folia Fac. Sci. nat. Univ. Purkynianae brunensis, 15, Biologia 43:77–81.

Dušek, J., and P. Láska. 1976. European species of *Metasyrphus*: key, descriptions and notes (Diptera, Syrphidae). Čas. čsl. Spol. ent. 73:263–282.

Dušek, J., and P. Láska. 1980. Species of *Metasyrphus* from Afghanistan and Kirghizia, with keys and descriptions of three new species (Diptera, Syrphidae). Čas. čsl. Spol. ent. 77:118–130.

Enderlein, G. 1938. Beiträge zur Kenntnis der Syrphiden. Sber. Ges. naturf. Freunde Berl. 1937:192–237.

Fabricius, J.C. 1775. Systema entomologiae, sistens insectorum classes, ordines, genera, species adiectis synonymis, locis, descriptionibus, observationibus. Korte, Flensburgi et Lipsiae [=Flensburg and Leipzig]. 832 pp.

Fabricius, J.C. 1781. Species insectorum exhibentes eorum differentias specificas, synonyma, auctorum, loca natalia, metamorphosin. Vol. 2. Bohn, Hamburgi et Kilonii [=Hamburg and Kiel]. 517 pp.

Fabricius, J.C. 1787. Mantissa insectorum sistens species nuper detectas. Vol. 2. Proft, Hafniae [=Copenhagen]. 382 pp.

Fabricius, J.C. 1794. Entomologia systematica emendata et aucta. Vol. 4. Proft, Hafniae [=Copenhagen]. 472 pp.

Fallén, C.F. 1816–1817. Syrphici Sveciae. Pages 1–22 [1816], 23–62 [1817]. Berling, Lundae [=Lund].

Fitzpatrick, S.M., and W.G. Wellington. 1983. Contrasts in the territorial behaviour of three species of hover flies (Diptera: Syrphidae). Can. Ent. 115:559–566.

Fluke, C.L. 1922. Syrphidae of Wisconsin. Trans. Wis. Acad. Sci. Arts Lett. 20:215–253, 2 pls.

Fluke, C.L. 1929. The known predacious and parasitic enemies of the pea aphid in North America. Wis. agric. Exp. Stn Res. Bull. 93:1–47, 6 pls.

Fluke, C.L. 1930. High-altitude Syrphidae with descriptions of new species (Diptera). Ann. ent. Soc. Am. 23:133–144.

Fluke, C.L. 1931. Notes on certain *Syrphus* flies related to *Xanthogramma* (Diptera Syrphidae) with descriptions of two new species. Trans. Wis. Acad. Sci. Arts Lett. 26:289–309, 2 pls.

Fluke, C.L. 1933. Revision of the *Syrphus* flies of America north of Mexico (Diptera, Syrphidae, *Syrphus* s. 1.). Part I. Trans. Wis. Acad. Sci. Arts Lett. 28:63–127, 3 pls.

Fluke, C.L. 1935. Revision of the *Epistrophe* flies of America north of Mexico (Diptera, Syrphidae). Entomologica am. 15:1–56, 3 pls.

Fluke, C.L. 1939. New Syrphidae (Diptera) from Central and North America. Ann. ent. Soc. Am. 32:365–375, 1 pl.

Fluke, C.L. 1949. Some Alaskan syrphid flies, with descriptions of new species. Proc. U.S. natn. Mus. 100:39–54.

Fluke, C.L. 1950. The male genitalia of *Syrphus*, *Epistrophe* and related genera (Diptera, Syrphidae). Trans. Wis. Acad. Sci. Arts Lett. 40:115–148, 10 pls.

Fluke, C.L. 1952. The *Metasyrphus* species of North America (Syrphidae). Am. Mus. Novit. 1590:1–27.

Fluke, C.L. 1954. Two new North American species of Syrphidae, with notes on *Syrphus* (Diptera). Am. Mus. Novit. 1690:1–10.

Fluke, C.L. 1958. A study of the male genitalia of the Melanostomini (Diptera-Syrphidae). Trans. Wis. Acad. Sci. Arts Lett. (1957) 46:261–279, 6 pls.

Folsom, J.W. 1909. The insect pests of clover and alfalfa. Illinois agric. Exp. Stn Bull. 134:144–152.

Forster, J.R. 1771. Novae species insectorum. Centuria I. London. 100 pp.

Frazer, B.D. 1972. A simple and efficient method of rearing aphidophagous hoverflies (Diptera: Syrphidae). J. ent. Soc. B.C. 69:23–24.

Gabritschevsky, E. 1924. Farbenpolymorphismus und Vererbung mimetischer Varietäten der Fliege *Volucella bombylans* und anderer "hummelähnlicher" Zweiflügler. Z. indukt. Abstamm.- u. VererbLehre 32:321–353.

Gabritschevsky, E. 1926. Convergence of coloration between American pilose flies and bumblebees (*Bombus*). Biol. Bull. mar. biol. Lab., Woods Hole 51:269–287.

Gilbert, F.S. 1981. Foraging ecology of hoverflies: morphology of the mouthparts in relation to feeding on nectar and pollen in some common urban species. Ecol. Ent. 6:245–262.

Goeldlin, P. 1974. Contribution a l'étude systématique et écologique des Syrphidae (Dipt.) de la Suisse occidentale. Mitt. schweiz. ent. Ges. 47:151–252.

Goeldlin, P. 1976. Révision du genre *Paragus* (Dipt., Syrphidae) de la région paléarctique occidentale. Mitt. schweiz. ent. Ges. 49:79–108.

Goot, V.S. van der. 1964. Fluke's catalogue of neotropical Syrphidae (Insecta, Diptera), a critical study with an appendix on new names in Syrphidae. Beaufortia 10:212–221.

Goot, V.S. van der. 1981. De zweefvliegen van Noordwest-Europa en Europees Rusland, in het bijzonder van der Benelux. Koninklijke Nederlandse Natuurhistorische Vereniging, Hoogwoud. 275 pp.

Goot, V.S. van der, and R.A.J. Grabandt. 1970. Some species of the genera *Melanostoma*, *Platycheirus* and *Pyrophaena* and their relation to flowers. Ent. Ber., Amst. 30:135–143.

Greene, C.T. 1918. Three new species of Diptera. Proc. ent. Soc. Wash. 20:69–71.

Hamrum, C.L. 1966. Food utilization of the common Minnesota Syrphinae species. Pages 71–73 *in* I. Hodek, ed. Ecology of aphidophagous insects. Academia, Prague. 360 pp., 11 pls.

Hardy, D.E. 1964. Syrphidae. Pages 380–419 *in* D.E. Hardy, M.A. Kohn, and E. Beyer, Insects of Hawaii. Vol. 11. Diptera. University of Hawaii Press, Honolulu.

Harris, T.L., and C.L. Hamrum. 1968. Notes on the biology of *Baccha fascipennis* Wied. (Diptera, Syrphidae). Ent. News 79:225–228.

Hartley, J.C. 1961. A taxonomic account of the larvae of some British Syrphidae. Proc. zool. Soc. Lond. 136:505–573.

Heiss, E.M. 1938. A classification of the larvae and puparia of the Syrphidae of Illinois exclusive of aquatic forms. Illinois biol. Monogr. 16(4):1–142, 17 pls.

Hine, J.S. 1922. Descriptions of Alaskan Diptera of the family Syrphidae. Ohio J. Sci. 22:143–147.

Hippa, H. 1968. A generic revision of the genus *Syrphus* and allied genera (Diptera, Syrphidae) in the Palaearctic region, with descriptions of the male genitalia. Acta ent. fenn. 25:1–94.

Holloway, B.A. 1976. Pollen-feeding in hover-flies (Diptera: Syrphidae). N.Z. Jl Zool. 3:339–350.

Hull, F.M. 1925. Notes on the North American species of the genus *Didea* with the description of a new species. Ann. ent. Soc. Am. 18:277–280, 1 pl.

Hull, F.M. 1930. Some new species of Syrphidae (Diptera) from North and South America. Trans. Am. ent. Soc. 56:139–148, 1 pl.

Hull, F.M. 1942. Some new species of Syrphidae from Florida, Cuba and Brazil (Syrphidae). Ent. News 52:278–283.

Hull, F.M. 1943*a*. Some new American syrphid flies (Diptera). Ent. News 54:29–37.

Hull, F.M. 1943*b*. New species of American syrphid flies. Bull. Brooklyn ent. Soc. 38:48–53.

Hull, F.M. 1944. Some new forms of *Platycheirus* of the family Syrphidae. Bull. Brooklyn ent. Soc. 39:75–79.

Hull, F.M. 1949. The morphology and inter-relationship of the genera of syrphid flies, recent and fossil. Trans. zool. Soc. Lond. 26:257–408.

Ide, F.P. 1926. Descriptions of two new species of *Platycheirus* (Syrphidae, Diptera). Can. Ent. 58:155–156.

Jaennicke, F. 1867. Neue exotische Dipteren. Abh. senckenb. naturforsch. Ges. 6:311–408, pls. 43, 44.

Johnson, C.W. 1898. Notes and descriptions of new Syrphidae from Mount St. Elias, Alaska. Ent. News 9:17–18.

Johnson, C.W. 1916. Some New England Syrphidae. Psyche, Camb. 23:75–80.

Johnson, C.W. 1919. The North American Diptera described by Nils S. Swederus. Can. Ent. 51:32.

Johnson, C.W. 1924. A review of the New England species of *Chrysotoxum*. Occ. Pap. Boston Soc. nat. Hist. 5:97–100.

Jones, C.R. 1917. New species of Colorado Syrphidae. Ann. ent. Soc. Am. 10:219–231.

Jones, C.R. 1922. A contribution to our knowledge of the Syrphidae of Colorado. Bull. Colo. St. Univ. agric. Exp. Stn 269:1–72, 5 pls.

Jones, P.R. 1907. A preliminary list of Nebraska Syrphidae with descriptions of new species. Jl N.Y. ent. Soc. 15:87–100.

Kanervo, E. 1934. Einige neue Syrphiden aus Petsamo. Suomal. eläin-ja kasvit. Seur. van. eläin. Julk. [=Ann. Soc. zool. bot. Vanamo, Helsingfors] 14:115–135.

Kanervo, E. 1946. Studien über die natürlichen Feinde des Erlenblattkäfers *Melasoma aenea* L. (Col. Chrysomelidae). Suomal. eläin-ja kasvit Seur. van. eläin. Julk. [=Ann. Soc. zool. bot. Vanamo, Helsinki] 12:1–206. [In Finnish with German summary].

Knab, F. 1914. Two North American Syrphidae. Insecutor Inscit. menstr. 2:151–153.

Knutson, L.V. 1973. Taxonomic revision of the aphid-killing flies of the genus *Sphaerophoria* in the Western Hemisphere. Misc. Publs. ent. Soc. Am. 9(1):1–50.

Kula, E. 1983. The larva and puparium of *Eriozona syrphoides* (Fallén) (Diptera, Syrphidae). Sb. ent. Odd. nár. Mus. Praze [=Acta ent. bohemoslovaca] 80:71–73, 2 pls.

Kuznetzov, S.Y. 1988. Morphology of the eggs of hover-flies (Diptera, Syrphidae). Ent. Obozr. 67:741–753, 10 pls.

Láska, P., and P. Starý. 1980. Prey records of aphidophagous syrphid flies from Czechoslovakia (Diptera, Syrphidae). Sb. ent. Odd. nár. Mus. Praze [=Acta ent. bohemoslovaca] 77:228–235.

Leereveld, H. 1982. Anthecological relations between reputedly anemophilous flowers and syrphid flies. III. Worldwide survey of crop and intestine contents of certain anthophilous syrphid flies. Tijdschr. Ent. 125:25–35.

Linnaeus, C. 1758. Systema naturae per regna tria naturae, secundum classes, ordines, genera, species, cum characteribus, differentiis, synonymis, locis. Ed. 10, Vol. 1. Laurentius Salvius, Holmiae [=Stockholm]. 824 pp.

Loew, H. 1860. Diptera Americana ab Osten-Sackenio collecta. Decas prima. Wien. ent. Monatschrift 4:79–84.

Loew, H. 1863*a*. Diptera Americae septentrionalis indigena. Centuria tertia. Berl. ent. Z. 7:1–55.

Loew, H. 1863*b*. Diptera Americae septentrionalis indigena. Centuria quarta. Berl. ent. Z. 7:275–326.

Loew, H. 1864. Diptera Americae septentrionalis indigena. Centurio quinta. Berl. ent. Z. 8:49–104.

Loew, H. 1866. Diptera Americae septentrionalis indigena. Centurio septa. Berl. ent. Z. 10:1–54.

Loew, H. 1871. Beschreibung europaischer Dipteren. Systematiche Beschreibung der bekannten europäischen zweiflugeligen Insecten, von Johann Wilhelm Meigen. Vol. 2: Neunter Theil oder dritter Supplementband. Schmidt, Halle. viii + 320 pp.

Lovett, A.L. 1919. Pages 238–251, pls. 18, 19 *in* F.R. Cole and A.L. Lovett, New Oregon Diptera. Proc. Calif. Acad. Sci. ser. 4, 9:221–255.

Lyon, J.P. 1965. Influence de quelques facteurs sur l'expression du potentiel de multiplication des syrphides aphidiphages. Annls Épiphyt. 16:397–398.

Macquart, J. 1829. Insectes Diptères du nord de la France. Syrphies. Mém. Soc. Sci. Agric. Lille 1827/1828:149–371, 4 pls.

Macquart, J. 1834. Histoire naturelle des Insectes—Diptères, Tome premier. Librairie encyclopédique de Roret, Paris. 578 pp., 12 pls.

Macquart, J. 1842. Diptères exotiques nouveaux ou peu connus. Mém. Soc. Sci. Agric. Lille 1841(1):65–200, 22 pls.

Macquart, J. 1847. Diptères exotiques nouveaux ou peu connus. 2e supplément. Mém. Soc. Sci. Agric. Lille 1846:21–120, 6 pls.

Macquart, J. 1850. Diptères exotiques nouveaux ou peu connus. 4e supplément [part]. Mém. Soc. Sci. Agric. Lille 1849:309–479, 14 pls.

Macquart, J. 1855. Diptères exotiques nouveaux ou peu connus. 5e supplément. Mém. Soc. Sci. Agric. Lille 1854:25–156, 7 pls.

Maier, C.T. 1978. The immature stages and biology of *Mallota posticata* (Fabricius) (Diptera: Syrphidae). Proc. ent. Soc. Wash. 80:424–440.

Maier, C.T. 1982. Larval habitats and mate-seeking sites of flower-flies (Diptera: Syrphidae, Eristalinae). Proc. ent. Soc. Wash. 84:603–609.

Maier, C.T., and G.P. Waldbauer. 1979. Dual mate-seeking strategies in male syrphid flies. Ann. ent. Soc. Am. 72:54–61.

Malloch, J.R. 1919. The Diptera collected by the Canadian Expedition, 1913–1918 (excluding the Tipulidae and Culicidae). Pages 34–90, pls. 7–10 *in* R.M. Anderson, ed. Report of the Canadian Arctic Expedition 1913–1918. Vol. 3: Insects, Pt. C.: Diptera. J. de Labroquerie Taché, Ottawa.

Matsumura, S. 1919. *In* S. Matsumura and J. Adachi. Synopsis of the economic Syrphidae of Japan. Pt. III [sic] [=IV]. Ent. Mag., Kyoto 3:128–144, 1 pl.

McAlpine, J.F. 1981. Morphology and terminology—adults. Pages 9–63 *in* J.F. McAlpine, B.V. Peterson, G.E. Shewell, H.J. Teskey, J.R. Vockeroth, and D.M. Wood, eds. Manual of Nearctic Diptera. Vol. 1. Agric. Can. Monogr. 27.

Meigen, J.W. 1822. Systematische Beschreibung der bekannten europaischen zweiflügeligen Insekten. Vol. 3. Schulz-Wundermann, Hamm. x + 416 pp., 11 pls.

Metcalf, C.L. 1916. Syrphidae of Maine. Bull. Me agric. Exp. Stn [ser. 2] 253:193–264.

Metcalf, C.L. 1917. Syrphidae of Maine. Second report. Bull. Me agric. Exp. Stn [ser. 2] 263:153–176.

Metcalf, C.L. 1921. The genitalia of male Syrphidae: their morphology, with especial reference to its taxonomic significance. Ann. ent. Soc. Am. 14:169–225.

Miller, R.L. 1929. A contribution to the biology and control of the green citrus aphid, *Aphis spiraecola* Patch. Bull. Fla. agric. Exp. Stn 203:431–476.

Müller, O.F. 1764. Fauna insectorium Fridrichsdalina, sive methodica descriptio insectorum agri Fridrichsdalinensis, cum characteribus genericis et specificis, nominibus trivialibus, locis natalibus, iconibus allegatis, novisque pluribus speciebus additis. Gleditsch, Hafniae et Lipsiae [=Copenhagen and Leipzig]. xxiv + 96 pp.

Nielsen, J.C. 1910. A catalogue of the insects of north-east Greenland with descriptions of some larvae. Pages 55–68, 2 pls., *in* F. Johansen and J.C. Nielsen. The insects of the “Danmark” expedition. Meddr Grønland (1911) 43.

Nielsen, T. 1971. Syrphidae (Dipt.) from Jaeren, Norway. I. With description of two new species. Norsk ent. Tidsskr. 18:53–73.

Nielsen, T. 1972. *Platycheirus monticolus* nov. sp., a northern species confused with *P. discimanus* Loew (Dipt., Syrphidae). Norsk ent. Tidsskr. 19:91–98.

Nielsen, T. 1974. Notes on two northern species of the genus *Platycheirus* St.-Farg. et Serv. (Dipt., Syrphidae). Norsk ent. Tidsskr. 21:167–171.

Nielsen, T. 1981. Studies on *Platycheirus* Lepeletier & Serville: *P. complicatus* Becker, *P. latimanus* (Wahlberg) and *P. boreomontanus* nom. nov. (Diptera: Syrphidae). Entomologica scand. 12:99–102.

Ninomiya, E. 1959. Further notes on the immature stages of aphidophagous syrphid flies of Japan. Sci. Bull. Fac. lib. Arts Educ. Nagasaki Univ. 10:3–52. [In Japanese with English summary].

Osburn, R.C. 1908. British Columbia Syrphidae, new species and additions to the list. Can. Ent. 40:1–14.

Osburn, R.C. 1910. Studies on Syrphidae. I. *Syrphus arcuatus* Fallén and a related new species; II. The invalidity of *Scaeva* (=*Catabomba*) as a genus; III. An interesting meristic variation in *Syrphus perplexus*. Jl N.Y. ent. Soc. 18:53–66, 3 pls.

Osten Sacken, C.R. 1875*a*. On the North American species of the genus *Syrphus* (in the narrowest sense). Proc. Boston Soc. nat. Hist. (1875–1876) 18:135–153.

Osten Sacken, C.R. 1875*b*. A list of the North American Syrphidae. Bull. Buffalo Soc. nat. Sci. 3:38–71.

Osten Sacken, C.R. 1877. Western Diptera: descriptions of new genera and species of Diptera from the region west of the Mississippi and especially from California. Bull. U.S. geol. geog. Surv. Terr. 3:189–354.

Rondani, C. 1857. Dipterologiae Italicae prodromus. Vol. 2: Species Italicae ordinis dipterorum in genera characteribus definita, ordinatim collectae, methodo analitica distinctae, et novis vel minus cognitis descriptis. Pars prima: Oestridae, Syrpfhidae [sic], Conopidae. Stocchi, Parmae [=Parma]. 264 pp.

Rotheray, G.E., and F.S. Gilbert. 1989. The phylogeny and systematics of European predacious Syrphidae (Diptera) based on larval and puparial stages. Zool. J. Linn. Soc. 95:29–70.

Sack, P. 1928–1932. Syrphidae. Fliegen palaearkt. Reg. 31:1–451, 18 pls.

Say, T. 1823. Descriptions of dipterous insects of the United States. J. Acad. nat. Sci. Philad. 3:9–54, 73–104.

Say, T. 1824. American entomology, or descriptions of the insects of North America [Vol. I]. Mitchell, Philadelphia. 101 pp., 18 pls.

Say, T. 1829. Descriptions of North American dipterous insects. J. Acad. nat. Sci. Philad. (1829–1830) 6:149–178.

Schneider, F. 1947. Zur Ueberwinterung von *Lasiopticus pyrastri* L. und *L. seleniticus* Meig. Mitt. schweiz. ent. Ges. 20:306–316.

Schneider, F. 1948. Beitrag zur Kenntnis der Generations—verhältnisse und Diapause räuberischer Schwebfliegen. Mitt. schweiz. ent. Ges. 21:249–285.

Schneider, F. 1950*a*. Die Abwehrreaktion des Insectenblutes und ihre Beeinflussung durch die Parasiten. Vjschr. naturf. Ges. Zürich 95:22–44.

Schneider, F. 1950*b*. Die Entwicklung des Syrphidenparasiten *Diplazon fissorius* Grav. in uni-, oligo- und polyvoltinen Wirten und sein Verhalten bei parasitärer Aktivierung der Diapauselarven durch *Diplazon pectoratorius* Grav. Mitt. schweiz. ent. Ges. 23:155–194.

Schneider, F. 1951. Einige physiologische Beziehungen zwischen Syrphidenlarven und ihre Parasiten. Z. angew. Ent. 33:150–162.

Schneider, F. 1953. *Syrphus nigritarsis* Zett., ein Ei- und Larvenrauber von *Melasoma* (Chrysom., Col.). Tijdschr. PlZiekt. 59:192–194, 1 pl.

Schneider, F. 1958. Kunstliche Blumen zum Nachweis von Winterquartieren, Futterpflanzen und Tageswanderungen von *Lasiopticus pyrastri* (L.) und andere Schwebfliegen. Mitt. schweiz. ent. Ges. 31:1–24.

Schneider, F. 1969. Biology and physiology of aphidophagous Syrphidae. A. Rev. Ent. 14:103–124.

Schneider, F., T. Wildbolz, and W. Vogel. 1957. Die Apfelgraslaus (*Rhopalosiphum insertum* Wlk.), eine sehr häufige, jedoch noch wenig bekannte Doppelgängerin der grunen Apfellaus (*Aphis pomi* De Geer). Schweiz. Z. Obst- u. Weinb. 66:8–17.

Scott, E.I. 1939. An account of the developmental stages of some aphidophagous Syrphidae and their parasites. Ann. appl. Biol. 26:509–532.

Sedman, Y.S. 1965. *In* W.W. Wirth, Y.S. Sedman, and H.V. Weems. Family Syrphidae. Pages 557–625 *in* A. Stone, C.W. Sabrosky, W.W. Wirth, et al., eds. A catalog of the Diptera of America north of Mexico. U.S. Dep. Agric., Agric. Handb. 276.

Séguy, E. 1961. Diptères Syrphides de l'Europe occidentale. Mém. Mus. natn. Hist. nat., Paris (series A. Zool.) 23:1–248.

Shannon, R.C. 1916. Appendix. Systematic and synonymic notes. Pages 195–203 *in* N. Banks, C.T. Greene, W.L. McAtee, and R.C. Shannon. District of Columbia Diptera: Syrphidae. Proc. biol. Soc. Wash. 29:173–203.

Shannon, R.C. 1926. The chrysotoxine syrphid-flies. Proc. U.S. natn. Mus. 69(11):1–20.

Shiraki, T. 1949. Studies on the Syrphidae. 1. The classification of the subfamilies. Mushi 20:59–73.

Snow, W.A. 1892. Notes and descriptions of Syrphidae. Kans. Univ. Quart. 1:33–38, 1 pl.

Snow, W.A. 1895. Diptera of Colorado and New Mexico. Syrphidae. Kans. Univ. Quart. 3:225–247.

Speight, M.C.D. 1978. *Melanostoma dubium* (Dipt. Syrphidae) in Britain and a key to the British Isles *Melanostoma* species. Ent. Rec. J. Var. 90:226–230.

Speight, M.C.D., P.J. Chandler, and R. Nash. 1975. Irish Syrphidae (Diptera): notes on the species and an account of their known distribution. Proc. R. Ir. Acad. 75(B):1–80.

Stackelberg, A.A. 1970. Syrphidae. Pages 11–96 *in* G.Y. Bey-Bienko. Classification key to the insects of the European part of USSR 5(2). Opred. Faune SSSR 103. [In Russian] [English Translation, 1988, Family Syrphidae. Pages 10–148 *in* Keys to the insects of the European part of the USSR 5(2): Diptera and Siphonaptera. Smithsonian Institution Libraries and National Science Foundation, Washington, D.C.].

Staeger, C. 1843. (Udsigt over den danske faunas arter af antliatslaegten *Platycheirus*). Pages 320–327 *in* J.C. Schiødte. Forhandlinger i det Skandinaviske Entomologiske Selskab. Naturh. Tidsskr. (1842–1843) 4:315–360.

Staeger, C. 1845. Grønlands antliater. Naturh. Tidsskr. (1844–1845) ser. 2, 1:346–369.

Stelleman, P., and A.D.J. Meeuse. 1976. Anthecological relations between reputedly anomophilous flowers and syrphid flies. I. The possible role of syrphid flies as pollinators of *Plantago*. Tijdschr. Ent. 119:15–31, 3 pls.

Stubbs, A.E., and S.J. Falk. 1983. British hoverflies. British Entomological and Natural History Society, London. 253 pp., 13 pls.

Stuckenberg, B.R. 1954. The immature stages of *Sphaerophoria quadrituberculata* Bezzi (Diptera, Syrphidae), a predator of *Cercothrips afer* Priesner. J. ent. Soc. sth. Afr. 17:58–61.

Sturken, K. 1964. Die Bedeutung der Imaginalernährung für das Reproduktionsvermögen der Syrphiden. Z. angew. Zool. 51:385–417.

Swezey, O.H. 1926. Note. Proc. Hawaii. ent. Soc. 6:226.

Tamaki, G., B.J. Landis, and R.E. Weeks. 1967. Autumn populations of green peach aphid on peach trees and the role of syrphid flies in their control. J. econ. Ent. 60:433–436.

Teskey, H.J. 1981. Key to families—larvae. Pages 125–147 *in* J.F. McAlpine, B.V. Peterson, G.E. Shewell, H.J. Teskey, J.R. Vockeroth, and D.M. Wood, eds. Manual of Nearctic Diptera. Vol. 1. Agric. Can. Monogr. 27.

Thompson, F.C. 1969. A new genus of microdontine flies (Diptera: Syrphidae) with notes on the placement of the subfamily. Psyche, Camb. 76:74–85.

Thompson, F.C. 1981. Revisionary notes on Nearctic *Microdon* flies (Diptera: Syrphidae). Proc. ent. Soc. Wash. 83:725–758.

Thompson, F.C., J.R. Vockeroth, and M.C.D. Speight. 1982. The Linnaean species of flower flies (Diptera: Syrphidae). Mem. ent. Soc. Wash. 10:150–165.

Thomson, G.C. 1869. 6. Diptera. Species nova descripsit. Pages 443–614, pl. 9 *in* K. Svenska Vetenskaps—Akademien, Kongliga svenska fregatten Eugenies resa omkring jorden. Pt. 2: Zoologie, [Sec.] 1: Insekter. Stockholm "1868."

Torp, E. 1984. De danske svirrefluer (Diptera: Syrphidae). Kendetegn, levevis og udbredelse. Danmarks Dyreliv 1:1–300.

Verrall, G.H. 1901. Platypezidae, Pipunculidae, and Syrphidae of Great Britain. Pages 1–691 *in* G.H. Verrall, ed. British flies, Vol. 8. Gurney and Jackson, London.

Vimmer, A. 1933. Larvae and puparia of the syrphid family from Bohemia, Čas. české Spol. ent. 30:181–187. [In Czech].

Violovitsh, N.A. 1983. Sirfidy Sibiri. 241 pp. Nauka, Novisibirsk [not seen]. [Foreword and keys translated into English by V.S. van der Goot and L. Verlinden, 1986, Versl. tech. Gegevens 43:1–228].

Vockeroth, J.R. 1966. A method of mounting insects from alcohol. Can. Ent. 98:69–70.

Vockeroth, J.R. 1969. A revision of the genera of the Syrphini (Diptera: Syrphidae). Mem. ent. Soc. Can. 62:1–176.

Vockeroth, J.R. 1980. A review of the Nearctic species of *Melangyna* (*Meligramma*) Frey (Diptera: Syrphidae). Can. Ent. 112:775–778.

Vockeroth, J.R. 1983. Nomenclatural notes on Nearctic Syrphinae, with descriptions of new species of *Syrphus* and keys to the species of *Didea*, *Epistrophe* s. str. and *Syrphus* (Diptera: Syrphidae). Can. Ent. 115:175–182.

Vockeroth, J.R. 1986*a*. Revision of the New World species of *Paragus* (Diptera: Syrphidae). Can. Ent. 118:183–198.

Vockeroth, J.R. 1986*b*. Nomenclatural notes on Nearctic *Eupeodes* (including *Metasyrphus*) and *Dasysyrphus*. Can. Ent. 118:199–204.

Vockeroth, J.R. 1990. Revision of the Nearctic species of *Platycheirus* (Diptera, Syrphidae). Can. Ent. 122:659–766.

Waldbauer, G.P. 1970. Mimicry of hymenopteran antennae by Syrphidae. Psyche, Camb. 77:45–49.

Waldbauer, G.P., and J.K. Sheldon. 1971. Phenological relationships of some aculeate Hymenoptera, their dipteran mimics, and insectivorous birds. Evolution 25:371–382.

Waldbauer, G.P., J.G. Sternburg, and C.T. Maier. 1977. Phenological relationships of wasps, bumblebees, their mimics, and insectivorous birds in an Illinois sand area. Ecology 58:583–591.

Walker, F. 1849. List of the specimens of dipterous insects in the collection of the British Museum. Vol. 3:485–687. British Museum, London.

Walker, F. 1852. Diptera. Vol. 1. Pages 157–414, 4 pls. *in* W.W. Saunders, ed. Insecta Saundersiana. van Voorst, London "1856."

Walton, W.R. 1911. Notes on Pennsylvanian Diptera, with two new species of Syrphidae. Ent. News 22:318–322, 1 pl.

Weems, H.V. 1954. Natural enemies and insecticides that are detrimental to beneficial Syrphidae. Ohio J. Sci. 54:45–54.

Wehr, E.E. 1922. A synopsis of the Syrphidae of Nebraska, with descriptions of new species from Nebraska and Colorado. Univ. Stud. Univ. Nebr. 22:119–162.

Wellington, W.G., and S.M. Fitzpatrick. 1981. Territoriality in the drone fly, *Eristalis tenax* (Diptera: Syrphidae). Can. Ent. 113:695–704.

Wiedemann, C.R.W. 1830. Aussereuropäische zweiflugelige Insekten. Vol. 2. Schulz, Hamm. xii + 684 pp., 5 pls.

Williston, S.W. 1882. Contribution to a monograph of the North American Syrphidae. Proc. Am. phil. Soc. 20:299–332.

Williston, S.W. 1887. Synopsis of the North American Syrphidae. Bull. U.S. natn. Mus. (1886) 31:i–xxx, 1–335.

Wirth, W.W., Y.S. Sedman, and H.V. Weems. 1965. Family Syrphidae. Pages 557–625 *in* A. Stone, C.W. Sabrosky, W.W. Wirth, et al., eds. A catalog of the Diptera of America north of Mexico. U.S. Dept. Agric., Agric. Handb. 276.

Wnuk, A., and R. Fuchs. 1977. The observations on the effectiveness of the limitation of *Brevicoryne brassicae* (L.) by Syrphidae. Polskie Pismo ent. 47:147–155.

Wulp, F.M. van der. 1867. Eenige Noord-Americaansche Diptera. Tijdschr. Ent. 10:125–164, pls. 3–5.

Yakhontov, V.V. 1966. Food specificity in Syrphidae and Cocinellidae of Central Asia. Pages 35–36 *in* I. Hodek, ed. Ecology of aphidophagous insects. Academia, Prague. 310 pp., 11 pls.

Zetterstedt, J.W. 1838. Dipterologis Scandinaviae. Sect. 3: Diptera. Pages 447–868 *in* Zetterstedt, J.W. Insecta Lapponica. Voss, Lipsiae [=Leipzig]. vi + 1140 pp.

Zetterstedt, J.W. 1843. Diptera Scandinaviae. Disposita et descripta. Vol. 2:441–894. Lundberg, Lundae [=Lund].

Zetterstedt, J.W. 1849. Diptera Scandinaviae. Disposita et descripta. Vol. 8:2935–3366. Lundberg, Lundae [=Lund].

Zetterstedt, J.W. 1859. Diptera Scandinaviae. Disposita et descripta. Vol. 13: Sup. 4, pp. v–xvi, 4943–6190. Lundberg, Lundae [=Lund].

Index

(Page numbers of principal entries are in boldface; synonyms are in italics.)